Convex and Complex: Perspectives on Positivity in Geometry

CONTEMPORARY MATHEMATICS

810

Convex and Complex: Perspectives on Positivity in Geometry

Conference in Honor of Bo Berndtsson's 70th Birthday
Convex and Complex: Perspectives on Positivity in Geometry
October 31 – November 4, 2022
Cetraro, Italy

Robert J. Berman
Yanir A. Rubinstein
Editors

AMERICAN MATHEMATICAL SOCIETY

2020 *Mathematics Subject Classification.* Primary 32L05, 14C20, 32Q15, 32Q20, 52A40, 32U05, 32P05, 14G22, 32Q25, 53A15.

For additional information and updates on this book, visit
www.ams.org/bookpages/conm-810

Library of Congress Cataloging-in-Publication Data

Names: Conference on Convex and Complex: perspectives on positivity in geometry (2022 : Cetraro, Italy), author. | Berman, Robert J., 1976– editor. | Rubinstein, Yanir A., editor. | Berndtsson, Bo, 1950– honoree.

Title: Convex and complex : perspectives on positivity in geometry / Robert J. Berman, Yanir A. Rubinstein, editors.

Description: Providence, Rhode Island : American Mathematical Society, [2025] | Series: Contemporary mathematics, 0271-4132 ; volume 810 | "Conference on Convex and Complex: perspectives on positivity in geometry, in honor of Bo Berndtsson's 70th birthday, October 31–November 4, 2022, Cetraro, Italy". | Includes bibliographical references.

Identifiers: LCCN 2024035670 | ISBN 9781470473389 (paperback) | ISBN 9781470478612 (ebook)

Subjects: Convex geometry Congresses. | Several complex variables and analytic spaces – Holomorphic fiber spaces – Holomorphic bundles and generalizations. | Algebraic geometry – Cycles and subschemes – Divisors, linear systems, invertible sheaves. | Several complex variables and analytic spaces – Complex manifolds – Kähler manifolds. | Several complex variables and analytic spaces – Complex manifolds – Kähler-Einstein manifolds. | Convex and discrete geometry – General convexity – Inequalities and extremum problems. | Several complex variables and analytic spaces – Pluripotential theory – Plurisubharmonic functions and generalizations. | Several complex variables and analytic spaces – Non-Archimedean analysis – Non-Archimedean analysis. | Algebraic geometry – Arithmetic problems. Diophantine geometry – Rigid analytic geometry. | Several complex variables and analytic spaces – Complex manifolds – Calabi-Yau theory. | Differential geometry – Classical differential geometry – Affine differential geometry.

Classification: LCC QA639.5 .C644 2025 | DDC 516/.08–dc23/eng/20241112

LC record available at https://lccn.loc.gov/2024035670

Contemporary Mathematics ISSN: 0271-4132 (print); 1098-3627 (online)

DOI: https://doi.org/10.1090/conm/810

∞ The paper used in this book is acid-free and falls within the guidelines
established to ensure permanence and durability.
Visit the AMS home page at `https://www.ams.org/`

10 9 8 7 6 5 4 3 2 1 30 29 28 27 26 25

Contents

Preface

This volume presents a collection of research articles arising from the workshop "Convex and Complex—Perspectives on Positivity in Geometry", held in Cetraro, Italy, on October 31–November 4, 2022. The workshop celebrated the 70th birthday of Bo Berndtsson.[1] The format of the workshop was innovative compared to standard conferences in mathematics, with focused 30-minute talks, aimed to stimulate lively discussions and foster new collaborations across the diverse themes of complex and convex geometry, all united by the overarching concept of positivity. We believe this volume reflects the spirit of the workshop, showcasing the vitality of current research in these areas.

This preface is organized as follows. First, we discuss the unique, and very successful, format the workshop followed. This is followed by an overview of the contributions to the volume. We end with a tribute to Bo Berndtsson, whom we consider one of the founding fathers of the "Convex–Complex exchange".

The Cetraro 30-minute format. It is perhaps worthwhile to dwell a bit on the format of the workshop, which was anything but standard.[2]

When we were graduate students, we recall our department elders recounting how when *they* were graduate students, there was just one department seminar: the *colloquium*. For us that was hard to fathom: already in the early 2000s each department seemed to have many seminars each week in addition to the colloquium. A similar statement could be made about conferences, with the regional meetings of societies (e.g., AMS and ICM) historically playing a much more central role due to the scarcity of other workshops.

Fast forward twenty years, and it seems that this trend is even more pronounced, with almost every tenured faculty having in their resumé some seminar and conference organization experience.

Perhaps it is worth pausing to wonder what the effects are of the ubiquity of seminars and conferences on the profession. For one, listening to talks has become a major time commitment. On the flip side, as seminars become ever more focused, many talks are barely understandable (barring the opening minutes) except to few experts in the room (or on Zoom).

Colloquium talks target diverse mathematical audiences, and it is understandably challenging to squeeze a presentation into the allotted 50-60 minutes. For

[1]The conference was originally scheduled for October 2020, and was postponed twice due to the pandemic.

[2]A more in-depth article focused on this new format for talks and conferences has recently been published, and we refer the interested reader to it: R.J. Berman, Y.A. Rubinstein, "Less is More: 30-minute talks", *Notices AMS*, December 2024, DOI: `https://url.us.m.mimecastprotect.com/s/clMVC31rx5UpN27gugfOFQOR1w?domain=doi.org`

seminar talks, which are typically more technical, there is no particular reason to stretch the talk over a whole hour. After all, what is the point of the speaker transmitting more information than the audience can absorb?

Another problem with many workshops is that while one of their declared goals is to foster collaboration, in practice there is relatively very little time remaining in the packed schedule, aside from numerous talks and seemingly ever shorter meal breaks. Perhaps this is related to the ubiquity of conferences: most faculty will not attend a conference unless they are an invited speaker, which results in squeezing more talks into the schedule.

With these conflicting priorities in mind, our goal in this workshop was to experiment with a different format. Our first idea was to limit both the number and the length of the talks, while simultaneously increasing the length and number of breaks. We insisted that speakers prepare a 30-minute talk, a rather drastic departure from the traditional model. We asked speakers to focus on the main ideas, challenges, and problems, and not to dwell on technical details of proofs. We also allotted an additional 30 minutes between talks. In practice what happened is similar to the well-known "flipped classroom" model: there were lively discussions that went on for about 20 minutes after the talk ended, and speakers engaged with participants (or, perhaps, participants engaged with speakers!) in a way that was absolutely inspiring, in effect transforming the (typically passive) audience into (active) *participants*. By limiting the schedule to two talks in the morning and two talks in the afternoon, we were able to have a 3.5-hour break in the middle of the day. In addition, on Wednesday and Friday we had only a morning session. As a result, there were several useful blocks of collaboration time as well as much-needed rest time (a side effect was that participants did not have to skip talks in order to find time to collaborate or rest, and this in turn benefited the lively discussions following the talks). This resulted in participants being less fatigued in the afternoon session and also able to be more fruitful in collaboration in the evening.

The format was a great success according to testimonials by many participants, and we truly hope it will become popular.

A bird's eye view of the contributions. Positivity plays a central role in complex and convex geometry. It arises from a range of complementary perspectives, as illustrated by the breadth of the papers appearing in this volume. Many of the positivity results in global complex geometry can be traced back to the basic fact that any complex projective manifold can be endowed with a *strictly positive* real closed 2-form; i.e., in local holomorphic coordinates this form is represented by a matrix-valued function that is positive definite. This means that the 2-form can be identified with a Riemannian metric preserving the complex structure, called a Kähler metric. The metrics with constant Ricci curvature, called Kähler–Einstein metrics, play a privileged role in Kähler geometry.

The paper of C. Jin considers the problem, posed by Cheltsov and Rubinstein, of classifying Kähler–Einstein metrics with positive Ricci curvature that have edge-cone singularities along a given smooth complex hypersurface with a small angle. Such spaces are called asymptotically log Fano manifolds, and the Cheltsov–Rubinstein problem can be thought of as a natural singular analogue of Calabi's problem for determining which Fano manifolds admit a Kähler–Einstein metric.

The problem is settled in the case of complex surfaces by treating the last remaining open case where the hypersurface could consist of two disconnected components and have different edge-cone singularities on each component.

The expository paper of J. Hultgren is inspired by the study of degenerations of Kähler–Einstein metrics with vanishing Ricci curvature. According to the conjectures of Kontsevich and Soibelman and of Gross and Wilson—originating in the phenomenon of mirror symmetry observed in the string theory literature—such Kähler metrics degenerate (in the large complex structure limit) to Hessian metrics on singular real affine spaces. The potentials of these Hessian metrics are convex functions, satisfying a Monge–Ampère equation, and the paper surveys the duality structure of such Hessian metrics. This duality has played an important role in recent developments, and the paper highlights new connections to optimal transport theory, which is a classical topic in convex analysis.

The subject of L. Lempert's contribution is the geometry of the space of all Kähler metrics on a given complex manifold. This space is infinite dimensional (assuming that it is non-empty) and it has previously been endowed with canonical Riemannian and Finsler metrics. In Lempert's paper a new measure of the distance between the between given Kähler potentials is introduced. More precisely, the setup of global pluripotential theory on compact Kähler manifolds is considered, where the role of Kähler potentials is played by potentials of positive closed currents with full mass.

In the contribution of S. Boucksom and M. Jonsson the focus is shifted from the space of all such potentials to the space of all probability measures. A synthetic approach is introduced, which applies in a very general framework, covering the case of a compact Kähler manifold and that of projective Berkovich spaces over a non-Archimedean field. A general free energy type functional is introduced, which, in the case of a compact Kähler manifold, may be identified with Mabuchi's classical K-energy functional. On the other hand, in the case of the trivial non-Archimedean absolute value on the complex number field, the free energy functional can be identified with the Donaldson–Futaki invariant of a test configuration, up to a simple error term that disappears after base change. It does, in particular, show that the coercivity of the free energy functional is, in general, an open condition with respect to the polarization.

The subject of M. Xia's contribution is a different transcendental framework for the non-Archimedean global pluripotential theory, recently introduced in a joint work with Darvas and Zhang. In particular, the analogue of Boucksom–Jonsson's envelope conjecture is established.

A positivity result that lies at the heart of many recent developments in complex geometry and analysis is Berndtsson's positivity of direct image bundle. This is the subject of several of the contributions of this volume. In D. Varolin's paper a new proof of Berndtsson's curvature formula is given that uses a generalization of the holomorphic Gauss formula to Berndtsson–Lempert–Szöke (BLS) fields, for locally trivial direct image bundles. In the contribution of F. Campana, J. Cao, and M. Paun new positivity results for direct images are established, with applications to holomorphic foliations. Furthermore, in the contribution of T. Nguyen and X. Wang, positivity properties of certain trivial Hilbert bundles are used to give a new proof of the sharp strong openness theorem for plurisubharmonic functions. This approach confirms a conjecture of Berndtsson and is based on an extra L^2-estimate for the $\bar{\partial}$-operator. In general, such estimates can be viewed as quantitative refinements of the Kodaira vanishing theorem for the higher adjoint Dolbeault cohomology groups attached to semi-positive line bundles. This vanishing theorem admits a generalization on complex projective manifolds, known as Kollar's injectivity theorem. A conjectural generalization of this injectivity theorem, for divisors with simple normal crossings on Kähler manifolds, was proposed by Fujino. It serves as a motivation for the contribution of T. Chan and Y. Choi, where the use of adjoint ideal sheaves and residue functions is explained in this context. Moreover, a new extension result is established.

Berndtsson's positivity of direct image bundles can be viewed as a generalization of the classical Brunn–Minkowski inequality for convex bodies. This inequality can be rephrased as a concavity result for the convex interpolation between two given convex bodies. In the contribution of J. Ross and D. Witt-Nyström a different type of interpolation is considered, called the harmonic interpolation, which is shown to be superharmonic. An application to determinant of random vectors is provided, through the connection to zonoids.

The contribution of V. Mastrantonis is concerned with a generalization of another classical inequality for convex bodies: the Santaló inequality for the Mahler volume of a convex body. In a very recent previous work of Berndtsson, Rubinstein, and Mastrantonis an L^p-type generalization of the Mahler volume was introduced, coinciding with the classical Mahler volume when $p = \infty$. In this collaboration, initiated during the conference at Cetraro, a generalized Santaló inequality for the L^p-Mahler volume was established for symmetric convex bodies. In the present paper of Mastrantonis the symmetry assumption is removed resulting in a generalization of Santaló's classical 1949 inequality for the Mahler volume.

Finally, as already pointed out, the workshop was originally scheduled for 2020 and postponed twice. Jean-Pierre Demailly and Steve Zelditch, two giants of complex geometry, were among the confirmed speakers for 2020, but tragically passed away before the workshop finally took place in October 2022. They were sorely missed during the meeting and their contributions would have no doubt made this volume even richer.

A tribute to Bo Berndtsson. (A speech given by Y. A. Rubinstein at the gala dinner). It is only fitting that a conference celebrating Bo Berndtsson's mathematics and birthday(s) is centered around the theme of positivity.

Bo is a rock star.

In fact, Bo's early career started as a rock star. In the post-Woodstock 1970s in Sweden, he was the lead singer for Love Explosion. I think you will all agree that Love Explosion falls into the realm of positivity.

Bo and I entered Kähler geometry around the same time.... While I was a graduate student, I took a class by Gang Tian (my advisor) at Princeton where he went over his 100+ page monograph with Chen on geodesics in the space of Kähler metrics. This was in Spring 2005. After tea-time one afternoon that semester, Tian asked me whether I could generalize his PhD thesis theorem on Fubini–Study approximation of Kähler metrics to geodesics in the space of Kähler metrics. After working on the problem for several weeks I realized I was missing one key estimate: a bound on the time-derivative of the approximating geodesics independent of the size of the change-of-basis matrix. Before the semester ended, I found out from Tian that Phong and Sturm had posted a paper which solved the problem. Because the Phong–Sturm proof was so similar, and, admittedly, better, than mine, I couldn't get myself to look at the paper for some time, but indeed their first lemma concerned that time-derivative estimate.

While I was still struggling as a graduate student trying to make my splash, in 2007 Bo made his splash in Kähler geometry when he obtained his beautiful generalization of the Phong–Sturm theorem. First, he twisted the line bundles with the canonical bundle to effectively eliminate the arbitrary volume form from the Hilbert norms. Second, he invoked an old interpolation inequality of Rochberg for matrices, a sort of a maximum principle for systems, to get that time-derivative estimate very naturally. And, finally, he used his subharmonicity result from 2005 to get an estimate in the other direction, so to speak. The proof was elegant, short, and gave a more powerful statement.

Of course, if you had asked Bo, he would have emphasized that this didn't quite generalize Phong–Sturm since he could not prove the convergence of the geodesics using the original sequence of line bundles.

At the end of the introduction to his paper, you will find the following paragraph:

> As is probably clear from the text, I am a novice in the study of extremal Kähler metrics, and I apologize for any omission in accrediting results properly. The motivation for this paper lies not so much in the particular results as in showing the link between $\bar\partial$-theory and this beautiful area.

For the next 15 years Bo continued to show us and teach us these links, and in doing so has really transformed and enriched Kähler geometry and revitalized $\bar\partial$ theory. (By the way, in 2009 Bo did provide a short, original, and masterful proof for the original sequence of line bundles.)

As a student, I remember trying to read Bo's 2007 preprint. Where is the "catch" (I asked myself)? I realized that it boiled down to his magic rod, a certain theorem that I have fondly referred to over the years as "Theorem 1.1."

In a 2005 preprint (published in 2009 in the *Annals*) Bo stated his Theorem 1.1 on the very first page. It might have seemed at the time a beautiful but innocent and esoteric theorem on Nakano positivity of certain vector bundles whose fibers consist of Bergman spaces.

The proof was difficult to read (in 2017 Dror Varolin agreed to give a week-long mini-course at Maryland, a sort of "Berndtsson for dummies," where he beautifully broke down the proof).

Proofs and style have a lasting impact on a field. For many years, starting with Aubin and Yau, Kähler geometry was strongly influenced by geometric analysis in the sense of Riemannian geometry and PDEs. Their breakthroughs in the 1970s had a strong psychological influence on a whole generation of geometers. Even though some of the key methods they introduced into Kähler geometry were standard in the PDE community (e.g., Moser iteration), they went much farther than the perceived softer methods of the older generation (e.g., Chern, Calabi, Lichnerowicz, Berger).

The US, France, and China largely followed suit, but in Japan Mabuchi started developing some independent ideas using infinite-dimensional metric geometry. Independently, Semmes, in the US, who much like Bo was an outsider, found a connection between Kähler geometry and interpolation. And, Donaldson, also independently, about a decade after Mabuchi and Semmes, came up with similar ideas and many of his own. Additionally, Kołodziej brought pluripotential theory into the mix, generalizing Aubin and Yau's estimates. But it really wasn't until Bo came along that the full force of $\bar{\partial}$ theory and Bergman kernels was unleashed on the hardest problems of Kähler geometry.

As far as I can tell, one difference between Bo's introduction into Kähler geometry of those methods coming from a different field and previous such attempts was that Bo was a geometer-waiting-to-be-discovered. Second, Bo was actually developing those methods on $\bar{\partial}$-equations for many years and had just proved the best result in the field ("Theorem 1.1"), perhaps since Hörmander's theorem in the 1960s. (Here, again, Bo used to say that his Theorem 1.1 was a sort of consequence of Hörmander's theorem.)

At the same time, Bo had the remarkable curiosity of a graduate student, even though he was in his 50s.

One of Theorem 1.1's most remarkable consequences was the Complex Bo–Minkowski Theory. This is the theory of how to apply Theorem 1.1 in Complex Geometry as a black box, without having to understand it. And, as in Convex Geometry, it is the theory of basically all that varies in families.

Moving away from Complex Geometry. I also want to mention that Bo is a man of principles and ideals. For instance, as a chair of the department of mathematics at Chalmers he resigned in order to protect the jobs of certain of his colleagues and stood up to the higher-up admins. He also dedicated one of his papers, in fact the one on his resolution of the Openness Conjecture, to whistleblowers.

As much as Bo's papers have influenced a whole generation, perhaps his outlook on life and mathematics have influenced so many of us as well. His kindness, sense of humor, and generosity with ideas are what I refer to as "Berndtsson positivity". They have influenced me deeply.

I'll end with a story as an illustratation. When I visited Bo and Robert in 2014, I brought along my puppy. Bo enjoyed her company very much, and he would kindly dog-sit on occasions, as he had a spacious office with a comfortable sofa that my dog liked, and so I equipped Bo's office with a dog water bowl. When I came to Chalmers again about three years later, that water bowl was still there. I asked Bo why, to which he replied, "this is for my visitors".

Finally, if I had to go to a desert island and could only bring one book of collected works, I would undoubtedly bring along that of ... Alexandre Dumas, Raymond Chandler, ... hmmmm .... But, if you force me to bring a mathematics author, I would bring along either that of Bo Berndtsson or of Blaine Lawson.

Thank you all for coming here to celebrate Berndtsson positivity.

Acknowledgments. First and foremost, this conference owes its existence to Bo Berndtsson, who inspired several generations of mathematicians with his original ideas and foundational contributions, and, importantly, his sense of humor and free exchange of ideas.

Second, this conference owes so much to the inspiring setting of the Grand Hotel San Michele at Cetraro and the unsurpassed hospitality and attention to detail of its staff, spearheaded by its General Manager, Dott. Andrea Peiser. We would also like to express our special gratitude to him for allowing us to hold the conference two years after its original scheduled date and for going out of his way to accommodate so many special requests en route to an unforgettable workshop.

We are also grateful to the Swedish Research Council for funding the conference.

Finally, we thank the participants and speakers in the workshop, the authors of this volume for their inspiring contributions, and the referees for their meticulous work.

ROBERT J. BERMAN
YANIR A. RUBINSTEIN

Conference Participants

Photo courtesy of Mingchen Xia

FIGURE 1. From left to right: Anders Krøger Evensen, Tai Nguyen, Young-Jun Choi, Yan He, Xu Wang, Sébastien Boucksom, Chenzi Ji, Remi Reboulet, Dror Varolin, Rolf Andreasson, Vlassis Mastrantonis, Julius Ross, Jakob Hultgren, László Lempert, Mattias Jonsson, Robert Berman, Tamas Darvas, Mingchen Xia, Zbigniew Blocki, David Witt-Nyström, Bo Berndtsson, and Yanir Rubinstein

Contemporary Mathematics
Volume **810**, 2025
https://doi.org/10.1090/conm/810/16211

The Cheltsov–Rubinstein problem for strongly asymptotically log del Pezzo surfaces

Chenzi Jin

ABSTRACT. The notion of (strongly) asymptotically log Fano varieties was introduced in 2013 by Cheltsov–Rubinstein, who posed the problem of classifying all strongly asymptotically log del Pezzo surfaces with smooth boundary that admit Kähler–Einstein edge metrics. Thanks to the Cheltsov–Rubinstein classification, this amounts to considering 10 families. In 8 families the problem has been solved by work of Cheltsov–Rubinstein, Fujita and Mazzeo–Rubinstein. The remaining 2 families are rational surfaces parameterized by the self-intersection of the 0-section n and the number of blow-ups m. By Cheltsov–Rubinstein, Cheltsov–Rubinstein–Zhang and Fujita, Kähler–Einstein edge metrics exist when either $m = 0$ or $m \geq 3$ for the first family, and the cases $m = 1, 2$ have been studied by Fujita–Liu–Süß–Zhang–Zhuang and Fujita. The final remaining family, denoted (II.6A.n.m) in the Cheltsov–Rubinstein classification, is more difficult as the boundary consists of two components, unlike any of the other 9 families. It is the generalization of the football to complex surfaces with the pair (II.6A.0.0) being exactly the football times $\mathbb{P}^1$. The pairs (II.6A.n.0) have been completely understood by the work of Rubinstein–Zhang using the $\mathbb{P}^1$-bundle structure of Hirzebruch surfaces. This article studies the family (II.6A.n.m) for $m \geq 1$. These pairs no longer have a $\mathbb{P}^1$-bundle structure and are therefore more difficult to tackle. The main result is a necessary and sufficient condition on the angles for the existence of Kähler–Einstein edge metrics, generalizing the Rubinstein–Zhang condition. Thus, we resolve the Cheltsov–Rubinstein problem for strongly asymptotically log del Pezzo surfaces.

CONTENTS

2020 *Mathematics Subject Classification.* Primary 32Q20; Secondary 14J45, 32Q26.

This is part of the author's PhD work. The author is grateful to his advisor Y.A. Rubinstein for suggesting this problem, and a careful reading and several corrections. The author is thankful to K. Fujita, I. Cheltsov and K. Zhang for their comments and informing the author of the results in [**1, 19**]. The author is also grateful to the referee for the careful reading and comments that improved the presentation of the paper. The research is supported by NSF grants DMS-1906370, 2204347, and University of Maryland Summer Research Fellowship for Summer 2022.

1. Introduction

The purpose of this article is to complete the solution of the Cheltsov–Rubinstein problem [**7**, **22**] on the classfication of strongly asymptotically log del Pezzo surfaces admitting a Kähler–Einstein edge metric. This has been a topic of intense research. See, e.g., [**7**–**9**, **14**, **22**–**24**]. As a consequence of these works, one (infinite) family of pairs remains to be understood: the pairs (II.6A.n.m) (see Definition 1.6). In this article we treat this family completely and hence solve the Cheltsov–Rubinstein problem for strongly asymptotically log del Pezzo surfaces. The Cheltsov–Rubinstein problem for general asymptotically log del Pezzo surfaces is very challenging and remains wide open.

To explain the problem and our work let us introduce some notation. Let X be a Kähler manifold. Fix a divisor $D = D_1 + \cdots + D_r$ with simple normal crossings on X, and angle vector $\beta = (\beta_1, \ldots, \beta_r) \in (0, 1]^r$. A Kähler–Einstein edge metric is a smooth Kähler–Einstein metric on $X \setminus \bigcup_i D_i$ that has a conical singularity of angle $2\pi\beta_i$ along D_i [**16**, **22**, **26**]. The existence of a Kähler–Einstein edge metric with positive Ricci curvature imposes the cohomological condition that

$$-K_X - \sum_{i=1}^{r} (1 - \beta_i) D_i$$

is ample. When $\beta_i = 1$, the metric reduces to the ordinary (smooth) Kähler–Einstein metric. On the other hand, the case when the β_i's are *small* is also of much interest; this corresponds to the notion of asymptotically log Fano varieties, introduced by Cheltsov–Rubinstein [**7**]:

Definition 1.1. $(X, D_1 + \cdots + D_r)$ is *(strongly) asymptotically log Fano* if $-K_X - \sum_{i=1}^{r} (1 - \beta_i) D_i$ is ample for (all) sufficiently small $\beta \in (0, 1]^r$. In dimension 2, these are also called *(strongly) asymptotically log del Pezzo surfaces.*

Strongly asymptotically log del Pezzo surfaces have been classified by Cheltsov–Rubinstein [**7**, **23**]. A natural question to ask is the existence of Kähler–Einstein edge metrics on such surfaces, namely the following uniformization problem posed by Cheltsov–Rubinstein [**22**, Problem 9.1].

Problem 1.2. *Determine which strongly asymptotically log del Pezzo surfaces admit Kähler–Einstein edge metrics for sufficiently small β.*

Assume the boundary is smooth. By Cheltsov–Rubinstein classification, we are reduced to the following 10 families.

Theorem 1.3 ([**7**, Theorems 2.1 and 3.1]). *Let S be a smooth surface, let $C_1, \ldots, C_r$ be disjoint smooth irreducible curves on S. Let Z_n denote the curve of self-intersection $-n$ on the Hirzebruch surface $\mathbb{F}_n$, and F a fiber on it. Then $(S, C_1 + \cdots + C_r)$ is strongly asymptotically log del Pezzo if and only if it is one of the following pairs.*

(I.3A) *$S \cong \mathbb{F}_1$, and $C_1 \in |2(Z_1 + F)|$;*
(I.4A) *$S \cong \mathbb{P}^1 \times \mathbb{P}^1$, and C_1 is a smooth elliptic curve of bi-degree $(2, 2)$;*
(I.5.m) *S is a blow-up of $\mathbb{P}^2$ at $0 \leq m \leq 8$ distinct points on a smooth cubic elliptic curve such that $-K_S$ is ample, and C_1 is the proper transform of the elliptic curve;*

(I.6B.m) *S is a blow-up of $\mathbb{P}^2$ at $m \geq 0$ distinct points on a smooth conic, and C_1 is the proper transform of the conic;*

(I.6C.m) *S is a blow-up of $\mathbb{P}^2$ at $m \geq 0$ distinct points on a line, and C_1 is the proper transform of the line;*

(I.7.n.m) *S is a blow-up of $\mathbb{F}_n$ at $m \geq 0$ distinct points on Z_n, and C_1 is the proper transform of Z_n, for any $n \geq 0$;*

(I.8B.m) *S is a blow-up of $\mathbb{F}_1$ at $m \geq 0$ distinct points on a smooth rational curve in the linear system $|Z_1 + F|$, and C_1 is the proper transform of the rational curve;*

(I.9B.m) *S is a blow-up of $\mathbb{P}^1 \times \mathbb{P}^1$ at $m \geq 0$ distinct points on a smooth rational curve of bi-degree $(2,1)$ with no two of them on a single curve of bi-degree $(0,1)$, and C_1 is the proper transform of the rational curve;*

(I.9C.m) *S is a blow-up of $\mathbb{P}^1 \times \mathbb{P}^1$ at $m \geq 0$ distinct points on a smooth rational curve of bi-degree $(1,1)$, and C_1 is the proper transform of the rational curve;*

(II.6A.n.m) *S is a blow-up of $\mathbb{F}_n$ at $m \geq 0$ distinct points on Z_n and a smooth rational curve in the linear system $|Z_n + nF|$ with no two of them on a single curve in the linear system $|F|$, C_1 is the proper transform of Z_n, and C_2 is the proper transform of the rational curve, for any $n \geq 0$.*

As a step towards Problem 1.2, Cheltsov–Rubinstein proposed the following conjecture [**7**, Conjecture 1.6].

CONJECTURE 1.4. *Let (S, C) be a strongly asymptotically log del Pezzo surface with $r = 1$. Then (S, C) admits a Kähler–Einstein edge metric for sufficiently small β if and only if $(K_S + C)^2 = 0$.*

The necessary part and all but 6 sub-families of the sufficient part were verified. Namely, the pairs (I.6B.m),(I.6C.m),(I.7.n.m),(I.8B.m),(I.9C.m) do not admit Kähler–Einstein edge metrics with small angles [**8**, Theorem 1.6], while the pairs (I.3A),(I.4A),(I.5.m), and (I.9B.m) for $m = 0$ or $m \geq 7$ admit Kähler–Einstein edge metrics with small angles [**7**, Propositions 7.4 and 7.5], [**9**, Theorem 1.3], [**16**, Corollary 1].

Let us now discuss the remaining cases, namely, the sub-families (I.9B.m) with $1 \leq m \leq 6$ and (II.6A.n.m).

For the sub-families (I.9B.m) with $1 \leq m \leq 6$, each sub-family is parameterized by the position of the m blown-up points on the smooth $(2,1)$-curve. Let p_0 and p_∞ denote the points where the $(2,1)$-curve on $\mathbb{P}^1 \times \mathbb{P}^1$ is tangent to a $(0,1)$-curve. Fujita–Liu–Süß–Zhang–Zhuang showed that the pairs (I.9B.m) contradict the conjecture if $m = 1, 2$ and exactly 1 of the points p_0 and p_∞ is blown-up, by computing an upper bound for the δ-invariant [**14**, Corollary 2.9]. Fujita verified that these 2 pairs are the only counterexamples in the 6 sub-families [**12**, Theorem 1.3]. Thus, Conjecture 1.4 of Cheltsov–Rubinstein holds for the vast majority of the cases, and the remaining cases are semi-stable degenerations that are now completely understood by the aforementioned works.

In this article we restrict to the remaining tenth and final family, (II.6A.n.m), where $\operatorname{Supp} C$ has multiple components, and hence is not within the scope of Conjecture 1.4.

PROBLEM 1.5. *Determine whether (II.6A.n.m) admit Kähler–Einstein edge metrics for sufficiently small β.*

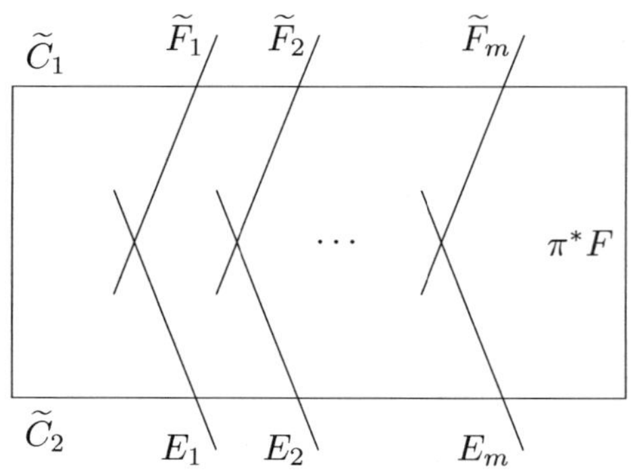

FIGURE 1. The pair (II.6A.n.m). The surface is given by blowing up m points $p_1, \dots, p_m$ on the curve $C_2 \subseteq \mathbb{F}_n$, where $\mathbb{F}_n$ is the Hirzebruch surface containing a curve C_1 whose self intersection number is $-n$ and a generic fiber F, and C_2 is a smooth element of $|C_1 + nF|$. The fiber through the point p_i is denoted by F_i, and the exceptional curve over p_i is denoted E_i. The curves $\widetilde{C}_1, \widetilde{C}_2, \widetilde{F}_1, \dots, \widetilde{F}_m$ are the proper transforms of $C_1, C_2, F_1, \dots, F_m$, respectively.

First we introduce some notation to describe the curves appearing in the study of (II.6A.n.m).

DEFINITION 1.6. Let $\mathbb{F}_n$ denote the Hirzebruch surface containing a curve C_1 whose self intersection number is $-n$, and pick a smooth curve C_2 in $|C_1 + nF|$ where F denotes a fiber. Let S be the blow-up of $\mathbb{F}_n$ at m points on $C_1 \cup C_2$ such that no two of them lie on the same fiber, and $C = \widetilde{C}_1 + \widetilde{C}_2$, where $\widetilde{C}_i \subseteq S$ is the proper transform of C_i. This family is named (II.6A.n.m). See Figure 1.

This family is the generalization of the football to complex surfaces. In particular, the pair (II.6A.0.0) can precisely be equipped with the standard conical metric on the football times $\mathbb{P}^1$, if $\beta_1 = \beta_2$. Generalizing the product football metric, using the $\mathbb{P}^1$-bundle structure of Hirzebruch surfaces, Rubinstein–Zhang constructed a family of Kähler–Einstein edge metrics using the Calabi ansatz [**24**, Theorem 1.2], and settled the case $m = 0$ for all n by finding an explicit relation between β_1 and β_2.

THEOREM 1.7 ([**24**, Theorem 1.2]). *Consider (II.6A.n.0). Namely, S is the Hirzebruch surface $\mathbb{F}_n$ containing a curve C_1 whose self intersection number is $-n$, and C_2 is a smooth curve in $|C_1 + nF|$ where F denotes a fiber. Let β_1 and β_2 be the cone angles along the boundary curves C_1 and C_2. Then Kähler–Einstein edge metrics exist if and only if*

$$\beta_1^2 - \frac{n}{3}\beta_1^3 = \beta_2^2 + \frac{n}{3}\beta_2^3. \tag{1.1}$$

It is difficult to generalize such a construction to the cases $m \geq 1$, as these pairs no longer have a $\mathbb{P}^1$-bundle structure. In this article, we study Problem 1.5 for $m \geq 1$ using algebraic methods, i.e., verifying log K-stability.

The notion of K-stability introduced by Tian [**27**] and refined by Donaldson [**10**], Berman [**3**], Fujita–Odaka [**15**] and Fujita [**12**] turned out to characterize the existence of Kähler–Einstein metrics. It was also extended to log setting [**11**], namely Kähler–Einstein edge metrics exist if and only if the pair is log K-polystable [**4**]. Moreover, Fujita–Odaka [**15**] introduced the δ-invariant (Definition 2.7) that characterizes K-stability. In fact, it has been shown by Blum–Jonsson [**6**] that $\delta > 1$ is equivalent to (uniform) K-stability, which is stronger than K-polystability, and $\delta \geq 1$ is equivalent to K-semistability, which is weaker than K-polystability. This was also extended to log setting [**5**].

The δ-invariant is the main technical tool used in this article. Our main result generalizes (1.1), imposing a relation on the cone angles.

THEOREM 1.8. *Consider (II.6A.n.m) as in Definition* 1.6. *We have*

$$\delta\left(S, (1-\beta_1)\widetilde{C}_1 + (1-\beta_2)\widetilde{C}_2\right) \leq 1,$$

with equality if and only if

$$\beta_1^2 + \frac{1}{3}\widetilde{C}_1^2\beta_1^3 = \beta_2^2 + \frac{1}{3}\widetilde{C}_2^2\beta_2^3, \quad m \neq 1. \tag{1.2}$$

The pair is log K-polystable, i.e., it admits a Kähler–Einstein edge metric, if and only if (1.2) *holds.*

The equivalence between stability and existence of Kähler–Einstein edge metrics used in the statement is due to the resolution of the logarithmic Yau–Tian–Donaldson conjecture [**20**, Theorem 1.6].

Theorem 1.8 generalizes Rubinstein–Zhang's Theorem 1.7 and completely solves Problem 1.5.

Organization. We first prove the necessary part of the main theorem by directly using the definition of the δ-invariant. Section 2 contains the preliminary. We explain the setup in Section 2.1. Then in Sections 2.2 and 2.3 we recall the definition of the δ-invariant and Zariski decomposition which serves as the main tool in the computation of the δ-invariant. The nef cone and the negative definite intersection matrices are computed in Sections 3 and 4, respectively, as preparation for the computation of the volumes of divisors. Finally, in Section 5, we use the computation above to give upper bounds for the δ-invariant.

Next we make use of the T-variety structure of the pair to give an alternative proof of the necessary part, as well as a proof of the sufficient part. In Section 6 we recall the relevant definitions and theorems. The proof is given in Section 7.

2. Preliminaries

2.1. The pairs (II.6A.n.m). In this section, we start by explaining the definition of the family (II.6A.n.m). As a preparation, we also compute the intersection numbers of curves as well as the log anti-canonical divisor of the surface S.

For $n \geq 0$, let $\mathbb{F}_n$ denote the Hirzebruch surface containing a curve C_1 whose self intersection number is $-n$. A fiber F is a smooth curve on $\mathbb{F}_n$ whose self intersection number is 0. Pick a smooth element of $|C_1 + nF|$ and call it C_2. Let S be the blow-up of $\mathbb{F}_n$ at m points $p_1, \ldots, p_m$ on $C_1 \cup C_2$ such that no two of them lie on the same fiber, and $\widetilde{C}_i \subseteq S$ the proper transform of C_i. Consider the pair $(S, \widetilde{C}_1 + \widetilde{C}_2)$. Let $\pi : S \to \mathbb{F}_n$ denote the blow-up morphism, E_i the exceptional divisor $\pi^{-1}(p_i)$, F a generic fiber, i.e., a fiber not through any p_i, and F_i the fiber through p_i. For an irreducible curve $C \subseteq \mathbb{F}_n$, let $\widetilde{C}$ denote its proper transform. For $i = 1, 2$, define

$$I_i := \{j \,|\, p_j \in C_i\}, \quad m_i := |I_i|.$$

REMARK 2.1. Recall that $\mathbb{F}_{n+1}$ can be obtained by blowing up a point on the $-n$-curve of $\mathbb{F}_n$ and then blow down the proper transform of the fiber through that point. In other words, if S is obtained from $\mathbb{F}_n$ by blowing up m_1 points on $C_1 \subseteq \mathbb{F}_n$ and m_2 points on $C_2 \subseteq \mathbb{F}_n$ such that no two of them lie on the same fiber, then it can also be obtained from $\mathbb{F}_{n+m_1}$ by blowing up $m_1 + m_2$ points on $C_2 \subseteq \mathbb{F}_{n+m_1}$. Because of this, we may assume

$$m_1 = 0, \quad m_2 = m.$$

The Picard group of S is generated by the linear equivalent classes of $\pi^* C_2$, $\pi^* F$ and E_i, giving intersection matrix

$$\begin{pmatrix} n & 1 & & & \\ 1 & 0 & & & \\ & & -1 & & \\ & & & \ddots & \\ & & & & -1 \end{pmatrix}.$$

Note that

$$\begin{aligned} \widetilde{C}_1 &\sim \pi^* (C_2 - nF), \\ \widetilde{C}_2 &\sim \pi^* C_2 - \sum_i E_i, \\ -K_S = \widetilde{C}_1 + \widetilde{C}_2 + 2\pi^* F &\sim \pi^* (2C_2 + (2-n) F) - \sum_i E_i. \end{aligned}$$

Let

$$\begin{aligned} \Delta_\beta := (1-\beta_1)\widetilde{C}_1 + (1-\beta_2)\widetilde{C}_2 \sim_{\mathbb{Q}} \pi^* ((2-\beta_1-\beta_2) C_2 - (n - n\beta_1) F) \\ - (1-\beta_2) \sum_i E_i. \end{aligned}$$

Then

$$-K_{S,\Delta_\beta} := -K_S - \Delta_\beta = \pi^* ((\beta_1 + \beta_2) C_2 + (2 - n\beta_1) F) - \beta_2 \sum_i E_i. \tag{2.1}$$

2.2. Nefness, bigness, volume and Zariski decomposition. In this section, we recall the definition of nefness, bigness, and volume [**18**, Deinition 1.4.1, Definition 2.2.1, Definition 2.2.31 and Remark 2.2.39]. We also recall Zariski decomposition, which we will use as the main tool to compute the volume of a divisor.

DEFINITION 2.2. Let X be a projective manifold of dimension n and D a divisor on X. D is called *numerically effective*, or *nef*, if

$$D \cdot C \geq 0$$

for any irreducible curve C. Define its *volume*

$$\operatorname{vol}(D) := \lim_{k\to\infty} \frac{n!}{k^n} \dim H^0 (X, kD).$$

In general, if D is a $\mathbb{Q}$-divisor, assume its multiple mD is an integral divisor. Then its volume is defined by scaling

$$\operatorname{vol}(D) := \frac{1}{m^n} \operatorname{vol}(mD).$$

We say D is *big* if $\operatorname{vol}(D) > 0$.

To compute the volume, the following standard result is useful [**18**, Corollary 1.4.41].

PROPOSITION 2.3. *Let X be a projective manifold of dimension n and D a nef divisor on X. Then*

$$\operatorname{vol}(D) = D^n.$$

When X is a surface, we can use Zariski decomposition to compute the volume [**28**, Theorems 7.7 and 8.1].

DEFINITION 2.4. Let D be an effective $\mathbb{Q}$-divisor on a smooth projective surface S. There is a unique effective $\mathbb{Q}$-divisor $N = \sum_i a_i N_i$ satisfying the following:

(1) $D - N$ is nef;
(2) The intersection matrix $(N_i \cdot N_j)$ is negative definite;
(3) For any i,

$$(D - N) \cdot N_i = 0. \tag{2.2}$$

The $\mathbb{Q}$-divisor N is called the *negative part* of D, and the $\mathbb{Q}$-divisor

$$P := D - N$$

is called the *positive part* of D.

PROPOSITION 2.5. *Let D be an effective divisor on a smooth projective surface S with Zariski decomposition $D = P + N$. Then*

$$\operatorname{vol}(D) = \operatorname{vol}(P) = P^2.$$

2.3. δ-invariant. In this section we recall the definition of the δ-invariant.

Let X be a projective manifold of dimension n, and $\Delta = \sum_i a_i D_i$ a simple normal crossing $\mathbb{Q}$-divisor on X with coefficient $a_i \in (0,1)$. Assume

$$K_{X,\Delta} := K_X + \Delta$$

is an anti-ample $\mathbb{Q}$-divisor.

Recall the following definitions [**4**, Theorem 3.3] [**17**, Definition 2.24] .

DEFINITION 2.6. Let X be a projective manifold. We call E *a prime divisor over* X if there is a projective birational morphism $f : Y \to X$ with normal variety Y, and E is a prime divisor on Y. For a big line bundle $L \to X$, define the *pseudo-effective threshold*

$$\tau_L(E) := \sup\{x \in \mathbb{Q}_{>0} \mid f^*(L) - xE \text{ is big}\},$$

and the *expected vanishing order*

$$S_L(E) := \frac{1}{\operatorname{vol}(L)} \int_0^{\tau_L(E)} \operatorname{vol}(f^*(L) - xE)\, dx.$$

Define its *log discrepancy*

$$A_{X,\Delta}(E) := 1 - \operatorname{ord}_E(f^*(K_{X,\Delta}) - K_Y). \tag{2.3}$$

Below is the definition of the δ-invariant [**4**, Theorem C] that we will use in this article.

DEFINITION 2.7. Let X be a projective manifold. For a big line bundle $L \to X$, define the *δ-invariant*

$$\delta(X, \Delta; L) := \inf_E \frac{A_{X,\Delta}(E)}{S_L(E)},$$

where E is taken to be prime divisors over X. In particular, if $L = -K_{X,\Delta}$, set

$$\delta(X, \Delta) := \delta(X, \Delta; -K_{X,\Delta}).$$

3. Nefness criterion

For the remainder of the article we let (S, C) be the pair (II.6A.n.m). See Section 2.1.

To compute Zariski decomposition, we need to check nefness of various divisors. To simplify this process, in this section we first compute the nef cone. The result is the following.

LEMMA 3.1. *A $\mathbb{Q}$-divisor*

$$D \sim_{\mathbb{Q}} \pi^*(aC_2 + bF) - \sum_i c_i E_i$$

is nef if and only if

$$\begin{aligned} 0 &\le D \cdot E_i = c_i, \\ 0 &\le D \cdot \widetilde{F}_i = a - c_i, \\ 0 &\le D \cdot \widetilde{C}_1 = b, \\ 0 &\le D \cdot \widetilde{C}_2 = na + b - \sum_i c_i. \end{aligned} \tag{3.1}$$

PROOF. Recall Definition 2.2. If D is nef, then (3.1) is necessary.

On the other hand, if (3.1) is satisfied, then the $\mathbb{Q}$-divisor

$$D' := a\widetilde{C}_2 + \sum_i (a - c_i) E_i + b\pi^* F \sim_{\mathbb{Q}} D$$

is effective. To prove that D is nef, it suffices to show that $D \cdot C \ge 0$ for each irreducible component C of D'. Since (3.1) is satisfied,

$$\begin{aligned} D \cdot \widetilde{C}_2 &= na + b - \sum_i c_i \ge 0, \\ D \cdot E_i &= c_i \ge 0, \\ D \cdot \pi^* F &= a \ge 0. \end{aligned}$$

This shows D is nef. □

COROLLARY 3.2. *The log anti-canonical divisor $-K_{S,\Delta_\beta}$ is ample if and only if*

$$\beta_1 \in \left(0, \frac{2}{n}\right), \quad \beta_2 \in \begin{cases} (0, +\infty), & n \ge m \\ \left(0, \frac{2}{m-n}\right), & n < m. \end{cases} \tag{3.2}$$

PROOF. Since the ample cone is the interior of the nef cone [**18**, Theorem 1.4.23], $-K_{S,\Delta_\beta}$ is ample if and only if the inequalities of (3.1) are strict. By (2.1),

$$\begin{aligned} \beta_2 &> 0, \\ \beta_1 &> 0, \\ 2 - n\beta_1 &> 0, \\ 2 + n\beta_2 &> m\beta_2. \end{aligned}$$

This is (3.2). □

4. Intersection matrices

Since the Zariski decomposition of a divisor also involves curves with negative definite intersection matrices, in this section we give some collections of curves that have negative definite intersection matrices. They will be used in the computation of Zariski decompositions.

CLAIM 4.1. *The following families of divisors have negative definite intersection matrices.*

(1) $\widetilde{F}_1, \dots, \widetilde{F}_m$;
(2) $E_1, \dots, E_m$;
(3) $\widetilde{C}_1, \widetilde{F}_i$, *where* $n > 1$ *and* $1 \le i \le m$;
(4) $\widetilde{C}_1, \widetilde{F}_1, \dots, \widetilde{F}_m$, *where* $n > m$;
(5) $\widetilde{C}_2, \widetilde{F}_i$, *where* $n < m$ *and* $1 \le i \le m$;
(6) $\widetilde{C}_1, \widetilde{C}_2, \widetilde{F}_i$, *where* $1 < n < m$ *and* $1 \le i \le m$;
(7) $\widetilde{C}_2, \widetilde{F}_i, E_1, \dots, \widehat{E}_i, \dots, E_m$, *where* $n = 0$ *and* $1 \le i \le m$.

PROOF. (1) The intersection matrix is $-I$, which is negative definite.
(2) The intersection matrix is $-I$, which is negative definite.
(3) The intersection matrix is

$$\begin{pmatrix} -n & 1 \\ 1 & -1 \end{pmatrix}.$$

Since

$$\begin{pmatrix} -n & 1 \\ 1 & -1 \end{pmatrix} = \begin{pmatrix} 1 & -1 \\ & 1 \end{pmatrix} \begin{pmatrix} -(n-1) & \\ & -1 \end{pmatrix} \begin{pmatrix} 1 & \\ -1 & 1 \end{pmatrix},$$

it is negative definite.
(4) The intersection matrix is

$$\begin{pmatrix} -n & 1 & \cdots & 1 \\ 1 & -1 & & \\ \vdots & & \ddots & \\ 1 & & & -1 \end{pmatrix}.$$

Since

$$
\begin{pmatrix} -n & 1 & \cdots & 1 \\ 1 & -1 & & \\ \vdots & & \ddots & \\ 1 & & & -1 \end{pmatrix}
= \begin{pmatrix} 1 & -1 & \cdots & -1 \\ & 1 & & \\ & & \ddots & \\ & & & 1 \end{pmatrix}
\begin{pmatrix} -(n-m) & & & \\ & -1 & & \\ & & \ddots & \\ & & & -1 \end{pmatrix}
\begin{pmatrix} 1 & & & \\ -1 & 1 & & \\ \vdots & & \ddots & \\ -1 & & & 1 \end{pmatrix},
$$

it is negative definite.

(5) The intersection matrix is

$$
\begin{pmatrix} -(m-n) & \\ & -1 \end{pmatrix},
$$

which is negative definite.

(6) The intersection matrix is

$$
\begin{pmatrix} -n & & 1 \\ & -(m-n) & \\ 1 & & -1 \end{pmatrix}.
$$

Since

$$
\begin{pmatrix} -n & & 1 \\ & -(m-n) & \\ 1 & & -1 \end{pmatrix}
= \begin{pmatrix} 1 & & -1 \\ & 1 & \\ & & 1 \end{pmatrix}
\begin{pmatrix} -(n-1) & & \\ & -(m-n) & \\ & & -1 \end{pmatrix}
\begin{pmatrix} 1 & & \\ & 1 & \\ -1 & & 1 \end{pmatrix},
$$

it is negative definite.

(7) We may assume $i = 1$. The intersection matrix is

$$
\begin{pmatrix} -m & 0 & 1 & \cdots & 1 \\ 0 & -1 & & & \\ 1 & & -1 & & \\ \vdots & & & \ddots & \\ 1 & & & & -1 \end{pmatrix}.
$$

Since

$$\begin{pmatrix} -m & 0 & 1 & \cdots & 1 \\ 0 & -1 & & & \\ 1 & & -1 & & \\ \vdots & & & \ddots & \\ 1 & & & & -1 \end{pmatrix} = \begin{pmatrix} 1 & 0 & -1 & \cdots & -1 \\ & 1 & & & \\ & & 1 & & \\ & & & \ddots & \\ & & & & 1 \end{pmatrix} \begin{pmatrix} -1 & & & & \\ & -1 & & & \\ & & -1 & & \\ & & & \ddots & \\ & & & & -1 \end{pmatrix} \begin{pmatrix} 1 & & & & \\ 0 & 1 & & & \\ -1 & & 1 & & \\ \vdots & & & \ddots & \\ -1 & & & & 1 \end{pmatrix},$$

it is negative definite. □

5. Upper bounds for the delta-invariant

In this section, we give some upper bounds for the δ-invariant by computing the ratio

$$\frac{A_{S,\Delta_\beta}(E)}{S_L(E)}$$

for various divisors E. In particular, by letting $E = \widetilde{C}_1, \widetilde{C}_2$ we prove Theorem 5.5, and by letting $E = E_i$ we prove Theorem 5.11. They combine to give the necessary part of the main theorem.

LEMMA 5.1. *The pseudo-effective threshold*

$$\tau_{-K_{S,\Delta_\beta}}\left(\widetilde{C}_1\right) = \beta_1 + \beta_2,$$

and

$$\operatorname{vol}\left(-K_{S,\Delta_\beta} - x\widetilde{C}_1\right) = \begin{cases} \operatorname{vol}\left(-K_{S,\Delta_\beta}\right) - 2\left(2 - n\beta_1\right)x - nx^2, & x \in [0, \beta_1], \\ \left(\beta_1 + \beta_2 - x\right)\left(4 - (m-n)\left(x - \beta_1 + \beta_2\right)\right), & x \in (\beta_1, \beta_1 + \beta_2]. \end{cases}$$

PROOF. By (3.1) and (3.2),

$$-K_{S,\Delta_\beta} - x\widetilde{C}_1 = \pi^*\left(\left(\beta_1 + \beta_2 - x\right)C_2 + \left(2 - n\beta_1 + nx\right)F\right) - \beta_2 \sum_i E_i$$

is nef when $x \in [0, \beta_1]$. Therefore by Proposition 2.3,

$$\operatorname{vol}\left(-K_{S,\Delta_\beta} - x\widetilde{C}_1\right) = \operatorname{vol}\left(-K_{S,\Delta_\beta}\right) - 2\left(2 - n\beta_1\right)x - nx^2.$$

When x reaches β_1, the intersection number $(-K_{S,\Delta_\beta} - x\widetilde{C}_1) \cdot \widetilde{F}_i$ decreases to 0. When $x > \beta_1$, assume there is an effective $\mathbb{R}$-divisor N supported on $\bigcup_i \widetilde{F}_i$ that satisfies (2.2). We can solve

$$N = (x - \beta_1)\sum_i \widetilde{F}_i.$$

By (3.1) and (3.2),

$$\begin{aligned}-K_{S,\Delta_\beta}-x\widetilde{C}_1-N=\pi^*\left(\left(\beta_1+\beta_2-x\right)C_2+\left(2+\left(m-n\right)\left(\beta_1-x\right)\right)F\right)\\-\left(\beta_1+\beta_2-x\right)\sum_i E_i\end{aligned}$$

is nef when $x\in(\beta_1,\beta_1+\beta_2]$. Also note Claim 4.1(1). By Definition 2.4, N is indeed the negative part of $-K_{S,\Delta_\beta}-x\widetilde{C}_1$ for $x\in(\beta_1,\beta_1+\beta_2]$. Therefore by Proposition 2.5,

$$\operatorname{vol}\left(-K_{S,\Delta_\beta}-x\widetilde{C}_1\right)=\left(\beta_1+\beta_2-x\right)\left(4-\left(m-n\right)\left(x-\beta_1+\beta_2\right)\right).$$

When x reaches $\beta_1+\beta_2$, the volume $\operatorname{vol}(-K_{S,\Delta_\beta}-x\widetilde{C}_1)$ decreases to 0. This implies

$$\tau_{-K_{S,\Delta_\beta}}\left(\widetilde{C}_1\right)=\beta_1+\beta_2.$$

□

LEMMA 5.2. *The expected vanishing order*

$$S_{-K_{S,\Delta_\beta}}\left(\widetilde{C}_1\right)=\beta_1-\frac{2}{\operatorname{vol}\left(-K_{S,\Delta_\beta}\right)}\left(\beta_1^2-\beta_2^2+\frac{1}{3}\left(-n\beta_1^3+\left(m-n\right)\beta_2^3\right)\right).$$

PROOF. By Lemma 5.1,

$$\int_0^{\beta_1}\operatorname{vol}\left(-K_{S,\Delta_\beta}-x\widetilde{C}_1\right)dx=\beta_1\operatorname{vol}\left(-K_{S,\Delta_\beta}\right)-2\beta_1^2+\frac{2}{3}n\beta_1^3,$$

$$\int_{\beta_1}^{\beta_1+\beta_2}\operatorname{vol}\left(-K_{S,\Delta_\beta}-x\widetilde{C}_1\right)dx=2\beta_2^2-\frac{2}{3}\left(m-n\right)\beta_2^3.$$

Therefore

$$\begin{aligned}S_{-K_{S,\Delta_\beta}}\left(\widetilde{C}_1\right)&=\frac{1}{\operatorname{vol}\left(-K_{S,\Delta_\beta}\right)}\int_0^{\beta_1+\beta_2}\operatorname{vol}\left(-K_{S,\Delta_\beta}-x\widetilde{C}_1\right)dx\\&=\beta_1-\frac{2}{\operatorname{vol}\left(-K_{S,\Delta_\beta}\right)}\left(\beta_1^2-\beta_2^2+\frac{1}{3}\left(-n\beta_1^3+\left(m-n\right)\beta_2^3\right)\right).\end{aligned}$$

□

LEMMA 5.3. *The pseudo-effective threshold*

$$\tau_{-K_{S,\Delta_\beta}}\left(\widetilde{C}_2\right)=\beta_1+\beta_2,$$

and

$$\begin{aligned}&\operatorname{vol}\left(-K_{S,\Delta_\beta}-x\widetilde{C}_2\right)\\&=\begin{cases}\operatorname{vol}\left(-K_{S,\Delta_\beta}\right)-2\left(2-\left(m-n\right)\beta_2\right)x-\left(m-n\right)x^2, & x\in[0,\beta_2],\\ \left(\beta_1+\beta_2-x\right)\left(4-n\left(x+\beta_1-\beta_2\right)\right), & x\in(\beta_2,\beta_1+\beta_2].\end{cases}\end{aligned}$$

PROOF. By (3.1) and (3.2),

$$-K_{S,\Delta_\beta}-x\widetilde{C}_2=\pi^*\left(\left(\beta_1+\beta_2-x\right)C_2+\left(2-n\beta_1\right)F\right)-\left(\beta_2-x\right)\sum_i E_i$$

is nef when $x\in[0,\beta_2]$. Therefore by Proposition 2.3,

$$\operatorname{vol}\left(-K_{S,\Delta_\beta}-x\widetilde{C}_2\right)=\operatorname{vol}\left(-K_{S,\Delta_\beta}\right)-2\left(2-\left(m-n\right)\beta_2\right)x-\left(m-n\right)x^2.$$

When x reaches β_2, the intersection number $(-K_{S,\Delta_\beta} - x\widetilde{C}_2)\cdot E_i$ decreases to 0. When $x > \beta_2$, assume there is an effective $\mathbb{R}$-divisor N supported on $\bigcup_i E_i$ that satisfies (2.2). We can solve

$$N = (x-\beta_2)\sum_i E_i.$$

By (3.1) and (3.2),

$$-K_{S,\Delta_\beta} - x\widetilde{C}_2 - N = \pi^*\left(\left(\beta_1+\beta_2-x\right)C_2 + \left(2-n\beta_1\right)F\right)$$

is nef when $x \in (\beta_2, \beta_1+\beta_2]$. Also note Claim 4.1(2). By Definition 2.4, N is indeed the negative part of $-K_{S,\Delta_\beta} - x\widetilde{C}_2$ for $x \in (\beta_2, \beta_1+\beta_2]$. Therefore by Proposition 2.5,

$$\operatorname{vol}\left(-K_{S,\Delta_\beta} - x\widetilde{C}_2\right) = \left(\beta_1+\beta_2-x\right)\left(4-n\left(x+\beta_1-\beta_2\right)\right).$$

When x reaches $\beta_1+\beta_2$, the volume $\operatorname{vol}(-K_{S,\Delta_\beta} - x\widetilde{C}_2)$ decreases to 0. This implies

$$\tau_{-K_{S,\Delta_\beta}}\left(\widetilde{C}_2\right) = \beta_1+\beta_2.$$

□

LEMMA 5.4. *The expected vanishing order*

$$S_{-K_{S,\Delta_\beta}}\left(\widetilde{C}_2\right) = \beta_2 + \frac{2}{\operatorname{vol}\left(-K_{S,\Delta_\beta}\right)}\left(\beta_1^2-\beta_2^2+\frac{1}{3}\left(-n\beta_1^3+\left(m-n\right)\beta_2^3\right)\right).$$

PROOF. By Lemma 5.3,

$$\int_0^{\beta_2}\operatorname{vol}\left(-K_{S,\Delta_\beta} - x\widetilde{C}_2\right)dx = \beta_2\operatorname{vol}\left(-K_{S,\Delta_\beta}\right) - 2\beta_2^2 + \frac{2}{3}\left(m-n\right)\beta_2^3,$$

$$\int_{\beta_2}^{\beta_1+\beta_2}\operatorname{vol}\left(-K_{S,\Delta_\beta} - x\widetilde{C}_2\right)dx = 2\beta_1^2 - \frac{2}{3}n\beta_1^3.$$

Therefore

$$\begin{aligned} S_{-K_{S,\Delta_\beta}}\left(\widetilde{C}_1\right) &= \frac{1}{\operatorname{vol}\left(-K_{S,\Delta_\beta}\right)}\int_0^{\beta_1+\beta_2}\operatorname{vol}\left(-K_{S,\Delta_\beta} - x\widetilde{C}_2\right)dx \\ &= \beta_2 + \frac{2}{\operatorname{vol}\left(-K_{S,\Delta_\beta}\right)}\left(\beta_1^2-\beta_2^2+\frac{1}{3}\left(-n\beta_1^3+\left(m-n\right)\beta_2^3\right)\right). \end{aligned}$$

□

THEOREM 5.5. *If*

$$\beta_1^2-\beta_2^2+\frac{1}{3}\left(-n\beta_1^3+\left(m-n\right)\beta_2^3\right) = 0,$$

then $\delta(S,\Delta_\beta) \le 1$. *Otherwise* $\delta(S,\Delta_\beta) < 1$.

PROOF. By (2.3), $A_{S,\Delta_\beta}(\widetilde{C}_i) = 1 - \operatorname{ord}_{\widetilde{C}_i}\Delta_\beta = \beta_i$, for $i = 1, 2$. Recall Lemmas 5.2 and 5.4,

$$S_{-K_{S,\Delta_\beta}}\left(\widetilde{C}_i\right) = \begin{cases} \beta_1 - \frac{2}{\operatorname{vol}\left(-K_{S,\Delta_\beta}\right)}\left(\beta_1^2-\beta_2^2+\frac{1}{3}\left(-n\beta_1^3+\left(m-n\right)\beta_2^3\right)\right), & i=1, \\ \beta_2 + \frac{2}{\operatorname{vol}\left(-K_{S,\Delta_\beta}\right)}\left(\beta_1^2-\beta_2^2+\frac{1}{3}\left(-n\beta_1^3+\left(m-n\right)\beta_2^3\right)\right), & i=2. \end{cases}$$

By Definition 2.7,

$$\delta\left(S,\Delta_\beta\right)\le \min_{i=1,2}\frac{A_{S,\Delta_\beta}\left(\widetilde{C}_i\right)}{S_{-K_{S,\Delta_\beta}}\left(\widetilde{C}_i\right)}. \tag{5.1}$$

When

$$\beta_1^2-\beta_2^2+\frac{1}{3}\left(-n\beta_1^3+(m-n)\,\beta_2^3\right)=0,$$

(5.1) becomes $\delta(S,\Delta_\beta)\le 1$. Otherwise if

$$\beta_1^2-\beta_2^2+\frac{1}{3}\left(-n\beta_1^3+(m-n)\,\beta_2^3\right)<0,$$

then (5.1) gives $\delta(S,\Delta_\beta)\le \frac{A_{S,\Delta_\beta}(\widetilde{C}_1)}{S_{-K_{S,\Delta_\beta}}(\widetilde{C}_1)}<1$; if

$$\beta_1^2-\beta_2^2+\frac{1}{3}\left(-n\beta_1^3+(m-n)\,\beta_2^3\right)>0,$$

then (5.1) gives $\delta(S,\Delta_\beta)\le \frac{A_{S,\Delta_\beta}(\widetilde{C}_2)}{S_{-K_{S,\Delta_\beta}}(\widetilde{C}_2)}<1$. □

Recall that we assume $m\ge 1$, i.e., S contains some exceptional curve E_i.

LEMMA 5.6. *Assume $n\ge m$. The pseudo-effective threshold*

$$\tau_{-K_{S,\Delta_\beta}}\left(E_i\right)=2+(n-m)\,\beta_2,$$

and

$$\operatorname{vol}\left(-K_{S,\Delta_\beta}-xE_i\right)$$
$$=\begin{cases}4\left(\beta_1+\beta_2\right)-n\beta_1^2+(n-m)\,\beta_2^2-2\beta_2x-x^2, & x\in\left[0,\beta_1\right],\\ 4\left(\beta_1+\beta_2\right)-(n-1)\,\beta_1^2+(n-m)\,\beta_2^2-2\left(\beta_1+\beta_2\right)x, & x\in\left(\beta_1,2-(n-1)\,\beta_1\right],\\ (n-1)\left(\beta_2+\frac{2-x}{n-1}\right)^2-(m-1)\,\beta_2^2, & x\in\left(2-(n-1)\,\beta_1,2\right],\\ \frac{(2+(n-m)\beta_2-x)^2}{n-m}, & x\in\left(2,2+(n-m)\,\beta_2\right].\end{cases}$$

PROOF. By (3.1) and (3.2),

$$-K_{S,\Delta_\beta}-xE_i=\pi^*\left(\left(\beta_1+\beta_2\right)C_2+\left(2-n\beta_1\right)F\right)-\left(\beta_2+x\right)E_i-\sum_{j\neq i}\beta_2E_j$$

is nef when $x\in[0,\beta_1]$. Therefore by Proposition 2.3,

$$\operatorname{vol}\left(-K_{S,\Delta_\beta}-xE_i\right)=4\left(\beta_1+\beta_2\right)-n\beta_1^2+(n-m)\,\beta_2^2-2\beta_2x-x^2.$$

When x reaches β_1, the intersection number $\left(-K_{S,\Delta_\beta}-xE_i\right)\cdot\widetilde{F}_i$ decreases to 0. When $x>\beta_1$, assume there is an effective $\mathbb{R}$-divisor N supported on $\widetilde{F}_i$ that satisfies (2.2). We can solve

$$N=\left(x-\beta_1\right)\widetilde{F}_i.$$

By (3.1) and (3.2),

$$\begin{aligned}-K_{S,\Delta_\beta}-xE_i-N=\pi^*&\left(\left(\beta_1+\beta_2\right)C_2+\left(2-(n-1)\,\beta_1-x\right)F\right)\\ &-\left(\beta_1+\beta_2\right)E_i-\sum_{j\neq i}\beta_2E_j\end{aligned}$$

is nef when $x \in (\beta_1, 2-(n-1)\beta_1]$. Also note Claim 4.1(1). By Definition 2.4, N is indeed the negative part of $-K_{S,\Delta_\beta} - xE_i$ for $x \in (\beta_1, 2-(n-1)\beta_1]$. Therefore by Proposition 2.5,

$$\operatorname{vol}\left(-K_{S,\Delta_\beta} - xE_i\right) = 4\left(\beta_1+\beta_2\right) - (n-1)\,\beta_1^2 + (n-m)\,\beta_2^2 - 2\left(\beta_1+\beta_2\right)x.$$

When x reaches $2-(n-1)\beta_1$, the intersection number $(-K_{S,\Delta_\beta} - xE_i)\cdot \widetilde{C}_1$ decreases to 0. If $n > 1$, when $x > 2-(n-1)\beta_1$, assume there is an effective $\mathbb{R}$-divisor N supported on $\widetilde{F}_i \cup \widetilde{C}_1$ that satisfies (2.2). We can solve

$$N = \left(\beta_1 - \frac{2-x}{n-1}\right)\widetilde{C}_1 + \frac{nx-2}{n-1}\widetilde{F}_i.$$

By (3.1) and (3.2),

$$-K_{S,\Delta_\beta} - xE_i - N = \pi^*\left(\left(\beta_2 + \frac{2-x}{n-1}\right)C_2\right) - \left(\beta_2 + \frac{2-x}{n-1}\right)E_i - \sum_{j\neq i}\beta_2 E_j$$

is nef when $x \in (2-(n-1)\beta_1, 2]$. Also note Claim 4.1(3). By Definition 2.4, N is indeed the negative part of $-K_{S,\Delta_\beta} - xE_i$ for $x \in (2-(n-1)\beta_1, 2]$. Therefore by Proposition 2.5,

$$\operatorname{vol}\left(-K_{S,\Delta_\beta} - xE_i\right) = (n-1)\left(\beta_2 + \frac{2-x}{n-1}\right)^2 - (m-1)\,\beta_2^2.$$

When x reaches 2, the intersection number $(-K_{S,\Delta_\beta} - xE_i)\cdot\widetilde{F}_j$ decreases to 0 for $j \neq i$. If $n > m$, when $x > 2$, assume there is an effective $\mathbb{R}$-divisor N supported on $\widetilde{C}_1 \cup \bigcup_j \widetilde{F}_j$ that satisfies (2.2). We can solve

$$N = \left(\beta_1 + \frac{x-2}{n-m}\right)\widetilde{C}_1 + x\widetilde{F}_i + \sum_j \frac{x-2}{n-m}\widetilde{F}_j.$$

By (3.1) and (3.2),

$$-K_{S,\Delta_\beta} - xE_i - N = \pi^*\left(\left(\beta_2 - \frac{x-2}{n-m}\right)C_2\right) - \sum_j\left(\beta_2 - \frac{x-2}{n-m}\right)E_i$$

is nef when $x \in (2, 2+(n-m)\beta_2]$. Also note Claim 4.1(4). By Definition 2.4, N is indeed the negative part of $-K_{S,\Delta_\beta} - xE_i$ for $x \in (2, 2+(n-m)\beta_2]$. Therefore by Proposition 2.5,

$$\operatorname{vol}\left(-K_{S,\Delta_\beta} - xE_i\right) = \frac{\left(2+(n-m)\,\beta_2 - x\right)^2}{n-m}.$$

When x reaches $2+(n-m)\beta_2$, the volume $\operatorname{vol}(-K_{S,\Delta_\beta} - xE_i)$ decreases to 0. This implies

$$\tau_{-K_{S,\Delta_\beta}}(E_i) = 2 + (n-m)\,\beta_2.$$

□

LEMMA 5.7. *Assume $n < m$ and $(n-1)\beta_1 \geq (m-n)\beta_2$. The pseudo-effective threshold*

$$\tau_{-K_{S,\Delta_\beta}}(E_i) = 2,$$

and

$$\operatorname{vol}\left(-K_{S,\Delta_\beta}-xE_i\right)$$
$$=\begin{cases}4\left(\beta_1+\beta_2\right)-n\beta_1^2+\left(n-m\right)\beta_2^2-2\beta_2x-x^2, & x\in\left[0,\beta_1\right],\\ 4\left(\beta_1+\beta_2\right)-\left(n-1\right)\beta_1^2+\left(n-m\right)\beta_2^2-2\left(\beta_1+\beta_2\right)x, & x\in\left(\beta_1,2-\left(n-1\right)\beta_1\right],\\ \left(n-1\right)\left(\beta_2+\frac{2-x}{n-1}\right)^2-\left(m-1\right)\beta_2^2, & x\in\left(2-\left(n-1\right)\beta_1,2-\left(m-n\right)\beta_2\right],\\ \frac{(m-1)}{(m-n)(n-1)}\left(x-2\right)^2, & x\in\left(2-\left(m-n\right)\beta_2,2\right].\end{cases}$$

PROOF. By (3.1) and (3.2),

$$-K_{S,\Delta_\beta}-xE_i=\pi^*\left(\left(\beta_1+\beta_2\right)C_2+\left(2-n\beta_1\right)F\right)-\left(\beta_2+x\right)E_i-\sum_{j\neq i}\beta_2E_j$$

is nef when $x\in[0,\beta_1]$. Therefore by Proposition 2.3,

$$\operatorname{vol}\left(-K_{S,\Delta_\beta}-xE_i\right)=4\left(\beta_1+\beta_2\right)-n\beta_1^2+\left(n-m\right)\beta_2^2-2\beta_2x-x^2.$$

When x reaches β_1, the intersection number $(-K_{S,\Delta_\beta}-xE_i)\cdot\widetilde{F}_i$ decreases to 0. When $x>\beta_1$, assume there is an effective $\mathbb{R}$-divisor N supported on $\widetilde{F}_i$ that satisfies (2.2). We can solve

$$N=\left(x-\beta_1\right)\widetilde{F}_i.$$

By (3.1) and (3.2),

$$\begin{aligned}-K_{S,\Delta_\beta}-xE_i-N=\pi^*\left(\left(\beta_1+\beta_2\right)C_2+\left(2-\left(n-1\right)\beta_1-x\right)F\right)\\ -\left(\beta_1+\beta_2\right)E_i-\sum_{j\neq i}\beta_2E_j\end{aligned}$$

is nef when $x\in(\beta_1,2-(n-1)\beta_1]$. Also note Claim 4.1(1). By Definition 2.4, N is indeed the negative part of $-K_{S,\Delta_\beta}-xE_i$ for $x\in(\beta_1,2-(n-1)\beta_1]$. Therefore by Proposition 2.5,

$$\operatorname{vol}\left(-K_{S,\Delta_\beta}-xE_i\right)=4\left(\beta_1+\beta_2\right)-\left(n-1\right)\beta_1^2+\left(n-m\right)\beta_2^2-2\left(\beta_1+\beta_2\right)x.$$

When x reaches $2-(n-1)\beta_1$, the intersection number $(-K_{S,\Delta_\beta}-xE_i)\cdot\widetilde{C}_1$ decreases to 0. When $x>2-(n-1)\beta_1$, assume there is an effective $\mathbb{R}$-divisor N supported on $\widetilde{F}_i\cup\widetilde{C}_1$ that satisfies (2.2). We can solve

$$N=\left(\beta_1-\frac{2-x}{n-1}\right)\widetilde{C}_1+\frac{nx-2}{n-1}\widetilde{F}_i.$$

By (3.1) and (3.2),

$$-K_{S,\Delta_\beta}-xE_i-N=\pi^*\left(\left(\beta_2+\frac{2-x}{n-1}\right)C_2\right)-\left(\beta_2+\frac{2-x}{n-1}\right)E_i-\sum_{j\neq i}\beta_2E_j$$

is nef when $x\in(2-(n-1)\beta_1,2-(m-n)\beta_2]$. Also note Claim 4.1(3). By Definition 2.4, N is indeed the negative part of $-K_{S,\Delta_\beta}-xE_i$ for $x\in(2-(n-1)\beta_1,2-(m-n)\beta_2]$. Therefore by Proposition 2.5,

$$\operatorname{vol}\left(-K_{S,\Delta_\beta}-xE_i\right)=\left(n-1\right)\left(\beta_2+\frac{2-x}{n-1}\right)^2-\left(m-1\right)\beta_2^2.$$

When x reaches $2-(m-n)\beta_2$, the intersection number $(-K_{S,\Delta_\beta}-xE_i)\cdot\widetilde{C}_2$ decreases to 0. When $x>2-(m-n)\beta_2$, assume there is an effective $\mathbb{R}$-divisor N supported on $\widetilde{F}_i\cup\widetilde{C}_1\cup\widetilde{C}_2$ that satisfies (2.2). We can solve

$$N=\left(\beta_1-\frac{2-x}{n-1}\right)\widetilde{C}_1+\frac{nx-2}{n-1}\widetilde{F}_i+\left(\beta_2+\frac{x-2}{m-n}\right)\widetilde{C}_2.$$

By (3.1) and (3.2),

$$\begin{aligned}-K_{S,\Delta_\beta}-xE_i-N=\pi^*&\left(\frac{m-1}{(n-1)(m-n)}(2-x)\,C_2\right)\\&-\frac{m-1}{(n-1)(m-n)}(2-x)\,E_i-\sum_{j\neq i}\frac{2-x}{m-n}E_j\end{aligned}$$

is nef when $x\in(2-(m-n)\beta_2,2]$. Also note Claim 4.1(6). By Definition 2.4, N is indeed the negative part of $-K_{S,\Delta_\beta}-xE_i$ for $x\in(2-(m-n)\beta_2,2]$. Therefore by Proposition 2.5,

$$\operatorname{vol}\left(-K_{S,\Delta_\beta}-xE_i\right)=\frac{(m-1)}{(m-n)(n-1)}(x-2)^2.$$

When x reaches 2, the volume $\operatorname{vol}(-K_{S,\Delta_\beta}-xE_i)$ decreases to 0. This implies

$$\tau_{-K_{S,\Delta_\beta}}(E_i)=2.$$

□

LEMMA 5.8. *Assume $(n-1)\beta_1<(m-n)\beta_2$ and $n\geq 1$. The pseudo-effective threshold*

$$\tau_{-K_{S,\Delta_\beta}}(E_i)=2,$$

and

$$\operatorname{vol}\left(-K_{S,\Delta_\beta}-xE_i\right)$$
$$=\begin{cases}4(\beta_1+\beta_2)-n\beta_1^2+(n-m)\beta_2^2-2\beta_2x-x^2, & x\in[0,\beta_1],\\ 4(\beta_1+\beta_2)-(n-1)\beta_1^2+(n-m)\beta_2^2-2(\beta_1+\beta_2)x, & x\in(\beta_1,2-(m-n)\beta_2],\\ \frac{4+(m-n)(4-(n-1)\beta_1)\beta_1-2(2+(m-n)\beta_1)x+x^2}{m-n}, & x\in(2-(m-n)\beta_2,2-(n-1)\beta_1],\\ \frac{(m-1)}{(m-n)(n-1)}(x-2)^2, & x\in(2-(n-1)\beta_1,2].\end{cases}$$

PROOF. By (3.1) and (3.2),

$$-K_{S,\Delta_\beta}-xE_i=\pi^*\left((\beta_1+\beta_2)\,C_2+(2-n\beta_1)\,F\right)-(\beta_2+x)\,E_i-\sum_{j\neq i}\beta_2E_j$$

is nef when $x\in[0,\beta_1]$. Therefore by Proposition 2.3,

$$\operatorname{vol}\left(-K_{S,\Delta_\beta}-xE_i\right)=4(\beta_1+\beta_2)-n\beta_1^2+(n-m)\beta_2^2-2\beta_2x-x^2.$$

When x reaches β_1, the intersection number $(-K_{S,\Delta_\beta}-xE_i)\cdot\widetilde{F}_i$ decreases to 0. When $x>\beta_1$, assume there is an effective $\mathbb{R}$-divisor N supported on $\widetilde{F}_i$ that satisfies (2.2). We can solve

$$N=(x-\beta_1)\,\widetilde{F}_i.$$

By (3.1) and (3.2),

$$-K_{S,\Delta_\beta} - xE_i - N = \pi^*\left(\left(\beta_1+\beta_2\right)C_2 + \left(2-(n-1)\beta_1 - x\right)F\right) - \left(\beta_1+\beta_2\right)E_i - \sum_{j\neq i}\beta_2 E_j$$

is nef when $x \in (\beta_1, 2-(m-n)\beta_2]$. Also note Claim 4.1(1). By Definition 2.4, N is indeed the negative part of $-K_{S,\Delta_\beta} - xE_i$ for $x \in (\beta_1, 2-(m-n)\beta_2]$. Therefore by Proposition 2.5,

$$\operatorname{vol}\left(-K_{S,\Delta_\beta} - xE_i\right) = 4\left(\beta_1+\beta_2\right) - (n-1)\beta_1^2 + (n-m)\beta_2^2 - 2\left(\beta_1+\beta_2\right)x.$$

When x reaches $2-(m-n)\beta_2$, the intersection number $(-K_{S,\Delta_\beta} - xE_i)\cdot \widetilde{C}_2$ decreases to 0. When $x > 2-(m-n)\beta_2$, assume there is an effective $\mathbb{R}$-divisor N supported on $\widetilde{F}_i \cup \widetilde{C}_2$ that satisfies (2.2). We can solve

$$N = \left(\beta_2 - \frac{2-x}{m-n}\right)\widetilde{C}_2 + (x-\beta_1)\widetilde{F}_i.$$

By (3.1) and (3.2),

$$-K_{S,\Delta_\beta} - xE_i - N = \pi^*\left(\left(\beta_1 + \frac{2-x}{m-n}\right)C_2 + \left(2-(n-1)\beta_1 - x\right)F\right) - \left(\beta_1 + \frac{2-x}{m-n}\right)E_i - \sum_{j\neq i}\frac{2-x}{m-n}E_j$$

is nef when $x \in (2-(m-n)\beta_2, 2-(n-1)\beta_1]$. Also note Claim 4.1(5). By Definition 2.4, N is indeed the negative part of $-K_{S,\Delta_\beta} - xE_i$ for $x \in (2-(m-n)\beta_2, 2-(n-1)\beta_1]$. Therefore by Proposition 2.5,

$$\operatorname{vol}\left(-K_{S,\Delta_\beta} - xE_i\right) = \frac{4 + (m-n)\left(4-(n-1)\beta_1\right)\beta_1 - 2\left(2+(m-n)\beta_1\right)x + x^2}{m-n}.$$

When x reaches $2-(n-1)\beta_1$, the intersection number $(-K_{S,\Delta_\beta} - xE_i)\cdot \widetilde{C}_1$ decreases to 0. If $n>1$, when $x > 2-(n-1)\beta_1$, assume there is an effective $\mathbb{R}$-divisor N supported on $\widetilde{F}_i \cup \widetilde{C}_1 \cup \widetilde{C}_2$ that satisfies (2.2). We can solve

$$N = \left(\beta_1 - \frac{2-x}{n-1}\right)\widetilde{C}_1 + \left(\beta_2 - \frac{2-x}{m-n}\right)\widetilde{C}_2 + \frac{nx-2}{n-1}\widetilde{F}_i.$$

By (3.1) and (3.2),

$$-K_{S,\Delta_\beta} - xE_i - N = \pi^*\left(\frac{m-1}{(n-1)(m-n)}(2-x)C_2\right) - \frac{m-1}{(n-1)(m-n)}(2-x)E_i - \sum_{j\neq i}\frac{2-x}{m-n}E_j$$

is nef when $x \in (2-(n-1)\beta_1, 2]$. Also note Claim 4.1(6). By Definition 2.4, N is indeed the negative part of $-K_{S,\Delta_\beta} - xE_i$ for $x \in (2-(n-1)\beta_1, 2]$. Therefore by Proposition 2.5,

$$\operatorname{vol}\left(-K_{S,\Delta_\beta} - xE_i\right) = \frac{(m-1)}{(m-n)(n-1)}(x-2)^2.$$

When x reaches 2, the volume $\mathrm{vol}(-K_{S,\Delta_\beta} - xE_i)$ decreases to 0. This implies

$$\tau_{-K_{S,\Delta_\beta}}(E_i) = 2.$$

□

LEMMA 5.9. *Assume $n = 0$. The pseudo-effective threshold*

$$\tau_{-K_{S,\Delta_\beta}}(E_i) = 2 + \beta_1,$$

and

$$\mathrm{vol}\left(-K_{S,\Delta_\beta} - xE_i\right) = \begin{cases} 4(\beta_1+\beta_2) - m\beta_2^2 - 2\beta_2 x - x^2, & x \in [0,\beta_1], \\ 4(\beta_1+\beta_2) + \beta_1^2 - m\beta_2^2 - 2(\beta_1+\beta_2)x, & x \in (\beta_1, 2-m\beta_2], \\ \frac{4+m(4+\beta_1)\beta_1 - 2(2+m\beta_1)x + x^2}{m}, & x \in (2-m\beta_2, 2], \\ (2+\beta_1-x)^2, & x \in (2, 2+\beta_1]. \end{cases}$$

PROOF. By (3.1) and (3.2),

$$-K_{S,\Delta_\beta} - xE_i = \pi^*\left((\beta_1+\beta_2)C_2 + 2F\right) - (\beta_2+x)E_i - \sum_{j\neq i}\beta_2 E_j$$

is nef when $x \in [0,\beta_1]$. Therefore by Proposition 2.3,

$$\mathrm{vol}\left(-K_{S,\Delta_\beta} - xE_i\right) = 4(\beta_1+\beta_2) - m\beta_2^2 - 2\beta_2 x - x^2.$$

When x reaches β_1, the intersection number $(-K_{S,\Delta_\beta} - xE_i)\cdot\widetilde{F}_i$ decreases to 0. When $x > \beta_1$, assume there is an effective $\mathbb{R}$-divisor N supported on $\widetilde{F}_i$ that satisfies (2.2). We can solve

$$N = (x-\beta_1)\widetilde{F}_i.$$

By (3.1) and (3.2),

$$-K_{S,\Delta_\beta} - xE_i - N = \pi^*\left((\beta_1+\beta_2)C_2 + (2+\beta_1-x)F\right) - (\beta_1+\beta_2)E_i - \sum_{j\neq i}\beta_2 E_j$$

is nef when $x \in (\beta_1, 2-m\beta_2]$. Also note Claim 4.1(1). By Definition 2.4, N is indeed the negative part of $-K_{S,\Delta_\beta} - xE_i$ for $x \in (\beta_1, 2-m\beta_2]$. Therefore by Proposition 2.5,

$$\mathrm{vol}\left(-K_{S,\Delta_\beta} - xE_i\right) = 4(\beta_1+\beta_2) + \beta_1^2 - m\beta_2^2 - 2(\beta_1+\beta_2)x.$$

When x reaches $2-m\beta_2$, the intersection number $(-K_{S,\Delta_\beta} - xE_i)\cdot\widetilde{C}_2$ decreases to 0. When $x > 2-m\beta_2$, assume there is an effective $\mathbb{R}$-divisor N supported on $\widetilde{F}_i \cup \widetilde{C}_2$ that satisfies (2.2). We can solve

$$N = \left(\beta_2 - \frac{2-x}{m}\right)\widetilde{C}_2 + (x-\beta_1)\widetilde{F}_i.$$

By (3.1) and (3.2),

$$\begin{aligned} -K_{S,\Delta_\beta} - xE_i - N = \pi^*&\left(\left(\beta_1 + \frac{2-x}{m}\right)C_2 + (2+\beta_1-x)F\right) \\ &- \left(\beta_1 + \frac{2-x}{m}\right)E_i - \sum_{j\neq i}\frac{2-x}{m}E_j \end{aligned}$$

is nef when $x \in (2-m\beta_2, 2]$. Also note Claim 4.1(5). By Definition 2.4, N is indeed the negative part of $-K_{S,\Delta_\beta} - xE_i$ for $x \in (2-m\beta_2, 2]$. Therefore by Proposition 2.5,

$$\operatorname{vol}\left(-K_{S,\Delta_\beta} - xE_i\right) = \frac{4 + m\left(4+\beta_1\right)\beta_1 - 2\left(2+m\beta_1\right)x + x^2}{m}.$$

When x reaches 2, the intersection number $(-K_{S,\Delta_\beta} - xE_i) \cdot E_j$ decreases to 0 for $j \neq i$. When $x > 2$, assume there is an effective $\mathbb{R}$-divisor N supported on $\widetilde{F}_i \cup \widetilde{C}_2 \cup \bigcup_{j\neq i} E_j$ that satisfies (2.2). We can solve

$$N = \left(\beta_2 + x - 2\right)\widetilde{C}_2 + \left(x - \beta_1\right)\widetilde{F}_i + \sum_{j\neq i}\left(x-2\right)E_j.$$

By (3.1) and (3.2),

$$-K_{S,\Delta_\beta} - xE_i - N = \pi^*\left(\left(2+\beta_1 - x\right)C_2 + \left(2+\beta_1-x\right)F\right) - \left(2+\beta_1-x\right)E_i$$

is nef when $x \in (2, 2+\beta_1]$. Also note Claim 4.1(7). By Definition 2.4, N is indeed the negative part of $-K_{S,\Delta_\beta} - xE_i$ for $x \in (2, 2+\beta_1]$. Therefore by Proposition 2.5,

$$\operatorname{vol}\left(-K_{S,\Delta_\beta} - xE_i\right) = \left(2+\beta_1-x\right)^2.$$

When x reaches $2+\beta_1$, the volume $\operatorname{vol}(-K_{S,\Delta_\beta} - xE_i)$ decreases to 0. This implies

$$\tau_{-K_{S,\Delta_\beta}}\left(E_i\right) = 2+\beta_1.$$

$\square$

LEMMA 5.10. *The expected vanishing order*

$$S_{-K_{S,\Delta_\beta}}\left(E_i\right) = 1 + \frac{1}{\operatorname{vol}\left(-K_{S,\Delta_\beta}\right)} \times \left(-\left(n-2\right)\beta_1^2 + \left(n-m\right)\beta_2^2 + \frac{n\left(n-2\right)\beta_1^3 + \left(n-m\right)^2\beta_2^3}{3}\right).$$

PROOF. By Lemmas 5.6, 5.7, 5.8 and 5.9,

$$\begin{aligned} S_{-K_{S,\Delta_\beta}}\left(E_i\right) &= \frac{1}{\operatorname{vol}\left(-K_{S,\Delta_\beta}\right)}\int_0^{\tau_{-K_{S,\Delta_\beta}}(E_i)} \operatorname{vol}\left(-K_{S,\Delta_\beta} - xE_i\right)dx \\ &= 1 + \frac{1}{\operatorname{vol}\left(-K_{S,\Delta_\beta}\right)} \\ &\quad \times \left(-\left(n-2\right)\beta_1^2 + \left(n-m\right)\beta_2^2 + \frac{n\left(n-2\right)\beta_1^3 + \left(n-m\right)^2\beta_2^3}{3}\right). \end{aligned}$$

$\square$

THEOREM 5.11. *Assume* $\beta_1^2 - \beta_2^2 - \frac{1}{3}\left(n\beta_1^3 + (n-m)\beta_2^3\right) = 0$. *If* $m = 1$, *then* $\delta(S, \Delta_\beta) < 1$.

PROOF. By (2.3), $A_{S,\Delta_\beta}(E_i) = 1 - \operatorname{ord}_{E_i}\Delta_\beta = 1$. Recall Lemma 5.10,

$$S_{-K_{S,\Delta_\beta}}\left(E_i\right) = 1 + \frac{1}{\operatorname{vol}\left(-K_{S,\Delta_\beta}\right)}\left(2-m\right)\beta_1^2\left(1 - \frac{n}{3}\beta_1\right).$$

By definition 2.7, when $m = 1$,

$$\delta\left(S, \Delta_\beta\right) \leq \frac{A_{S,\Delta_\beta}\left(E_i\right)}{S_{-K_{S,\Delta_\beta}}\left(E_i\right)} < 1.$$

□

Theorems 5.5 and 5.11 combined imply the necessary part of the main theorem.

THEOREM 5.12. *Consider (II.6A.n.m) as in Definition* 1.6. *We have*

$$\delta\left(S, \left(1-\beta_1\right)\widetilde{C}_1 + \left(1-\beta_2\right)\widetilde{C}_2\right) \leq 1,$$

with equality only if

$$\beta_1^2 + \frac{1}{3}\widetilde{C}_1^2\beta_1^3 = \beta_2^2 + \frac{1}{3}\widetilde{C}_2^2\beta_2^3$$

and

$$m \neq 1.$$

PROOF. By Theorem 5.5,

$$\delta\left(S, \Delta_\beta\right) \leq 1.$$

If equality holds, by Theorem 5.5,

$$\beta_1^2 + \frac{1}{3}\widetilde{C}_1^2\beta_1^3 = \beta_2^2 + \frac{1}{3}\widetilde{C}_2^2\beta_1^3;$$

by Theorem 5.11,

$$m \neq 1.$$

□

6. Rational T-varieties of complexity 1

In this section we collect some definitions and properties of rational T-varieties of complexity 1 that will be used in the proof of the main theorem.

6.1. Construction of rational T-varieties of complexity 1. We first recall some definitions for T-varieties [**2**, Sections 2.2 and 2.3], in particular, rational T-varieties of complexity 1 [**29**, Section 1].

DEFINITION 6.1. Let N be a lattice and M its dual lattice. We say $\Delta \subseteq N_{\mathbb{Q}} := N \otimes_{\mathbb{Z}} \mathbb{Q}$ is a *polyhedron* if it is a finite intersection of half-spaces. Its *tailcone* is defined by

$$\operatorname{tail}\left(\Delta\right) := \left\{a \in N_{\mathbb{Q}} \,|\, a + \Delta \subseteq \Delta\right\}.$$

A cone is always assumed to be *pointed*, i.e., it does not contain non-trivial linear subspaces. Fix a cone σ, a formal sum

$$\mathcal{D} = \sum_{p \in \mathbb{P}^1} \mathcal{D}_p \otimes p$$

is called a *polyhedral divisor* on $\mathbb{P}^1$ with $\operatorname{tail}(\mathcal{D}) := \sigma$ if each coefficient $\mathcal{D}_p$ is either a polyhedron with $\operatorname{tail}(\mathcal{D}_p) = \sigma$, or the empty set, and all but finitely many coefficients are σ. Define its *locus*

$$\operatorname{Loc}\mathcal{D} := \mathbb{P}^1 \setminus \bigcup_{\mathcal{D}_p = \varnothing} p$$

and *degree*

$$\deg \mathcal{D} := \sum_{p \in \mathbb{P}^1} \mathcal{D}_p.$$

Here the sum is in the sense of Minkowski sum, i.e.,

$$\sum_{i=1}^{N} \Delta_i = \left\{ \sum_{i=1}^{N} a_i \,\middle|\, a_i \in \Delta_i \right\}.$$

Note that if the tailcone of a polyhedron Δ is σ, then $\Delta \subseteq \Delta + \sigma \subseteq \Delta$, i.e., $\Delta + \sigma = \Delta$. Hence the definition of $\deg \mathcal{D}$ makes sense. Indeed,

$$\sum_{p \in \mathbb{P}^1} \mathcal{D}_p = \sum_{\mathcal{D}_p \neq \sigma} \mathcal{D}_p$$

is a finite sum, as any polyhedron $\mathcal{D}_p$ equaling σ can be ignored. By convention, if $\mathcal{D}_p = \sigma$ for all p, then $\deg \mathcal{D} = \sigma$. We assume $\mathcal{D}$ is a *p-divisor*, that is, $\deg \mathcal{D} \subsetneq \sigma$. A ray ρ of σ is called *extremal* if $\rho \cap \deg \mathcal{D} = \varnothing$. For $u \in \sigma^\vee \cap M$, the evaluation

$$\mathcal{D}(u) := \sum_{p \in \operatorname{Loc} \mathcal{D}} \min_{v \in \mathcal{D}_p} \langle u, v \rangle \cdot p$$

is a $\mathbb{Q}$-divisor on $\operatorname{Loc} \mathcal{D}$. Consider the graded $\mathcal{O}_{\operatorname{Loc} \mathcal{D}}$-algebra

$$\mathbb{A}(\mathcal{D}) := \bigoplus_{u \in \sigma^\vee \cap M} \mathcal{O}_{\operatorname{Loc} \mathcal{D}}(\mathcal{D}(u)) \chi^u.$$

We can define two T-varieties

$$\widetilde{\operatorname{TV}}(\mathcal{D}) := \mathbf{Spec}\, \mathbb{A}(\mathcal{D}), \quad \operatorname{TV}(\mathcal{D}) := \operatorname{Spec} H^0(\operatorname{Loc} \mathcal{D}, \mathbb{A}(\mathcal{D})),$$

equipped with an action of $T := \operatorname{Spec} \mathbb{C}[M]$ induced by the action of $\mathbb{C}[M]$ on $\mathbb{A}(\mathcal{D})$.

THEOREM 6.2 ([**21**, Proposition 3.13], [**29**, Section 1]). *There are two types of T-invariant prime divisors on $\widetilde{\operatorname{TV}}(D)$: vertical and horizontal divisors. Vertical divisors are classified by pairs (p, v), where v is a vertex of $\mathcal{D}_p$. Such divisor is denoted $D_{p,v}$. Horizontal divisors are classified by rays ρ of σ. Such divisor is denoted D_ρ. The natural morphism $r : \widetilde{\operatorname{TV}}(D) \to \operatorname{TV}(D)$ is a proper birational T-equivariant contraction morphism which contracts any horizontal divisor D_ρ where ρ is not extremal. The other divisors survive and are precisely the T-invariant prime divisors on* $\operatorname{TV}(D)$.

Similar to the toric case, general T-varieties are obtained by gluing affine T-varieties.

DEFINITION 6.3 ([**2**, Definition 13], [**29**, Definition 1.4]). For two p-divisors $\mathcal{D} = \sum_{p \in \mathbb{P}^1} \mathcal{D}_p \otimes p$ and $\mathcal{D}' = \sum_{p \in \mathbb{P}^1} \mathcal{D}'_p \otimes p$, define their intersection

$$\mathcal{D} \cap \mathcal{D}' := \sum_{p \in \mathbb{P}^1} \left(\mathcal{D}_p \cap \mathcal{D}'_p\right) \otimes p.$$

If $\mathcal{D}' \subseteq \mathcal{D}$, then there is an induced map $\operatorname{TV}(\mathcal{D}') \to \operatorname{TV}(\mathcal{D})$. If this map is an open embedding, then $\mathcal{D}'$ is called a *face* of $\mathcal{D}$. A finite set $\mathcal{S} = \{\mathcal{D}^i\}$ of p-divisors on $\mathbb{P}^1$ is called a *divisorial fan* if the intersection of two p-divisors from $\mathcal{S}$ is a common face of both and also belongs to $\mathcal{S}$. For $p \in \mathbb{P}^1$, define a *slice* $\mathcal{S}_p := \{\mathcal{D}^i_p\}$. $\mathcal{S}$ is called *complete* if each slice is a complete polyhedral subdivision of $N_\mathbb{Q}$.

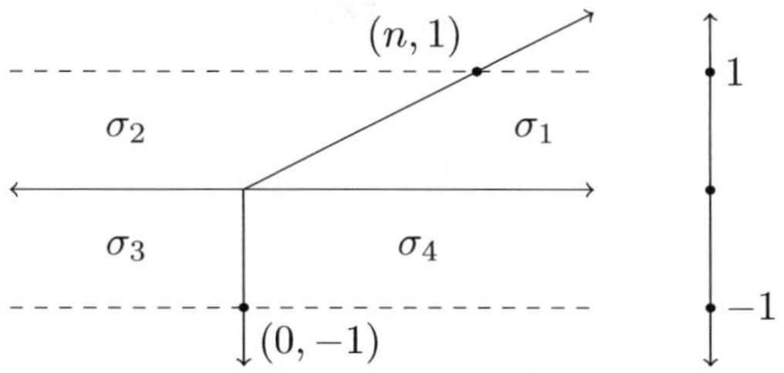

FIGURE 2. The downgrading of the torus action on $\mathbb{F}_n$. The projection to the second variable maps the fan corresponding to $\mathbb{F}_n$ to the fan corresponding to the base $\mathbb{P}^1$. Let $0, \infty \in \mathbb{P}^1$ denote the torus-invariant divisors corresponding to the rays with primitive generators 1 and -1, respectively. Then each 2-dimensional cone σ_i gives rise to a polyhedral divisor $\mathcal{D}_i$ whose coefficient over a torus-invariant divisor is the intersection of that cone and its preimage under the projection map.

EXAMPLE 6.4. Consider the toric variety $\mathbb{F}_n$. We restrict its torus action to that of a subtorus, consisting of the actions that fix the fibers. This induces the structure of a rational T-variety of complexity 1. For such T-varieties, we can use toric downgrade to obtain its divisorial fan [**21**, Remark 2.8] (see also [**29**, Section 4.1]). See Figure 2.

Start from the fan corresponding to the toric variety $\mathbb{F}_n$. As explained in Figure 2, let $pr_k : N_{\mathbb{Q}} \to N_{\mathbb{Q}}$ be the projection map onto the k-th variable, $k = 1, 2$. The fan corresponding to $\mathbb{F}_n$ is mapped via pr_2 to the fan corresponding to $\mathbb{P}^1$, which has two rays corresponding to the two torus-invariant divisors $D_1 = 0$ and $D_{-1} = \infty$, with primitive generators 1 and -1, respectively. For each 2-dimensional cone σ_i, define a polyhedral divisor $\mathcal{D}_i$, whose coefficient over D_j is the polyhedron $pr_1(\sigma_i \cap pr_2^{-1}(D_j))$, $j = \pm 1$. That is,

$$\begin{aligned}
\mathcal{D}_1 &= [n, +\infty) \otimes 0 + \varnothing \otimes \infty, \\
\mathcal{D}_2 &= (-\infty, n] \otimes 0 + \varnothing \otimes \infty, \\
\mathcal{D}_3 &= \varnothing \otimes 0 + (-\infty, 0] \otimes \infty, \\
\mathcal{D}_4 &= \varnothing \otimes 0 + [0, +\infty) \otimes \infty.
\end{aligned}$$

Note that $\deg \mathcal{D}_i = \varnothing$ for any i. These polyhedral divisors now give rise to a divisorial fan corresponding to $\mathbb{F}_n$.

EXAMPLE 6.5. Recall how we blow up a torus-invariant point on a smooth toric variety. We can also interpret it under toric downgrade. See Figure 3. The cone $\sigma = \sigma_1 \cup \sigma_2$ corresponds to $\mathbb{C}^2$, while the cones σ_1 and σ_2 give rise to a fan that corresponds to $Bl_0\mathbb{C}^2$. After the blow-up, we replace the polyhedral divisor corresponding to σ by two polyhedral divisors corresponding to σ_1 and σ_2, whose coefficients over $0 \in \mathbb{C}$ are $[0, 1]$ and $[1, +\infty)$, respectively. Note that since blow-ups are local, similar construction applies if we replace the base to some open neighborhood $U \subseteq \mathbb{C}$ of 0.

REMARK 6.6 ([**2**, Section 4.4]). In general, the slices of a divisorial fan $\mathcal{S}$ do not capture the entire information of $\mathcal{S}$, since the information of which polyhedra

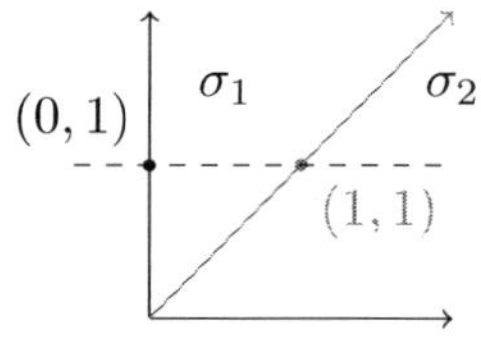

FIGURE 3. The blow-up of $\mathbb{C}^2$ at 0. Since the cone $\sigma = \sigma_1 \cup \sigma_2$ is replaced by the cones σ_1 and σ_2, we replace the polyhedral divisor corresponding to σ with two polyhedral divisors, whose coefficients over 0 are $[0,1]$ and $[1,+\infty)$, respectively.

from different slices belong to the same p-divisor is lost. However, in complexity 1, with the additional information of the degree, this problem can be overcome, which leads to the definition of f-divisors.

DEFINITION 6.7 ([**2**, Definition 19]). Consider a pair $(\sum_{p\in\mathbb{P}^1} \mathcal{S}_p \otimes p, \mathfrak{deg})$ where each $\mathcal{S}_p$ is a polyhedral subdivision with common tailfan Σ and $\mathfrak{deg} \subseteq |\Sigma|$. For any full-dimensional $\sigma \in \operatorname{tail}(\mathcal{S})$ that intersects $\mathfrak{deg}$, let Δ_p^σ be the unique polyhedron in $\mathcal{S}_p$ with tailcone σ, and $\mathcal{D}^\sigma := \sum_{p\in\mathbb{P}^1} \Delta_p^\sigma$. The pair $(\sum_{p\in\mathbb{P}^1} \mathcal{S}_p \otimes p, \mathfrak{deg})$ is called an *f-divisor* if each $\mathcal{D}^\sigma$ is a p-divisor with $\deg \mathcal{D}^\sigma = \mathfrak{deg} \cap \sigma$.

THEOREM 6.8 ([**2**, Proposition 20]). *For any f-divisor* $(\sum_{p\in\mathbb{P}^1} \mathcal{S}_p \otimes p, \mathfrak{deg})$, *there is a divisorial fan* $\mathcal{S}$ *with slices* $\mathcal{S}_p$ *and degree* $\mathfrak{deg}$. *Moreover, the T-variety constructed from* $\mathcal{S}$ *is independent of the choice of* $\mathcal{S}$.

Let X be a rational T-variety of complexity 1, characterized by a complete divisorial fan $\mathcal{S}$. By [**21**, Section 3.1], a T-invariant Cartier divisor is uniquely characterized by its divisorial support function $h = \sum_{p\in\mathbb{P}^1} h_p \otimes p$, where each h_p is defined on the slice $\mathcal{S}_p$ and is affine on each polyhedron. Moreover, the linear part h_p on a polyhedron depends only on its tailcone but not on p. The corresponding Weil divisor is given by the following theorem.

THEOREM 6.9 ([**21**, Corollary 3.19]). *The Weil divisor corresponding to the divisorial support function* h *is given by*

$$-\sum_{\rho} h_t(n_\rho) D_\rho - \sum_{(p,v)} \mu(v) h_p(v) D_{p,v},$$

where h_t *is the linear part of* h, n_ρ *is the primitive generator of the extremal ray* ρ, *and* $\mu(v)$ *is the smallest natural number such that* $\mu(v)v$ *is a lattice point.*

THEOREM 6.10 ([**21**, Theorem 3.21]). *Fix an anti-canonical divisor* $-K_{\mathbb{P}^1} = \sum_{p\in\mathbb{P}^1} a_p \cdot p$. *Then the anti-canonical divisor*

$$-K_X = \sum_{\rho} D_\rho + \sum_{(p,v)} (\mu(v) a_p + 1 - \mu(v)) D_{p,v}.$$

DEFINITION 6.11 ([**21**, Definition 3.22]). For a divisorial support function h with linear part h_t, define

$$\square_h := \bigcap_{\rho} \{u \in M_{\mathbb{R}} \mid \langle u, n_\rho \rangle \geq h_t(n_\rho)\},$$

where ρ runs through all rays in tail($\mathcal{S}$). The *dual* function of h is defined to be

$$\Psi := \sum_{p \in \mathbb{P}^1} \Psi_p \otimes p,$$

where

$$\Psi_p(u) := h_p^*(u) = \min_v \left(\langle u, v\rangle - h_p(v)\right\}$$

is the Legendre transform of h_p on $\square_h$. Note that it suffices to let v run through all vertices in $\mathcal{S}_p$. We also define its degree

$$\deg \Psi := \sum_{p \in \mathbb{P}^1} \Psi_p.$$

6.2. δ-invariant and K-stability. In this section, we provide results relevant to the δ-invariant and K-stability.

THEOREM 6.12 ([**29**, Section 3], [**25**, Theorem 3.15]). *Let X be a rational Fano T-variety of complexity 1 with log terminal singularities, characterized by a complete divisorial fan $\mathcal{S}$. Let h be the divisorial support function corresponding to $-K_X$ specified by Theorem* 6.10*, and Ψ its dual function. Define*

$$\begin{aligned}
\operatorname{vol}(\Psi) &:= \int_{\square_h} \deg \Psi(u)\, du, \\
\operatorname{bc}(\Psi) &:= \frac{1}{\operatorname{vol}(\Psi)} \int_{\square_h} u \cdot \deg \Psi(u)\, du, \qquad (6.1)\\
\operatorname{bc}_p &:= \frac{1}{\operatorname{vol}(\Psi)} \int_{\square_h} \left(\frac{1}{2}\deg \Psi(u) - \Psi_p(u)\right) \deg \Psi(u)\, du.
\end{aligned}$$

Then

$$\delta(X) = \min \left\{ \frac{-h_t(n_\rho)}{\langle \operatorname{bc}(\Psi), n_\rho \rangle - h_t(n_\rho)} \right\}_\rho \cup \left\{ \frac{1}{\mu(v)\left(\operatorname{bc}_p + \langle \operatorname{bc}(\Psi), v\rangle - h_p(v)\right)} \right\}_{(p,v)},$$

where each term corresponds to the valuation given by D_ρ and $D_{p,v}$.

The Futaki character vanishes precisely when $\operatorname{bc}(\Psi) = 0$.

THEOREM 6.13 ([**1**, Theorem 1.31], [**19**, Theorem 3.2]). *Let (X, Δ) be log Fano with Kawamata log terminal singularities. Let T be the maximal torus in* $\operatorname{Aut}(X, \Delta)$. *Suppose* $\dim T = \dim X - 1$. *Then (X, Δ) is log K-polystable if and only if the Futaki character vanishes, and for any T-invariant prime divisor E on X,*

$$\frac{A_{X,\Delta}(E)}{S_{-K_{X,\Delta}}(E)} \geq 1,$$

with equality if and only if E is horizontal.

If (X, Δ) is toric, recall the following theorem [**5**, **6**].

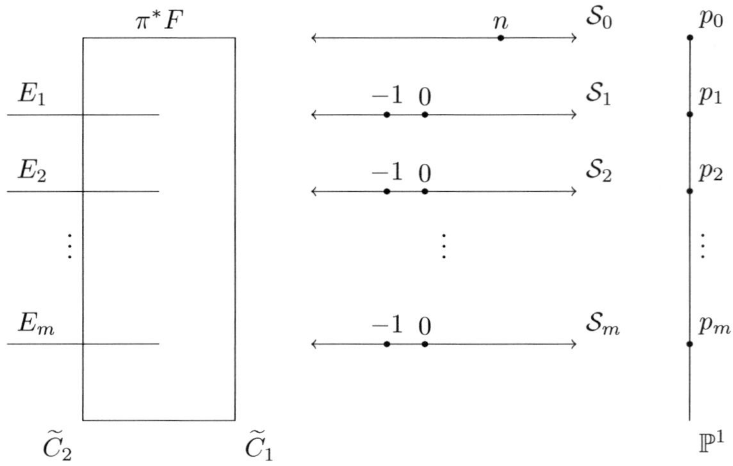

FIGURE 4. An f-divisor corresponding to S. The rays $[0,+\infty)$ and $(-\infty,0]$ correspond to the horizontal divisors $\widetilde{C}_1$ and $\widetilde{C}_2$, respectively. The vertices correspond to vertical divisors. The vertex n on $\mathcal{S}_0$ correspond to the fiber over p_0. The vertices -1 and 0 on $\mathcal{S}_i$ correspond to E_i and $\widetilde{F}_i$, respectively. The vertex 0 on any trivial slice over $p \in \mathbb{P}^1$ correspond to the fiber over p.

THEOREM 6.14. *Let (X,Δ) be log Fano and toric. Then it is log K-polystable if and only if for any torus-invariant prime divisor E on X,*

$$\frac{A_{X,\Delta}(E)}{S_{-K_{X,\Delta}}(E)} = 1.$$

7. Proof of main theorem

In this section we make use of the T-variety structure of the pairs to compute the δ-invariant, independent of the computation from Section 5, and prove Theorem 1.8.

Start from the toric variety $\mathbb{F}_n$. The $\mathbb{C}^*$-action along the fibers can be lifted to S, equipping the surface with the structure of a T-variety of complexity 1. We choose the f-divisor $(\sum_{i=0}^{m} \mathcal{S}_i \otimes p_i, \mathfrak{deg})$ on $\mathbb{P}^1$, where

$$\mathcal{S}_i = \begin{cases} \{(-\infty, n], [n,+\infty)\}, & i=0, \\ \{(-\infty,-1], [-1,0], [0,+\infty)\}, & i \geq 1, \end{cases}$$

(intersections omitted), and $\mathfrak{deg} = \varnothing$, as explained in Figure 4. By Example 6.4, when $m=0$, this f-divisor corresponds to $\mathbb{F}_n$. By Example 6.5, for general m, this f-divisor corresponds to S.

Recall Theorem 6.10. Consider

$$-K_S = \widetilde{C}_1 + \widetilde{C}_2 + 2\pi^*F,$$
$$-K_{S,\Delta_\beta} = \beta_1\widetilde{C}_1 + \beta_2\widetilde{C}_2 + 2\pi^*F.$$

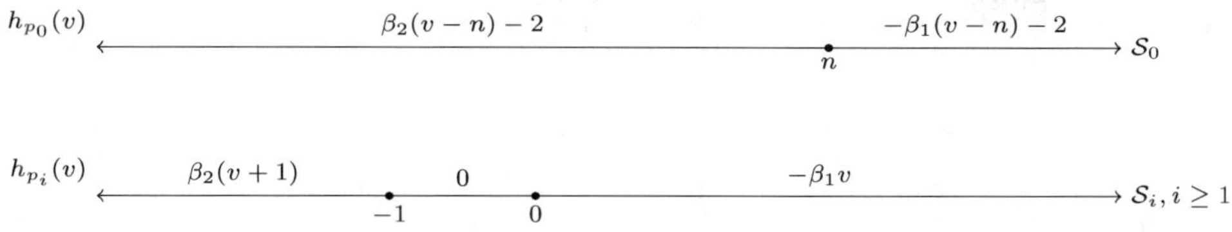

FIGURE 5. The divisorial support function h corresponding to the log anti-canonical divisor $-K_{S,\Delta_\beta}$.

By Theorem 6.9, $-K_{S,\Delta_\beta}$ corresponds to the divisorial support function h with linear part

$$h_t(v) = \begin{cases} \beta_2 v, & v < 0, \\ -\beta_1 v, & v \geq 0, \end{cases}$$

and satisfies

$$h_{p_0}(n) = -2, \quad h_{p_1}(-1) = h_{p_1}(0) = \cdots = h_{p_m}(-1) = h_{p_m}(0) = 0,$$

as shown in Figure 5.

By Definition 6.11, the dual function Ψ of h is given by

$$\Psi_{p_i}(u) = \begin{cases} nu + 2, & i = 0, \\ \min\{0, -u\}, & i \geq 1, \end{cases}$$

where the domain $\square_h = [-\beta_1, \beta_2]$, and its degree

$$\deg \Psi(u) = \begin{cases} nu + 2, & u < 0, \\ (n-m)u + 2, & u \geq 0. \end{cases}$$

By (6.1), we compute its volume

$$\operatorname{vol}(\Psi) = 2(\beta_1 + \beta_2) - \frac{n}{2}\beta_1^2 + \frac{n-m}{2}\beta_2^2,$$

and barycenter

$$\operatorname{bc}(\Psi) = -\frac{2}{\operatorname{vol}\left(-K_{S,\Delta_\beta}\right)}\left(\beta_1^2 - \beta_2^2 - \frac{1}{3}\left(n\beta_1^3 + (n-m)\beta_2^3\right)\right),$$

$$\operatorname{bc}_P(\Psi) = \begin{cases} -\frac{2}{\operatorname{vol}\left(-K_{S,\Delta_\beta}\right)}\left((2 - n\beta_1 + n\beta_2)(\beta_1+\beta_2) + \frac{1}{6}\left(n^2\beta_1^3 + (n^2-m^2)\beta_2^3\right)\right), & i = 0, \\ \frac{2}{\operatorname{vol}\left(-K_{S,\Delta_\beta}\right)}\left(2(\beta_1+\beta_2) - n\beta_1^2 + (n-m+1)\beta_2^2 + \frac{n^2\beta_1^3 + (n-m+2)(n-m)\beta_2^3}{6}\right), & i \geq 1. \end{cases}$$

Recall that by Theorem 6.12, the Futaki character vanishes precisely when $\operatorname{bc}(\Psi) = 0$, i.e.,

$$\beta_1^2 - \beta_2^2 - \frac{1}{3}\left(n\beta_1^3 + (n-m)\beta_2^3\right) = 0. \tag{7.1}$$

We are now ready to compute the δ-invariant.

(7.2)

$$\frac{A_{S,\Delta_\beta}\left(\widetilde{C}_1\right)}{S_{-K_{S,\Delta_\beta}}\left(\widetilde{C}_1\right)} = \frac{\beta_1}{\beta_1 - \frac{2}{\mathrm{vol}\left(-K_{S,\Delta_\beta}\right)}\left(\beta_1^2 - \beta_2^2 - \frac{1}{3}\left(n\beta_1^3 + (n-m)\,\beta_2^3\right)\right)},$$

$$\frac{A_{S,\Delta_\beta}\left(\widetilde{C}_2\right)}{S_{-K_{S,\Delta_\beta}}\left(\widetilde{C}_2\right)} = \frac{\beta_2}{\beta_2 + \frac{2}{\mathrm{vol}\left(-K_{S,\Delta_\beta}\right)}\left(\beta_1^2 - \beta_2^2 - \frac{1}{3}\left(n\beta_1^3 + (n-m)\,\beta_2^3\right)\right)},$$

$$\frac{A_{S,\Delta_\beta}(E_i)}{S_{-K_{S,\Delta_\beta}}(E_i)} = \frac{1}{1 + \frac{1}{\mathrm{vol}\left(-K_{S,\Delta_\beta}\right)}\left(-(n-2)\,\beta_1^2 + (n-m)\,\beta_2^2 + \frac{n(n-2)\beta_1^3 + (n-m)^2\beta_2^3}{3}\right)},$$

$$\frac{A_{S,\Delta_\beta}(F_i)}{S_{-K_{S,\Delta_\beta}}(F_i)} = \frac{1}{1 + \frac{1}{\mathrm{vol}\left(-K_{S,\Delta_\beta}\right)}\left(-n\beta_1^2 + (n-m+2)\,\beta_2^2 + \frac{n^2\beta_1^3 + (n-m+2)(n-m)\beta_2^3}{3}\right)}.$$

This reproduces the computation in Section 5 and hence proves Theorem 5.12, the necessary part of Theorem 1.8. We now prove the sufficient part of Theorem 1.8.

PROOF OF THEOREM 1.8. Assume

$$\beta_1^2 - \beta_2^2 - \frac{1}{3}\left(n\beta_1^3 + (n-m)\,\beta_2^3\right) = 0$$

and $m \geq 2$. By (7.2),

$$\frac{A_{S,\Delta_\beta}\left(\widetilde{C}_1\right)}{S_{-K_{S,\Delta_\beta}}\left(\widetilde{C}_1\right)} = \frac{A_{S,\Delta_\beta}\left(\widetilde{C}_2\right)}{S_{-K_{S,\Delta_\beta}}\left(\widetilde{C}_2\right)} = 1,$$

$$\frac{A_{S,\Delta_\beta}(E_i)}{S_{-K_{S,\Delta_\beta}}(E_i)} = \frac{A_{S,\Delta_\beta}(F_i)}{S_{-K_{S,\Delta_\beta}}(F_i)} = \frac{1}{1 + \frac{1}{\mathrm{vol}\left(-K_{S,\Delta_\beta}\right)}(2-m)\,\beta_1^2\left(1 - \frac{n}{3}\beta_1\right)}.$$

When $m \geq 3$,

$$\frac{A_{S,\Delta_\beta}(E_i)}{S_{-K_{S,\Delta_\beta}}(E_i)} = \frac{A_{S,\Delta_\beta}(F_i)}{S_{-K_{S,\Delta_\beta}}(F_i)} > 1.$$

Moreover, since $\pi^*F - F_i \geq 0$,

$$\frac{A_{S,\Delta_\beta}(\pi^*F)}{S_{-K_{S,\Delta_\beta}}(\pi^*F)} \geq \frac{A_{S,\Delta_\beta}(F_i)}{S_{-K_{S,\Delta_\beta}}(F_i)} > 1.$$

By (7.1), the Futaki character vanishes. Therefore by Theorem 6.13, the pair (S, Δ_β) is K-polystable.

When $m = 2$, for $i = 1, 2$,

$$\frac{A_{S,\Delta_\beta}\left(\widetilde{C}_i\right)}{S_{-K_{S,\Delta_\beta}}\left(\widetilde{C}_i\right)} = \frac{A_{S,\Delta_\beta}(E_i)}{S_{-K_{S,\Delta_\beta}}(E_i)} = \frac{A_{S,\Delta_\beta}(F_i)}{S_{-K_{S,\Delta_\beta}}(F_i)} = 1.$$

Therefore by Theorem 6.14, the pair (S, Δ_β) is K-polystable. ∎

REMARK 7.1. By [**13**, Theorem 5.1], the vanishing of the Futaki invariant also follows directly from the fact that $A_{S,\Delta_\beta}(\widetilde{C}_i) = S_{-K_{S,\Delta_\beta}}(\widetilde{C}_i)$.

References

[1] C. Araujo, A.-M. Castravet, I. Cheltsov, K. Fujita, A.-S. Kaloghiros, J. Martinez-Garcia, C. Shramov, H. Süß, and N. Viswanathan, *The Calabi problem for Fano threefolds*, London Mathematical Society Lecture Note Series, vol. 485, Cambridge University Press, Cambridge, 2023. MR4590444

[2] K. Altmann, N. O. Ilten, L. Petersen, H. Süß, and R. Vollmert, *The geometry of T-varieties*, Contributions to algebraic geometry, EMS Ser. Congr. Rep., Eur. Math. Soc., Zürich, 2012, pp. 17–69, DOI 10.4171/114-1/2. MR2975658

[3] R. J. Berman, *K-polystability of $\mathbb{Q}$-Fano varieties admitting Kähler-Einstein metrics*, Invent. Math. **203** (2016), no. 3, 973–1025, DOI 10.1007/s00222-015-0607-7. MR3461370

[4] R. J. Berman, S. Boucksom, and M. Jonsson, *A variational approach to the Yau-Tian-Donaldson conjecture*, J. Amer. Math. Soc. **34** (2021), no. 3, 605–652, DOI 10.1090/jams/964. MR4334189

[5] H. Blum, *Singularities and K-stability*, Diss., 2018.

[6] H. Blum and M. Jonsson, *Thresholds, valuations, and K-stability*, Adv. Math. **365** (2020), 107062, 57, DOI 10.1016/j.aim.2020.107062. MR4067358

[7] I. A. Cheltsov and Y. A. Rubinstein, *Asymptotically log Fano varieties*, Adv. Math. **285** (2015), 1241–1300, DOI 10.1016/j.aim.2015.08.001. MR3406526

[8] I. A. Cheltsov and Y. A. Rubinstein, *On flops and canonical metrics*, Ann. Sc. Norm. Super. Pisa Cl. Sci. (5) **18** (2018), no. 1, 283–311. MR3783790

[9] I. A. Cheltsov, Y. A. Rubinstein, and K. Zhang, *Basis log canonical thresholds, local intersection estimates, and asymptotically log del Pezzo surfaces*, Selecta Math. (N.S.) **25** (2019), no. 2, Paper No. 34, 36, DOI 10.1007/s00029-019-0473-z. MR3945265

[10] S. K. Donaldson, *Scalar curvature and stability of toric varieties*, J. Differential Geom. **62** (2002), no. 2, 289–349. MR1988506

[11] S. K. Donaldson, *Kähler metrics with cone singularities along a divisor*, Essays in mathematics and its applications, Springer, Heidelberg, 2012, pp. 49–79, DOI 10.1007/978-3-642-28821-0_4. MR2975584

[12] K. Fujita, *Optimal bounds for the volumes of Kähler-Einstein Fano manifolds*, Amer. J. Math. **140** (2018), no. 2, 391–414, DOI 10.1353/ajm.2018.0009. MR3783213

[13] K. Fujita, *A valuative criterion for uniform K-stability of $\mathbb{Q}$-Fano varieties*, J. Reine Angew. Math. **751** (2019), 309–338, DOI 10.1515/crelle-2016-0055. MR3956698

[14] K. Fujita, Y. Liu, H. Süss, K. Zhang, and Z. Zhuang, *On the Cheltsov-Rubinstein conjecture*, Birational geometry, Kähler-Einstein metrics and degenerations, Springer Proc. Math. Stat., vol. 409, Springer, Cham, [2023], pp. 865–883, DOI 10.1007/978-3-031-17859-7_43. MR4606670

[15] K. Fujita and Y. Odaka, *On the K-stability of Fano varieties and anticanonical divisors*, Tohoku Math. J. (2) **70** (2018), no. 4, 511–521, DOI 10.2748/tmj/1546570823. MR3896135

[16] T. Jeffres, R. Mazzeo, and Y. A. Rubinstein, *Kähler-Einstein metrics with edge singularities*, Ann. of Math. (2) **183** (2016), no. 1, 95–176, DOI 10.4007/annals.2016.183.1.3. MR3432582

[17] J. Kollár and S. Mori, *Birational geometry of algebraic varieties*, Cambridge Tracts in Mathematics, vol. 134, Cambridge University Press, Cambridge, 1998. With the collaboration of C. H. Clemens and A. Corti; Translated from the 1998 Japanese original, DOI 10.1017/CBO9780511662560. MR1658959

[18] R. Lazarsfeld, *Positivity in algebraic geometry. I: Classical setting: line bundles and linear series*, Ergebnisse der Mathematik und ihrer Grenzgebiete. 3. Folge. A Series of Modern Surveys in Mathematics [Results in Mathematics and Related Areas. 3rd Series. A Series of Modern Surveys in Mathematics], vol. 48, Springer-Verlag, Berlin, 2004, DOI 10.1007/978-3-642-18808-4. MR2095471

[19] Y. Liu, *K-stability of Fano threefolds of rank 2 and degree 14 as double covers*, Math. Z. **303** (2023), no. 2, Paper No. 38, 9, DOI 10.1007/s00209-022-03192-4. MR4530187

[20] Y. Liu, C. Xu, and Z. Zhuang, *Finite generation for valuations computing stability thresholds and applications to K-stability*, Ann. of Math. (2) **196** (2022), no. 2, 507–566, DOI 10.4007/annals.2022.196.2.2. MR4445441

[21] L. Petersen and H. Süss, *Torus invariant divisors*, Israel J. Math. **182** (2011), 481–504, DOI 10.1007/s11856-011-0039-z. MR2783981

[22] Y. A. Rubinstein, *Smooth and singular Kähler-Einstein metrics*, Geometric and spectral analysis, Contemp. Math., vol. 630, Amer. Math. Soc., Providence, RI, 2014, pp. 45–138, DOI 10.1090/conm/630/12665. MR3328541

[23] Y. A. Rubinstein, *Classification of strongly asymptotically log del Pezzo flags and surfaces*, Internat. J. Math. **33** (2022), no. 10-11, Paper No. 2250077, 27, DOI 10.1142/S0129167X2250077X. MR4514297

[24] Y. A. Rubinstein and K. Zhang, *Angle deformation of Kähler-Einstein edge metrics on Hirzebruch surfaces*, Pure Appl. Math. Q. **18** (2022), no. 1, 343–366, DOI 10.4310/PAMQ.2022.v18.n1.a11. MR4381856

[25] H. Süss, *Fano threefolds with 2-torus action: a picture book*, Doc. Math. **19** (2014), 905–940. MR3262075

[26] G. Tian, *Kähler-Einstein metrics on algebraic manifolds*, Transcendental methods in algebraic geometry (Cetraro, 1994), Lecture Notes in Math., vol. 1646, Springer, Berlin, 1996, pp. 143–185, DOI 10.1007/BFb0094304. MR1603624

[27] G. Tian, *Kähler-Einstein metrics with positive scalar curvature*, Invent. Math. **130** (1997), no. 1, 1–37, DOI 10.1007/s002220050176. MR1471884

[28] O. Zariski, *The theorem of Riemann-Roch for high multiples of an effective divisor on an algebraic surface*, Ann. of Math. (2) **76** (1962), 560–615, DOI 10.2307/1970376. MR141668

[29] K. Zhang, *Stability thresholds on rational Fano T-varieties of complexity one*, `https://keweizhang666.github.io/kwzhang.github.io/Stability_thresholds_on_Fano_T_varieties-2.pdf`

Department of Mathematics, University of Maryland, College Park, Maryland

Contemporary Mathematics
Volume **810**, 2025
https://doi.org/10.1090/conm/810/16213

A Santaló inequality for the L^p-polar body

Vlassis Mastrantonis

Dedicated to Bo Berndtsson on the occasion of his 71st birthday

ABSTRACT. In recent joint work with Berndtsson and Rubinstein, the polar of a convex body was realized as the $p = \infty$ case in a broader class of L^p-polar bodies. A Santaló inequality for the L^p-polar was proved for symmetric convex bodies. The aim of this article is to remove the symmetry assumption and to show that an L^p-Santaló inequality holds for any convex body after translation by the L^p-Santaló point.

1. Introduction

In this article, we prove a Santaló inequality for the L^p-polar body, recently introduced by Berndtsson, Rubinstein, and the author [**4**]. For $p \in (0, \infty)$, the *L^p-support function*,

$$h_{p,K}(y) := \log \left(\int_K e^{p\langle x,y\rangle} \frac{\mathrm{d}x}{|K|} \right)^{\frac{1}{p}} \tag{1.1}$$

of a convex body K, generalizes the classical support function

$$h_K(y) := \sup_{x \in K} \langle x, y \rangle, \tag{1.2}$$

which is obtained by (1.1) for $p = \infty$. Indeed, $h_{p,K}$ increases in p [**4**, Lemma 2.2 (v)] and converges pointwise to (1.2), thus $h_{\infty,K} := \lim_{p\to\infty} h_{p,K} = h_K$ [**4**, Corollary 2.7]. In contrast to h_K, for finite p, $h_{p,K}$ is not homogeneous. Nonetheless, it is always convex [**4**, Lemma 2.1] thus, by a theorem of Ball [**1**, Theorem 5] (also Klartag in the non-even case [**8**, Theorem 2.2]), it induces a near norm (that is, non-negative and sub-additive, but only positively 1-homogeneous: $\|\lambda y\| = \lambda\|y\|$ for $\lambda \geq 0$)

$$\|y\|_{K^{\circ,p}} := \left(\frac{1}{(n-1)!} \int_0^\infty r^{n-1} e^{-h_{p,K}(ry)} \mathrm{d}r \right)^{-\frac{1}{n}}. \tag{1.3}$$

The *L^p-polar* is the induced convex body [**4**, Definition 1.1]

$$K^{\circ,p} := \{y \in \mathbb{R}^n : \|y\|_{K^{\circ,p}} \leq 1\}.$$

2020 *Mathematics Subject Classification.* Primary.

The author is grateful to Y. A. Rubinstein for his support and guidance. Research supported through NSF grants DMS-1906370, 2204347.

By the homogeneity of h_K, $\|y\|_{K^{\circ,\infty}} = h_K(y)$, thus

$$K^{\circ,\infty} = K^\circ := \{y \in \mathbb{R}^n : \langle x, y\rangle \le 1, \text{ for all } x \in K\},$$

coincides with the polar of K.

The L^p*-Mahler volume* is defined via [**4**, (1.9), Theorem 1.2]

$$\mathcal{M}_p(K) := |K| \int_{\mathbb{R}^n} e^{-h_{p,K}(y)} \mathrm{d}y = n!|K||K^{\circ,p}|. \tag{1.4}$$

Denote by

$$B_2^n := \{x \in \mathbb{R}^n : x_1^2 + \ldots + x_n^2 \le 1\}.$$

Santaló's inequality [**15**, (3.12)] was generalized to the L^p-polar for symmetric convex bodies.

THEOREM 1.1 ([**4**, Theorem 1.6]). *Let $p \in (0, \infty]$. For a symmetric convex body $K \subset \mathbb{R}^n$, $\mathcal{M}_p(K) \le \mathcal{M}_p(B_2^n)$.*

The main result of the present paper is to remove the symmetry assumption:

THEOREM 1.2. *Let $p \in (0, \infty]$. For a convex body $K \subset \mathbb{R}^n$,*

$$\inf_{x \in \mathbb{R}^n} \mathcal{M}_p(K - x) \le \mathcal{M}_p(B_2^n). \tag{1.5}$$

Theorem 1.2 is a strengthening of Theorem 1.1 because the infimum on the left-hand side of (1.5) is attained at the origin for symmetric convex bodies. As we show in our previous work, the infimum in (1.5) is always a minimum attained at a unique point that lies in the interior of K.

DEFINITION 1.3 ([**4**, Proposition 1.5]). Let $p \in (0, \infty]$, and $K \subset \mathbb{R}^n$ a convex body. Denote by $s_p(K)$ the unique point in $\operatorname{int} K$ for which

$$\mathcal{M}_p(K - s_p(K)) = \inf_{x \in \mathbb{R}^n} \mathcal{M}_p(K - x), \tag{1.6}$$

called the L^p-Santaló point of K.

The L^p-Santaló point is uniquely characterized by the vanishing of the barycenter of $h_{p,K}$. Denote the *barycenter* of a convex body K by

$$b(K) := \int_K x \frac{\mathrm{d}x}{|K|}.$$

For a convex function $\phi : \mathbb{R}^n \to \mathbb{R} \cup \{\infty\}$, denote by

$$V(\phi) := \int_{\mathbb{R}^n} e^{-\phi(x)} \mathrm{d}x, \quad \text{and} \quad b(\phi) := \int_{\mathbb{R}^n} x e^{-\phi(x)} \frac{\mathrm{d}x}{V(\phi)},$$

its volume and barycenter respectively. For the convex indicator function of K, $\mathbf{1}_K^\infty$ being 0 for $x \in K$ and ∞ otherwise, $V(\mathbf{1}_K^\infty) = |K|$ and $b(\mathbf{1}_K^\infty) = b(K)$.

LEMMA 1.4. [**4**, Proposition 1.5] *Let $p \in (0, \infty]$, and $K \subset \mathbb{R}^n$ a convex body. For $x \in \mathbb{R}^n$, $x = s_p(K)$ if and only if $b(h_{p,K-x}) = 0$.*

For $p = \infty$, Theorem 1.2 is due to Santaló [**15**, (3.12)] and, in fact, $s_\infty(K)$ is the Santaló point of K. The maximzers were uniquely characterized as ellipsoids (affine images of B_2^n) by Petty [**13**, Theorem 3.2]. In future work we hope to characterize the equality case for finite p. A proof using Steiner symmetrization was later given by Saint-Raymond for symmetric convex bodies [**14**]. Meyer–Pajor adapted the proof to any convex body [**10**]. Our proof is mainly inspired by their work.

Finally, Theorem 1.2 can be used to obtain an upper bound on the Bergman kernels of tube domains evaluated on the diagonal. The connection between $\mathcal{M}$ and Bergman kernels goes back to Nazarov [**12**], Błocki [**5**], and Berndtsson [**3**] who used Bergman kernels to obtain lower bounds on $\mathcal{M}$. For a convex body $K \subset \mathbb{R}^n$, let

$$T_K := \mathbb{R}^n + \sqrt{-1}\,\mathrm{int}\, K,$$

and

$$A^2(T_K) := \{f : T_K \to \mathbb{C} : \text{ holomorphic with } \int_{T_K} |f(z)|^2 \mathrm{d}\lambda\,(z) < \infty\},$$

be the Bergman space of T_K, where λ denotes the Lebesgue measure. As observed in our previous work, $\mathcal{M}_1$ and the Bergman kernel of $A^2(T_K)$ evaluated at the diagonal are related by the following [**9**, (42)],

$$\mathcal{M}_1(K-x) = (4\pi)^n |K|^2 \sup_{\substack{f\in A^2(T_K)\\ f\neq 0}} \frac{|f(\sqrt{-1}x)|^2}{\|f\|^2_{L^2(T_K)}}, \qquad \text{for all } x \in \mathrm{int}\, K.$$

Therefore, by Theorem 1.2 for $p = 1$, we obtain the following estimate.

COROLLARY 1.5. *Let $K \subset \mathbb{R}^n$ be a convex body. For $f \in A^2(T_K)$,*

$$|f(\sqrt{-1}s_1(K))|^2 \leq \frac{\mathcal{M}_1(B_2^n)}{(4\pi)^n|K|^2} \int_{T_K} |f(z)|^2 \mathrm{d}\lambda\,(z).$$

Organization. In §2.1, an outline of the proof of Theorem 1.2 is given, stating without proof a proposition and three lemmas necessary for its proof. In §2.2, a few facts about Steiner symmetrization are recalled. §2.3 is dedicated to demonstrating that the L^p-Santaló point (Definition 1.3) of a convex body symmetric with respect to hyperplane lies on the hyperplane (Lemma 2.17). In §2.4, starting with a convex body K that contains the origin in the interior, it is shown that it can be translated in any direction u so that $u^\perp$ λ-separates (Definition 2.4) the L^p-polar (Lemma 2.8). A necessary generalization of a theorem of Ball is proved in §2.5. In §2.6, it is shown that starting with a convex body K and a direction u so that $u^\perp$ 1/2-separates $K^{\circ,p}$, then Steiner symmetrization increases the L^p-Mahler volume (Lemma 2.6). An essential step in the proof of Lemma 2.6, an inequality between the L^p-support functions of K and its Steiner symmetrization (Lemma 2.26), is detailed in §2.7. Finally, Theorem 1.2 is proved in §2.8.

2. Inequality

2.1. Outline of the proof of Theorem 1.2. The proof of Theorem 1.2 is by Steiner symmetrization. For a unit vector $u \in \partial B_2^n$, denote by

$$u^\perp := \{x \in \mathbb{R}^n : \langle x, u\rangle = 0\},$$

the hyperplane through the origin perpendicular to u. Let, also,

$$\pi_{u^\perp} : \mathbb{R}^n \ni x \mapsto x - \langle x, u\rangle u \in u^\perp,$$

the projection onto $u^\perp$.

DEFINITION 2.1. For a convex body $K \subset \mathbb{R}^n$ and $u \in \partial B_2^n$, the Steiner symmetral of K in the u direction is given by

$$\sigma_u K := \{x + tu \in \mathbb{R}^n : x \in \pi_{u^\perp}(K) \text{ and } |t| \leq \frac{1}{2}|K \cap (x + \mathbb{R}u)|\}.$$

We know that Steiner symmetrization always increases the L^p-Mahler volume of a symmetric convex body [**4**, Proposition 5.1]. This is not true without the symmetry assumption. Indeed, recall the following.

LEMMA 2.2. [**4**, Lemma 4.2] *Let $p \in (0, \infty]$. For a convex body $K \subset \mathbb{R}^n$, $\mathcal{M}_p(K)$ is finite if and only if $0 \in \operatorname{int} K$.*

Therefore, given an appropriately misplaced convex body $K \subset \mathbb{R}^n$, say with $0 \notin \operatorname{int} K$, and $u \in \partial B_2^n$ such that $0 \in \operatorname{int} \sigma_u K$, then

$$\infty = \mathcal{M}_p(K) > \mathcal{M}_p(\sigma_u K).$$

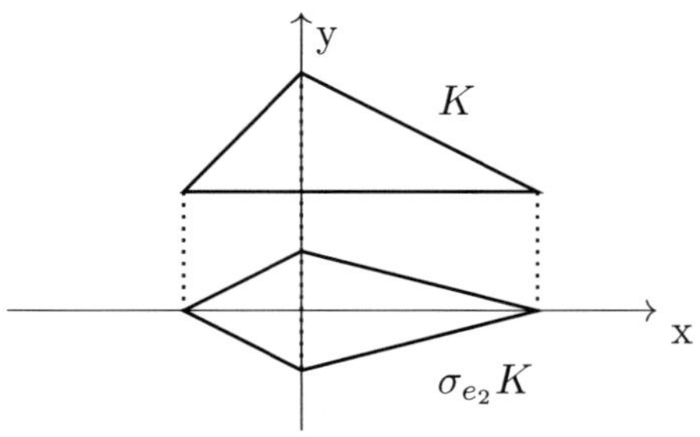

FIGURE 1. Steiner symmetrization could potentially decrease L^p-Mahler volume.

It is, thus, necessary to reposition K before applying Steiner symmetrization. Our proof is inspired by the works of Meyer–Pajor [**10**] and Meyer–Reisner [**11**]. However, the construction in the proof of their main theorem of Meyer–Pajor [**10**, Theorem] cannot be entirely generalized in our setting; duality of the polar body is heavily used, but for finite p, L^p-polarity is no longer an involution on convex bodies.

Even though Steiner symmetrization does not always increase $\mathcal{M}_p$ (Figure 1), it, nonetheless, increases (1.6).

PROPOSITION 2.3. *Let $p \in (0, \infty]$. For a convex body $K \subset \mathbb{R}^n$ and $u \in \partial B_2^n$,*

$$\inf_{x \in \mathbb{R}^n} \mathcal{M}_p(K - x) \leq \inf_{x \in \mathbb{R}^n} \mathcal{M}_p((\sigma_u K) - x).$$

The main ingredient for the proof of Proposition 2.3 is a generalization of a lemma of Meyer–Pajor [**10**, Lemma 7] to the L^p-polar (Lemma 2.6 below). First, a couple of definitions. For a unit vector $u \in \partial B_2^n$, let

$$(u^\perp)^+ := \{x \in \mathbb{R}^n : \langle x, u \rangle \geq 0\},$$
$$(u^\perp)^- := \{x \in \mathbb{R}^n : \langle x, u \rangle \leq 0\},$$

the two closed half-spaces defined by $u^\perp$.

DEFINITION 2.4. Let $K \subset \mathbb{R}^n$ be a convex body, $u \in \partial B_2^n$, and $\lambda \in (0, 1)$. The hyperplane $u^\perp$ λ-separates K if

$$\frac{|K \cap (u^\perp)^+|}{|K|} \frac{|K \cap (u^\perp)^-|}{|K|} = \lambda(1 - \lambda).$$

REMARK 2.5. Definition 2.4 is equivalent to having either $|K \cap (u^\perp)^+| = \lambda |K|$ and $|K \cap (u^\perp)^-| = (1-\lambda)|K|$, or $|K \cap (u^\perp)^+| = (1-\lambda)|K|$ and $|K \cap (u^\perp)^-| = \lambda |K|$. To see why, it is enough to note that for $x, \lambda \in (0,1)$, $x(1-x) = \lambda(1-\lambda)$ if and only if $x = \lambda$ or $x = 1-\lambda$.

LEMMA 2.6. *Let $p \in (0, \infty]$. For $u \in \partial B_2^n$ and $\lambda \in (0,1)$ such that $u^\perp$ λ-separates $K^{\circ,p}$,*

$$|(\sigma_u K)^{\circ,p}| \geq 4\lambda(1-\lambda)|K^{\circ,p}|.$$

The proof of Lemma 2.6 is rather technical and so its proof is delayed until §2.6. Given Lemma 2.6, to prove Proposition 2.3 we capitalize on the following two observations. First, the Steiner symmetral of a convex body K translated by a vector x perpendicular to the direction of the symmetrization u, is the Steiner symmetral of K with respect to u translated by x,

$$\sigma_u(K - x) = (\sigma_u K) - x, \qquad \text{for all } x \in u^\perp. \tag{2.1}$$

Indeed, it is enough to note that since $x \in u^\perp$, $\pi_{u^\perp}(K-x) = \pi_{u^\perp}(K) - x$. Therefore, $z + tu \in \sigma_u(K-x)$ if and only if $z \in \pi_{u^\perp}(K-x) = \pi_{u^\perp}(K) - x$ and $2|t| \leq |(K-x) \cap (z + \mathbb{R}u)|$. Or, equivalently, $z + x \in \pi_{u^\perp}(K)$ and $2|t| \leq |K \cap (z + x + \mathbb{R}u)|$, because $(K-x) \cap (z + \mathbb{R}u) = \big(K \cap (z + x + \mathbb{R}u)\big) - x$, i.e., $z + x + tu \in \sigma_u K$ proving (2.1).

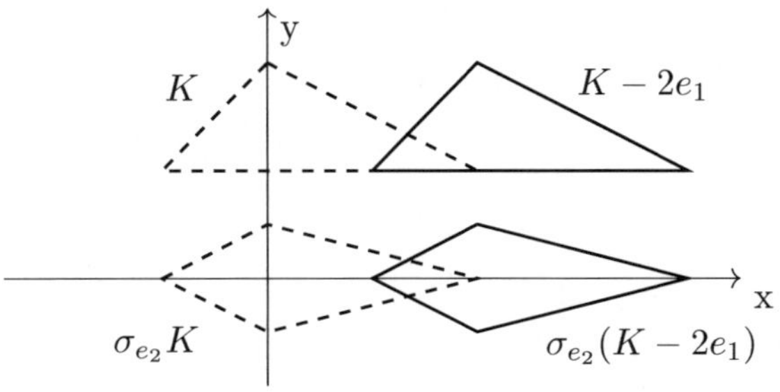

FIGURE 2. Translating K in a direction normal to u has the same effect on the Steiner symmetral.

Second, the Steiner symmetral of a convex body K with respect to a unit vector u remains invariant under translation of the body in the direction of u, i.e.,

$$\sigma_u(K - tu) = \sigma_u K, \qquad \text{for all } t \in \mathbb{R}. \tag{2.2}$$

This follows directly from $\pi_{u^\perp}(K - tu) = \pi_{u^\perp}(K)$ and $(K - tu) \cap (x + \mathbb{R}u) = \big(K \cap (x + \mathbb{R}u)\big) - tu$.

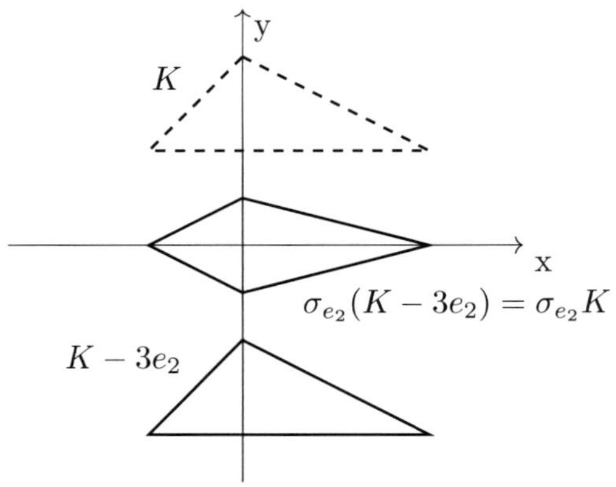

FIGURE 3. Translating K in a direction parallel to u has no effect on the Steiner symmetral.

The first observation (2.1), will be used to place K in a position such that the L^p-Santaló point of the Steiner symmetral lies at the origin, $s_p(\sigma_u K) = 0$, by translating K in a direction perpendicular to u. This is always possible because the L^p-Santaló point of $\sigma_u K$ lies on $u^\perp$. The proof occupies §2.3.

LEMMA 2.7. *Let $p \in (0, \infty]$. For a convex body $K \subset \mathbb{R}^n$ and $u \in \partial B_2^n$, $s_p(\sigma_u K) \in u^\perp$.*

The second observation (2.2) is used to further translate K in a direction parallel to u, ensuring that the L^p-Santaló point of the Steiner symmetral remains at the origin, so that, in addition, $u^\perp$ 1/2-separates $K^{\circ,p}$. Consequently, by Lemma 2.6, Steiner symmetrization can be used to increase $\mathcal{M}_p$.

LEMMA 2.8. *Let $p \in (0, \infty]$, and $K \subset \mathbb{R}^n$ a convex body with $0 \in \operatorname{int} K$. For $u \in \partial B_2^n$ and $\lambda \in (0,1)$, there is $t \in \mathbb{R}$ such that $u^\perp$ λ-separates $(K - tu)^{\circ,p}$.*

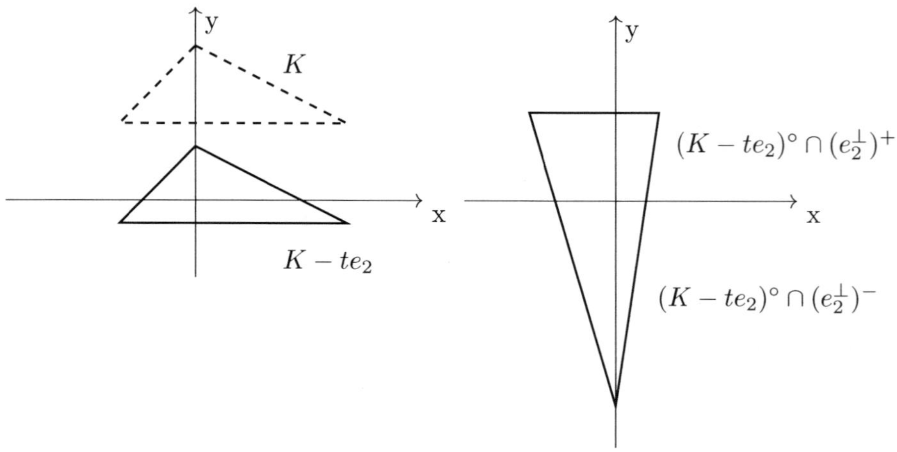

FIGURE 4. Translating K in a direction parallel to u so that $u^\perp$ 1/2-separates the L^p-polar.

Given a direction $u \in \partial B_2^n$, Lemmas 2.6, 2.7, and 2.8, allow us to reposition K in a way that achieves two objectives simultaneously: first, $u^\perp$ 1/2-separates its L^p-polar, resulting in an increased L^p-Mahler volume of the Steiner symmetral, and second, the L^p-Mahler volume of its Steiner symmetral attains its minimum value over translations. This is sufficient to prove Proposition 2.3. Theorem 1.2 then follows by repeated applications of Steiner symmetrization.

2.2. Steiner symmetrization. We recall a few facts about Steiner symmetrization [**16**, pp. 286–287] without proof.

DEFINITION 2.9. Let $u \in \partial B_2^n$. A convex body $K \subset \mathbb{R}^n$ is symmetric with respect to a hyperplane $u^\perp$ if for all $x \in K$, $x - 2\langle x, u\rangle u \in K$.

Steiner symmetrization produces a convex body that is symmetric with respect to $u^\perp$. It also preserves convexity and volume [**7**, Proposition 9.1].

LEMMA 2.10. *For a convex body $K \subset \mathbb{R}^n$ and $u \in \partial B_2^n$, $\sigma_u K$ is a convex body, symmetric with respect to $u^\perp$, with $|\sigma_u K| = |K|$.*

Since L^p-polarity preserves the symmetries of the convex body with respect to hyperplanes [**4**, Lemma 5.17], by Lemma 2.10, $(\sigma_u K)^{\circ,p}$ is symmetric with respect to $u^\perp$.

LEMMA 2.11. *Let $p \in (0,\infty]$. For a convex body $K \subset \mathbb{R}^n$ and $u \in \partial B_2^n$, $(\sigma_u K)^{\circ,p}$ is symmetric with respect to $u^\perp$.*

For a matrix $A \in GL(n,\mathbb{R})$, A^T denotes the transpose of A, and $O(n) := \{A \in GL(n,\mathbb{R}) : A^T = A^{-1}\}$ the orthogonal group. The next lemma allows for working with e_n instead of an arbitrary unit vector [**4**, Lemma 5.8].

LEMMA 2.12. *For a convex body $K \subset \mathbb{R}^n$, $u \in \partial B_2^n$, and $A \in O(n)$,*

$$\sigma_u(K) = A^{-1}\sigma_{Au}(AK).$$

In particular, $|\sigma_u(K)| = |\sigma_{Au}(AK)|$.

Finally, Steiner symmetrization preserves symmetries in orthogonal directions [**17**, Lemma 3.1].

LEMMA 2.13. *Let $K \subset \mathbb{R}^n$ be a convex body, and $u \in \partial B_2^n$ such that K is symmetric with respect to $u^\perp$. For $v \in \partial B_2^n \cap u^\perp$, $\sigma_v K$ is symmetric with respect to $u^\perp$.*

2.3. Symmetries of the L^p-Santaló point. In this section, we prove Lemma 2.7.

A measure μ on $\mathbb{R}^n$ is called *absolutely continuous* with respect to another measure ν, if $\mu(A) = 0$ for all measurable $A \subset \mathbb{R}^n$ with $\nu(A) = 0$. As a consequence, if for a positive function $f : \mathbb{R}^n \to [0,\infty)$,

$$\int_{\mathbb{R}^n} f(x) \mathrm{d}\mu(x) = 0,$$

then f is zero μ-almost everywhere, and hence f is zero ν-almost everywhere.

LEMMA 2.14. *Let $p \in (0,\infty]$. For a convex body $K \subset \mathbb{R}^n$, the Lebsegue measure λ is absolutely continuous with respect to the measure $e^{-h_{p,K}(y)}\mathrm{d}y$.*

PROOF. Let us start with $p=\infty$. Let $A \subset \mathbb{R}^n$ be a measurable set such that $\int_A e^{-h_K(y)}\mathrm{d}y = 0$. For $r>0$,

$$0 = \int_A e^{-h_K(y)}\mathrm{d}y \geq \int_{A\cap\{h_K\leq r\}} e^{-h_K(y)}\mathrm{d}y \geq \lambda(A\cap(rK^\circ))e^{-r} \geq 0,$$

because $\{h_K \leq r\} = rK^\circ$, by the homogeneity of h_K. Therefore, $\lambda(A\cap(rK^\circ))=0$ for all $r>0$. Taking $r\to\infty$, $\lambda(A)=0$.

For $p\in(0,\infty)$, note

$$h_{p,K}(y) = \frac{1}{p}\log\int_K e^{p\langle x,y\rangle}\frac{\mathrm{d}x}{|K|} \leq \frac{1}{p}\log\int_K e^{ph_K(y)}\frac{\mathrm{d}x}{|K|} = h_K(y).$$

Therefore, for measurable $A\subset\mathbb{R}^n$ with $\int_A e^{-h_{p,K}(y)}\mathrm{d}y = 0$,

$$0 = \int_A e^{-h_{p,K}(y)}\mathrm{d}y \geq \int_A e^{-h_K(y)}\mathrm{d}y \geq 0,$$

and hence $\int_A e^{-h_K(y)}\mathrm{d}y = 0$. By the $p=\infty$ case, $\lambda(A)=0$. □

The L^p-support function inherits symmetries with respect to hyperplanes.

CLAIM 2.15. *Let $p\in(0,\infty]$, $K\subset\mathbb{R}^n$ a convex body, and $u\in\mathbb{R}^n$. If K is symmetric with respect to $u^\perp$, then $h_{p,K}$ is symmetric with respect to $u^\perp$, i.e.,*

$$h_{p,K}(y+su) = h_{p,K}(y-su), \qquad \text{for all } y\in u^\perp, s\in\mathbb{R}.$$

PROOF. Since K is symmetric with respect to $u^\perp$, there is $f: K\cap u^\perp \to [0,\infty)$ such that

$$K = \{x+tu : x\in K\cap u^\perp, |t|\leq f(x)\}.$$

For $y\in u^\perp, s\in\mathbb{R}$,

$$\begin{aligned}
h_{p,K}(y+su) &= \frac{1}{p}\log\int_{K\cap u^\perp}\int_{-f(x)}^{f(x)} e^{p\langle x+tu,y+su\rangle}\frac{\mathrm{d}t\,\mathrm{d}x}{|K|}\\
&= \frac{1}{p}\log\int_{K\cap u^\perp}\int_{-f(x)}^{f(x)} e^{p\langle x,y\rangle}e^{pts}\frac{\mathrm{d}t\,\mathrm{d}x}{|K|}\\
&= \frac{1}{p}\log\int_{K\cap u^\perp}\int_{-f(x)}^{f(x)} e^{p\langle x,y\rangle}e^{-p\tau s}\frac{\mathrm{d}\tau\,\mathrm{d}x}{|K|}\\
&= \frac{1}{p}\log\int_{K\cap u^\perp}\int_{-f(x)}^{f(x)} e^{p\langle x+\tau u,y-su\rangle}\frac{\mathrm{d}\tau\,\mathrm{d}x}{|K|}\\
&= h_{p,K}(y-su),
\end{aligned}$$

by the change of variables $\tau=-t$. □

The L^p-polar behaves exactly like the polar under the action of $GL(n,\mathbb{R})$ [**4**, (4.7)], that is for a convex body $K\subset\mathbb{R}^n$ and $A\in GL(n,\mathbb{R})$,

$$(AK)^{\circ,p} = (A^{-1})^T K^{\circ,p}. \tag{2.3}$$

To see why, by a simple change of variables, $h_{p,AK}(y) = h_{p,K}(A^T y)$ [**4**, Lemma 2.2 (iii)]. Therefore,

$$\|y\|_{(AK)^{\circ,p}} := \left(\int_0^\infty r^{n-1} e^{-h_{p,AK}(ry)} \mathrm{d}r\right)^{-\frac{1}{n}}$$
$$= \left(\int_0^\infty r^{n-1} e^{-h_{p,K}(rA^T y)} \mathrm{d}r\right)^{-\frac{1}{n}} = \|A^T y\|_{K^{\circ,p}},$$

from which (2.3) follows.

LEMMA 2.16. *Let $p \in (0,\infty]$. For a convex body $K \subset \mathbb{R}^n$, $A \in GL(n,\mathbb{R})$, $s_p(AK) = As_p(K)$.*

PROOF. By Lemma 1.4, the vanishing of the barycenter of the L^p-support function uniquely characterizes the L^p-Santaló point. By changing variables,

$$h_{p,AK-As_p(K)}(y) = h_{p,A(K-s_p(K))}(y) = h_{p,K-s_p(K)}(A^T y),$$

[**4**, Lemma 2.2 (iii)], and $V(h_{p,AK-As_p(K)}) = |\det A|^{-1} V(h_{p,K-s_p(K)})$. For the barycenter,

$$b(h_{p,AK-As_p(K)}) = \int_{\mathbb{R}^n} y e^{-h_{p,AK-As_p(K)}(y)} \frac{\mathrm{d}y}{V(h_{p,AK-As_p(K)})}$$
$$= \int_{\mathbb{R}^n} y e^{-h_{p,K-s_p(K)}(A^T y)} \frac{\mathrm{d}y}{|\det A|^{-1} V(h_{p,K-s_p(K)})}$$
$$= (A^T)^{-1} \int_{\mathbb{R}^n} z e^{-h_{p,K-s_p(K)}(z)} \frac{|\det A|^{-1}\mathrm{d}z}{|\det A|^{-1} V(h_{p,K-s_p(K)})}$$
$$= (A^T)^{-1} b(h_{p,K-s_p(K)}).$$

Therefore, $b(h_{p,K-s_p(K)}) = 0$ if and only if $b(h_{p,AK-As_p(K)}) = 0$, from which the claim follows. □

For Lemma 2.7, let us prove something slightly more general.

LEMMA 2.17. *Let $p \in (0,\infty]$, $K \subset \mathbb{R}^n$ a convex body, and $u \in \partial B_2^n$. If K is symmetric with respect to $u^\perp$, then $s_p(K) \in u^\perp$.*

PROOF. To begin with, take $u = e_n$. Since $K \subset \mathbb{R}^n$ is a convex body symmetric with respect to $e_n^\perp$ there is $f : K \cap e_n^\perp \to [0,\infty)$ such that

$$K = \{(\xi, x_n) \in \mathbb{R}^{n-1} \times \mathbb{R} : (\xi, 0) \in K \cap u^\perp \text{ and } |x_n| \le f(x)\}.$$

Let

$$s_p(K) = (\xi_0, t_0), \quad (\xi_0, 0) \in K \cap u^\perp, t_0 \in \mathbb{R},$$

be the L^p-Santaló point of K. The claim is that $t_0 = 0$. Indeed, by Lemma 1.4, the barycenter of $h_{p,K-s_p(K)}$ lies at the origin. In particular, the n-th coordinate is 0,

$$\begin{aligned}
0 &= \int_{\mathbb{R}^n} y_n e^{-h_{p,K-s_p(K)}(y)} \mathrm{d}y \\
&= \int_{\mathbb{R}^{n-1}} \int_{\mathbb{R}} y_n e^{-h_{p,K}(\eta,y_n)} e^{\langle(\xi_0,t_0),(\eta,y_n)\rangle} \mathrm{d}y_n \mathrm{d}\eta \\
&= \int_{\mathbb{R}^{n-1}} \int_{\mathbb{R}} y_n e^{-h_{p,K}(\eta,y_n)} e^{\langle\xi_0,\eta\rangle} e^{t_0 y_n} \mathrm{d}y_n \mathrm{d}\eta \\
&= \int_{\mathbb{R}^{n-1}} \int_0^\infty y_n e^{-h_{p,K}(\eta,y_n)} e^{\langle\xi_0,\eta\rangle} e^{t_0 y_n} \mathrm{d}y_n \mathrm{d}\eta \\
&\quad + \int_{\mathbb{R}^{n-1}} \int_{-\infty}^0 y_n e^{-h_{p,K}(\eta,y_n)} e^{\langle\xi_0,\eta\rangle} e^{t_0 y_n} \mathrm{d}y_n \mathrm{d}\eta \\
&= \int_{\mathbb{R}^{n-1}} \int_0^\infty y_n e^{-h_{p,K}(\eta,y_n)} e^{\langle\xi_0,\eta\rangle} e^{t_0 y_n} \mathrm{d}y_n \mathrm{d}\eta \\
&\quad - \int_{\mathbb{R}^{n-1}} \int_0^\infty y_n e^{-h_{p,K}(\eta,-y_n)} e^{\langle\xi_0,\eta\rangle} e^{-t_0 y_n} \mathrm{d}y_n \mathrm{d}\eta \\
&= \int_{\mathbb{R}^{n-1}} \int_0^\infty y_n e^{-h_{p,K}(\eta,y_n)} e^{\langle\xi_0,\eta\rangle} (e^{t_0 y_n} - e^{-t_0 y_n}) \mathrm{d}y_n \mathrm{d}\eta \\
&= \int_{\mathbb{R}^{n-1}} \int_0^\infty y_n e^{-h_{p,K-(\xi_0,0)}(\eta,y_n)} 2\sinh(t_0 y_n) \mathrm{d}y_n \mathrm{d}\eta\,,
\end{aligned}$$

because K is symmetric with respect to e_n, thus by Lemma 2.14, $h_{p,K}(\eta,-y_n) = h_{p,K}(\eta,y_n)$. By Lemma 2.14, $2y_n \sinh(t_0 y_n)$ vanishes almost everywhere in $\mathbb{R}^{n-1} \times (0,\infty)$, hence $t_0 = 0$.

In general, for $u \in \partial B_2^n$, let $A \in O(n)$ such that $Au = e_n$. If K is symmetric with respect to u, then AK is symmetric with respect to $Au = e_n$. By Lemma 2.16, $As_p(K) = s_p(AK) \in e_n^\perp = (Au)^\perp$. To finish the proof, since for $A \in O(n)$, $(A^T)^{-1} = A$, is remains to show

$$(Au)^\perp = (A^T)^{-1} u^\perp, \tag{2.4}$$

Since A is invertible, it is enough to show $(Au)^\perp \subset (A^T)^{-1} u^\perp$. For $x \in (Au)^\perp$, $0 = \langle x, Au\rangle = \langle A^T x, u\rangle$, thus $A^T x \in u^\perp$, or $x \in (A^T)^{-1} u$. □

PROOF OF LEMMA 2.7. By Lemma 2.10, $\sigma_u K$ is symmetric with respect to $u^\perp$, thus by Lemma 2.17, $s_p(\sigma_u K) \in u^\perp$. □

2.4. Separating the L^p-polar. We now turn to the proof of Lemma 2.8.

After proving Lemma 2.6 for $p = \infty$, in order to show the classical Santaló inequality, Meyer–Pajor translate K so that $u^\perp$ 1/2-separates it. Since $(K^\circ)^\circ = K$, they switch to working with K°, so that $u^\perp$ 1/2-separates its polar $(K^\circ)^\circ$ [**10**, pp. 88–89]. Then, by Lemma 2.6 for $\lambda = 1/2$,

$$\inf_{x\in\mathbb{R}^n} \mathcal{M}(K - x) \le \mathcal{M}(K) = \mathcal{M}(K^\circ) \le \mathcal{M}(\sigma_u(K^\circ)).$$

Doing this n times for n orthogonal directions produces a symmetric convex body S with $\inf_{x\in\mathbb{R}^n} \mathcal{M}(K - x) \le \mathcal{M}(S)$. It is, therefore, enough to prove Santaló's inequality for symmetric convex bodies.

L^p-polarity is not a duality operation, so a similar argument would not work in our case. However, even in the Meyer–Pajor argument, it is not necessary to switch to K°. Let us explain why. One may translate K in the direction of u such that $u^\perp$ 1/2-separates K°. Take $u = e_n$ for a moment. Starting with a convex body $K \subset \mathbb{R}^n$ such that $0 \in \operatorname{int} K$, the volume of the polar K° is finite (Lemma 2.2). In particular,

$$|K^\circ \cap (e_n^\perp)^+| < \infty \quad \text{and} \quad |K^\circ \cap (e_n^\perp)^-| < \infty.$$

Moving K in the direction of e_n will always keep one of the two finite,

$$|(K - te_n)^\circ \cap (e_n^\perp)^-| < \infty \quad \text{for all } t \in \mathbb{R} \text{ with } K - te_n \subset (e_n^\perp)^-, \tag{2.5}$$

even though, by Lemma 2.2, when the origin is no longer in the interior their sum will become infinite. In other words, if K mostly lies in $(e_n^\perp)^-$, then K° mostly lies in $(e_n^\perp)^+$.

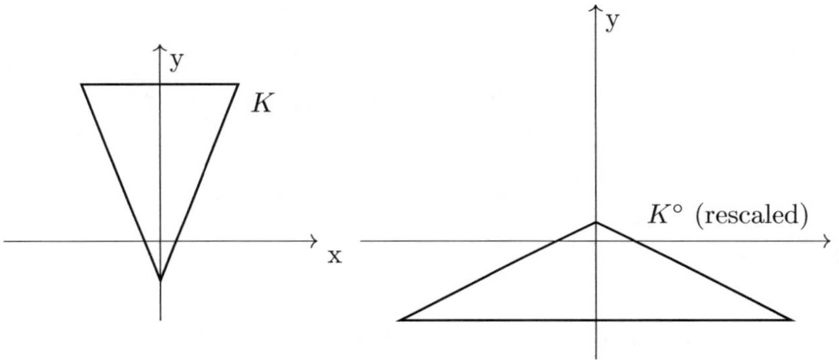

FIGURE 5. If K mostly lies in $(e_n^\perp)^+$, then K° mostly lies in $(e_n^\perp)^-$.

To prove (2.5), it is necessary to start with K with $0 \in \operatorname{int} K$, so that $|K^\circ|$ is finite (see Remark 2.22 below). In particular, K° is bounded, and hence, there is $M > 0$ such that

$$K^\circ \subset MB_2^n.$$

Then, for $t \in \mathbb{R}$ with $K - te_n \subset (e_n^\perp)^-$,

$$(K - te_n)^\circ \subset \{(\eta, y_n) \in \mathbb{R}^{n-1} \times \mathbb{R} : y_n > -1/t, |\eta| \leq M\sqrt{1 + y_n t}\}, \tag{2.6}$$

i.e., $(K - te_n)^\circ$ is contained in a cone whose intersection with $(e_n^\perp)^-$ is of finite volume. Indeed, for $(\eta, y_n) \in (K - te_n)^\circ$,

$$\langle(\xi, x_n - t), (\eta, y_n)\rangle = \langle(\xi, x_n), (\eta, y_n)\rangle - ty_n \leq 1, \quad \text{for all } (\xi, x_n) \in K, \tag{2.7}$$

or, equivalently, $\langle(\xi, x_n), (\eta, y_n)\rangle \leq 1 + ty_n$. This forces $y_n > -1/t$, because if $y_n \leq -1/t$, or equivalently, $1 + ty_n \leq 0$, choosing $\varepsilon > 0$ such that $(0, -\varepsilon) \in K$ (such an ε exists because $0 \in \operatorname{int} K$), (2.7) gives $-\varepsilon y_n \leq 1 + ty_n \leq 0$, which in turn implies $y_n \geq 0$, a contradiction to $y_n \leq -1/t$. Thus, $1 + ty_n > 0$. As a result, by (2.7),

$$\frac{(\eta, y_n)}{1 + ty_n} \in K^\circ,$$

which is bounded by M, thus

$$\frac{|\eta|}{\sqrt{1 + ty_n}} \leq \frac{\sqrt{|\eta|^2 + |y_n|^2}}{\sqrt{1 + ty_n}} \leq M.$$

proving (2.6). The part of the cone of the right-hand side of (2.6) that lies in $(e_n^\perp)^-$ has finite volume. Indeed, it is contained in the cylinder $(MB_2^{n-1}) \times [-1/t, 0]$.

In sum, we have verified the following claim for $p = \infty$.

CLAIM 2.18. *Let $p \in (0, \infty]$. For a convex body $K \subset \mathbb{R}^n$ with $0 \in \operatorname{int} K$, and $t \in \mathbb{R}$ such that $(K - te_n) \subset (e_n^\perp)^-$,*

$$|(K - te_n)^{\circ,p} \cap (e_n^\perp)^-| < \infty.$$

To prove Claim 2.18 for finite p, we need an approximation of $K^{\circ,p}$ by K°.

LEMMA 2.19 ([**4**, Lemma 2.6]). *Let $p \in (0, \infty)$. For a convex body $K \subset \mathbb{R}^n$ with $b(K) = 0$, and $\lambda \in (0,1)$,*

$$h_K(y) \leq h_{p,K}(y/\lambda) - \frac{n}{p}\log(1-\lambda),$$

for all $y \in \mathbb{R}^n$.

COROLLARY 2.20. *Let $p \in (0, \infty]$. For a convex body $K \subset \mathbb{R}^n$,*

$$K^{\circ,p} \subset \frac{1}{\lambda(1-\lambda)^p}\left(K - \left(1 - \frac{1}{\lambda}\right)b(K)\right)^\circ,$$

for all $\lambda \in (0,1)$.

PROOF OF COROLLARY 2.20. Let $\lambda \in (0,1)$. By Lemma 2.19,

$$h_{K-b(K)}(y) \leq h_{p,K-b(K)}(y/\lambda) - \frac{n}{p}\log(1-\lambda), \qquad \text{for all } y \in \mathbb{R}^n. \tag{2.8}$$

Since $h_{p,K-b(K)}(y) = h_{p,K}(y) - \langle y, b(K)\rangle$ and $h_{K-b(K)}(y) = h_K(y) - \langle y, b(K)\rangle$ [**4**, Lemma 2.2 (ii)], (2.8) gives

$$h_{K-(1-\frac{1}{\lambda})b(K)}(y) = h_K(y) - \left(1 - \frac{1}{\lambda}\right)\langle y, b(K)\rangle \leq h_{p,K}(y/\lambda) - \frac{n}{p}\log(1-\lambda).$$

So, for the norm (1.3),

$$\|y\|_{(K-(1-\frac{1}{\lambda})b(K))^\circ} \leq \frac{1}{\lambda(1-\lambda)^p}\|y\|_{K^{\circ,p}},$$

from which the claim follows. □

PROOF OF CLAIM 2.18. The $p = \infty$ case follows from (2.6). Let $p \in (0, \infty)$, and

$$a := \max\{t \in \mathbb{R} : K - te_n \subset (e_n^\perp)^+\},$$
$$b := \min\{t \in \mathbb{R} : K - te_n \subset (e_n^\perp)^-\}.$$

Since $0 \in \operatorname{int} K$, $a < 0$ and $b > 0$.

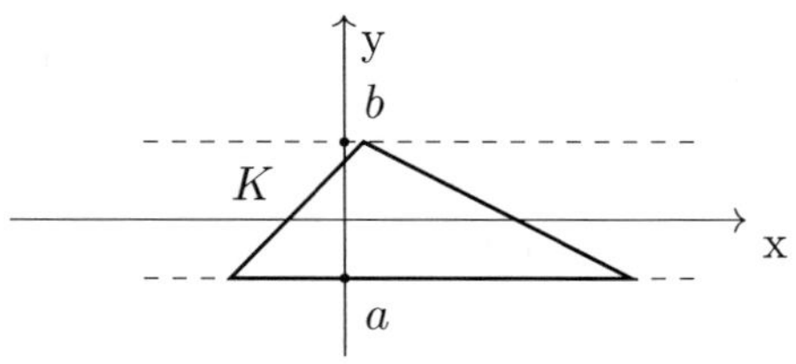

FIGURE 6. a and b.

For $t \in (a,b)$, $te_n \in \operatorname{int} K$, thus $0 \in \operatorname{int}(K - te_n)$. By Lemma 2.2, $\mathcal{M}_p(K - te_n) < \infty$ for all $t \in (a,b)$. In particular, $|(K - te_n)^{\circ,p} \cap (e_n^\perp)^-| < \infty$ for all $t \in (a,b)$. By continuity, it is then enough to prove the claim for $t > b$. In that case, since $t > b$, for $\lambda_0 \in (0,1)$ close enough to 1, if

$$L := (K - te_n) - \left(1 - \frac{1}{\lambda_0}\right) b(K - te_n),$$

then $L \subset (e_n^\perp)^-$. By the $p = \infty$ case, $|L^\circ \cap (e_n^\perp)^-| < \infty$. By Corollary 2.20,

$$(K - te_n)^{\circ,p} \cap (e_n^\perp)^- \subset \frac{L^\circ \cap (e_n^\perp)^-}{\lambda_0(1-\lambda_0)^p},$$

from which the claim follows. □

Replacing e_n by $-e_n$ proves the following.

CLAIM 2.21. *Let $p \in (0,\infty]$. For a convex body $K \subset \mathbb{R}^n$ with $0 \in \operatorname{int} K$, and $t \in \mathbb{R}$ such that $(K - te_n) \subset (e_n^\perp)^+$,*

$$|(K - te_n)^{\circ,p} \cap (e_n^\perp)^+| < \infty.$$

REMARK 2.22. For Claims 2.18 and 2.21, it is important to start with a convex body K with $0 \in \operatorname{int} K$. Otherwise, the claim does not hold. For example, the polar body of the standard simplex in $\mathbb{R}^2$,

$$\Delta_2 := \{(x,y) \in [0,\infty)^2 : x + y \leq 1\},$$

translated in the direction of the y-axis, for $t \in (0,1)$,

$$(\Delta_2 - te_2)^\circ = \left\{(x,y) \in \mathbb{R}^2 : -\frac{1}{t} \leq y \leq \frac{1}{1-t} \text{ and } x - ty \leq 1\right\}.$$

Both $(\Delta_2 - te_2)^\circ \cap (e_2^\perp)^+$ and $(\Delta_2 - te_2)^\circ \cap (e_2^\perp)^-$ has infinite volume.

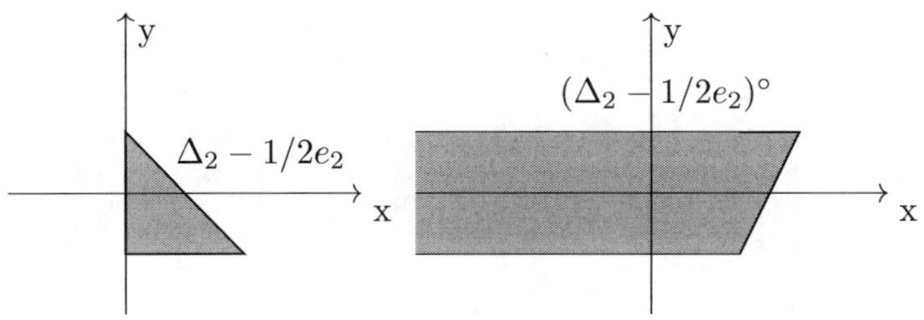

FIGURE 7. $\Delta_2 - 1/2e_2$ and its polar $(\Delta_2 - 1/2e_2)^\circ$.

PROOF OF LEMMA 2.8. Let

$$a := \max\{t < 0 : K - te_n \subset (e_n^\perp)^+\},$$
$$b := \min\{t > 0 : K - te_n \subset (e_n^\perp)^-\}$$

(Figure 6). Consider

$$f(t) := \frac{|(K - te_n)^{\circ,p} \cap (e_n^\perp)^+|}{|(K - te_n)^{\circ,p} \cap (e_n^\perp)^-|}, \quad t \in [a,b].$$

By [**4**, Lemma 4.3], f is finite valued and C^2 for $t \in (a,b)$. By Lemma 2.2, $|(K - ae_n)^{\circ,p}| = \infty$ and $|(K - be_n)^{\circ,p}| = \infty$. In addition, by Claim 2.21,

$|(K-ae_n)^{\circ,p}\cap(e_n^\perp)^+|<\infty$, forcing $|(K-ae_n)^{\circ,p}\cap(e_n^\perp)^-|=\infty$ thus $f(a)=0$. Similarly, by Claim 2.18, $|(K-be_n)^{\circ,p}\cap(e_n^\perp)^-|<\infty$ and $|(K-be_n)^{\circ,p}\cap(e_n^\perp)^+|=\infty$, thus $f(b)=\infty$. By the intermediate valued theorem, there is $t_0\in(a,b)$ such that $f(t_0)=\frac{\lambda}{1-\lambda}$ (could also choose $\frac{1-\lambda}{\lambda}$), i.e., $e_n^\perp$ λ-separates $(K-t_0e_n)^{\circ,p}$. □

2.5. A generalization Ball's Brunn–Minkowski inequality for the harmonic mean. For $p=\infty$, $\|\cdot\|_{K^\circ}=h_K$, thus going from inequalities on h_K to inclusions on K° is immediate. A similar equality does not hold for finite p, so a little more work is required. The following theorem due to Ball is very useful for this purpose [**2**, Theorem 4.10].

THEOREM 2.23. *Let $F,G,H:(0,\infty)\to[0,\infty)$ be measurable functions, not almost everywhere 0, so that*

$$H(r)\geq F(t)^{\frac{s}{t+s}}G(s)^{\frac{t}{t+s}},\qquad \textit{for all}\quad \frac{2}{r}=\frac{1}{t}+\frac{1}{s}.$$

Then, for $q\geq 1$,

$$\left(\int_0^\infty r^{q-1}H(r)\mathrm{d}r\right)^{-\frac{1}{q}}\leq\frac{1}{2}\left(\int_0^\infty t^{q-1}F(t)\mathrm{d}t\right)^{-\frac{1}{q}}+\frac{1}{2}\left(\int_0^\infty s^{q-1}G(s)\mathrm{d}s\right)^{-\frac{1}{q}}.$$

As a corollary we obtain the following.

COROLLARY 2.24. *Let $\lambda\in[0,1]$, and $F,G,H:(0,\infty)\to[0,\infty)$ be measurable functions, not almost everywhere 0, so that*

$$H(r)\geq F(t)^{\frac{(1-\lambda)s}{\lambda t+(1-\lambda)s}}G(s)^{\frac{\lambda t}{\lambda t+(1-\lambda)s}},\qquad \textit{for all}\quad \frac{1}{r}=\frac{1-\lambda}{t}+\frac{\lambda}{s}.$$

Then,

$$\left(\int_0^\infty r^{q-1}H(r)\mathrm{d}r\right)^{-\frac{1}{q}}\leq(1-\lambda)\left(\int_0^\infty t^{q-1}F(t)\mathrm{d}t\right)^{-\frac{1}{q}}+\lambda\left(\int_0^\infty s^{q-1}G(s)\mathrm{d}s\right)^{-\frac{1}{q}}.$$

PROOF. When $\lambda=0$ or 1, the statement follows from the monotonicity of the integral. For $\lambda\in(0,1)$, let

$$\tau=\frac{t}{2(1-\lambda)},\qquad\text{and}\qquad\sigma=\frac{s}{2\lambda}.$$

Then,

$$\frac{(1-\lambda)s}{\lambda t+(1-\lambda)s}=\frac{(1-\lambda)2\lambda\sigma}{\lambda 2(1-\lambda)\tau+(1-\lambda)2\lambda\sigma}=\frac{\sigma}{\tau+\sigma},$$

and

$$\frac{\lambda t}{\lambda t+(1-\lambda)s}=\frac{\lambda 2(1-\lambda)\tau}{\lambda 2(1-\lambda)\tau+(1-\lambda)2\lambda\sigma}=\frac{\tau}{\tau+\sigma}.$$

Therefore, if

$$f(\tau):=F(2(1-\lambda)\tau),\qquad\text{and}\qquad g(\sigma):=G(2\lambda\sigma),$$

then

$$H(r)\geq f(\tau)^{\frac{\sigma}{\tau+\sigma}}g(\sigma)^{\frac{\tau}{\tau+\sigma}},\ \text{for all}\quad\frac{2}{r}=\frac{1}{\tau}+\frac{1}{\sigma}.$$

By Theorem 2.23, for all $q \geq 1$,

$$\begin{aligned}
&\left(\int_0^\infty r^{q-1}H(r)\mathrm{d}r\right)^{-\frac{1}{q}}\\
&\leq \frac{1}{2}\left(\int_0^\infty \tau^{q-1}f(\tau)\mathrm{d}\tau\right)^{-\frac{1}{q}} + \frac{1}{2}\left(\int_0^\infty \sigma^{q-1}g(\sigma)\mathrm{d}\sigma\right)^{-\frac{1}{q}}\\
&= \frac{1}{2}\left(\int_0^\infty \tau^{q-1}F(2(1-\lambda)\tau)\mathrm{d}\tau\right)^{-\frac{1}{q}} + \frac{1}{2}\left(\int_0^\infty \sigma^{q-1}G(2\lambda\sigma)\mathrm{d}\sigma\right)^{-\frac{1}{q}}\\
&= \frac{1}{2}\left(\int_0^\infty \left(\frac{t}{2(1-\lambda)}\right)^{q-1}F(t)\frac{\mathrm{d}t}{2(1-\lambda)}\right)^{-\frac{1}{q}} + \frac{1}{2}\left(\int_0^\infty \left(\frac{s}{2\lambda}\right)^{q-1}G(s)\frac{\mathrm{d}s}{2\lambda}\right)^{-\frac{1}{q}}\\
&= (1-\lambda)\left(\int_0^\infty t^{q-1}F(t)\mathrm{d}t\right)^{-\frac{1}{q}} + \lambda\left(\int_0^\infty s^{q-1}G(s)\mathrm{d}s\right)^{-\frac{1}{q}},
\end{aligned}$$

concluding the proof. □

2.6. Monotonicity of the volume of the slices of the L^p-polar under Steiner symmetrization. The method of proof for Lemma 2.6 is standard: compare the volume of the slices of K° and $(\sigma_u K)^{\circ,p}$ normal to the direction of the symmetrization. For a convex body $K \subset \mathbb{R}^n$ let

$$K(x_n) := \{\xi \in \mathbb{R}^{n-1} : (\xi, x_n) \in K\}, \tag{2.9}$$

the 'slice' of K at height x_n. By Tonelli's theorem [**6**, §2.37],

$$|K| = \int_K \mathrm{d}x = \int_{\{(\xi,x_n)\in\mathbb{R}^{n-1}\times\mathbb{R}:\xi\in K(x_n)\}} \mathrm{d}\xi\,\mathrm{d}x_n = \int_{-\infty}^\infty |K(x_n)|\mathrm{d}x_n. \tag{2.10}$$

Therefore, to obtain Lemma 2.6, we will use the following.

LEMMA 2.25. *Let $p \in [0,\infty)$. Let, also, $\lambda \in (0,1)$ and $K \subset \mathbb{R}^n$ a convex body for which $e_n^\perp$ λ-separates $K^{\circ,p}$. Then,*

$$|(\sigma_{e_n}K)^{\circ,p}(r)| \geq |K^{\circ,p}(t)|^{\frac{s}{t+s}}|K^{\circ,p}(-s)|^{\frac{t}{t+s}},$$

for all $r,t,s > 0$ with $\frac{2}{r} = \frac{1}{t} + \frac{1}{s}$.

To prove Lemma 2.25 we apply the multiplicative Brunn–Minkowski inequality on

$$\frac{s}{t+s}K^{\circ,p}(t) + \frac{t}{t+s}K^{\circ,p}(-s) \subset (\sigma_{e_n}K)^{\circ,p}(r), \quad \text{for all} \quad \frac{2}{r} = \frac{1}{t} + \frac{1}{s}, \tag{2.11}$$

Before proving (2.11), note that the symmetry of $((\sigma_{e_n}K)^{\circ,p}$ with respect to $e_n^\perp$ (Lemma 2.11), allows for the right-hand side of (2.11) to be written as

$$(\sigma_{e_nK})^{\circ,p}(r) = \frac{s}{t+s}(\sigma_{e_n}K)^{\circ,p}(t) + \frac{t}{t+s}(\sigma_{e_n}K)^{\circ,p}(-s),$$

because $(\sigma_{e_n}K)^{\circ,p}(-s) = (\sigma_{e_n}K)^{\circ,p}(s)$, and $r = \frac{2ts}{t+s}$. Thus, (2.11) is a monotonicity property for the slices of $K^{\circ,p}$ under Steiner symmetrization.

The proof of (2.11) follows from an inequality of norms

$$\left\|\left(\frac{s}{t+s}\xi + \frac{t}{t+s}\xi', r\right)\right\|_{(\sigma_{e_n}K)^{\circ,p}} \leq \frac{s}{t+s}\|(\xi,t)\|_{K^{\circ,p}} + \frac{t}{t+s}\|(\xi',-s)\|_{K^{\circ,p}}, \tag{2.12}$$

for all $r,t,s>0$ with $\frac{2}{r}=\frac{1}{t}+\frac{1}{s}$. Now, to prove (2.12), we use Corollary 2.24 on the L^p-support functions, hence we require the next lemma proved in §2.7 below.

LEMMA 2.26. *Let $p\in(0,\infty]$, $\xi,\xi'\in\mathbb{R}^{n-1}$, $t,s,r>0$ with $\frac{2}{r}=\frac{1}{t}+\frac{1}{s}$, and $\tau:=\frac{t}{t+s}$. For all $\alpha,\beta,\gamma>0$ with $\frac{1}{\gamma}=\frac{1-\tau}{\alpha}+\frac{\tau}{\beta}$,*

$$\begin{aligned}&h_{p,\sigma_{e_n}K}\Big(\gamma((1-\tau)\xi+\tau\xi',r)\Big)\\&\leq\frac{(1-\tau)\beta}{\tau\alpha+(1-\tau)\beta}h_{p,K}(\alpha(\xi,t))+\frac{\tau\alpha}{\tau\alpha+(1-\tau)\beta}h_{p,K}(\beta(\xi',-s)).\end{aligned}$$

Indeed, applying Corollary 2.24 with $q=n$ on Lemma 2.26 gives

$$\begin{aligned}&\Big((n-1)!\Big)^{-\frac{1}{n}}\|((1-\tau)\xi+\tau\xi',r)\|_{(\sigma_{e_n}K)^{\circ,p}}\\&=\left(\int_0^\infty\gamma^{n-1}e^{-h_{p,\sigma_{e_n}K}(\gamma((1-\tau)\xi+\tau\xi',r))}\mathrm{d}\gamma\right)^{-\frac{1}{n}}\\&\leq(1-\tau)\left(\int_0^\infty\alpha^{n-1}e^{-h_{p,K}(\alpha(\xi,t))}\mathrm{d}\alpha\right)^{-\frac{1}{n}}+\tau\left(\int_0^\infty\beta^{n-1}e^{-h_{p,K}(\beta(\xi',s))}\mathrm{d}\beta\right)^{-\frac{1}{n}}\\&=\Big((n-1)!\Big)^{-\frac{1}{n}}(1-\tau)\|(\xi,t)\|_{K^{\circ,p}}+\Big((n-1)!\Big)^{-\frac{1}{n}}\tau\|(\xi',s)\|_{K^{\circ,p}},\end{aligned}$$

proving (2.12), because $\tau=\frac{t}{t+s}$.

For (2.11), given $\xi\in K^{\circ,p}(t)$ and $\xi'\in K^{\circ,p}(-s)$, by definition (2.9), $(\xi,t)\in K^{\circ,p}$ and $(\xi',-s)\in K^{\circ,p}$, i.e., $\|(\xi,t)\|_{K^{\circ,p}}\leq 1$ and $\|(\xi',-s)\|_{K^{\circ,p}}\leq 1$. By (2.12), since $r=\frac{2ts}{t+s}$,

$$\begin{aligned}\left\|\left(\frac{s}{t+s}\xi+\frac{t}{t+s}\xi',r\right)\right\|_{(\sigma_{e_n}K)^{\circ,p}}&\leq\frac{s}{t+s}\|(\xi,t)\|_{K^{\circ,p}}+\frac{t}{t+s}\|(\xi',-s)\|_{K^{\circ,p}}\\&\leq\frac{s}{t+s}+\frac{t}{t+s}=1\end{aligned}$$

thus $(\frac{s}{t+s}\xi+\frac{t}{t+s}\xi',r)\in(\sigma_{e_n}K)^{\circ,p}$ and $\frac{s}{t+s}\xi+\frac{t}{t+s}\xi'\in(\sigma_{e_n}K)^{\circ,p}(r)$.

PROOF OF LEMMA 2.25. By the multiplicative Brunn–Minkowski inequality applied on (2.11),

$$|(\sigma_{e_n}K)^{\circ,p}(r)|\geq\left|\frac{s}{t+s}K^{\circ,p}(t)+\frac{t}{t+s}K^{\circ,p}(-s)\right|\geq|K^{\circ,p}(t)|^{\frac{s}{t+s}}|K^{\circ,p}(-s)|^{\frac{t}{t+s}}.$$

□

PROOF OF LEMMA 2.6. To begin with, take $u=e_n$. Let

$$\begin{aligned}F(t)&:=|K^{\circ,p}(t)|,\quad t\in(0,\infty),\\G(s)&:=|K^{\circ,p}(-s)|,\quad s\in(0,\infty),\\H(r)&:=|(\sigma_{e_n}K)^{\circ,p}(r)|,\quad r\in(0,\infty).\end{aligned}$$

By Lemma 2.25, $H(r) \geq F(t)^{\frac{s}{t+s}} G(s)^{\frac{t}{t+s}}$, for all $t, s, r > 0$ with $\frac{2}{r} = \frac{1}{t} + \frac{1}{s}$. Therefore, by Theorem 2.23 for $q = 1$, and (2.10),

$$\begin{aligned}
\frac{4}{|(\sigma_{e_n} K)^{\circ,p}|} = \frac{2}{\int_0^\infty |(\sigma_{e_n} K)^{\circ,p}(r)| \mathrm{d}r} &\leq \frac{1}{\int_0^\infty |K^{\circ,p}(t)| \mathrm{d}t} + \frac{1}{\int_0^\infty |K^{\circ,p}(-s)| \mathrm{d}s} \\
&= \frac{1}{\int_0^\infty |K^{\circ,p}(t)| \mathrm{d}t} + \frac{1}{\int_{-\infty}^0 |K^{\circ,p}(s)| \mathrm{d}s} \\
&= \frac{1}{|K^{\circ,p} \cap (e_n^\perp)^+|} + \frac{1}{|K^{\circ,p} \cap (e_n^\perp)^-|} \\
&= \frac{|K^{\circ,p} \cap (e_n^\perp)^+| + |K^{\circ,p} \cap (e_n^\perp)^-|}{|K^{\circ,p} \cap (e_n^\perp)^+||K^{\circ,p} \cap (e_n^\perp)^-|} \\
&= \frac{|K^{\circ,p}|}{|K^{\circ,p} \cap (e_n^\perp)^+||K^{\circ,p} \cap (e_n^\perp)^-|} \\
&= \frac{1}{\lambda(1-\lambda)|K^{\circ,p}|},
\end{aligned}$$

the last equality is because, by assumption, $e_n^\perp$ λ-separates $K^{\circ,p}$.

In general, let $A \in O(n)$ such that $u = Ae_n$. If u λ-separates $K^{\circ,p}$, then $Au = e_n$ λ-separates $(AK)^{\circ,p}$. Indeed, if $|K^{\circ,p} \cap (u^\perp)^+| = \lambda|K^{\circ,p}|$, then, by (2.3) and (2.4),

$$(AK)^{\circ,p} \cap ((Au)^\perp)^+ = \Big((A^{-1})^T K^{\circ,p}\Big) \cap \Big((A^{-1})^T (u^\perp)^+\Big) = (A^{-1})^T (K^{\circ,p} \cap (u^\perp)^+).$$

Thus, $|(AK)^{\circ,p} \cap ((Au)^\perp)^+| = |\det A||K^{\circ,p} \cap (u^\perp)^+| = |\det A|\lambda|K^{\circ,p}| = \lambda|(AK)^{\circ,p}|$, i.e., Au λ-separates $(AK)^{\circ,p}$. By Lemma 2.12, since $Au = e_n$, and e_n λ-separates $(AK)^{\circ,p}$,

$$\begin{aligned}
|(\sigma_u K)^{\circ,p}| &= |(A^{-1}\sigma_{e_n}(AK))^{\circ,p}| \\
&= |A^T (\sigma_{e_n}(AK))^{\circ,p}| \\
&= |\det A||(\sigma_{e_n}(AK))^{\circ,p}| \\
&\geq 4\lambda(1-\lambda)|\det A||(AK)^{\circ,p}| \\
&= 4\lambda(1-\lambda)|K^{\circ,p}|,
\end{aligned}$$

because by (2.3), $|\det A||(AK)^{\circ,p}| = |\det A||(A^{-1})^T K^{\circ,p}| = |K^{\circ,p}|$. □

2.7. An inequality for the L^p-support functions.

PROOF OF LEMMA 2.26. Let $f, g : \pi_{e_n^\perp}(K) \to \mathbb{R}, g \leq f$, such that

$$K = \{(\xi, x_n) \in \mathbb{R}^{n-1} \times \mathbb{R} : (\xi, 0) \in \pi_{e_n^\perp}(K) \text{ and } g(\xi) \leq x_n \leq f(\xi)\}.$$

Then,

$$\sigma_{e_n^\perp} K = \left\{ (\xi, x_n) \in \mathbb{R}^{n-1} \times \mathbb{R} : (\xi, 0) \in \pi_{e_n^\perp}(K) \text{ and } |x_n| \leq \frac{f(\xi) - g(\xi)}{2} \right\}.$$

Let

$$J(\eta, y_n) := \frac{2\sinh(py_n \frac{f(\eta)-g(\eta)}{2})}{py_n}, \quad \eta \in \mathbb{R}^{n-1}, y_n \in \mathbb{R}.$$

Compute,

(2.13)

$$\begin{aligned}
&h_{p,\sigma_{e_n} K}(\gamma((1-\tau)\xi + \tau\xi', r)) \\
&= \frac{1}{p} \log \int_{\pi_{e_n^\perp}(K)} \int_{-\frac{f(\eta)-g(\eta)}{2}}^{\frac{f(\eta)-g(\eta)}{2}} e^{p\gamma\langle(1-\tau)\xi+\tau\xi',\eta\rangle} e^{p\gamma y_n r} \frac{\mathrm{d}y_n \mathrm{d}\eta}{|\sigma_{e_n} K|} \\
&= \frac{1}{p} \log \int_{\pi_{e_n^\perp}(K)} e^{p\gamma(1-\tau)\langle\xi,\eta\rangle + p\gamma\tau\langle\xi',\eta\rangle} \frac{1}{p\gamma r} (e^{p\gamma r \frac{f(\eta)-g(\eta)}{2}} - e^{-p\gamma r \frac{f(\eta)-g(\eta)}{2}}) \frac{\mathrm{d}\eta}{|K|} \\
&= \frac{1}{p} \log \int_{\pi_{e_n^\perp}(K)} e^{p\gamma(1-\tau)\langle\xi,\eta\rangle + p\gamma\tau\langle\xi',\eta\rangle} \frac{2\sinh(p\gamma r \frac{f(\eta)-g(\eta)}{2})}{p\gamma r} \frac{\mathrm{d}\eta}{|K|} \\
&= \frac{1}{p} \log \int_{\pi_{e_n^\perp}(K)} e^{p\gamma(1-\tau)\langle\xi,\eta\rangle + p\gamma\tau\langle\xi',\eta\rangle} J(\eta, \gamma r) \frac{\mathrm{d}\eta}{|K|}.
\end{aligned}$$

Also,

$$\begin{aligned}
h_{p,K}(\alpha(\xi, t)) &= \frac{1}{p} \log \int_{\pi_{e_n^\perp}(K)} \int_{g(\eta)}^{f(\eta)} e^{p\alpha\langle\xi,\eta\rangle} e^{p\alpha t y_n} \frac{\mathrm{d}y_n \mathrm{d}\eta}{|K|} \\
&= \frac{1}{p} \log \int_{\pi_{e_n^\perp}(K)} e^{p\alpha\langle\xi,\eta\rangle} \frac{1}{p\alpha t} (e^{p\alpha t f(\eta)} - e^{p\alpha t g(\eta)}) \frac{\mathrm{d}\eta}{|K|} \\
&= \frac{1}{p} \log \int_{\pi_{e_n^\perp}(K)} e^{p\alpha\langle\xi,\eta\rangle} e^{p\alpha t \frac{f(\eta)+g(\eta)}{2}} \frac{2\sinh(p\alpha t \frac{f(\eta)-g(\eta)}{2})}{p\alpha t} \frac{\mathrm{d}\eta}{|K|} \\
&= \frac{1}{p} \log \int_{\pi_{e_n^\perp}(K)} e^{p\alpha\langle\xi,\eta\rangle} e^{p\alpha t \frac{f(\eta)+g(\eta)}{2}} J(\eta, \alpha t) \frac{\mathrm{d}\eta}{|K|}.
\end{aligned} \tag{2.14}$$

Similarly,

$$h_{p,K}(\beta(\xi', -s)) = \frac{1}{p} \log \int_{\pi_{e_n^\perp}(K)} e^{p\beta\langle\xi',\eta\rangle} e^{-p\beta s \frac{f(\eta)+g(\eta)}{2}} J(\eta, \beta s) \frac{\mathrm{d}\eta}{|K|}, \tag{2.15}$$

because J is even in the second variable. By (2.14), (2.15), and Hölder's inequality,
(2.16)

$$\begin{aligned}
&\frac{(1-\tau)\beta}{\tau\alpha+(1-\tau)\beta}h_{p,K}(\alpha(\xi,t))+\frac{\tau\alpha}{\tau\alpha+(1-\tau)\beta}h_{p,K}(\beta(\xi',-s))\\
&=\frac{1}{p}\log\Bigg[\left(\int_{\pi_{e_n^\perp}(K)}e^{p\alpha\langle\xi,\eta\rangle}e^{p\alpha t\frac{f(\eta)+g(\eta)}{2}}J(\eta,\alpha t)\frac{d\eta}{|K|}\right)^{\frac{(1-\tau)\beta}{\tau\alpha+(1-\tau)\beta}}\\
&\qquad\left(\int_{\pi_{e_n^\perp}(K)}e^{p\beta\langle\xi',\eta\rangle}e^{-p\beta s\frac{f(\eta)+g(\eta)}{2}}J(\eta,\beta s)\frac{d\eta}{|K|}\right)^{\frac{\tau\alpha}{\tau\alpha+(1-\tau)\beta}}\Bigg]\\
&\geq\frac{1}{p}\log\int_{\pi_{e_n^\perp}(K)}e^{p\frac{(1-\tau)\alpha\beta}{\tau\alpha+(1-\tau)\beta}\langle\xi,\eta\rangle}e^{p\frac{(1-\tau)\alpha\beta}{\tau\alpha+(1-\tau)\beta}t\frac{f(\eta)+g(\eta)}{2}}J(\eta,\alpha t)^{\frac{(1-\tau)\beta}{\tau\alpha+(1-\tau)\beta}}\\
&\qquad e^{p\frac{\tau\alpha\beta}{\tau\alpha+(1-\tau)\beta}\langle\xi',\eta\rangle}e^{-p\frac{\tau\alpha\beta}{\tau\alpha+(1-\tau)\beta}s\frac{f(\eta)+g(\eta)}{2}}J(\eta,\beta s)^{\frac{\tau\alpha}{\tau\alpha+(1-\tau)\beta}}\frac{d\eta}{|K|}\\
&=\frac{1}{p}\log\int_{\pi_{e_n^\perp}(K)}e^{p\gamma(1-\tau)\langle\xi,\eta\rangle}e^{p\gamma(1-\tau)t\frac{f(\eta)+g(\eta)}{2}}J(\eta,\alpha t)^{\frac{(1-\tau)\beta}{\tau\alpha+(1-\tau)\beta}}\\
&\qquad e^{p\gamma\tau\langle\xi',\eta\rangle}e^{-p\gamma\tau s\frac{f(\eta)+g(\eta)}{2}}J(\eta,\beta s)^{\frac{\tau\alpha}{\tau\alpha+(1-\tau)\beta}}\frac{d\eta}{|K|}\\
&=\frac{1}{p}\log\int_{\pi_{e_n^\perp}(K)}e^{p\gamma(1-\tau)\langle\xi,\eta\rangle+p\gamma\tau\langle\xi',\eta\rangle}e^{p\gamma((1-\tau)t-\tau s)\frac{f(\eta)+g(\eta)}{2}}\\
&\qquad J(\eta,\alpha t)^{\frac{(1-\tau)\beta}{\tau\alpha+(1-\tau)\beta}}J(\eta,\beta s)^{\frac{\tau\alpha}{\tau\alpha+(1-\tau)\beta}}\frac{d\eta}{|K|}\\
&=\frac{1}{p}\log\int_{\pi_{e_n^\perp}(K)}e^{p\gamma(1-\tau)\langle\xi,\eta\rangle+p\gamma\tau\langle\xi',\eta\rangle}J(\eta,\alpha t)^{\frac{(1-\tau)\beta}{\tau\alpha+(1-\tau)\beta}}J(\eta,\beta s)^{\frac{\tau\alpha}{\tau\alpha+(1-\tau)\beta}}\frac{d\eta}{|K|}
\end{aligned}$$

because

$$\gamma=\frac{\alpha\beta}{\tau\alpha+(1-\tau)\beta},\tag{2.17}$$

and

$$(1-\tau)t-\tau s=\frac{s}{t+s}t-\frac{t}{t+s}s=0.$$

By the log-concavity of J in the second variable (Claim 2.27 below), and (2.17),
(2.18)

$$\begin{aligned}
J(\eta,\alpha t)^{\frac{(1-\tau)\beta}{\tau\alpha+(1-\tau)\beta}}J(\eta,\beta s)^{\frac{\tau\alpha}{\tau\alpha+(1-\tau)\beta}}&\geq J\left(\eta,\frac{(1-\tau)\alpha\beta}{\tau\alpha+(1-\tau)\beta}t+\frac{\tau\alpha\beta}{\tau\alpha+(1-\tau)\beta}s\right)\\
&=J(\eta,(1-\tau)\gamma t+\tau\gamma s)\\
&=J\left(\eta,\gamma\frac{ts}{t+s}+\gamma\frac{ts}{t+s}\right)\\
&=J\left(\eta,\gamma\frac{2ts}{t+s}\right)\\
&=J(\eta,\gamma r),
\end{aligned}$$

because $(1-\tau)=\frac{s}{t+s}, \tau=\frac{t}{t+s}$ and $r=\frac{2ts}{t+s}$. Finally, by (2.16), (2.18), and (2.13),

$$\begin{aligned}&\frac{(1-\tau)\beta}{\tau\alpha+(1-\tau)\beta}h_{p,K}(\alpha(\xi,t))+\frac{\tau\alpha}{\tau\alpha+(1-\tau)\beta}h_{p,K}(\beta(\xi',-s))\\&\geq\frac{1}{p}\log\int_{\pi_{e_n^\perp}(K)}e^{p\gamma(1-\tau)\langle\xi,\eta\rangle+p\gamma\tau\langle\xi',\eta\rangle}J(\eta,\gamma r)\frac{\mathrm{d}\eta}{|K|}\\&=h_{p,\sigma_{e_n}K}(\gamma((1-\tau)\xi+\tau\xi',r)),\end{aligned}$$

as desired. □

Claim 2.27. *For any $x>0$, $t\mapsto\log\left(\frac{1}{t}\sinh(tx)\right)$, $t>0$, is convex.*

Proof. Write

$$f(t):=\log\left(\frac{1}{t}\sinh(tx)\right)=\log(\sinh(tx))-\log t.$$

Compute the derivatives, $f'(t)=x\frac{\cosh(tx)}{\sinh(tx)}-\frac{1}{t}$ and

$$\begin{aligned}f''(t)&=x^2\frac{\sinh(tx)}{\sinh(tx)}-x^2\frac{(\cosh(tx))^2}{(\sinh(tx))^2}+\frac{1}{t^2}=x^2\left(1-\frac{(\cosh(tx))^2}{(\sinh(tx))^2}+\frac{1}{(tx)^2}\right)\\&=x^2\left(1-\frac{1+(\sinh(tx))^2}{(\sinh(tx))^2}+\frac{1}{(tx)^2}\right)=x^2\left(\frac{1}{(tx)^2}-\frac{1}{(\sinh(tx))^2}\right)\geq 0,\end{aligned}$$

because $\sinh(y)\geq y$, for all $y\geq 0$. □

2.8. Finishing the proof.

Proof of Proposition 2.3. Let

$$L:=K-s_p(\sigma_uK).$$

By Lemma 2.7, $s_p(\sigma_uK)\in u^\perp$, thus, by (2.1), $\sigma_uL=(\sigma_uK)-s_p(\sigma_uK)$. Therefore, $s_p(\sigma_uL)=0$, and hence, by (1.6),

$$\mathcal{M}_p(\sigma_uL)=\inf_{x\in\mathbb{R}^n}\mathcal{M}_p((\sigma_uL)-x). \tag{2.19}$$

By Lemma 2.8, there is $t\in\mathbb{R}$ such that $u^\perp$ 1/2-separates $(L-tu)^{\circ,p}$. By Lemma 2.6 for $\lambda=1/2$,

$$|(\sigma_u(L-tu))^{\circ,p}|\geq|(L-tu)^{\circ,p}|. \tag{2.20}$$

By (2.2), $\sigma_u(L-tu)=\sigma_uL$, and hence,

$$(\sigma_u(L-tu))^{\circ,p}=(\sigma_uL)^{\circ,p}. \tag{2.21}$$

By (1.4), (2.20), and (2.21),

$$\begin{aligned}\mathcal{M}_p(L-tu)&=n!|L||(L-tu)^{\circ,p}|\leq n!|L||(\sigma_u(L-tu))^{\circ,p}|\\&=n!|L||(\sigma_uL)^{\circ,p}|=\mathcal{M}_p(\sigma_uL),\end{aligned} \tag{2.22}$$

because $|L|=|\sigma_uL|$ (Lemma 2.10 below). Note that $L-tu=K-tu-s_p(\sigma_uK)$ is a translate of K, thus

$$\inf_{x\in\mathbb{R}^n}\mathcal{M}_p(K-x)\leq\mathcal{M}_p(L-tu). \tag{2.23}$$

Combining (2.23), (2.22), and (2.19) proves the claim. □

PROOF OF THEOREM 1.2. Denote by $e_1, \ldots, e_n$ the standard basis in $\mathbb{R}^n$. Let $K_0 := K - s_p(K)$, and

$$K_i := (\sigma_{e_i} K_{i-1}) - s_p(K_{i-1})$$

for $i = 1, \ldots, n$. By (1.6) and Proposition 2.3,

$$\inf_{x \in \mathbb{R}^n} \mathcal{M}_p(K - x) = \mathcal{M}_p(K_0) \leq \mathcal{M}_p(K_1) \leq \ldots \leq \mathcal{M}_p(K_n). \tag{2.24}$$

By Lemma 2.13 below, since $e_1, \ldots, e_n$ are all orthogonal to each other, $\sigma_{e_i} K_{i-1}$ is symmetric with respect to $e_1, \ldots, e_i$. By Lemma 2.7, $s_p(\sigma_{e_i} K_{i-1}) \in e_1^\perp \cap \ldots \cap e_i^\perp$, thus $(\sigma_{e_i} K_{i-1}) - s_p(K_{i-1})$ is also symmetric with respect to $e_1, \ldots, e_i$. Therefore, K_n is symmetric with respect to all $e_1, \ldots, e_n$, and hence it is a symmetric convex body. By (2.24) and Theorem 1.1, $\inf_{x \in \mathbb{R}^n} \mathcal{M}_p(K) \leq \mathcal{M}_p(K_n) \leq \mathcal{M}_p(B_2^n)$. □

REMARK 2.28. For a short proof of Theorem 1.1, note that for a symmetric convex body K, $K^{\circ,p}$ is also symmetric [**4**, Theorem 1.2], thus $u^\perp$ 1/2-separates it. By Lemma 2.6, Steiner symmetrization increases its L^p-Mahler volume, $\mathcal{M}_p(K) \leq \mathcal{M}_p(\sigma_u K)$. Repeated applications of Steiner symmetrization converge to B_2^n [**7**, Theorem 9.1], and hence, by the continuity of $\mathcal{M}_p$ [**4**, Lemma 5.11], $\mathcal{M}_p(K) \leq \mathcal{M}_p(B_2^n)$.

References

[1] K. Ball, *Logarithmically concave functions and sections of convex sets in* $\mathbf{R}^n$, Studia Math. **88** (1988), no. 1, 69–84, DOI 10.4064/sm-88-1-69-84. MR932007

[2] K. Ball, *PhD Thesis*, University of Cambridge, 1986.

[3] B. Berndtsson, *Bergman kernels for Paley-Wiener spaces and Nazarov's proof of the Bourgain-Milman theorem*, Pure Appl. Math. Q. **18** (2022), no. 2, 395–409. MR4429213

[4] B. Berndtsson, V. Mastrantonis, Y. A. Rubinstein, *L^p-polarity, Mahler volumes, and the isotropic constant*, 2023, Preprint, `arXiv:2304.14363`.

[5] Z. Błocki, *On Nazarov's complex analytic approach to the Mahler conjecture and the Bourgain-Milman inequality*, Complex analysis and geometry, Springer Proc. Math. Stat., vol. 144, Springer, Tokyo, 2015, pp. 89–98, DOI 10.1007/978-4-431-55744-9_6. MR3446749

[6] G. B. Folland, *Real analysis*, 2nd ed., Pure and Applied Mathematics (New York), John Wiley & Sons, Inc., New York, 1999. Modern techniques and their applications; A Wiley-Interscience Publication. MR1681462

[7] P. M. Gruber, *Convex and discrete geometry*, Grundlehren der mathematischen Wissenschaften [Fundamental Principles of Mathematical Sciences], vol. 336, Springer, Berlin, 2007. MR2335496

[8] B. Klartag, *On convex perturbations with a bounded isotropic constant*, Geom. Funct. Anal. **16** (2006), no. 6, 1274–1290, DOI 10.1007/s00039-006-0588-1. MR2276540

[9] F. Nazarov, *The Hörmander proof of the Bourgain-Milman theorem*, Geometric aspects of functional analysis, Lecture Notes in Math., vol. 2050, Springer, Heidelberg, 2012, pp. 335–343, DOI 10.1007/978-3-642-29849-3_20. MR2985302

[10] M. Meyer and A. Pajor, *On the Blaschke-Santaló inequality*, Arch. Math. (Basel) **55** (1990), no. 1, 82–93, DOI 10.1007/BF01199119. MR1059519

[11] M. Meyer and S. Reisner, *Shadow systems and volumes of polar convex bodies*, Mathematika **53** (2006), no. 1, 129–148 (2007), DOI 10.1112/S0025579300000061. MR2304056

[12] F. Nazarov, *The Hörmander proof of the Bourgain-Milman theorem*, Geometric aspects of functional analysis, Lecture Notes in Math., vol. 2050, Springer, Heidelberg, 2012, pp. 335–343, DOI 10.1007/978-3-642-29849-3_20. MR2985302

[13] C. M. Petty, *Affine isoperimetric problems*, Discrete geometry and convexity (New York, 1982), Ann. New York Acad. Sci., vol. 440, New York Acad. Sci., New York, 1985, pp. 113–127, DOI 10.1111/j.1749-6632.1985.tb14545.x. MR809198

[14] J. Saint-Raymond, *Sur le volume des corps convexes symétriques* (French), Initiation Seminar on Analysis: G. Choquet-M. Rogalski-J. Saint-Raymond, 20th Year: 1980/1981, Publ. Math. Univ. Pierre et Marie Curie, vol. 46, Univ. Paris VI, Paris, 1981, pp. Exp. No. 11, 25. MR670798

[15] L. A. Santaló, *An affine invariant for convex bodies of n-dimensional space* (Spanish), Portugal. Math. **8** (1949), 155–161. MR39293

[16] J. Steiner, *Einfache Beweise der isoperimetrischen Hauptsätze* (German), J. Reine Angew. Math. **18** (1838), 281–296, DOI 10.1515/crll.1838.18.281. MR1578194

[17] D. Xi and Y. Zhao, *General affine invariances related to Mahler volume*, Int. Math. Res. Not. IMRN **18** (2022), 14151–14180, DOI 10.1093/imrn/rnab118. MR4485955

Department of Mathematics, University of Maryland, College Park, Maryland
Email address: vmastr@umd.edu

Contemporary Mathematics
Volume **810**, 2025
https://doi.org/10.1090/conm/810/16217

Subharmonicity of direct images and applications

Frédéric Campana, Junyan Cao, and Mihai Păun

1. Introduction and main results

The positivity properties of direct images are at the heart of many recent developments in complex geometry. In the present article our main goal is to explore further some of these topics. Actually the main results we obtain here concern the *subharmonicity properties of direct images.* As a consequence we answer partially to a conjecture formulated by Pereira-Touzet, cf. [**PeTou13**] and the references therein. We start with the application as follows.

Let X be a compact Kähler manifold whose canonical bundle K_X is pseudo-effective. Let $\mathcal{F} \subset T_X$ be a holomorphic foliation whose first Chern class $c_1(\mathcal{F})$ is trivial. Then Pereira-Touzet show (cf. [**PeTou13**]) that $\mathcal{F}$ is non-singular, and they conjecture that if moreover $c_2(\mathcal{F}) \neq 0$ and $\mathcal{F}$ is stable with respect to some Kähler metric ω_X on X, then the leaves of $\mathcal{F}$ are algebraic.

In this article we obtain the following partial confirmation of their conjecture.

THEOREM 1.1. *Let (X, ω_X) be a projective manifold. Let $\mathcal{F} \subset T_X$ be a holomorphic foliation such that the following hold.*

(i) *The first Chern class of $\mathcal{F}$ is zero, i.e. $c_1(\mathcal{F}) = 0$ and $c_2(\mathcal{F}) \neq 0$.*

(ii) *The sheaf $\mathrm{Sym}^k \mathcal{F}^\star$ is ω_X-stable for some $k \geqslant r$, where r is the rank of $\mathcal{F}$.*

Then the leaves of $\mathcal{F}$ are algebraic.

We refer to [**HP19**], [**Dr18**], [**LPT18**] and the references therein for other particular cases of the aforementioned conjecture and its relevance in the context of singular Calabi-Yau manifolds.

Now the main new ingredient in the proof of Theorem 1.1 is Theorem 1.2 below that we next discuss. We recall that if $\mathcal{F}$ is any torsion free, coherent sheaf on X then there exists a modification $\pi : \widehat{X} \to X$ such that the inverse image $\pi^\star(\mathcal{F})$ modulo torsion is a vector bundle which we denote by E. Let $\mathcal{O}(1) \to \mathbb{P}(E)$ be the (dual of the) tautological bundle over $\mathbb{P}(E)$ (our convention here is that a point of $\mathbb{P}(E)$ corresponds to a line of the fiber of E at some point of X).

Then we have the following statement.

THEOREM 1.2. *Let (X, ω_X) be a compact Kähler manifold, and let $\mathcal{F}$ be a coherent reflexive sheaf such that the following hold.*

(i) *The first Chern class of $\mathcal{F}$ is zero, i.e. $c_1(\mathcal{F}) = 0$.*

2020 *Mathematics Subject Classification.* Primary 14D06, 14C30, 14F18, 32U05, 32U40.

(ii) *The sheaf* $\widehat{\mathrm{Sym}}^k\mathcal{F}^\star$ *is* ω_X*-stable for some* $k \geqslant r+1$*, where* r *is the rank of* $\mathcal{F}$ *and* $\widehat{\mathrm{Sym}}^k\mathcal{F}^\star := (\mathrm{Sym}^k\mathcal{F}^\star)^{\star\star}$.

Then the bundle $\mathcal{O}(1)$ *on* $\mathbb{P}(E)$ *is either not pseudo-effective, or* $\mathcal{F}$ *is Hermitian flat. Moreover, if* X *is projective then we can derive the same conclusion provided that* $\widehat{\mathrm{Sym}}^k\mathcal{F}^\star$ *is* ω_X*-stable for some* $k \geqslant r$ *instead of* (ii).

REMARK 1.3. As already mentioned, Theorem 1.1 follows from Theorem 1.2. Indeed, if the foliation $\mathcal{F}$ is not algebraic, then the bundle $\mathcal{O}(1)$ is pseudo-effective –cf. Theorem 3.6 in Section 3– so Theorem 1.2 implies that $\mathcal{F}$ is Hermitian flat. The hypothesis $c_2(\mathcal{F}) \neq 0$ prevents this from happening.

Unlike the other articles [**HP19**], [**Dr18**], [**LPT18**] mentioned above in connection with Pereira-Touzet conjecture, our methods here are relying on new positivity results for direct images that we are now introducing.

Let $p : Y \to X$ be a holomorphic proper map between two Kähler manifolds, and let (L, h_L) be a Hermitian line bundle on the total space Y. The metric h_L could be singular, but we assume that

$$\Theta_{h_L}(L) \geqslant -Cp^*\omega_X \tag{1}$$

in the sense of currents on Y for some constant $C > 0$, where ω_X is some smooth hermitian form on X. In this context, if $C = 0$, it is then established that the direct image sheaf

$$\mathcal{E} := p_\star\left((K_{Y/X} + L) \otimes \mathcal{I}(h_L)\right) \tag{2}$$

is semi-positively curved in the sense of Griffiths when endowed with the natural L^2 metric denoted by $h_{Y/X}$ cf. [**HPS18**], [**PT18**].

The next result we are presenting here concerns the positivity of $\det \mathcal{E}$ for an arbitrary constant C as follows.

THEOREM 1.4. *Let* $p : Y \to X$ *be a holomorphic surjective and proper map, where* X *and* Y *are Kähler manifolds. Moreover, we assume that* p *is locally projective.*[1] *Let* (L, h_L) *be a Hermitian line bundle over* Y *such that* (1) *holds true, and such that*

$$\Theta_{h_L}(L) \wedge p^\star\omega_X^{n-1} \geqslant \varepsilon_0 p^\star\omega_X^n \tag{3}$$

where ω_X *is a Hermitian metric on* X, $n = \dim X$, *and* $\varepsilon_0 \geqslant 0$ *is a non-negative real number. Then we have*

$$\Theta_{\det h_{Y/X}}(\det(\mathcal{E})) \wedge \omega_X^{n-1} \geqslant r\varepsilon_0\omega_X^n \tag{4}$$

where r *is the rank of the direct image* $\mathcal{E}$.

REMARK 1.5. If we replace the hypothesis (3) with the following

$$\Theta_{h_L}(L) \geqslant \varepsilon_0 p^\star\omega_X, \tag{5}$$

then it is well-known that the curvature of $(\mathcal{E}, h_{Y/X})$ is greater than $\varepsilon_0\omega_X \otimes \mathrm{Id}_{\mathrm{End}(\mathcal{E})}$. Thus our statement 1.4 can be seen as a stronger version of this result.

[1] This means that each point x of the base X has a neighbourhood U whose p-inverse image admits an embedding in the product $U \times \mathbb{P}^N$ over U.

Very roughly, the proof for Theorem 1.4 goes as follows: by techniques due to [**Ber06**], [**BP08**], we show that it is enough to show the inequality (4) in the particular case of map $p : U \times V \to U$ which is simply the projection on the first factor, where U is an open ball in some Euclidean space and V is a Stein manifold. In this case the statement to prove looks quite different from (4) and it will be established by a (long and) direct computation.

An immediate consequence of Theorem 1.4 is the following.

THEOREM 1.6. *Let $p : Y \to X$ be a holomorphic surjective and proper map, where X and Y are Kähler manifolds. Moreover, we assume that p is locally projective. Let (L, h_L) be a Hermitian holomorphic line bundle over Y such that*

$$\Theta_{h_L}(L) \geqslant -Cp^\star\omega_X, \qquad \Theta_{h_L}(L) \wedge p^\star\omega_X^{n-1} \geqslant 0 \tag{6}$$

where ω_X is a Hermitian metric on X, $n = \dim X$, and $C > 0$ is a positive real number. Then we have

$$\Theta_{\det h_{Y/X}}(\det(\mathcal{E})) \wedge \omega_X^{n-1} \geqslant 0. \tag{7}$$

A last result we mention here is the following statement, cf. Theorem 5.4, section 5.

THEOREM 1.7. *Let $p : Y \to X$ be a holomorphic surjective and proper map, where X and Y are Kähler manifolds. Moreover, we assume that p is locally projective. Let L be a line bundle over the total space Y endowed with a possible singular metric h_L such that $\Theta_{h_L}(L) \geqslant 0$, and*

$$\Theta_{h_L}(L) \wedge p^\star\omega_X^{n-1} \geqslant \varepsilon_0 p^\star\omega_X^n \tag{8}$$

where ω_X is a Hermitian metric on X, $n = \dim X$, and $\varepsilon_0 > 0$ is a positive real number. We assume that the space of fiberwise $L^{\frac{2}{m}}$-holomorphic sections (with respect to h_L) of $p_\star(mK_{Y/X} + L)$ is non zero. Then there exists a metric h on the bundle $mK_{Y/X} + L$ such that we have

$$\Theta_h(mK_{Y/X} + L) \wedge p^\star\omega_X^{n-1} \geqslant \varepsilon_0 \cdot p^\star\omega_X^n \tag{9}$$

in the sense of currents on Y.

REMARK 1.8. We see that Theorem 1.7 has the same flavor as the usual results concerning direct images: the positivity of the curvature of (L, h_L) induces similar properties for the twisted pluricanonical bundle $mK_{Y/X} + L$. More precisely, if $\Theta_{h_L}(L) \geqslant 0$ then we already know that the bundle $mK_{Y/X} + L$ is pseudo-effective. We prove here that the alternative positivity requirement (8) is inherited by the twisted pluricanonical bundle of the map p.

Organization of the paper. The remaining part of this article will unfold as follows. In the first section we recall a few technical statements which are playing an important role in our arguments. Then as a *warm-up*, we prove Theorem 1.2 in the projective case; the main ideas are as follows. Assume that $\mathcal{O}(1)$ is pseudo-effective. Then it admits a singular metric $e^{-\varphi}$ whose curvature is semi-positive.

The stability condition (ii) implies that the multiplier ideal sheaf

$$\mathcal{I}(e^{-(r+1-\varepsilon)\varphi}|_{\mathbb{P}(E_x^\star)})$$

is trivial for any $\varepsilon > 0$, and for all $x \in X$ generic. In order to prove that this is the case (i.e. the multiplier ideal sheaf is trivial), we are using the positivity of direct images, combined with the restriction theorem of Mehta-Ramanathan.

Now if the multiplier ideal above is trivial, then we can construct a singular Hermitian metric on $\mathcal{F}^\star$ with positive curvature in the sense of Griffiths. Together with the fact that $c_1(\mathcal{F}) = 0$, this implies that $\mathcal{F}$ is Hermitian flat by a result in [**CP17**].

The general case of Theorem 1.2 is much more subtle, since one has to find an alternative argument in order to compensate the absence of Mehta-Ramanathan theorem. It is at this point that Theorem 1.6 comes into the picture. Thanks to this result we can still analyse the singularities of the metric on $\mathcal{O}(1)$ as discussed above.

On the other hand, Theorem 1.4 is interesting in its own right: it can be seen as very precise analysis of the positivity properties of the determinant of direct images. Further results and applications of this statement (and its proof) are discussed in the last section of this article. We can interpret them as a first step towards the analysis of the positivity of direct images on currents of bi-dimension $(1,1)$.

2. Stable reflexive sheaves and singular Hermitian metrics

In this preliminary section we collect a few definitions and properties concerning the stability of reflexive sheaves on compact Kähler manifolds, as well as the notion of singular Hermitian metric on a vector bundle.

2.1. Stable reflexive sheaves. We will mostly follow the article by Bando-Siu, cf. [**BS94**], combined with important clarifications communicated to us by M. Toma, cf. [**T19**].

Let X be a compact Kähler manifold, and let $\mathcal{F}$ be a reflexive coherent subsheaf. Then there exists a sequence of blow-up maps of non-singular centres whose composition is denoted by

$$\pi : \widehat{X} \to X \tag{10}$$

such that the inverse image $\pi^\star(\mathcal{F})$ modulo torsion becomes a vector bundle denoted by $E \to \widehat{X}$ in [**BS94**]. We call E a desingularization of the sheaf $\mathcal{F}$.

Let ω be a Kähler form on X. Then the inverse image $\widehat{\omega} := \pi^\star(\omega)$ *is not* Kähler in general, but nevertheless the class corresponding to $\widehat{\omega}^{n-1}$ is movable. In this context, we have the following result cf. e.g. [**GKP16**].

LEMMA 2.1. *The sheaf $\mathcal{F}$ is stable with respect to ω if and only if E is stable with respect to $\widehat{\omega}$.*

For the basic facts concerning the stability with respect to a movable class we refer to [**GKP16**].

The important result of Bando-Siu states as follows.

THEOREM 2.2 ([**BS94**]). *Let $\mathcal{F}$ be a coherent reflexive sheaf on a compact Kähler manifold, which moreover is stable with respect to a Kähler metric ω. Then $\mathcal{F}$ has an admissible Hermite-Einstein metric $h_{\mathcal{F}}$.*

We do not recall here the notion of *admissible metric* of a vector bundle because we do not need it; we will rather work with its regularization defined as follows. Let

$$K_{\widehat{X}/X} = \sum_{i=1}^{N} e_i E_i \tag{11}$$

be the relative canonical divisor of the map π. Then there exist a set of non-singular representatives $\theta_i \in c_1(E_i)$ such that the form

$$\widehat{\omega}_0 := \widehat{\omega} - \varepsilon_0 \sum_{i=1}^{N} \varepsilon_i \theta_i \tag{12}$$

is Kähler for any positive and small enough $0 < \varepsilon_0 \ll 1$, provided that the coefficients ε_i are carefully chosen.

By Lemma 2.1 the vector bundle E is stable with respect to $\widehat{\omega}$. It is stated as a remark in [**BS94**] that E is stable with respect to the perturbation $\widehat{\omega}_0$ of $\widehat{\omega}$, as soon as the ε_0 in (12) is small enough. We were not able to find a reference/proof for this assertion (which is most likely true) so we will rather use the following result established in the recent article [**T19**].

THEOREM 2.3 ([**T19**]). *The bundle E is stable with respect to the degree function induced by the form*

$$(1-\varepsilon)\widehat{\omega}_0^{n-1} + \varepsilon\widehat{\omega}^{n-1} \tag{13}$$

for any $1 \geqslant \varepsilon \geqslant 0$, provided that $\varepsilon_0 \ll 1$.

It then follows that there exists a metric $h_{E,\varepsilon}$ on E which verifies the Hermite-Einstein condition with respect to g_ε^{n-1}, where g_ε is the unique Gauduchon metric on $\widehat{X}$ such that

$$g_\varepsilon^{n-1} = (1-\varepsilon)\widehat{\omega}_0^{n-1} + \varepsilon\widehat{\omega}^{n-1}. \tag{14}$$

Let us fix a coordinate system (z_j) centered at some point x_0 of $\widehat{X}$. We denote by G_ε the matrix corresponding to g_ε and we define

$$(1-\varepsilon)\widehat{\omega}_0^{n-1} + \varepsilon\widehat{\omega}^{n-1}|_U = (n-1)!(-1)^{\frac{(n-1)^2}{2}} \sum_{\alpha,\beta} (-1)^{\alpha+\beta} \Omega_{\alpha\overline{\beta},\varepsilon} \widehat{dz_\alpha} \wedge \widehat{d\overline{z}_\beta}. \tag{15}$$

We have the formula $\Omega_\varepsilon^t G_\varepsilon = \det(G_\varepsilon) I_n$ which implies that we have the equality $\det G_\varepsilon = (\det \Omega_\varepsilon)^{\frac{1}{n-1}}$. We therefore obtain

$$\int_{\widehat{X}} dV_{g_\varepsilon} \leqslant C \tag{16}$$

for some constant $C > 0$ independent of ε.

In conclusion, a coherent reflexive sheaf $\mathcal{F}$ which is stable with respect to a Kähler metric ω has an admissible Hermite-Einstein metric $h_{\mathcal{F}}$ which is limit of non-singular Hermite-Einstein metrics on a modification $\widehat{X}$ of X with respect to a Gauduchon metric g_ε.

2.2. Singular Hermitian metrics and positivity of direct images. For the convenience of the reader, we collect here a few basic notions and results concerning the positivity of push-forward of relative canonical bundles, cf. [**PT18**] [**HPS18**], [**Pa15**] and the references therein. As we will next see, the proof of the projective case of Theorem 1.2 is relying heavily on them.

The following notion appeared naturally in [**BP08**], and it was subsequently studied in [**Rau15**], [**PT18**], [**HPS18**]. Let $E \to X$ be a holomorphic vector bundle of rank r on a complex manifold X. We denote by

$$H_r := \{A = (a_{i\overline{j}})\} \tag{17}$$

the set of $r \times r$, semi-positive definite Hermitian matrices. The manifold X is endowed with the Lebesgue measure. We recall next the following notion.

DEFINITION 2.4. A *singular Hermitian metric* h on E is given locally by a measurable map with values in H_r such that

$$0 < \det h < +\infty \tag{18}$$

almost everywhere.[2]

Let (E, h) be a vector bundle endowed with a singular Hermitian metric h. Given a local section v of E, i.e. an element $v \in H^0(U, E)$ defined on some open subset $U \subset X$, the function $|v|_h^2 : U \to \mathbb{R}_{\geqslant 0}$ is measurable, given by

$$|v|_h^2 = {}^t v h \overline{v} = \sum h_{i\overline{j}} v^i \overline{v^j} \tag{19}$$

where $v = {}^t(v^1, \ldots, v^r)$ is a column vector.

Following [**BP08**, p. 357], we recall next the notion of positivity/negativity of a singular Hermitian vector bundle as follows.

DEFINITION 2.5. Let h be a singular Hermitian metric on E.

(1) The metric h is *negatively curved in the sense of Griffiths* if the function

$$x \to \log |v|^2_{h,x} \tag{20}$$

is psh for any local section v of E.

(2) The metric h is *positively curved in the sense of Griffiths* if the dual singular Hermitian metric $h^\star := {}^t h^{-1}$ on the dual vector bundle $E^\star$ is negatively curved.

We recall the following important property of the notion Definition 2.5.

PROPOSITION 2.6 ([**CP17**, Cor 2.8]). *Let (E, h) be a vector bundle on a compact manifold X, endowed with a possible singular Hermitian metric h positively curved in the sense of Griffiths. If the first Chern class $c_1(E) = 0$, then the metric h is non-singular and the curvature of (E, h) is equal to zero.*

If E is replaced by a coherent, torsion-free sheaf $\mathcal{E}$ then the conclusion of the statement 2.6 still holds provided that we restrict to the open subset on which $\mathcal{E}$ is a vector bundle.

The notion of positively curved singular Hermitian metric is the natural property of push-forwards of relative canonical bundles as we see from the next results.

THEOREM 2.7 ([**PT18, Pa15, HPS18**]). *Let $p : Y \to X$ be an algebraic fiber space, and let (L, h_L) be a Hermitian holomorphic line bundle over Y. The metric h_L could be singular, and the corresponding curvature current $\Theta(L, h_L)$ is semi-positive in the sense of current. Then the singular Hermitian metric $h_{Y/X}$ on the push-forward sheaf (which is torsion-free)*

$$p_\star \left(\mathcal{O}_Y(K_{Y/X} + L) \otimes \mathcal{I}(h_L)\right) \tag{21}$$

[2]As it was pointed out to us by the referee, the RHS inequality in (18) is a consequence of the fact that h has values in H_r, (so all in all the definition above could be a bit confusing). We certainly agree with that, but since it seems to late to change it, we prefer to leave it as it is...

is positively curved in the sense of Griffiths. If moreover $\mathcal{I}(h_L|_{Y_x}) = \mathcal{O}_{Y_x}$ for a "Lebesgue-general" fiber Y_x[3]*, then $h_{Y/X}$ extends naturally as a singular Hermitian metric for $p_\star\left(K_{Y/X}+L\right)$, which is also positively curved in the sense of Griffiths.*

We also recall the following variant of Theorem 2.7.

COROLLARY 2.8 ([**CP17**, Lemma 5.25]). *Let $p : Y \to X$ be an algebraic fiber space, and let (L, h_L) be a Hermitian line bundle. The metric h_L could be singular and the corresponding curvature current $\Theta(L, h_L) + Cp^\star\omega_X$ is semi-positive for some $C > 0$. Let ξ be a local holomorphic section of the dual of $p_\star\left(\mathcal{O}_Y(K_{Y/X} + L) \otimes \mathcal{I}(h_L)\right)$. Then we have*

$$\sqrt{-1}\partial\bar{\partial}\log|\xi|^2_{h_{Y/X}} \geqslant -C\omega_X. \tag{22}$$

The property (22) will be formally denoted by

$$\Theta_{h_{Y/X}}(p_\star\left(\mathcal{O}_Y(K_{Y/X} + L) \otimes \mathcal{I}(h_L)\right)) \succeq -C\omega_X \otimes \mathrm{Id}. \tag{23}$$

In other words, if the curvature of the bundle L is bounded from below by the p-pull-back of a smooth form on the base X, then the function $\log|\xi|^2_{h_{Y/X}}$ is quasi-psh, for any local holomorphic section ξ as in the Corollary 2.8.

We will equally need the following generalization of Proposition 2.6 proved in [**CCM21**] and later generalized to the Kähler setting in [**W22**].

THEOREM 2.9 ([**CCM21, W22**]). *Let X be a compact Kähler manifold and $\mathcal{F}$ be a reflexive sheaf on X. Let X_0 be the locally free locus of $\mathcal{F}$. If $c_1(\mathcal{F}) = 0$ and for every $\varepsilon > 0$, there exsits a possible singular hermitian metric h_ε on $\mathcal{F}|_{X_0}$ such that*

$$\Theta_{h_\varepsilon}(\mathcal{F}) \succeq -\varepsilon\omega_X \otimes \mathrm{Id} \qquad \text{on } X_0, \tag{24}$$

then $\mathcal{F}$ is a numerically flat vector bundle on X, namely both $\mathcal{F}$ and $\mathcal{F}^\star$ are nef vector bundles.

3. Proof of the main result: the projective case

In this section we will prove our main result under the assumption that X is projective. Indeed, in this case all the necessary tools we will be using in our arguments are already available: we show that Theorem 1.2 follows from the positivity of direct images, combined with the existence of Hermite-Einstein metric on stable vector bundles.

To make things precise, the result we establish in this first section is the following.

THEOREM 3.1. *Let (X, H) be a smooth polarized projective manifold. We consider a coherent reflexive sheaf $\mathcal{F}$ with the following properties.*

(i) *The first Chern class of $\mathcal{F}$ is zero, i.e. $c_1(\mathcal{F}) = 0$.*

(ii) *The sheaf $\widehat{\mathrm{Sym}}^k\mathcal{F}^\star$ is H-stable for some $k \geqslant r$, where r is the rank of $\mathcal{F}$.*

Let E be a desingularisation bundle of $\mathcal{F}$, cf. section 1. Then the bundle $\mathcal{O}(1)$ on $\mathbb{P}(E)$ is not pseudo-effective, or $\mathcal{F}$ is Hermitian flat.

[3]By this we mean that the point $x \in X$ belongs to the complement of a subset of X of Lebesgue measure 0.

As an application of Theorem 3.1, together with [**CP19, Dr18**], we obtain the following statement.

COROLLARY 3.2. *Let (X, H) be a smooth polarized projective manifold. We consider a saturated coherent subsheaf $\mathcal{F} \subset T_X$ satisfying the properties in Theorem 3.1 and moreover we assume that $\mathcal{F}$ is closed under the Lie bracket. Then $\mathcal{F}$ is either Hermitian flat or all of its leaves are algebraic.*

We assume that the tautological bundle $\mathcal{O}(1) \to \mathbb{P}(E)$ is pseudo-effective. Then we obtain a singular metric $h = e^{-\varphi}$ on $\mathcal{O}(1)$ whose curvature current Θ is positive,

$$\Theta \geqslant 0.$$

The main point in our arguments is the following statement concerning the singularities of the metric h.

LEMMA 3.3. *In the setting of Theorem 3.1, we consider the map $p : \mathbb{P}(E) \to X$ obtained by composing the natural projection $\mathbb{P}(E) \to \widehat{X}$ with π. If $\mathrm{Sym}^k \mathcal{F}^\star$ is H-stable for some $k \in \mathbb{N}$, then the multiplier sheaf corresponding to the restriction*

$$e^{-(k+1-\varepsilon)\varphi}|_{\mathbb{P}(E)_x} \tag{25}$$

is trivial for any $\varepsilon > 0$. Here $x \in X$ is any point in the complement of a Lebesgue measure 0 subset in X, and we denote by $\mathbb{P}(E)_x$ the p–inverse image of the point $x \in X$.

REMARK 3.4. In this lemma it is not necessary to require $k \geqslant r$. Ideally, we want to consider the case $k = 1$. The bundle $\mathcal{O}(r + 1)$ is endowed with the metric

$$e^{-(2-\varepsilon)\varphi-(r-1+\varepsilon)\varphi_{HE}}$$

where we assume that we are in the projective case and we restrict $\mathcal{F}$ to a complete intersection curve C in the linear system $|mH|$ for m sufficiently large. We denote by $e^{-\varphi_{HE}}$ the metric on $\mathcal{O}(1)$ induced by the flat Hermite-Einstein metric on $\mathcal{F}|_C$. Of course, the direct image is just $\mathcal{F}^\star|_C$, but it is endowed with a metric depending on $e^{-\varphi}$. Since the determinant is cohomologically trivial, the hope is that this should have important consequences on the Hessian of φ "in the base directions", cf. calculations section 4.

The rest of this section is divided into three parts. We will first show that Theorem 3.1 follows from Lemma 3.3. Then we prove the lemma, and finally we give a quick argument for Corollary 3.2.

3.1. Lemma 3.3 implies Theorem 3.1. Now we can prove Theorem 3.1: since $k \geqslant r$, thanks to the Lemma 3.3, we know that $\mathcal{I}(e^{-(r+1-\varepsilon)\varphi}|_{\mathbb{P}(E)_x}) = \mathcal{O}_{\mathbb{P}(E)_x}$ for a generic $x \in X$. The inclusion

$$p_\star\left(K_{\mathbb{P}(E)/X} \otimes \mathcal{O}(r+1) \otimes \mathcal{I}(e^{-(r+1-\varepsilon)\varphi})\right) \subset p_\star\left(K_{\mathbb{P}(E)/X} \otimes \mathcal{O}(r+1)\right) = \mathcal{F}^\star \tag{26}$$

is thus generically isomorphic.

As $\mathcal{O}(1)$ is relatively ample with respect to $\mathbb{P}(E) \to \widehat{X}$, there exists a metric h_0 on $\mathcal{O}(1)$ such that $\Theta_{h_0}\mathcal{O}(1) \geqslant -p^\star\omega_X$ for some Kähler metric ω_X on X, and h_0 is smooth outside the exceptional locus of the birational morphism $\widehat{X} \to X$. Set $h_\varepsilon := e^{-(r+1-\varepsilon)\varphi} \cdot h_0^\varepsilon$. It is a metric on $\mathcal{O}(r+1)$ with $\Theta_{h_\varepsilon}(\mathcal{O}(r+1)) \geqslant -\varepsilon p^\star\omega_X$. By Corollary 2.8, h_ε induces a metric $h_{\varepsilon,\mathcal{F}^\star}$ on $\mathcal{F}^\star$ such that

$$i\Theta_{h_{\varepsilon,\mathcal{F}^\star}}(\mathcal{F}^\star) \succeq -\varepsilon\omega_X \otimes \mathrm{Id}.$$

Together with the fact that $c_1(\mathcal{F}^\star) = 0$, it follows that $\mathcal{F}^\star$ is numerically flat by Theorem 2.9. Note that the stability of $\widehat{\mathrm{Sym}}^k \mathcal{F}^\star$ implies that $\mathcal{F}^\star$ is stable. $\mathcal{F}^\star$ is thus hermitian flat by using [**DPS94**, Thm 1.18].

3.2. Proof of Lemma 3.3. By the restriction theorem of Mehta-Ramanathan, the bundles

$$\mathrm{Sym}^k \mathcal{F}^\star|_C \tag{27}$$

are still stable provided that C is a generic complete intersection of $n-1$ divisors in a large enough multiple of H.

The curves "C" for which the restriction (27) is stable cover a non-empty Zariski open subset of X. We assume that our point x belongs to the union of these curves, so $x \in C$ for some complete intersection curve which we fix for the rest of the proof.

Let h_C be the Hermite-Einstein metric on the bundle $\mathcal{F}|_C$. It is of course smooth, and thanks to the first hypothesis (i) the curvature tensor $\Theta(\mathcal{F}|_C, h_C)$ is identically equal to zero. In particular, h_C induces a smooth metric on $\mathcal{O}(1)$ whose curvature is semi-positive. Roughly speaking, this is so because the curvature of $\mathcal{O}(1)$ at a point $(y, [v])$ has two components: the Fubini-Study form in the directions of the fibers of the map

$$\pi : \mathbb{P}(\mathcal{F}|_C) \to C \tag{28}$$

plus the curvature $\Theta(\mathcal{F}|_C, h_C)$ evaluated in the direction v.

We assume that x is such that the multiplier ideal $\mathcal{I}(e^{-(k+1-\varepsilon)\varphi}|_{\mathbb{P}(\mathcal{F}_x)})$ coincides with $\mathcal{I}(e^{-(k+1-\varepsilon)\varphi})|_{\mathbb{P}(\mathcal{F}_x)}$. Again, this is a genericness requirement which holds true in the complement of some Lebesgue measure 0 subset of C.

Consider the bundle

$$K_{\mathbb{P}(\mathcal{F})/C} + L. \tag{29}$$

We endow $L := \mathcal{O}(k+r)$ with a the metric whose local weights are

$$\psi_L := (k+1-\varepsilon)\varphi + (r-1+\varepsilon)\varphi_C; \tag{30}$$

the corresponding curvature curent is semi-positive. By standard L^2 estimates cf. Lemma 4.4, the vector space

$$\begin{aligned} &H^0\left(\mathbb{P}(\mathcal{F}_x), \mathcal{O}(K_{\mathbb{P}(\mathcal{F})/C} + L) \otimes \mathcal{I}(e^{-\psi_L})\right) \\ &\quad = H^0\left(\mathbb{P}(\mathcal{F}_x), \mathcal{O}(k) \otimes \mathcal{I}(e^{-(k+1-\varepsilon)\varphi}|_{\mathbb{P}(\mathcal{F}_x)})\right) \end{aligned} \tag{31}$$

is non-trivial. Moreover, if the multiplier ideal sheaf of the metric $e^{-\psi_L}$ is non-trivial, since $H^0(\mathbb{P}(\mathcal{F}_x), \mathcal{O}(k))$ is globally generated, (31) is a strict subspace of $H^0(\mathbb{P}(\mathcal{F}_x), \mathcal{O}(k))$.

By Theorem 2.7 we infer that the direct image

$$\mathcal{G} := \pi_\star\left((K_{\mathbb{P}(\mathcal{F})/C} + L) \otimes \mathcal{I}(e^{-\psi_L})\right) \tag{32}$$

is semi-positively curved. In particular, its degree with respect to C is semi-positive.

On the other hand, we have

$$\mathcal{G} \subset \pi_\star\left(K_{\mathbb{P}(\mathcal{F})/C} + L\right) = \mathrm{Sym}^k \mathcal{F}^\star \otimes \det \mathcal{F}^* \tag{33}$$

and as explained before, the inclusion is strict. Hence the degree of $\mathcal{G}$ is *strictly smaller* than the degree of $\mathrm{Sym}^k \mathcal{F}^\star$, which is zero. The lemma is proved in the projective case.

QUESTION 3.5. We see that the stability condition allow us to deduce regularity properties of the metric $e^{-\varphi}$ in case $c_1(\mathcal{F}) = 0$. Can one formulate (and eventually prove) similar results if $c_1(\mathcal{F})$ contains a negative representative?

3.3. Proof of Corollary 3.2. The statement 3.2 follows immediately as combination of Theorem 3.1 and the following algebraicity criteria which is a direct consequence of [**CP19**, Thm 4.2], [**Dr18**, Prop 8.4].

THEOREM 3.6. *Let X be a projective manifold and let $\mathcal{F} \subset T_X$ be a holomorphic foliation. If $\mathcal{F}$ is not algebraic, then the tautological line bundle $\mathcal{O}(1)$ over $\mathbb{P}(E)$ is pseudo-effective. Here we denote by E any desingularization of the sheaf $\mathcal{F}$.*

Indeed, this result does not appear explicitly in [**CP19**] but it is quickly deduced from the proof of [**CP19**, Thm 1.1]. More precisely this is proved in section 4.1, pages 14-18 of *loc.cit.*

4. Proof of the main result: the general case

We establish in the next subsection Theorem 1.4. This statement can be seen as a subharmonicity property of direct images and it represents the new technical result in the present article. Once this is done we show that the Kähler version of Lemma 3.3 follows. Further results based on Theorem 1.4 will be given in the next section.

A good point to start with is the remark that if h_L is non-singular, p is a submersion and any section $u \in H^0(Y_x, K_{Y_x} + L|_{Y_x})$ can be extended holomorphically to a neighbourhood of Y_x, then the curvature formula [**Ber09**, (4.8)] still holds. Then (4) follows directly from the curvature formula [**Ber09**, (4.8)].

Indeed, let $x \in X$ be an arbitrary point and let $u \in H^0(Y_x, K_{Y_x} + L|_{Y_x})$ be a holomorphic section. Since $i\Theta_{h_L}(L) \geqslant -Cp^\star\omega_X$, by Ohsawa-Takegoshi theorem, we can extend u to be a section in $H^0\left(U, p_\star(K_{Y/X} + L)\right)$. Here we denote by U a small coordinate set centred at x. Assume first that the constant "C" above is equal to zero, so that the curvature current of (L, h_L) is semi-positive. As h_L and p are smooth, we can find a section

$$\widehat{u} \in H^0\left(U, p_\star(K_{Y/X} + L)\right)$$

extending u as in *loc. cit.*, [**Ber09**, Prop 4.2].

The equality [**Ber09**, (4.8)] implies that we have

$$\langle \Theta_{h_{Y/X}}\left(p_\star(K_{Y/X} + L)\right) u, u\rangle \wedge \omega_X^{n-1} \geqslant c_n \int_{Y_x} \widehat{u} \wedge \overline{\widehat{u}} \wedge \Theta_{h_L}(L) \wedge p^\star(\omega_X^{n-1}) e^{-\phi_L}. \tag{34}$$

Since by hypothesis (3) we have,

$$\Theta_{h_L}(L) \wedge p^\star\omega_X^{n-1} \geqslant \varepsilon_0 p^\star\omega_X^n$$

it follows that

$$\langle \Theta_{h_{Y/X}}\left(p_\star(K_{Y/X} + L)\right) u, u\rangle \wedge \omega_X^{n-1} \geqslant \varepsilon_0 \|u\|^2_{h_{Y/X}} \omega_X^n$$

at x. We thus obtain (4) in this case by taking the trace.

In general, we consider a real function $\phi_U \in \mathcal{C}^\infty(U)$ such that $i\partial\bar\partial\varphi_U = \omega_X|_U$ as well as the induced metric $e^{-C\phi_U \circ p} h_L$ on the bundle $L|_{p^{-1}(U)}$. We apply [**Ber09**, (4.8)] for this data, and we see that the inequality (34) still holds true for (L, h_L). Indeed, the main observation is that the curvature of $(L, e^{-C\phi_U \circ p} h_L)$ is simply $i\Theta_{h_L}(L) + Cp^\star(\omega_X)$, and the curvature of the direct image has an additional term

$Cp^\star(\omega_X)\otimes \mathrm{Id}$ (the identity endomorphism of the direct image sheaf). As a result, the termes in (34) induced by the presence of the additional weight $e^{-C\phi_U}$ cancel, which proves our claim.

4.1. A subharmonicity property of direct images. In general case the metric h_L is not necessary smooth, but nevertheless the function $\log|\sigma|^2_{h^\star_{Y/X}}$ is still quasi-psh for any holomorphic section σ of $\mathcal{E}^\star$ defined on some open subset U of X. This is the content of Corollary 2.8 we have recalled in Section 2. Alternatively, the claim can be directly deduced from Theorem 2.7. Indeed, remark that we can modify the metric h_L by the pull-back of a function defined on U, such that the curvature of the resulting metric is semi-positive. By Theorem 2.7, the direct image metric is semi-positively curved, and its expression differs from $h_{Y/X}$ by the (exponential of the) function we have "corrected" h_L with. The assertion follows.

The next step is to show that under the assumptions of Theorem 1.4 the following holds.

CLAIM 4.1. Let σ be a local holomorphic section of the dual bundle $\mathcal{E}^\star$. Then we have

$$dd^c\log|\sigma|^2_{h^\star_{Y/X}}\wedge\omega_X^{n-1}\geqslant\varepsilon_0\omega_X^n \tag{35}$$

in the sense of currents on X.

Prior to explaining the proof of this claim, we see here that it implies the inequality (4). Toward this end, we are using a trick from the article [**HPS18**], which consists in considering the relative adjoint bundle

$$K_{\mathbb{P}(\mathcal{E}^\star)/X_0}+\mathcal{O}(r) \tag{36}$$

on $\mathbb{P}(\mathcal{E}^\star)$, where $X_0\subset X$ is the maximal subset of X such that the restriction $\mathcal{E}|_{X_0}$ is a vector bundle. The bundle $\mathcal{O}(1)$ is endowed with the metric h induced from $h_{Y/X}$. Then the relation (35) implies that we have

$$\Theta_h(\mathcal{O}(1))\wedge\pi^\star\omega_X^{n-1}\geqslant\varepsilon_0\pi^\star\omega_X^n. \tag{37}$$

Now we apply the Claim one more time, for the direct image sheaf

$$\pi_\star\left(K_{\mathbb{P}(\mathcal{E}^\star)/X}+\mathcal{O}(r)\right) \tag{38}$$

which coincides with $\det\mathcal{E}|_{X_0}$. Since h is induced by the metric $h_{Y/X}$, it follows by the definition of a singular metric that the corresponding multiplier sheaf is trivial over a large open subset of $\mathbb{P}(\mathcal{E}^\star)$. Then we have

$$\chi_{X_0}\Theta_{\det h_{Y/X}}(\det\mathcal{E})\wedge\omega_X^{n-1}\geqslant r\varepsilon_0\omega_X^n. \tag{39}$$

in the sense of currents on X_0 as consequence of (35), where χ_{X_0} is the characteristic function of the set $X_0\subset X$. Indeed this follows from the fact that

$$T:=\Theta_{\det h_{Y/X}}(\det\mathcal{E})\wedge\omega_X^{n-1} \tag{40}$$

is a closed positive current on X, and we clearly have

$$T\geqslant\lim_{k\to\infty}v_k\Theta_{\det h_{Y/X}}(\det\mathcal{E})\wedge\omega_X^{n-1} \tag{41}$$

for a sequence (v_k) converging to χ_{X_0}, cf. [**Dem1**]. For each k we have

$$v_k\Theta_{\det h_{Y/X}}(\det\mathcal{E})\wedge\omega_X^{n-1}\geqslant v_k r\varepsilon_0\omega_X^n \tag{42}$$

as currents of bi-degree (n,n) on X. Summing up, we have

$$T \geqslant \lim_{k\to\infty} v_k r\varepsilon_0\omega_X^n \tag{43}$$

and this is equivalent to the inequality (4).

REMARK 4.2. It is absolutely crucial that the lower bound (35) is the same as the lower bound of the trace of the curvature of (L, h_L) we have by hypothesis in Theorem 1.6.

We detail now the proof of Claim 4.1. An important point in our argument here is that the function $\log|\sigma|^2_{h^\star_{Y/X}}$ is *already* known to be almost psh. It is therefore enough to show that the inequality (35) holds true in any small Stein open set in the complement of an analytic subset of X. In particular, we can assume that p is smooth, and that $\mathcal{E}$ is a trivial vector bundle of rank r.

4.1.1. *Reduction to a local statement.* We invoke here the regularization arguments in [**BP08**] in order to reduce our claim to a subharmonicity property of fiberwise Bergman kernels for Stein submanifolds in $\mathbb{C}^n$.

Let $s_1, s_2, \cdots, s_r$ be a set of sections of the fibration $p: p^{-1}(U)\to U$ and let a_i be local sections of $(K_{Y/X}+L)^\star$ defined locally near the image of s_i. We define the Bergman kernel-type function on U as follows

$$B\langle a, a\rangle := \sum_{j,k} a_j\overline{a}_k B(s_j, s_k), \tag{44}$$

where

$$B(s_j(x), s_k(x)) = \sum_i u_i(s_j(x))\otimes \overline{u}_i(s_k(x)), \tag{45}$$

and $\{u_1,\dots,u_r\}$ is an orthogonormal basis of $H^0\left(Y_x, \mathcal{O}_Y(K_{Y/X}+L)\otimes\mathcal{I}(h_{\phi_L})\right)$. Also, we are using the same notation a_i for the section a_i evaluated at s_i. Therefore the expression (44) is indeed a function on U.

The link between (44) and our problem is as follows. We consider the expression

$$\xi_x := \sum a_j(s_j(x))\mathrm{ev}_{s_j(x)} \tag{46}$$

which is a local holomorphic section of the dual bundle $\mathcal{E}^\star$. Conversely, as observed in [**BP08**] any local section of $\mathcal{E}^\star$ can be obtained in this manner. Moreover, the norm of the section (46) with respect to the metric $h^\star_{Y/X}$ is precisely $B\langle a,a\rangle$. This is very important, because it shows that $B\langle a,a\rangle$ is an *extremal function.*

It is therefore sufficient to show that for every such a we have

$$dd^c\log B\langle a,a\rangle\wedge\omega_X^{n-1}\geqslant\varepsilon_0\omega_X^n \tag{47}$$

on U. Actually, thanks to a standard trick which we recall in Lemma 4.3 below it is enough to show the -equivalent- inequality

$$dd^c B\langle a,a\rangle\wedge\omega_X^{n-1}\geqslant\varepsilon_0 B\langle a,a\rangle\omega_X^n \tag{48}$$

It is at this point that we need our map p to be locally projective. In this case we can use the regularization procedure in [**Ber06**], [**BP08**] and reduce our problem (i.e. the inequality (48) above) to a local situation in which we can solve it by a direct computation. We will therefore recall the main steps of the relevant part of [**BP08**] and explain the way in which this is implemented in our case.

- According to the hypothesis, given any point $t_0 \in X$ there exists an open subset $U \subset Y$ containing x_0 such that

(49)
$$D := p^{-1}(U) \setminus H$$

is a Stein manifold, where H is a hypersurface. We can therefore replace our initial map with a Stein fibration

(50)
$$p : D \to U.$$

The corresponding Bergman kernel $B\langle a, a\rangle$ is defined as in (44) except that we replace the finite dimensional basis $u_1, \dots, u_r$ with a Hilbert basis of the space of holomorphic L^2 top forms on $D_t := D \cap p^{-1}(t)$ with respect to the L^2 norm

(51)
$$\|u\|_t^2 := \int_{D_t} c_n u \wedge \overline{u} e^{-\varphi_L}.$$

- Let ρ be a psh exhaustion function for the Stein manifold D. For c large enough, the image of the sections s_i will be contained in $D^c := (\rho < c)$. Let $B_c\langle a, a\rangle$ be the function (44) associated to the domain D^c. By the extremal characterization of $B\langle a, a\rangle$ and $B_c\langle a, a\rangle$ respectively, we see that it is enough to show that we have

(52)
$$dd^c B_c\langle a, a\rangle \wedge \omega_X^{n-1} \geqslant \varepsilon_0 B_c\langle a, a\rangle \omega_X^n$$

in the sense of currents on D^c. The inequality (48) would follow as $c \to \infty$.
- Let c be a regular value for the function ρ. There exists a holomorphic embedding

(53)
$$\Phi : D_{t_0}^{c+1/2} \times U \to D$$

such that

(54)
$$D_t^c \subset \Phi\big(D_{t_0}^{c+1/2} \times \{t\}\big) \subset D_t^{c+1}.$$

This map is obtained as usual, by considering the flow associated to the holomophic lifting of vector fields of U (we might be forced to shrink U, but this is fine since the estimates we have to prove are independent of the size of this set).

We can assume that the bundle L is trivial when restricted to D, and that the weight of the metric h_L is given by the psh function φ. Via the map (53), the curvature hypothesis (3) becomes

(55)
$$dd^c\phi \wedge \pi^\star \omega^{n-1} \geqslant \varepsilon_0 \pi^\star \omega^n$$

where

(56)
$$\phi := \varphi \circ \Phi, \qquad \omega := \omega_X|_U.$$

and π is the projection on the second factor of $D_{t_0}^{c+1/2} \times U$.

Let $\big(B(x_i, \varepsilon)\big)_{i=1,\dots N_\varepsilon}$ be a covering of the base X with balls of radius ε. On each such ball we certainly have

(57)
$$(1 - \delta_\varepsilon)\omega_i \leqslant \omega_X \leqslant (1 + \delta_\varepsilon)\omega_i$$

where ω_i is a flat metric on $B(x_i, \varepsilon)$, and $\delta_\varepsilon \to 0$ as $\varepsilon \to 0$.

Then we have

(58)
$$dd^c\phi \wedge \pi^\star \omega_i^{n-1} \geqslant \varepsilon_0 \frac{(1 - \delta_\varepsilon)^{n-1}}{(1 + \delta_\varepsilon)^n} \pi^\star \omega_i^n$$

for each index i such that $x_i \in U$ (so we assume that Φ is defined on a domain slightly bigger, which is again harmless).

Now the point is that the inequality (58) behaves well with respect to convolutions on the inverse image of $B(x_i, \varepsilon)$ –since ω_i is flat– in the sense that we have

$$dd^c\phi_\delta \wedge \pi^\star\omega_i^{n-1} \geqslant \varepsilon_0\frac{(1-\delta_\varepsilon)^{n-1}}{(1+\delta_\varepsilon)^n}\pi^\star\omega_i^n \tag{59}$$

for a monotonic sequence of smooth psh functions (ϕ_δ) converging to ϕ. This can be seen e.g. as follows. We first use the fact that the Stein manifold $D_{t_0}^{c+1/2}$ can be embedded in an Euclidean space $\mathbb{C}^N$ (in our case this is much simpler, since it is a domain in a projective manifold). The image of the embedding

$$D_{t_0}^{c+1/2} \to \mathbb{C}^N \tag{60}$$

has a Stein neighborhood W by a result of Siu, and moreover we have a holomorphic retract $\mu : W \to D_{t_0}^{c+1/2}$. We can regularize the function

$$\widetilde{\phi} : U \times W \to \mathbb{R} \cup \{-\infty\} \qquad \widetilde{\phi}(t, w) := \phi\big(t, \mu(w)\big) \tag{61}$$

by the usual convolution kernel, and ϕ_δ is the restriction of the convolution to $U \times D_{t_0}^{c+1/2}$. Given this explicit definition of ϕ_δ, the inequality (59) becomes obvious.

- It is proved in [**BP08**] that there exists a sequence of smooth psh functions ψ_j such that if we denote by $B_{j,\delta}\langle a, a\rangle$ the Bergman kernel-type function induced fiberwise by $e^{-\varphi_\delta - \psi_j}$, then the limit

$$\lim_j B_{j,\delta}\langle a, a\rangle|_{D_{t_0}^{c+1/2}\times\{t\}} \tag{62}$$

is *equal* to the Bergman kernel on the fiber D_t of our initial map p with respect to the weight $\phi_\delta \circ \Phi^{-1}$. The role of the sequence (ψ_j) is thus to "erase" the difference between the domains in (54). This is first proved for points t in the complement of measure zero set, and then in general, cf. pages 355-356 in [**BP08**]. A last observation at this point is that we also have

$$dd^c(\phi_\delta + \psi_j) \wedge \pi^\star\omega_i^{n-1} \geqslant \varepsilon_0\frac{(1-\delta_\varepsilon)^{n-1}}{(1+\delta_\varepsilon)^n}\pi^\star\omega_i^n \tag{63}$$

for every j, since ψ_j is psh.

- We show in the next subsection that we have

$$dd^c B_{j,\delta}\langle a, a\rangle \wedge \omega_i^{n-1} \geqslant \varepsilon_0\frac{(1-\delta_\varepsilon)^{n-1}}{(1+\delta_\varepsilon)^n}B_{j,\delta}\langle a, a\rangle\omega_i^n \tag{64}$$

on $B(x_i, \varepsilon)$. By using (57) and the fact that $B\langle a, a\rangle$ is psh, we infer that we have

$$dd^c B\langle a, a\rangle \wedge \omega_X^{n-1} \geqslant \varepsilon_0\frac{(1-\delta_\varepsilon)^{2n-2}}{(1+\delta_\varepsilon)^{2n}}B\langle a, a\rangle\omega_X^n \tag{65}$$

and we are done, provided that we are able to establish (64). This will be done in the next subsection.

4.1.2. *The local computation.* As consequence of the discussion in the previous subsection, it is enough to consider the following set-up:

(1) The map $p : U \times V \to U$ is simply the projection on the 1st factor. Here U is the unit ball in $\mathbb{C}^n$ and V is a Stein manifold.
(2) We denote by $t_1, \dots, t_n$ the coordinate functions on U, and we use the notation $z_1, \dots, z_d$ for coordinates on V at some point.
(3) We have a smooth, psh function φ on $U \times V$ such that
$$dd^c\varphi \wedge p^\star\omega^{n-1} \geqslant \varepsilon_0 p^\star\omega^n$$
point-wise on $U \times V$. In addition, φ is strictly psh when restricted to $\{t\} \times V$ for each $t \in U$.
(4) For each $t \in U$ we denote by B_t the Bergman kernel on
$$\left(\{t\} \times V, e^{-\varphi(t,\cdot)}\right)$$
corresponding to $(d, 0)$ holomorphic forms.
(5) Let $s_i : U \to V$ be a set of holomorphic functions (i.e. sections of the projection map p), where $i = 1, \dots r$. We also consider the holomorphic sections a_i of $s_i^\star \Lambda^d T_V$ and we define

$$B_t\langle a, a\rangle := \sum_{k,p=1}^{r} a_k(t)\overline{a_p(t)} B_t\big(s_k(t), s_p(t)\big) \tag{66}$$

Now we will show that we have

$$dd^c B_t\langle a, a\rangle \wedge \omega^{n-1} \geqslant \varepsilon_0 B_t\langle a, a\rangle\omega^n \tag{67}$$

at each point $t \in U$. This would be indeed sufficient, thanks to the following standard fact.

LEMMA 4.3. *Assume that* (67) *holds true. Then we have*

$$dd^c \log B_t\langle a, a\rangle \wedge \omega^{n-1} \geqslant \varepsilon_0\omega^n \tag{68}$$

PROOF. The argument is well-know in the case of positive psh functions, cf. e.g. [**LG86**, Lemma 3.46]. Strictly the same proof goes through in our situation, as we will see in what follows.

The main observation is that if we replace the function φ with

$$\varphi_\lambda := \varphi + \Re\left(\sum_{i=1}^{n} \alpha_i t_i\right) \tag{69}$$

then the hypothesis (3) above is still verified. The corresponding fiberwise Bergman kernel becomes

$$e^{\Re\left(\sum_{i=1}^n \alpha_i t_i\right)} B_t\langle a, a\rangle. \tag{70}$$

Now let $t_0 \in U$ be an arbitrary point. We assume that the coordinates (t_i) are such that ω_{t_0} is the flat metric. Since the inequality (67) is true for the Bergman kernel (70), we infer that we have

$$\sum_i \frac{\partial^2 B}{\partial t_i \partial \bar{t}_i} + \frac{1}{2}\left(\alpha_i \frac{\partial B}{\partial \bar{t}_i} + \overline{\alpha}_i \frac{\partial B}{\partial t_i}\right) + \frac{1}{4}|\alpha_i|^2 B \geqslant n\varepsilon_0 B \tag{71}$$

where we denote by $B := B_t\langle a, a\rangle$, and the quantities above are evaluated at $t = t_0$. The inequality (71) holds true for any choice of the coefficients α_i. We then choose

$$\alpha_i := -\frac{2}{B(t_0)}\frac{\partial B}{\partial t_i}(t_0) \tag{72}$$

and the lemma follows. □

The inequality (67) will be established by a direct computation detailed along the next lines. The case $n = 1$ was treated by [**Ber06**], and our arguments below represent a generalization of his approach.

By the reproducing property of Bergman kernels, we have

$$B_t\langle a, a\rangle = \sum_{i,j} a_i(t)\overline{a_j(t)}\int_V B_t(\xi, s_i(t))\overline{B_t(\xi, s_j(t))}e^{-\varphi(t,\xi)} \tag{73}$$

We first take the anti-holomorphic derivative with respect to t_α.

$$\begin{aligned}\frac{\partial}{\partial \bar t_\alpha}B_t\langle a, a\rangle =&\sum_{i,j} a_i(t)\overline{\frac{\partial a_j(t)}{\partial t_\alpha}}\int_V B_t(\xi, s_i(t))\overline{B_t(\xi, s_j(t))}e^{-\varphi(t,\xi)}\\ &+\sum_{i,j} a_i(t)\overline{a_j(t)}\int_V \frac{\partial B_t}{\partial \bar t_\alpha}(\xi, s_i(t))\overline{B_t(\xi, s_j(t))}e^{-\varphi(t,\xi)}\\ &+\sum_{i,j} a_i(t)\overline{a_j(t)}\int_V \frac{\partial B_t}{\partial \overline{w}_\gamma}(\xi, s_i(t))\overline{\frac{\partial s_i^\gamma}{\partial t_\alpha}}\overline{B_t(\xi, s_j(t))}e^{-\varphi(t,\xi)}\\ &+\sum_{i,j} a_i(t)\overline{a_j(t)}\int_V B_t(\xi, s_i(t))\overline{\frac{\partial^\varphi B_t}{\partial t_\alpha}(\xi, s_j(t))}e^{-\varphi(t,\xi)}.\\ =:&I_1 + I_2 + I_3 + I_4.\end{aligned} \tag{74}$$

Given any holomorphic top form h which is L^2 on V we have

$$h(s_j(t)) = \int_V h(\xi)\overline{B_t(\xi, s_j(t))}e^{-\varphi(t,\xi)}. \tag{75}$$

We apply the operator $\dfrac{\partial}{\partial \bar t_\alpha}$ to (75) and it follows that we have

$$0 = \int_V h(\xi)\overline{\frac{\partial^\varphi B_t}{\partial t_\alpha}(\xi, s_j(t))}e^{-\varphi(t,\xi)}. \tag{76}$$

Therefore the last term I_4 of (74) is equal to zero.

We evaluate next the quantity

$$\sum_{\alpha,\beta}\omega^{\overline{\alpha}\beta}\frac{\partial^2}{\partial t_\beta \partial \bar t_\alpha}B_t\langle a, a\rangle \tag{77}$$

where $\left(\omega^{\overline{\alpha}\beta}\right)$ be the inverse of the coefficients of the metric ω at some point $t_0 \in U$.

The $\dfrac{\partial}{\partial t_\beta}$ of the term I_1 gives

$$\begin{aligned}
\frac{\partial I_1}{\partial t_\beta} =& \sum_{i,j} \frac{\partial a_i(t)}{\partial t_\beta}\overline{\frac{\partial a_j(t)}{\partial t_\alpha}} \int_V B_t(\xi, s_i(t))\overline{B_t(\xi, s_j(t))}e^{-\varphi(t,\xi)} \\
&+ \sum_{i,j} a_i(t)\overline{\frac{\partial a_j(t)}{\partial t_\alpha}} \int_V \frac{\partial^\varphi B_t}{\partial t_\beta}(\xi, s_i(t))\overline{B_t(\xi, s_j(t))}e^{-\varphi(t,\xi)} \\
&+ \sum_{i,j} a_i(t)\overline{\frac{\partial a_j(t)}{\partial t_\alpha}} \int_V B_t(\xi, s_i(t))\overline{\frac{\partial B_t}{\partial \bar{t}_\beta}(\xi, s_j(t))}e^{-\varphi(t,\xi)} \\
&+ \sum_{i,j} a_i(t)\overline{\frac{\partial a_j(t)}{\partial t_\alpha}} \int_V B_t(\xi, s_i(t))\overline{\frac{\partial B_t}{\partial \overline{w}_\gamma}(\xi, s_j(t))\overline{\frac{\partial s_j^\gamma}{\partial t_\beta}}}e^{-\varphi(t,\xi)}
\end{aligned} \tag{78}$$

We compute similar derivative for the other terms:

$$\begin{aligned}
\frac{\partial I_2}{\partial t_\beta} =& \sum_{i,j} \frac{\partial a_i(t)}{\partial t_\beta}\overline{a_j(t)} \int_V \frac{\partial B_t}{\partial \bar{t}_\alpha}(\xi, s_i(t))\overline{B_t(\xi, s_j(t))}e^{-\varphi(t,\xi)} \\
&+ \sum_{i,j} a_i(t)\overline{a_j(t)} \int_V \frac{\partial^\varphi}{\partial t_\beta}\Big(\frac{\partial B_t}{\partial \bar{t}_\alpha}\Big)(\xi, s_i(t))\overline{B_t(\xi, s_j(t))}e^{-\varphi(t,\xi)} \\
&+ \sum_{i,j} a_i(t)\overline{a_j(t)} \int_V \frac{\partial B_t}{\partial \bar{t}_\alpha}(\xi, s_i(t))\overline{\frac{\partial B_t}{\partial \bar{t}_\beta}(\xi, s_j(t))}e^{-\varphi(t,\xi)} \\
&+ \sum_{i,j} a_i(t)\overline{a_j(t)} \int_V \frac{\partial B_t}{\partial \bar{t}_\alpha}(\xi, s_i(t))\overline{\frac{\partial B_t}{\partial \overline{w}_\gamma}\overline{\frac{\partial s_j^\gamma}{\partial t_\beta}}(\xi, s_j(t))}e^{-\varphi(t,\xi)}
\end{aligned} \tag{79}$$

and

$$\begin{aligned}
\frac{\partial I_3}{\partial t_\beta} =& \sum_{i,j} \frac{\partial a_i(t)}{\partial t_\beta}\overline{a_j(t)} \int_V \frac{\partial B_t}{\partial \overline{w}_\gamma}(\xi, s_i(t))\overline{\frac{\partial s_i^\gamma}{\partial t_\alpha}}\overline{B_t(\xi, s_j(t))}e^{-\varphi(t,\xi)} \\
&+ \sum_{i,j} a_i(t)\overline{a_j(t)} \int_V \frac{\partial^\varphi}{\partial t_\beta}\Big(\frac{\partial B_t}{\partial \overline{w}_\gamma}\Big)(\xi, s_i(t))\overline{\frac{\partial s_i^\gamma}{\partial t_\alpha}}\overline{B_t(\xi, s_j(t))}e^{-\varphi(t,\xi)} \\
&+ \sum_{i,j} a_i(t)\overline{a_j(t)} \int_V \frac{\partial B_t}{\partial \overline{w}_\gamma}(\xi, s_i(t))\overline{\frac{\partial s_i^\gamma}{\partial t_\alpha}}\overline{\frac{\partial B_t}{\partial \bar{t}_\beta}(\xi, s_j(t))}e^{-\varphi(t,\xi)} \\
&+ \sum_{i,j} a_i(t)\overline{a_j(t)} \int_V \frac{\partial B_t}{\partial \overline{w}_\gamma}(\xi, s_i(t))\overline{\frac{\partial s_i^\gamma}{\partial t_\alpha}}\overline{\frac{\partial B_t}{\partial \overline{w}_\mu}\overline{\frac{\partial s_j^\mu}{\partial t_\beta}}(\xi, s_j(t))}e^{-\varphi(t,\xi)}.
\end{aligned} \tag{80}$$

In order to arrange a bit the terms above we make the following observations

(i) The second term in the rhs of (78) equals zero, by the formula (76).

(ii) We introduce the notation

$$\Xi_{i\overline{\alpha}}(t,\xi) := \overline{\frac{\partial a_i(t)}{\partial t_\alpha}} + a_i(t)\left(\frac{\partial B_t}{\partial \overline{t}_\alpha}(\xi, s_i(t)) + \frac{\partial B_t}{\partial \overline{w}_\mu}(\xi, s_i(t))\overline{\frac{\partial s_i^\mu}{\partial t_\alpha}}\right).$$

Then we have

$$\begin{aligned}\frac{\partial^2 B_t\langle a,a\rangle}{\partial t_\beta \partial \overline{t}_\alpha} &= \sum_{i,j}\int_V \Xi_{i\overline{\alpha}}(t,\xi)\overline{\Xi_{i\overline{\beta}}(t,\xi)}e^{\varphi} \\ &+ \sum_{i,j} a_i(t)\overline{a_j(t)}\int_V \frac{\partial^\varphi}{\partial t_\beta}\Big(\frac{\partial B_t}{\partial \overline{t}_\alpha}\Big)(\xi, s_i(t))\overline{B_t(\xi, s_j(t))}e^{-\varphi(t,\xi)} \\ &+ \sum_{i,j} a_i(t)\overline{a_j(t)}\int_V \frac{\partial^\varphi}{\partial t_\beta}\Big(\frac{\partial B_t}{\partial \overline{w}_\gamma}\Big)(\xi, s_i(t))\overline{\frac{\partial s_i^\gamma}{\partial t_\alpha}}\,\overline{B_t(\xi, s_j(t))}e^{-\varphi(t,\xi)}\end{aligned} \tag{81}$$

and since the following holds

$$\sum_{i,j,\alpha,\beta}\omega^{\overline{\alpha}\beta}\int_V \Xi_{i\overline{\alpha}}(t,\xi)\overline{\Xi_{j\overline{\beta}}(t,\xi)}e^{\varphi} \geqslant 0$$

for each $t \in U$ we only have to deal with the last two terms of the previous equality (81) in order to obtain a lower bound of (77).

(iii) The last two terms of (81) are obtained by applying the operator $\frac{\partial^\varphi}{\partial t_\beta}\circ\frac{\partial}{\partial \overline{t}_\alpha}$ to the function $t \to B_t(\xi, s_i(t))$. The commutation formula reads as

$$\left[\frac{\partial^\varphi}{\partial t_\beta}, \frac{\partial}{\partial \overline{t}_\alpha}\right] = \frac{\partial^2 \varphi}{\partial t_\beta \partial \overline{t}_\alpha}$$

so all in all we infer that we have

$$\begin{aligned}\sum \omega^{\overline{\alpha}\beta}\frac{\partial^2 B_t\langle a,a\rangle}{\partial t_\beta \partial \overline{t}_\alpha} &\geqslant \sum a_i(t)\overline{a_j(t)}\omega^{\overline{\alpha}\beta}\int_V \frac{\partial^2\varphi}{\partial t_\beta \partial \overline{t}_\alpha}B_t(\xi, s_j(t))\overline{B_t(\xi, s_j(t))}e^{-\varphi(t,\xi)} \\ &+ \sum a_i(t)\overline{a_j(t)}\omega^{\overline{\alpha}\beta}\int_V \frac{\partial}{\partial \overline{t}_\alpha}\Big(\frac{\partial^\varphi B_t}{\partial t_\beta}\Big)(\xi, s_i(t))\overline{B_t(\xi, s_j(t))}e^{-\varphi(t,\xi)}\end{aligned}$$

By the equation (76) we infer that

$$\begin{aligned}&\int_V \frac{\partial}{\partial \overline{t}_\alpha}\Big(\frac{\partial^\varphi B_t}{\partial t_\beta}\Big)(\xi, s_i(t))\overline{B_t(\xi, s_j(t))}e^{-\varphi(t,\xi)} \\ &= -\int_V \frac{\partial^\varphi B_t}{\partial t_\beta}(\xi, s_i(t))\overline{\frac{\partial^\varphi B_t}{\partial t_\alpha}(\xi, s_j(t))}e^{-\varphi(t,\xi)}\end{aligned} \tag{82}$$

The computations above were done with respect to an arbitrary coordinate system, so we can as well assume that $\omega^{\overline{\alpha}\beta} = \delta_{\alpha\beta}$ at some point $t_0 \in U$. We therefore get

$$\sum \frac{\partial^2 B_t\langle a,a\rangle}{\partial t_\alpha \partial \bar{t}_\alpha}(t_0) \geqslant \sum a_i(t_0)\overline{a_j(t_0)} \int_V \frac{\partial^2 \varphi}{\partial t_\alpha \partial \bar{t}_\alpha} B_{t_0}(\xi, s_i(t_0)) \overline{B_{t_0}(\xi, s_j(t_0))} e^{-\varphi(t_0,\xi)}$$
$$- \sum a_i(t_0)\overline{a_j(t_0)} \int_V \frac{\partial^\varphi B_{t_0}}{\partial t_\beta}(\xi, s_i(t_0)) \overline{\frac{\partial^\varphi B_{t_0}}{\partial t_\beta}(\xi, s_j(t_0))} e^{-\varphi(t_0,\xi)}$$
$$= \sum_\alpha \int_V \left(\frac{\partial^2 \varphi}{\partial t_\alpha \partial \bar{t}_\alpha}(t_0,\xi)|\Gamma(\xi)|^2 - |\Lambda_\alpha(\xi)|^2 \right) e^{-\varphi(t_0,\xi)},$$

where the notations used are as follows

$$\Gamma(\xi) := \sum_i a_i(t_0) B_{t_0}(\xi, s_i(t_0)) \tag{83}$$

and

$$\Lambda_\alpha(\xi) := \sum_i a_i(t_0) \frac{\partial^\varphi B_{t_0}}{\partial t_\alpha}(\xi, s_i(t_0)) \tag{84}$$

In order to evaluate the norm of Λ_α we compute its $\bar{\partial}$, and obtain

$$\bar{\partial}\Lambda_\alpha(\xi) = -\sum_{i,k} a_i(t_0) B_{t_0}(\xi, s_i(t_0)) \frac{\partial^2 \varphi}{\partial t_\alpha \partial \bar{\xi}_k} d\bar{\xi}_k \tag{85}$$

and we see that this can be rewritten as

$$\bar{\partial}\Lambda_\alpha(\xi) = -\Gamma(\xi) \sum_k \frac{\partial^2 \varphi}{\partial t_\alpha \partial \bar{\xi}_k} d\bar{\xi}_k. \tag{86}$$

Moreover, Λ_α is perpendicular on the space of holomorphic L^2 functions –by the property (76)–, and Hörmander estimates show that we must have

$$\int_V |\Lambda_\alpha(\xi)|^2 e^{-\varphi(t_0,\xi)} \leqslant \int_V |\Gamma|^2 \frac{\partial^2 \varphi}{\partial t_\alpha \partial \bar{\xi}_k} \frac{\partial^2 \varphi}{\partial \bar{t}_\alpha \partial \xi_m} \varphi^{\bar{k}m} e^{-\varphi(t_0,\xi)}, \tag{87}$$

where $(\varphi^{\bar{k}m})_{k,m}$ is the inverse of the Hessian matrix of φ in the fiber directions. We thus get the inequality

$$\sum_\alpha \frac{\partial^2 B_t\langle a,a\rangle}{\partial t_\alpha \partial \bar{t}_\alpha}(t_0) \geqslant \sum_\alpha \int_V |\Gamma|^2 \left(\frac{\partial^2 \varphi}{\partial t_\alpha \partial \bar{t}_\alpha} - \frac{\partial^2 \varphi}{\partial t_\alpha \partial \bar{\xi}_k} \frac{\partial^2 \varphi}{\partial \bar{t}_\alpha \partial \xi_m} \varphi^{\bar{k}m} \right) e^{-\varphi(t_0,\xi)}. \tag{88}$$

It turns out that the quantity

$$\frac{\partial^2 \varphi}{\partial t_\alpha \partial \bar{t}_\alpha} - \frac{\partial^2 \varphi}{\partial t_\alpha \partial \bar{\xi}_k} \frac{\partial^2 \varphi}{\partial \bar{t}_\alpha \partial \xi_m} \varphi^{\bar{k}m} \tag{89}$$

has an intrinsic interpretation: it is equal to

$$\frac{n}{d+1} \frac{(dd^c\varphi)^{d+1} \wedge p^\star \omega^{n-1}}{(dd^c\varphi)^d \wedge p^\star \omega^n} \tag{90}$$

evaluated at the point t_0. By hypothesis, φ is psh, which combined with the lower bound for the trace of $dd^c\varphi$ with respect to $p^\star\omega$ shows that

$$\frac{n}{d+1} \frac{(dd^c\varphi)^{d+1} \wedge p^\star \omega^{n-1}}{(dd^c\varphi)^d \wedge p^\star \omega^n} \geqslant n\varepsilon_0. \tag{91}$$

The inequality (91) is easy to justify in our situation, since φ is strictly psh, and we can choose the coordinates in such a way that (91) becomes trivial. It holds however true in more general circumstances, as we will now see: it is enough to assume that φ is strictly psh on the fibers of p.

Toward this end, let (z_i) be local coordinates centred at the origin, such that

$$dd^c\varphi = \sqrt{-1}\Big(\sum_{\alpha,\beta}\varphi_{\alpha\overline{\beta}}dt_\alpha\wedge d\overline{t}_\beta + \sum_{\alpha,a}\varphi_{\alpha\overline{a}}dt_\alpha\wedge d\overline{z}_a + \sum_{a,\beta}\varphi_{a\overline{\beta}}dz_a\wedge d\overline{t}_\alpha\Big) + \sqrt{-1}\sum_a dz_a\wedge d\overline{z}_a. \tag{92}$$

By hypothesis we have

$$\begin{aligned} n\varepsilon_0\prod_\alpha\sqrt{-1}dt_\alpha d\overline{t}_\alpha \leqslant{} & \sum_\alpha\varphi_{\alpha\overline{\alpha}}\prod_\alpha\sqrt{-1}dt_\alpha d\overline{t}_\alpha \\ & + \sqrt{-1}\sum_{\alpha,a}\varphi_{\alpha\overline{a}}dt_\alpha\wedge d\overline{z}_a\prod\widehat{dt_\alpha\wedge d\overline{t}_\alpha} \\ & + \sqrt{-1}\sum_{a,\alpha}\varphi_{a\overline{\alpha}}dz_a\wedge d\overline{t}_\alpha\prod\widehat{dt_\alpha\wedge d\overline{t}_\alpha} \\ & + \sqrt{-1}\Big(\sum_a dz_a\wedge d\overline{z}_a\Big)\prod\widehat{dt_\alpha\wedge d\overline{t}_\alpha}, \end{aligned}$$

and therefore for any choice of complex numbers $(\lambda_{\alpha\overline{a}})$ we have

$$\sum_\alpha\varphi_{\alpha\overline{\alpha}} - n\varepsilon_0 - 2\Re\big(\sum_{\alpha,a}\varphi_{\alpha\overline{a}}\overline{\lambda_{\alpha\overline{a}}}\big) + \sum|\lambda_{\alpha\overline{a}}|^2 \geqslant 0. \tag{93}$$

This now implies that

$$\sum_\alpha\varphi_{\alpha\overline{\alpha}} - \sum_{\alpha,a}|\varphi_{\alpha\overline{a}}|^2 \geqslant n\varepsilon_0 \tag{94}$$

which is what we wanted to prove.

The quantity we have on the LHS of (88) reads as

$$n\frac{dd^cB_t\langle a,a\rangle\wedge\omega^{n-1}}{\omega^n} \tag{95}$$

and we get

$$dd^cB_t\langle a,a\rangle\wedge\omega^{n-1} \geqslant \varepsilon_0B_t\langle a,a\rangle\omega^n. \tag{96}$$

In conclusion, the local version of our result holds true and Theorem 1.4 is proved.

4.2. Proof of Theorem 1.6. We first assume that ω_X is Kähler. Then Theorem 1.6 is easily reduced to Theorem 1.4 as follows.

Let $U\subset X$ be an open subset of X, such that

$$\omega_X|_U = dd^c\varphi_U \tag{97}$$

for some function smooth φ_U defined on U. We then consider the restriction of L on the p-inverse image of U denoted by Y_U and we endow it with the metric

$$h_1 := e^{-C\cdot\phi}h_L \tag{98}$$

where $\phi := \varphi_U \circ p$. The hypothesis of Theorem 1.4 are fulfilled. The constant ε_0 is in this case equal to C. Then we have

$$\Theta_{\det h_{1,Y_U/U}}(\det \mathcal{E}) \wedge \omega_X^{n-1} \geqslant rC\omega_X^n, \tag{99}$$

which is equivalent to

$$\Theta_{\det h_{Y/X}}(\det \mathcal{E}) \wedge \omega_X^{n-1} \geqslant 0, \tag{100}$$

by using the relation $h_{1,Y_U/U} = h_{Y/X} \cdot e^{-C\cdot\varphi_U}$.

Now for the case of an arbitrary metric ω we argue as in the proof of Theorem 1.4. Let $B(x_i, \varepsilon)$ be a covering of X with balls of radius ε, such that the condition (57) is satisfied. Then on each ball we consider the metric

$$h_{i,L} := e^{-C(1+\delta_\varepsilon)|t_i|^2} h_L \tag{101}$$

where (t_i) are coordinates on the ball $B(x_i, \varepsilon)$. By the Kähler case already discussed, we obtain

$$\big(\Theta_{\det h_{Y/X}}(\det \mathcal{E}) + rC(1+\delta_\varepsilon)\omega_i\big) \wedge \omega_i^{n-1} \geqslant rC\big(1 + \mathcal{O}(\delta_\varepsilon)\big)\omega_i^n \tag{102}$$

on each ball $B(x_i, \varepsilon)$. This inequality is basically unchanged if we replace ω_i by the global metric ω, again thanks to (57). We therefore obtain

$$\big(\Theta_{\det h_{Y/X}}(\det \mathcal{E}) + rC(1+\delta_\varepsilon)\omega\big) \wedge \omega^{n-1} \geqslant rC\big(1 + \mathcal{O}(\delta_\varepsilon)\big)\omega^n \tag{103}$$

Theorem 1.6 is therefore established by letting $\varepsilon \to 0$.

4.3. Kähler version of Lemma 3.3. We will establish here the Kähler version of Lemma 3.3. The statement is the same as in the projective case, except that we only assume the manifold X to be compact Kähler. Also, the arguments are similar to those used in the projective case but we apply Theorem 1.6 instead of Theorem 2.7. There is however an additional slight complication due to the singularities of $\mathcal{F}$, so we will provide a complete treatment in what follows.

Recall that we have the bundle $E \to \widehat{X}$ which admits a sequence of smooth metrics $h_{E,\varepsilon}$ such that

$$\Theta(E, h_{E,\varepsilon}) \wedge g_\varepsilon^{n-1} = \delta_\varepsilon g_\varepsilon^n \otimes \mathrm{Id}_{\mathrm{End}(E)} \tag{104}$$

for some constant δ_ε such that $\lim_{\varepsilon\to 0} \delta_\varepsilon = 0$. Let h be a metric on $\mathcal{O}(1)$ which could be singular, but such that

$$\Theta(\mathcal{O}(1), h) \geqslant 0 \tag{105}$$

in the sense of currents on $\mathbb{P}(E)$.

Assume by contradiction that the multiplier sheaf $\mathcal{I}((k+1-\delta_0)h|_{\mathbb{P}(E^\star)_x})$ is non-trivial on the generic fiber $\mathbb{P}(E)_x$ for some $\delta_0 > 0$. Then we consider the relative adjoint bundle

$$K_{\mathbb{P}(E)/\widehat{X}} + \mathcal{O}(r+k). \tag{106}$$

By a standard L^2-estimate (cf. Lemma 4.4), the direct image

$$\mathcal{G} := p_\star\Big((K_{\mathbb{P}(E)/\widehat{X}} + \mathcal{O}(r+k)) \otimes \mathcal{I}((k+1-\delta_0)h)\Big) \tag{107}$$

is non zero. We define the metric

$$h_\varepsilon := h_{E,\varepsilon}^{\otimes(r-1+\delta_0)} \otimes h^{\otimes(k+1-\delta_0)} \tag{108}$$

on the bundle $\mathcal{O}(r+k)$ –here we denote with the same symbol $h_{E,\varepsilon}$ the metric on E and the metric induced by it on $\mathcal{O}(1)$.

Then a direct computation shows that we have

$$\Theta(\mathcal{O}(r+k),h_\varepsilon)\wedge\pi^\star g_\varepsilon^{n-1}\geqslant(r-1)\delta_\varepsilon\pi^\star g_\varepsilon^n. \tag{109}$$

Here we are using (105) combined with the Hermite-Einstein identity (104) and the explicit expression of the curvature of $\mathcal{O}(1)$.

We therefore get a *proper* subsheaf $\mathcal{G}$ of $\operatorname{Sym}^k E^\star\otimes\det E^\star$ for which we have

$$\int_{\widehat{X}} c_1(\mathcal{G})\wedge(\pi^\star g_\varepsilon)^{n-1}\geqslant r(r-1)\delta_\varepsilon\int_{\widehat{X}}(\pi^\star g_\varepsilon)^n \tag{110}$$

by Theorem 1.6. Moreover, $\mathcal{G}$ is independent of ε. As $\varepsilon\to 0$ we obtain

$$\int_{\widehat{X}} c_1(\mathcal{G})\wedge\pi^\star\omega^{n-1}\geqslant 0 \tag{111}$$

since the volume of $\widehat{X}$ with respect to g_ε is uniformly bounded.

As in the projective case, the stability hypothesis forbids the existence of such subsheaf $\mathcal{G}$. In conclusion, we have

$$\mathcal{I}((k+1-\varepsilon)h|_{\mathbb{P}(E)_x})=\mathcal{O}_{\mathbb{P}(E)_x}, \tag{112}$$

for any $\varepsilon>0$, where x is a generic point (in the sense that outside some measure 0 set).

Finally, we prove Theorem 1.2 in the case when X is compact Kähler. Thanks to the stability condition (ii) of Theorem 1.2, (112) implies that

$$\mathcal{I}(h^{\otimes(r+1)}|_{\mathbb{P}(E)_x})=\mathcal{O}_{\mathbb{P}(E)_x}.$$

By applying Theorem 2.7 to

$$p_\star\left((K_{\mathbb{P}(E)/X}+\mathcal{O}(r+1))\right)=\mathcal{F}^\star,$$

$\mathcal{F}^\star$ is positively curved. Together with $c_1(\mathcal{F}^\star)=0$, Proposition 2.6 implies that $\mathcal{F}^\star$ is hermitian flat.

In our previous arguments we have used the following statement, which is a consequence of L^2-estimates.

LEMMA 4.4. *Let $\mathbb{P}^n$ be the n-dimensional projective space, and let $m\in\mathbb{N}^\star$. Let h be a singular metric on $\mathcal{O}_{\mathbb{P}^n}(m)$ such that $i\Theta_h(\mathcal{O}_{\mathbb{P}^n}(m))\geqslant 0$ on X. Then the space $H^0(\mathbb{P}^n,\mathcal{O}_{\mathbb{P}^n}(m-1)\otimes\mathcal{I}((1-\varepsilon)h))$ is non zero for every $\varepsilon>0$.*

PROOF. Set $L:=\mathcal{O}_{\mathbb{P}^n}(m+n)$. Then we have

$$\mathcal{O}_{\mathbb{P}^n}(m-1)=K_{\mathbb{P}^n}+L.$$

Let $x\in\mathbb{P}^n\setminus V(\mathcal{I}((1-\varepsilon)h))$ and let $\{s_1,s_2,\ldots,s_n\}$ be a basis of $H^0(\mathbb{P}^n,\mathcal{O}(1))$ vanishing on x. The local weight

$$\varphi:=n\log(\sum_i|s_i|^2)$$

defines a metric on $\mathcal{O}(n)$ of isolated singularity at x with Lelong number n. We define the metric

$$h_L:=h_\varphi+(1-\varepsilon)h+\varepsilon h_{FS}$$

on L, where h_{FS} is the Fubini-Study metric. Then the L^2-estimates (cf. [**Dem2**, Cor 5.12]) implies that

$$H^0(\mathbb{P}^n,\mathcal{O}_{\mathbb{P}^n}(K_{\mathbb{P}^n}+L)\otimes\mathcal{I}((1-\varepsilon)h))\to(\mathcal{O}_{\mathbb{P}^n}(K_{\mathbb{P}^n}+L))_x$$

is surjective. The lemma is proved. □

5. A few comments about the general conjecture and other results

We recall that the conjecture of Pereira-Touzet cf. [**PeTou13**] states that if $\mathcal{F} \subset T_X$ is a holomorphic foliation such that $c_1(\mathcal{F}) = 0$, $c_2(\mathcal{F}) \neq 0$ and $\mathcal{F}$ is ω_X-stable and K_X is pseudo-effective, then $\mathcal{F}$ is algebraic.

As we have already mentioned, under the conditions above $\mathcal{F}$ is a sub-bundle of T_X. If we assume (by contradiction) that $\mathcal{F}$ is not algebraic, then the bundle $\mathcal{O}(1) \to \mathbb{P}(\mathcal{F})$ is pseudo-effective. We can therefore find a singular metric $e^{-\varphi}$ on it such that the curvature $\Theta = dd^c\varphi \geqslant 0$ in the sense currents. The Siu decomposition of this current writes as

$$\Theta = \sum a_i[E_i] + T, \tag{113}$$

where $\sum a_i[E_i]$ is the divisorial part and T is a positive current with the property that $\operatorname{codim}_X E_c(T) \geqslant 0$ for every $c > 0$. Here $E_c(T)$ is the locus where the Lelong number of T is at least c.

We show now the following result concerning the divisorial part of the current Θ.

THEOREM 5.1. *Let X be a projective manifold, such that K_X is pseudo-effective. We assume that $\mathcal{F}$ is a holomorphic foliation on X such that $c_1(\mathcal{F}) = 0$. Assume moreover that $\mathcal{F}$ is H–stable, where H is an ample divisor on X. If $\mathcal{O}(1)$ is pseudo-effective and E is one component of the divisorial part of* (113), *then the map*

$$p_E : E \to X \tag{114}$$

induced by the restriction of p to E is a locally trivial fibration.

PROOF. We first remark that for each j we have

$$E_j \equiv m_j\mathcal{O}(1) + p^\star(L_j) \tag{115}$$

where m_j is a positive integer, L_j is a line bundle on X and $p : \mathbb{P}(T_\mathcal{F}) \to X$ is the projection map. We equally have

$$T \equiv m_T\mathcal{O}(1) + p^\star(L_T) \tag{116}$$

where $m_T \geqslant 0$ is a real number and L_T is a $\mathbb{R}$-bundle.

The hypothesis $c_1(\mathcal{F}) = 0$ together with the stability condition shows that for each j we have

$$L_j \cdot H^{n-1} \geqslant 0 \tag{117}$$

as soon as $m_j \geqslant 1$, i.e. if E_j is horizontal with respect to p. Now if E_j projects into a proper subvariety of X, then we also have $L_j \cdot H^{n-1} \geqslant 0$ since it means that the corresponding m_j equals zero. Moreover, T is a closed positive current, so a quick approximation argument shows that $L_T \cdot H^{n-1} \geqslant 0$ as well. On the other hand, the equality (113) shows that numerically we have

$$\sum a_j L_j + L_T \equiv 0. \tag{118}$$

Hence we infer that $L_j \cdot H^{n-1} = L_T \cdot H^{n-1} = 0$ for every j. Therefore, we obtain that $m_j \geqslant 1$ for every j, namely, every component E_j is horizontal with respect to p.

Consider the exact sequence

$$0 \to L_j^{-1} \to \operatorname{Sym}^{m_j}\mathcal{F}^\star \to Q_j \to 0 \tag{119}$$

induced by the section E_j. By taking the determinants and using the previous considerations we obtain $\det(Q_j) \wedge \omega^{n-1} = 0$. On the other hand, the foliation $\mathcal{F}$ is smooth so it follows that $\det Q_j$ is pseudo-effective, cf. [**CP19**]. In conclusion, $c_1(Q_j) = 0$, which in turn implies that $c_1(L_j) = 0$.

After these preliminary considerations, we are ready to prove that p_E is locally trivial. Let

$$S := H_1 \cap H_2 \cap \cdots \cap H_{n-1}$$

be the complete intersection of some smooth hypersurfaces $H_i \in |kH|$ for k large enough. To prove that p_E is locally trivial, it is sufficient to prove that

$$p_E|_{p_E^{-1}(S)} : p^{-1}(S) \cap E \to S$$

is locally trivial.

Thanks to [**Lan04**, Thm 5.2], $\mathcal{F}|_S$ is stable if k is large enough (note that we don't need ask that H_i is a generic hypersurface here). Since $c_1(\mathcal{F}|_S) = 0$, it follows that $\mathcal{O}_{\mathbb{P}(\mathcal{F})}(1)|_S$ admits a smooth metric h of semi-positive curvature. As we proved that $L_j \cdot H^{n-1} = 0$ for every j, we get

$$c_1(E) = c_1(\mathcal{O}(m)) \in H^{1,1}(\mathbb{P}(\mathcal{F}), \mathbb{R})$$

for some $m \in \mathbb{N}$. Let $a \geqslant m$ be some number large enough. $\mathcal{O}(a)|_S$ be can equipped with the metric

$$h_a := h^{a-m} \cdot e^{-\log |s_E|^2},$$

whose curvature is semipositive. Here s_E is the canonical section of $\mathcal{O}(E)$. We consider the direct image of $p : p^{-1}(S) \to S$

$$p_\star(K_{p^{-1}(S)/S} \otimes \mathcal{O}_{\mathbb{P}(\mathcal{F})}(a) \otimes \mathcal{I}(h_a)) \subset p_\star(K_{p^{-1}(S)/S} \otimes \mathcal{O}_{\mathbb{P}(\mathcal{F})}(a)) \qquad \text{on } S.$$

Then RHS is just $\mathrm{Sym}^{a-2}\mathcal{F}^\star|_S$, which is hermitian flat. The LHS is Griffiths-semi-positive by [**PT18**]. Therefore LHS is also hermitian flat, and the flat connection is compatible with the flat connection on $\mathcal{F}^\star|_S$. Note also that the germs of $p_\star(K_{p^{-1}(S)/S} \otimes \mathcal{O}_{\mathbb{P}(\mathcal{F})}(a) \otimes \mathcal{I}(h_a))_s$ is just the degree $(a-r)$-polynomials vanishing on E. For a large enough, the flatness $p_\star(K_{p^{-1}(S)/S} \otimes \mathcal{O}_{\mathbb{P}(\mathcal{F})}(a) \otimes \mathcal{I}(h_a))$ implies that the common zero locus of these polynomials (which is E) is also invariant by the flat connection on $\mathcal{F}|_S$. It means that $E|_S$ is invariant by the parallel tranport of the flat connection on $\mathcal{F}|_S$. Then $p_E^{-1}(S) \to S$ is locally trivial. As S is the complete intersection of any smooth hypersurfaces in $|kH|$, p_E is thus locally trivial. □

If the rank of $\mathcal{F}$ is equal to two we obtain the following result, which is already known cf. [**Dr17**], but our arguments here are different.

COROLLARY 5.2. *Let X be a projective manifold such that K_X is pseudo-effective and let $\mathcal{F} \subset T_X$ be a holomorphic subsheaf of rank 2 such that the following hold.*

(i) *The first Chern class of $\mathcal{F}$ is zero, i.e. $c_1(\mathcal{F}) = 0$.*
(ii) *The sheaf $\mathcal{F}^\star$ is ω_X-strongly stable, meaning that for any finite étale cover $\pi : X' \to X$, $\pi^\star\mathcal{F}^\star$ is $\pi^\star\omega_X$-stable.*

Then the bundle $\mathcal{O}(1)$ on $\mathbb{P}(\mathcal{F})$ is not pseudo-effective, or $\mathcal{F}$ is Hermitian flat.

PROOF. If the bundle $\mathcal{O}(1)$ on $\mathbb{P}(\mathcal{F})$ is pseudo-effective, we consider Siu's decomposition (113). If there is no divisorial part, namely $\sum a_i[E_i] = 0$, then the Lelong number of T vanishes over the generic fiber on the projection p (since the fibers are of dimension 1). We have already explained that in this case it follows that $\mathcal{F}$ is hermitian flat.

If we assume moreover that $\mathcal{F}$ is not hermitian flat, we have to have a component E of the divisor-like part of the current Θ. Thanks to Theorem 5.1, the map

$$p_E : E \to X$$

is an étale cover. We prove that $p_E^\star \mathcal{F}^\star$ is not stable and this contradicts with our strongly stable condition. Indeed, we have the arrows

$$0 \to \mathcal{O}(-1)|_E \to p_E^\star(\mathcal{F}^\star) \tag{120}$$

and a quick computation shows that we have

$$\int_E c_1(\mathcal{O}(-1)) \wedge p_E^\star(\omega_H^{n-1}) = 0, \tag{121}$$

because $c_1(\mathcal{F}) = 0$. This is the end of the proof. □

REMARK 5.3. Actually Claim 4.1 (on page 7) concerns the positivity of the curvature form of $\mathcal{E}$ on the current ω_X^{n-1}. It would be nice to extend this as initiated by Berndtsson-Sibony in [**BeSib02**] i.e. replace ω_X^{n-1} with more general currents of bidimension $(1,1)$ on X. The following two results are pointing in this direction.

THEOREM 5.4. *Let $p : Y \to X$ be a holomorphic surjective and proper map, where X and Y are Kähler manifolds. Moreover, we assume that p is locally projective. Let L be a line bundle over Y with a possible singular metric h_L such that*

$$i\Theta_{h_L}(L) \geqslant -Cp^*\omega_X$$

for some constant C, and

$$\Theta_{h_L}(L) \wedge p^\star\omega_X^{n-1} \geqslant \varepsilon_0 p^\star\omega_X^n \tag{122}$$

where ω_X is a Hermitian metric on X, $n = \dim X$, and $\varepsilon_0 > 0$ is a positive real number. We assume that the space of fiberwise $L^{\frac{2}{m}}$ sections (with respect to h_L) of $p_\star(mK_{Y/X}+L)$ is non zero. Then there exists a metric h on the bundle $mK_{Y/X}+L$ such that we have

$$\Theta_h(mK_{Y/X} + L) \wedge p^\star\omega_X^{n-1} \geqslant \varepsilon_0 \cdot p^\star\omega_X^n \tag{123}$$

in the sense of currents on X.

PROOF. We first show that this is true for $m = 1$. In this case the metric h in our statement 5.4 is precisely the fiberwise Bergman metric, as we will now see.

Actually, this a consequence of the proof of Theorem 1.6. We follow the same approximation process for the map p restricted to one of the balls $B(x_i, \varepsilon)$, on which the distortion between ω_X and the flat metric in coordinates (t_i) is $1 + \mathcal{O}(\delta_\varepsilon)$. The inequality (62) applied for the sections $s_1 = s_2 = \cdots = s_r = s$ shows that the regularized version of (9) holds true. Indeed (62) holds true when restricted to the image of any section s, hence point-wise by [**Dem2**, III Criteria 1.6].

In conclusion, the inequality

$$\Theta_{h_B}(K_{Y/X} + L) \wedge p^\star\omega_X^{n-1} \geqslant \varepsilon_0 p^\star\omega_X^n \tag{124}$$

follows by taking the limit of several parameters involved.

If $m \geqslant 2$ we argue as follows. Let h_B be the relative m-Bergman kernel metric on $mK_{Y/X} + L$. By our hypothesis, this is not identically $+\infty$. We define the bundle $F := (m-1)K_{Y/X} + L$. Then the local weights

$$\varphi_1 := (1 - \frac{1}{m})\varphi_B + \frac{1}{m}\varphi_L \tag{125}$$

define a metric on F satisfying

$$\Theta_{h_1}(F) \geqslant 0, \qquad \Theta_{h_1}(F) \wedge p^\star \omega_X^{n-1} \geqslant \frac{\varepsilon_0}{m} p^\star \omega_X^n. \tag{126}$$

By applying the above $m = 1$ case to $(K_{Y/X} + F, h_1)$, we obtain a Bergman type metric $h_{B,1}$ on $K_{Y/X} + F$ such that

$$\Theta_{h_{B,1}}(K_{Y/X} + F) \geqslant 0, \qquad \Theta_{h_{B,1}}(K_{Y/X} + F) \wedge p^\star \omega_X^{n-1} \geqslant \frac{\varepsilon_0}{m} p^\star \omega_X^n. \tag{127}$$

We now iterate this process, namely we define the metric

$$\varphi_2 := (1 - \frac{1}{m})\varphi_{B,1} + \frac{1}{m}\varphi_L \tag{128}$$

on F, and let $h_{B,2}$ be the relative Bergman kernel metric corresponding to the data $(K_{Y/X} + F, h_2)$. Note that we have

$$\Theta_{h_2}(F) \wedge p^\star \omega_X^{n-1} \geqslant (\frac{1}{m} + \frac{1}{m}(1 - \frac{1}{m}))\varepsilon_0 p^\star \omega_X^n, \tag{129}$$

we get thus $\Theta_{h_{B,2}}(K_{Y/X} + F) \geqslant 0$ and

$$\Theta_{h_{B,2}}(K_{Y/X} + F) \wedge p^\star \omega_X^{n-1} \geqslant \Big(\frac{1}{m} + \frac{1}{m}(1 - \frac{1}{m})\Big)\varepsilon_0 p^\star \omega_X^n. \tag{130}$$

Then for every $k \in \mathbb{N}$, we obtain a metric $h_{B,k}$ on $K_{Y/X} + F$ such that $\Theta_{h_{B,k}}(K_{Y/X} + F) \geqslant 0$ together with

$$\Theta_{h_{B,k}}(K_{Y/X} + F) \wedge p^\star \omega_X^{n-1} \geqslant \frac{1}{m}(\sum_{i=0}^{k-1}(1 - \frac{1}{m})^i)\varepsilon_0 p^\star \omega_X^n. \tag{131}$$

By normalization (and eventually taking a subsequence), $h_{B,k}$ converges to a metric denoted by h on $K_{Y/X} + F = mK_{Y/X} + L$ which satisfies (9). □

We remark that the limit metric h is not exactly the m-Bergman kernel metric, but nevertheless it has the following important property. Let D be some divisor such that $p^\star(D) = \sum a_i E_i$. By the construction, for every $k \in \mathbb{N}$, we have

$$\Theta_{h_{B,k}}(mK_{Y/X} + L) \geqslant \sum_i m(a_i - 1)^+[E_i] \tag{132}$$

Then the limit metric h satisfies a similar inequality

$$\Theta_h(mK_{Y/X} + L) \geqslant \sum_i m(a_i - 1)^+[E_i] \tag{133}$$

as well as

$$(\Theta_h(mK_{Y/X} + L) - \sum_i m(a_i - 1)^+[E_i]) \wedge p^\star \omega_X^{n-1} \geqslant \varepsilon_0 \cdot p^\star \omega_X^n. \tag{134}$$

We also have the following version of Theorem 5.4.

COROLLARY 5.5. *Let $p : Y \to X$ be a holomorphic surjective and proper map, where X and Y are Kähler manifolds. Moreover, we assume that p is locally projective. Let L be a line bundle over Y with a possible singular metric h_L such that we have*

$$\Theta_{h_L}(L) \geqslant -Cp^\star\omega_X, \qquad \Theta_{h_L}(L) \wedge p^\star\omega_X^{n-1} \geqslant 0 \tag{135}$$

where ω_X is a Hermitian metric on X, $n = \dim X$, and $C > 0$ is a positive real number. We assume that the space of fiberwise $L^{\frac{2}{m}}$ sections (with respect to h_L) of $p_\star(mK_{Y/X}+L)$ is non zero. Then there exists a metric h on the bundle $mK_{Y/X}+L$ such that we have

$$\Theta_h(mK_{Y/X} + L) \wedge p^\star\omega_X^{n-1} \geqslant 0 \tag{136}$$

in the sense of currents on X.

The proof of this statement follows in the same way we have obtained Theorem 1.6 as consequence of Theorem 1.4, so we provide no further explanations about it.

We note that as a direct consequence of the arguments we use in the proof of 5.4, we obtain the existence of a sequence of m-Bergman metrics on $mK_{Y/X} + L$ such that

$$\Theta_{h_{B,k}}(K_{Y/X} + F) \wedge p^\star\omega_X^{n-1} \geqslant -\delta_k p^\star\omega_X^n \tag{137}$$

where the sequence (δ_k) converges to zero.

Thanks to Corollary 5.5, we can prove the following variant of [**CP17**, Thm 3.4] (which is using the fundamental contributions of Viehweg, Tsuji... [**V95, T10**]).

COROLLARY 5.6. *Let $p : Y \to X$ be a holomorphic surjective map between two Kähler manifolds. We assume that p is locally projective. Let $\Sigma \subset X$ be the singular locus of p and we assume that both Σ and $p^{-1}(\Sigma)$ are normal crossing. Let (L, h_L) be a hermitian line bundle over X such that*

$$\Theta_{h_L}(L) \geqslant -Cp^\star\omega_X, \qquad \Theta_{h_L}(L) \wedge p^\star\omega_X^{n-1} \geqslant 0 \tag{138}$$

where ω_X is a Kähler metric on X, $n = \dim X$, and $C > 0$ is a positive real number. If $\mathcal{I}(h_L|_{Y_x}) = \mathcal{O}_{Y_x}$ for a generic $x \in X$ and $\mathcal{E} := p_\star(K_{Y/X} + L)$ is non zero, then there exists a divisor F in Y satisfying $\operatorname{codim}_X p_\star(F) \geqslant 2$, such that

$$\int_Y c_1\left(K_{Y/X} + L + F - \delta_0 p^\star(\det\mathcal{E})\right) \wedge p^\star\omega_X^{n-1} \geqslant 0. \tag{139}$$

for some positive δ_0.

PROOF. The proof is a linear combination of [**CP17**, Thm 3.4] with the arguments above. We sketch the proof here for the convenience of readers.

By hypothesis, $p^\star\Sigma$ is normal crossing, and it can be written as

$$p^\star\Sigma = \sum W_i + \sum a_i V_i,$$

where $\sum W_i + \sum V_i$ is snc and $a_i \geqslant 2$. Let r be the rank of $\mathcal{E}$. Let Y^r be the fiberwise product of p, and let $\mathrm{pr}_i : Y^r \to Y$ be the i-directional projection. Let $\varphi : Y^{(r)} \to Y^r$ be a desingularisation, and $p^{(r)} : Y^{(r)} \to X$ be the natural morphism. We set $L^{(r)} := \varphi^\star(\sum_i \mathrm{pr}_i^\star L)$.

We have the canonical morphism $\det \mathcal{E} \to \otimes^r \mathcal{E}$ over the locally free locus of p. This induces a section

$$s \in H^0(Y^{(r)}, K_{Y^{(r)}/Y} + L^{(r)} - (p^{(r)})^\star \det \mathcal{E} + E_1 + E_2),$$

where E_1 and E_2 are effective divisors such that $\mathrm{codim}_Y p^{(r)}_\star(E_2) \geqslant 2$ and

$$E_1 \leqslant C \sum_{i,k} (\varphi \circ \mathrm{pr}_i)^\star V_k$$

for some constant C. Let $m \in \mathbb{N}$ sufficiently large. We consider the line bundle

$$F := mL^{(r)} + (K_{Y^{(r)}/X} + L^{(r)} - (p^{(r)})^\star \det \mathcal{E} + E_1 + E_2)$$

with the metric $h_F := e^{-\log |s|} h_{L^{(r)}}^{\otimes m}$. Let h_k be the m-relative Bergman kernel metric on $mK_{Y^{(r)}/F} + F$ constructed in (137) with respect to h_F. They satisfy the inequality

$$\Theta_{h_k}(mK_{Y^{(r)}/X} + F) \wedge (p^{(r)})^\star \omega_X^{n-1} \geqslant -\delta_k (p^{(r)})^\star \omega_X^n \tag{140}$$

on $Y^{(r)}$.

Let $\pi : Y \to Y^{(r)}$ be the fiberwise diagonal embedding. Note that if $\mathcal{I}(h_L) = \mathcal{O}_X$ and if m large enough with respect to $\mathrm{Div}(s)$, we have

$$\mathcal{I}(h_F^{\frac{1}{m}}|_{Y_x^{(r)}}) = \mathcal{O}_{Y_x^{(r)}}$$

for a generic $x \in X$. As a consequence, $\pi^\star h_k$ *is not* identically $+\infty$. Therefore $\pi^\star \Theta_{h_k}(mK_{Y^{(r)}/X} + F)$ is well defined (thus quasi-psh) on Y. We have

$$\pi^\star(\Theta_{h_k}) \wedge p^\star \omega_X^{n-1} = \pi^\star \left(\Theta_{h_k} \wedge (p^r)^\star \omega_X^{n-1}\right) \geqslant 0 \qquad \text{on } Y,$$

as a consequence of (140). Note that $\pi^\star(\Theta_{h_k})$ belongs to the class of

$$\pi^\star(c_1(mK_{Y^{(r)}/X} + F)) = c_1((mr + r)(K_{Y/X} + L) - \pi^\star \det \mathcal{E} + E_3 + F),$$

where E_3 is supported $\sum_i V_i$ and $\mathrm{codim}_X p_\star(F) \geqslant 2$. Therefore we have

$$\int_Y c_1((mr + r)(K_{Y/X} + L) - \pi^\star \det \mathcal{E} + E_3 + F) \wedge p^\star \omega_X^{n-1} \geqslant -\delta_k p^\star \omega_X^n.$$

Finally, thanks to (134), we know that

$$\sum_i \int_Y c_1(V_i) \wedge p^\star \omega_X^{n-1} \leqslant \int_Y c_1(K_{Y/X} + L) \wedge p^\star \omega_X^{n-1}$$

Therefore (139) is proved as $k \to \infty$. □

Acknowledgment

We would like to thank B. Berndtsson, A. Höring and M. Toma for useful discussions and suggestions about the topics in this article. Many thanks to the anonymous referees, whose suggestions and constructive criticism helped us to improve our work. Bo Berndtsson's results and ideas represent a model, an inexhaustible source of inspiration for us, and it is a great honour to contribute to the current volume.

References

[BS94] Shigetoshi Bando and Yum-Tong Siu, *Stable sheaves and Einstein-Hermitian metrics*, Geometry and analysis on complex manifolds, World Sci. Publ., River Edge, NJ, 1994, pp. 39–50. MR1463962

[Ber06] Bo Berndtsson, *Subharmonicity properties of the Bergman kernel and some other functions associated to pseudoconvex domains* (English, with English and French summaries), Ann. Inst. Fourier (Grenoble) **56** (2006), no. 6, 1633–1662. MR2282671

[Ber09] Bo Berndtsson, *Curvature of vector bundles associated to holomorphic fibrations*, Ann. of Math. (2) **169** (2009), no. 2, 531–560, DOI 10.4007/annals.2009.169.531. MR2480611

[BP08] Bo Berndtsson and Mihai Păun, *Bergman kernels and the pseudoeffectivity of relative canonical bundles*, Duke Math. J. **145** (2008), no. 2, 341–378, DOI 10.1215/00127094-2008-054. MR2449950

[BeSib02] Bo Berndtsson and Nessim Sibony, *The $\overline{\partial}$-equation on a positive current*, Invent. Math. **147** (2002), no. 2, 371–428, DOI 10.1007/s002220100178. MR1881924

[CP19] Frédéric Campana and Mihai Păun, *Foliations with positive slopes and birational stability of orbifold cotangent bundles*, Publ. Math. Inst. Hautes Études Sci. **129** (2019), 1–49, DOI 10.1007/s10240-019-00105-w. MR3949026

[CH19] Junyan Cao and Andreas Höring, *A decomposition theorem for projective manifolds with nef anticanonical bundle*, J. Algebraic Geom. **28** (2019), no. 3, 567–597, DOI 10.1090/jag/715. MR3959071

[CCM21] Frédéric Campana, Junyan Cao, and Shin-ichi Matsumura, *Projective klt pairs with nef anti-canonical divisor*, Algebr. Geom. **8** (2021), no. 4, 430–464, DOI 10.14231/ag-2021-013. MR4371536

[CP17] Junyan Cao and Mihai Păun, *Kodaira dimension of algebraic fiber spaces over abelian varieties*, Invent. Math. **207** (2017), no. 1, 345–387, DOI 10.1007/s00222-016-0672-6. MR3592759

[Dem1] J.-P. Demailly, *Complex analytic and differential geometry* `https://www-fourier.ujf-grenoble.fr/~demailly/documents.html`, 2012.

[Dem2] Jean-Pierre Demailly, *Analytic methods in algebraic geometry*, Surveys of Modern Mathematics, vol. 1, International Press, Somerville, MA; Higher Education Press, Beijing, 2012. MR2978333

[DPS94] Jean-Pierre Demailly, Thomas Peternell, and Michael Schneider, *Compact complex manifolds with numerically effective tangent bundles*, J. Algebraic Geom. **3** (1994), no. 2, 295–345. MR1257325

[Dr17] Stéphane Druel, *Some remarks on regular foliations with numerically trivial canonical class*, Épijournal Géom. Algébrique **1** (2017), Art. 4, 20, DOI 10.46298/epiga.2017.volume1.2057. MR3743107

[Dr18] Stéphane Druel, *A decomposition theorem for singular spaces with trivial canonical class of dimension at most five*, Invent. Math. **211** (2018), no. 1, 245–296, DOI 10.1007/s00222-017-0748-y. MR3742759

[GKP16] Daniel Greb, Stefan Kebekus, and Thomas Peternell, *Movable curves and semistable sheaves*, Int. Math. Res. Not. IMRN **2** (2016), 536–570, DOI 10.1093/imrn/rnv126. MR3493425

[LPT18] Frank Loray, Jorge Vitório Pereira, and Frédéric Touzet, *Singular foliations with trivial canonical class*, Invent. Math. **213** (2018), no. 3, 1327–1380, DOI 10.1007/s00222-018-0806-0. MR3842065

[LG86] Lawrence Gruman and Pierre Lelong, *Entire functions of several complex variables*, Grundlehren der mathematischen Wissenschaften [Fundamental Principles of Mathematical Sciences], vol. 282, Springer-Verlag, Berlin, 1986, DOI 10.1007/978-3-642-70344-7. MR837659

[HPS18] Christopher Hacon, Mihnea Popa, and Christian Schnell, *Algebraic fiber spaces over abelian varieties: around a recent theorem by Cao and Păun*, Local and global methods in algebraic geometry, Contemp. Math., vol. 712, Amer. Math. Soc., Providence, RI, 2018, pp. 143–195, DOI 10.1090/conm/712/14346. MR3832403

[HP19] Andreas Höring and Thomas Peternell, *Algebraic integrability of foliations with numerically trivial canonical bundle*, Invent. Math. **216** (2019), no. 2, 395–419, DOI 10.1007/s00222-018-00853-2. MR3953506

[Lan04] Adrian Langer, *Semistable sheaves in positive characteristic*, Ann. of Math. (2) **159** (2004), no. 1, 251–276, DOI 10.4007/annals.2004.159.251. MR2051393

[Pa15] Mihai Păun, *Singular Hermitian metrics and positivity of direct images of pluricanonical bundles*, Algebraic geometry: Salt Lake City 2015, Proc. Sympos. Pure Math., vol. 97, Amer. Math. Soc., Providence, RI, 2018, pp. 519–553, DOI 10.1090/pspum/097.1/18. MR3821161

[PT18] Mihai Păun and Shigeharu Takayama, *Positivity of twisted relative pluricanonical bundles and their direct images*, J. Algebraic Geom. **27** (2018), no. 2, 211–272, DOI 10.1090/jag/702. MR3764276

[PeTou13] Jorge Vitório Pereira and Frédéric Touzet, *Foliations with vanishing Chern classes*, Bull. Braz. Math. Soc. (N.S.) **44** (2013), no. 4, 731–754, DOI 10.1007/s00574-013-0032-8. MR3167130

[Rau15] Hossein Raufi, *Singular hermitian metrics on holomorphic vector bundles*, Ark. Mat. **53** (2015), no. 2, 359–382, DOI 10.1007/s11512-015-0212-4. MR3391176

[T10] Tsuji, Hajime *Global generation of the direct images of relative pluricanonical systems*, `arXiv:1012.0884`, 2010.

[T19] Matei Toma, *Bounded sets of sheaves on Kähler manifolds*, J. Reine Angew. Math. **710** (2016), 77–93, DOI 10.1515/crelle-2013-0093. MR3437560

[V95] Eckart Viehweg, *Quasi-projective moduli for polarized manifolds*, Ergebnisse der Mathematik und ihrer Grenzgebiete (3) [Results in Mathematics and Related Areas (3)], vol. 30, Springer-Verlag, Berlin, 1995, DOI 10.1007/978-3-642-79745-3. MR1368632

[W22] Xiaojun Wu, *Strongly pseudo-effective and numerically flat reflexive sheaves*, J. Geom. Anal. **32** (2022), no. 4, Paper No. 124, 61, DOI 10.1007/s12220-021-00865-0. MR4375833

Université de Lorraine, Institut de Mathématiques Élie Cartan, B.P. 70239, 54506 Vandoeuvre-les-Nancy Cedex, France

Email address: `frederic.campana@univ-lorraine.fr`

Université Côte d'Azur, Laboratoire de Mathématiques J.A. Dieudonné, Parc Valrose 06108 Nice cedex 02, France

Email address: `junyan.cao@unice.fr`

Universität Bayreuth, Mathematisches Institut, Lehrstuhl Mathematik VIII, Universitätsstrasse 30, D-95447, Bayreuth, Germany

Email address: `mihai.paun@uni-bayreuth.de`

Contemporary Mathematics
Volume **810**, 2025
https://doi.org/10.1090/conm/810/16208

An application of adjoint ideal sheaves to injectivity and extension theorems

Tsz On Mario Chan and Young-Jun Choi

In honor of Professor Bo Benrdtsson on the occasion of his 70th birthday.

ABSTRACT. This note reviews the authors' approach to Fujino's conjecture, i.e. the injectivity theorem for lc pairs on compact Kähler manifolds, via the use of adjoint ideal sheaves coupled with the associated residue computations in their previous work. Using only the techniques and results obtained from that study and under a slightly stronger positivity assumption, a "qualitative" extension result is obtained. Such extension result guarantees the existence of a global holomorphic extension F of any holomorphic section f on some σ-lc centres of the given lc pair. The extension F can be shown to take values in the corresponding adjoint ideal sheaf, even though it comes without any L^2 estimate of F in terms of f. Moreover, the proof invokes and implies no vanishing theorem in general.

The version of the analytic adjoint ideal sheaf introduced in [**4**] is regarded by the authors as an algebraic manifestation of the properties exhibited when the residue function developed in [**6**] and [**3**], together with the corresponding techniques of residue computation, is applied to (germs of) holomorphic functions in multiplier ideal sheaves. In [**7**] (as well as the upcoming joint work with Shin-ichi Matsumura), it is shown that the adjoint ideal sheaves and the residue computation fit in so well in the setup for solving Fujino's conjecture, an injectivity theorem for log-canonical (lc) pairs on compact Kähler manifolds. The authors believe that the machinery can be useful for some other problems involving lc pairs.

In this note, some properties of adjoint ideal sheaves and some results of residue computations used in [**7**] are reviewed. Using only these results, a new "qualitative" extension result is obtained in Section 3, which is essentially the existence of holomorphic extensions suggested in [**5**, Conj. 2.2.3], except that the result here comes with no estimates.

0.1. Notation and conventions. Unless stated otherwise, the following notations (which follow most of the conventions in [**7**] and [**4**]) are used throughout

2020 *Mathematics Subject Classification.* Primary 32J25, 32Q15; Secondary 14B05.

Key words and phrases. L^2 injectivity, vanishing theorem, extension theorem, adjoint ideal sheaf, multiplier ideal sheaf.

This work was supported by the National Research Foundation of Korea (NRF) grant funded by the Korea government (MSIT) (Nos. 2023R1A2C1007227 and 2021R1A4A1032418).

this paper.

- X is a compact Kähler manifold of dimension n.
- $D = \sum_{i \in I_D} D_i$ is a simple normal crossing (snc) reduced divisor on X, in which each D_i is an irreducible component. Then, around every point in D, there is an open set $V \Subset X$ inside some coordinate chart with a holomorphic coordinate system $(z_1, \ldots, z_n)$ such that

$$D \cap V = \{z_1 z_2 \cdots z_{\sigma_V} = 0\} \quad \text{for some integer } \sigma_V \leq n\,. \tag{eq 0.1}$$

- F is a holomorphic line bundle equipped with a (possibly singular) potential φ_F (of the curvature of a metric $e^{-\varphi_F}$) which is *quasi-plurisubharmonic (quasi-psh)* and has *neat analytic singularities* such that $D \cup P_F$, where $P_F := \varphi_F^{-1}(-\infty)$, is an snc divisor and P_F contains no irreducible components of D (hence no lc centres of (X, D)). In this case, around any point in P_F, there is an open set $V \Subset X$ on which (eq 0.1) holds for some $\sigma_V \in \{0, \ldots, n\}$ and

$$\varphi_F|_V \equiv \sum_{k=\sigma_V+1}^{n} b_k \log |z_k|^2 \mod \mathscr{C}_X^\infty(V) \quad \text{for some constants } b_k \geq 0\,. \tag{eq 0.2}$$

- $X^\circ := X \setminus P_F$ and ω is the *complete* Kähler form on X° of form given in [**7**, §2.2 item (4)].
- $\mathscr{I}(\varphi) := \mathscr{I}_X(\varphi)$ is the multiplier ideal sheaf of a potential φ, which is given at each $x \in X$ by

$$\mathscr{I}(\varphi)_x := \left\{ f \in \mathscr{O}_{X,x} \;\middle|\; \begin{array}{l} f \text{ is defined on a coord. neighbourhood } V_x \ni x \\ \text{and } |f|^2 e^{-\varphi} \in L^1(V_x) \end{array} \right\}.$$

- s_i is a canonical section of the irreducible component D_i for each $i \in I_D$, which gives a canonical section $\mathsf{s}_D := \prod_{i \in I_D} \mathsf{s}_i$ of D.
- $\psi_D := \phi_D - \varphi_D^{\mathrm{sm}} \leq -1$ is a global function on X bounded above by -1, in which $\phi_D := \log |\mathsf{s}_D|^2$ is the potential on D induced from the canonical section s_D and φ_D^{sm} is a fixed smooth potential on D.
- For each $\sigma \in \{0, 1, \ldots, n\}$, $\mathrm{lc}_X^\sigma(D) := \bigcup_{p \in I_D^\sigma} \mathtt{D}_p^\sigma$ is the union of σ-*lc centres of* (X, D), i.e. the σ-codimensional irreducible components of any intersections of irreducible components of D (under the assumption (X, D) being log-smooth and lc), indexed by I_D^σ. Here $\mathrm{lc}_X^0(D) := X$ and I_D^0 being a singleton are set for convenience.
- $\mathrm{Diff}_p^* D$ is the effective divisor on $\mathtt{D}_p^\sigma$ defined by the adjunction formula

$$K_{\mathtt{D}_p^\sigma} \otimes \mathrm{Diff}_p^* D = (K_X \otimes D)|_{\mathtt{D}_p^\sigma}$$

  such that the restriction of $\mathsf{s}_{(p)} := \prod_{i \in I_D^\sigma \colon D_i \not\supset \mathtt{D}_p^\sigma} \mathsf{s}_i$ to $\mathtt{D}_p^\sigma$ is a canonical section of $\mathrm{Diff}_p^* D$.
- $\psi_{(p)} := \phi_{(p)} - \varphi_{(p)}^{\mathrm{sm}} \leq -1$ is a global function on X bounded above by -1 defined analogously to ψ_D, in which $\phi_{(p)} := \log \left|\mathsf{s}_{(p)}\right|^2$ is the potential induced from $\mathsf{s}_{(p)}$ and $\varphi_{(p)}^{\mathrm{sm}}$ is a fixed smooth potential on the corresponding line bundle.

An open set $V \subset X$ is said to be *admissible with respect to* (φ_F, ψ_D) if V is biholomorphic to a polydisc centred at the origin under a holomorphic coordinate system $(z_1, z_2, \cdots, z_n)$ such that the residue computation in [**7**, Thm. 2.6.1] is valid. In particular, both (eq 0.1) and (eq 0.2) are satisfied on an admissible open set V.

When an admissible set V is considered, an index $p \in I_D^\sigma$ such that $\mathtt{D}_p^\sigma \cap V \neq \emptyset$ is interpreted as a permutation representing a choice of σ elements from the set $\{1, 2, \ldots, \sigma_V\}$ such that

$$\mathtt{D}_p^\sigma \cap V = \left\{z_{p(1)} = z_{p(2)} = \cdots = z_{p(\sigma)} = 0\right\} \quad \text{and} \quad \mathtt{s}_{(p)}\big|_V = z_{p(\sigma+1)} \cdots z_{p(\sigma_V)}$$

(cf. the definition of the set $\mathfrak{C}_\sigma^{\sigma_V}$ in [**4**, §3.1]). Note also that the defining ideal sheaf $\mathcal{I}_{\mathrm{lc}_X^{\sigma+1}(D)}$ of $\mathrm{lc}_X^{\sigma+1}(D)$ is locally generated by $\mathtt{s}_{(p)}\big|_V = z_{p(\sigma+1)} \cdots z_{p(\sigma_V)}$ for all $p \in I_D^\sigma$.

1. Properties of adjoint ideal sheaves

In [**4**], the first author introduces the following version of analytic adjoint ideal sheaves.

DEFINITION 1.1 ([**4**, Def. 1.2.1]). Given φ_F and ψ_D described as in Section 0.1, the adjoint ideal sheaf $\mathscr{J}_\sigma(\varphi_F; \psi_D) := \mathscr{J}_{X,\sigma}(\varphi_F; \psi_D)$ of index σ is given at each $x \in X$ by

$$\mathscr{J}(\varphi_F; \psi_D)_x = \left\{ f \in \mathscr{O}_{X,x} \;\middle|\; \exists \text{ open set } V_x \ni x\,,\ \forall\, \varepsilon > 0, \right.$$
$$\left. \frac{|f|^2\, e^{-\varphi_F - \psi_D}}{|\psi_D|^\sigma \left(\log|e\psi_D|\right)^{1+\varepsilon}} \in L^1(V_x) \right\}.$$

According to [**4**, Thm. 1.2.3], under the assumption that φ_F and $\varphi_F + \psi_D$ have only neat analytic singularities with snc, one has

$$\mathscr{J}_\sigma(\varphi_F; \psi_D) = \mathscr{I}(\varphi_F) \cdot \mathcal{I}_{\mathrm{lc}_X^{\sigma+1}(D)} \qquad \text{(eq 1.1)}$$

for all integers $\sigma \geq 0$, which fit into the chain of natural inclusions

$$\mathscr{I}(\varphi_F + \phi_D) = \mathscr{J}_0(\varphi_F; \psi_D) \subset \mathscr{J}_1(\varphi_F; \psi_D) \subset \cdots \subset \mathscr{J}_{\sigma_{\mathrm{mlc}}}(\varphi_F; \psi_D) = \mathscr{I}(\varphi_F)\,,$$

where σ_{mlc} is the codimension of the minimal lc centres (mlc) of (X, D). Note that $\mathscr{J}_\sigma(\varphi_F; \psi_D) = \mathscr{J}_{\sigma_{\mathrm{mlc}}}(\varphi_F; \psi_D)$ for all integers $\sigma \geq \sigma_{\mathrm{mlc}}$.

Moreover, since $\varphi_F^{-1}(-\infty)$ contains no lc centres of (X, D), the analytic adjoint ideal sheaves fit into the residue short exact sequence
(eq 1.2)

$$0 \longrightarrow \mathscr{J}_{\sigma-1}(\varphi_F; \psi_D) \longrightarrow \mathscr{J}_\sigma(\varphi_F; \psi_D) \xrightarrow{\mathrm{Res}^\sigma} \mathscr{R}_\sigma(\varphi_F; \psi_D) \longrightarrow 0\ ,$$

in which the *residue sheaf* $\mathscr{R}_\sigma(\varphi_F; \psi_D)$ *of index* σ is given in this case as

$$\mathscr{R}_\sigma(\varphi_F; \psi_D) = \bigoplus_{p \in I_D^\sigma} \left(\mathrm{Diff}_p^* D\right)^{-1} \otimes \mathscr{I}_{\mathtt{D}_p^\sigma}(\varphi_F)\,.$$

Note that the adjunction formula implies that

$$K_X \otimes D \otimes F \otimes \mathscr{R}_\sigma(\varphi_F;\psi_D) = \bigoplus_{p\in I_D^\sigma} K_{\mathtt{D}_p^\sigma} \otimes F|_{\mathtt{D}_p^\sigma} \otimes \mathscr{I}_{\mathtt{D}_p^\sigma}(\varphi_F) \tag{eq 1.3}$$

(see [**4**, §4.2] for the precise construction of $\mathscr{R}_\sigma(\varphi_F;\psi_D)$). After tensoring with $K_X \otimes D \otimes F$, the *residue morphism* $\operatorname{Res}^\sigma$ can be described in terms of the Poincaré residue map $\mathcal{R}_{\mathtt{D}_p^\sigma}$ given in [**17**, §4.18] as follows. The Poincaré residue map $\mathcal{R}_{\mathtt{D}_p^\sigma}$ from X to each $\mathtt{D}_p^\sigma$ is uniquely determined after an orientation on the conormal bundle of $\mathtt{D}_p^\sigma$ in X is fixed. For an admissible open set $V \subset X$, one has $\operatorname{lc}_X^\sigma(D) \cap V = \bigcup_{p\in I_D^\sigma} \mathtt{D}_{V,p}^\sigma$ (where $\mathtt{D}_{V,p}^\sigma := \mathtt{D}_p^\sigma \cap V$, which is connected by the definition of the admissible open set, and possibly empty) and $\mathtt{D}_{V,p}^\sigma = \{z_{p(1)} = z_{p(2)} = \cdots = z_{p(\sigma)} = 0\}$ when non-empty. Under such coordinate system, a section f of $K_X \otimes D \otimes F \otimes \mathscr{J}_\sigma(\varphi_F;\psi_D)$ on $V \subset X$ can be written as

$$f = \sum_{p\in I_D^\sigma \,:\, \mathtt{D}_{V,p}^\sigma \neq \emptyset} dz_{p(1)} \wedge\cdots\wedge dz_{p(\sigma)} \wedge g_p \mathsf{s}_{(p)} = \sum_{p\in I_D^\sigma \,:\, \mathtt{D}_{V,p}^\sigma \neq \emptyset} \frac{dz_{p(1)}}{z_{p(1)}} \wedge\cdots\wedge \frac{dz_{p(\sigma)}}{z_{p(\sigma)}} \wedge g_p \mathsf{s}_D \quad \text{on } V ,$$

where g_p is a local holomorphic $(n-\sigma,0)$-form on V. It therefore follows that

$$\mathcal{R}_{\mathtt{D}_p^\sigma}\left(\frac{f}{\mathsf{s}_D}\right) = g_p|_{\mathtt{D}_p^\sigma} \in K_{\mathtt{D}_p^\sigma} \otimes F|_{\mathtt{D}_p^\sigma} \otimes \mathscr{I}_{\mathtt{D}_p^\sigma}(\varphi_F) \quad \text{on } \mathtt{D}_{V,p}^\sigma$$

under the assumption that the orientation on the conormal bundle of $\mathtt{D}_p^\sigma$ in X on V is given by $(dz_{p(1)}, dz_{p(2)}, \dots, dz_{p(\sigma)})$. The residue morphism $\operatorname{Res}^\sigma$ is then (according to [**4**, §4.2]) given by

$$\begin{array}{ccc} K_X \otimes D \otimes F \otimes \mathscr{J}_\sigma(\varphi_F;\psi_D) & \xrightarrow{\operatorname{Res}^\sigma} & K_X \otimes D \otimes F \otimes \mathscr{R}_\sigma(\varphi_F;\psi_D) \\ \text{\rotatebox{90}{$\in$}} & & \text{\rotatebox{90}{$\in$}} \\ f & \longmapsto & \left(g_p|_{\mathtt{D}_p^\sigma}\right)_{p\in I_D^\sigma} \end{array} .$$

In what follows, write

$$\mathscr{J}_\sigma := \mathscr{J}_\sigma(\varphi_F;\psi_D) \quad \text{and} \quad \mathscr{R}_\sigma := \mathscr{R}_\sigma(\varphi_F;\psi_D)$$

for convenience when there is no risk of confusion.

Here is an explicit example which may help to illustrate the notion described above.

EXAMPLE 1.2. Let $X = \Delta^n$ and $D = D_1 + D_2$, where D_1 and D_2 are hyperplanes defined by $D_1 := \{z_1 = 0\}$ and $D_2 := \{z_2 = 0\}$. Suppose that φ_F is smooth, which implies that $\mathscr{I}(\varphi_F) = \mathscr{O}_X$. Then, it is obvious that $\operatorname{lc}_X^1(D) = D_1 \cup D_2$,

$\operatorname{lc}^2_X(D) = D_1 \cap D_2$, and $\sigma_{\mathrm{mlc}} = 2$. By (eq 1.1), the adjoint ideal sheaves are given by

$$\mathscr{J}_0 = \mathcal{I}_{\operatorname{lc}^1_X(D)} = \mathcal{I}_{D_1 \cup D_2} = \langle z_1 z_2 \rangle \ , \quad \mathscr{J}_1 = \mathcal{I}_{\operatorname{lc}^2_X(D)} = \mathcal{I}_{D_1 \cap D_2} = \langle z_1, z_2 \rangle \quad \text{and} \quad \mathscr{J}_2 = \mathscr{O}_X \ ,$$

where $\mathscr{J}_\sigma = \mathscr{J}_\sigma(\varphi_F; \psi_D)$ for $\sigma = 0, 1$. Moreover, the residue sheaves are given by

$$\mathscr{R}_1 = \left(\operatorname{Diff}^*_{D_1} D\right)^{-1} \oplus \left(\operatorname{Diff}^*_{D_2} D\right)^{-1} = D_2^{-1}\big|_{D_1} \oplus D_1^{-1}\big|_{D_2} \quad \text{and} \quad \mathscr{R}_2 = \mathscr{O}_{D_1 \cap D_2}$$

(where D_i^{-1} is the dual of D_i as a line bundle). For any local section $f \in K_X \otimes D \otimes F \otimes \mathscr{J}_1$, it can be written as

$$f = (z_2 g_1 + z_1 g_2)\, dz_1 \wedge \cdots \wedge dz_n \ .$$

for some $g_1, g_2 \in \mathscr{O}_X$. The residue morphism Res^1 is then given by

$$\operatorname{Res}^1(f) = \left((g_1\, dz_2 \wedge \cdots \wedge dz_n)|_{D_1} \ , \ -(g_2\, dz_1 \wedge dz_3 \wedge \cdots dz_n)|_{D_2} \right) \ .$$

Most properties of the adjoint ideal sheaves are derived from the residue computations developed in [**6**, Prop. 2.2.1] and [**7**, Thm. 2.6.1] (see also [**3**, Prop. 2.2.1]). The computations are applicable even to smooth (n,q)-forms. To state the result, let $\boldsymbol{\Omega}_X$ be the holomorphic cotangent bundle of X and $\overline{\boldsymbol{\Omega}}_X$ be its complex conjugate. Write $\overline{\boldsymbol{\Omega}}^q_X := \bigwedge^q \overline{\boldsymbol{\Omega}}_X$. Extend each of the Poincaré residue maps $\mathcal{R}_{\mathrm{D}^\sigma_p}$ (hence $\operatorname{Res}^\sigma$) so that it sends sections of $K_X \otimes \overline{\boldsymbol{\Omega}}^q_X$ (which are identified with (n,q)-forms in $\mathscr{A}^{n,q}_X$) to $K_{\mathrm{D}^\sigma_p} \otimes \overline{\boldsymbol{\Omega}}^q_X\Big|_{\mathrm{D}^\sigma_p}$. Let $D + P_F = \sum_{k \in I_{D+F}} E_k$ be the decomposition of the polar set $D + P_F$ into irreducible components and let e_k be a canonical section of E_k for each $k \in I_{D+F}$. Set

$$\begin{aligned} \mathscr{C}^\infty_{X*} &:= \left(\mathscr{C}^\infty_X \left[\frac{1}{|\mathsf{e}_k|} : k \in I_{D+F} \right] \right)_{\mathrm{b}} \qquad (\mathsf{e}_k \text{ treated as a local defining function of } E_k) \\ &:= \left\{ \text{locally bounded elements in the } \mathscr{C}^\infty_X\text{-algebra generated by } \frac{1}{|\mathsf{e}_k|} \text{ for } k \in I_{D+F} \right\} . \end{aligned}$$ [1]

Set also

$$\mathscr{A}^{p,q}_{X*} := \{(p,q)\text{-forms with coefficients in } \mathscr{C}^\infty_{X*}\}.$$

The following proposition recalls the main result of the residue computations.

PROPOSITION 1.3 ([**7**, Thm. 2.6.1]). *Let $f \in \mathscr{A}^{n,q}_{X*}(X; D \otimes F \otimes \mathscr{J}_\sigma)$, one has*

$$\lim_{\varepsilon \to 0^+} \varepsilon \int_X \frac{|f|^2_\omega \, e^{-\phi_D - \varphi_F}}{|\psi_D|^{\sigma+\varepsilon}} = \sum_{p \in I^\sigma_D} \frac{\pi^\sigma}{(\sigma-1)!} \int_{\mathrm{D}^\sigma_p} \left| \mathcal{R}_{\mathrm{D}^\sigma_p}\left(\frac{f}{\mathsf{s}_D} \right) \right|^2_\omega e^{-\varphi_F} < +\infty \ .$$

---

[1] On an admissible open set V under the holomorphic coordinate system $(z_1, \dots, z_n)$ such that $D \cap V = \{z_1 z_2 \cdots z_{\sigma_V} = 0\}$ and $P_F \cap V = \left\{ z_{\sigma_V+1} \cdots z_{\tau_V} = 0 \right\}$, one has

$$\mathscr{C}^\infty_{X*}(V) = \mathscr{C}^\infty_X(V) \left[e^{\pm\sqrt{-1}\theta_1}, \dots, e^{\pm\sqrt{-1}\theta_{\tau_V}} \right]$$

where (r_k, θ_k) is the polar coordinate system of the z_k-plane for $k = 1, \dots, \tau_V$ in V, which is (almost) the same as the ad hoc definition of $\mathscr{C}^\infty_{X*}(V)$ given in [**7**, §2.6] (in which $e^{\pm\sqrt{-1}\theta_k}$ for $k \geq \tau_V + 1$ are also included in the set of generators of the algebra). The definition given here is independent of coordinates and its sheaf structure can be seen easily.

2. Injectivity theorem for lc pairs

Kollár's injectivity theorem ([**16**, Thm. 2.2]) can be considered as a generalization of the Kodaira vanishing theorem (see, for example, [**11**, Cor. 5.2] or [**18**, Remark 4.3.8]). It was first proved in [**16**] via algebraic method and later its generalization to the setting on compact Kähler manifolds was proved via harmonic theory by Enoki ([**10**]). The theorem is further generalized to lc pairs in the algebraic setting using the theory of (mixed) Hodge structure (see, for example, [**11**, §5], [**12**, §5] and [**1**]). Fujino formulated the following conjecture, expecting that it could be proved following the line of arguments in Enoki's proof.

CONJECTURE 2.1 (Fujino's conjecture, [**14**, Conj. 2.21]). *Given F as described in Section 0.1, which is semi-positive in particular, suppose that there exists a holomorphic section s of $F^{\otimes m}$ on X for some positive integer m such that s does not vanish identically on any lc centres of (X, D). Then, the multiplication map induced by $\otimes s$,*

$$H^q(X, K_X \otimes D \otimes F) \xrightarrow{\otimes s} H^q\Big(X, K_X \otimes D \otimes F^{\otimes(m+1)}\Big) ,$$

is injective for every $q \geq 0$.

In the spirit of Enoki's arguments, Fujino ([**13**]), Matsumura ([**19**]) and Gongyo–Matsumura ([**15**]) prove the corresponding injectivity theorem for Kawamata log-terminal (klt) pairs while Matsumura also obtains in [**20**] the result for purely log-terminal (plt) pairs (see these references for more information on the development of the injectivity theorem).

Inspired by the idea of Matsumura for the plt case in [**20**], the authors introduce the use of adjoint ideal sheaves to the study of Fujino's conjecture in [**7**], which leads to the scheme of a proof by induction on the codimension σ of the lc centres of (X, D) (see (eq 2.1) and Theorem 2.2). Moreover, a slightly more general statement, namely, Fujino's conjecture with multiplier ideal sheaves, will follow naturally under this framework. It will be shown in our upcoming article with Shin-ichi Matsumura that a proof of Fujino's conjecture in the spirit of Enoki's arguments can be obtained by following this scheme.

A solution to Fujino's conjecture is announced recently by Cao and Păun ([**2**]), in which a different approach is used. Their approach, instead of working within the framework of L^2 theory, makes use of a version of the Hodge decomposition on currents induced via duality from the Hodge decomposition of L^2 forms with respect to a Kähler form with conic singularities (along D).

The rest of this section is devoted to reviewing the statements in [**7**] which bring adjoint ideal sheaves and certain residue computations into play in the study of Fujino's conjecture.

For the sake of convenience, set

$$\mathbb{H}^q(\mathscr{F}) := H^q(X, K_X \otimes D \otimes F \otimes \mathscr{F})$$

for any sheaf $\mathscr{F}$ on X and any integer $q = 0, \dots, n$. Let $s \in H^0(X, M)$ be a non-trivial holomorphic section of a line bundle M (where $M := F^{\otimes m}$ in Fujino's conjecture) and φ_M a potential on M with the same regularity property as φ_F described in Section 0.1 such that $\sup_X |s|^2_{\varphi_M} < \infty$. From the residue short exact sequence (eq 1.2) and the multiplication map

$$K_X \otimes D \otimes F \otimes \mathscr{J}_\sigma(\varphi_F; \psi_D) \xrightarrow{\otimes s} K_X \otimes D \otimes F \otimes M \otimes \mathscr{J}_\sigma(\varphi_F + \varphi_M; \psi_D) ,$$

one obtains the following commutative diagram of cohomology groups:
(eq 2.1)

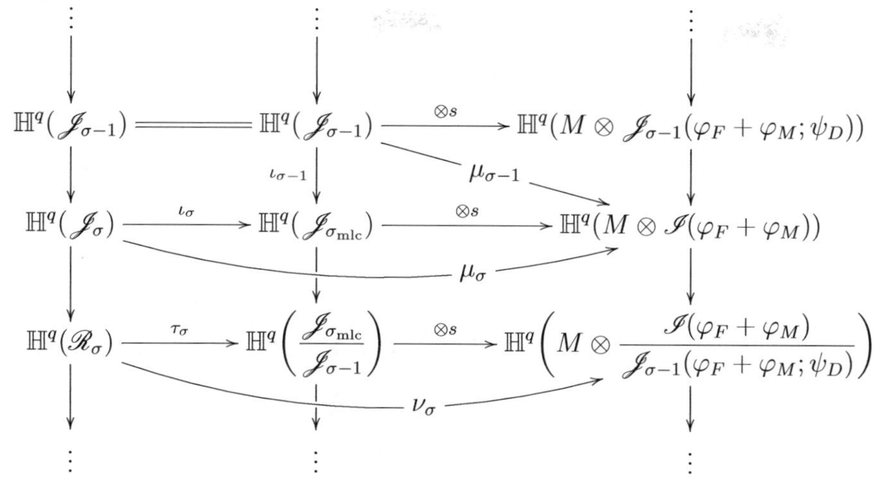

Note that the columns are all exact. The map ι_σ is induced from the natural inclusion $\mathscr{J}_\sigma \subset \mathscr{J}_{\sigma_{\mathrm{mlc}}} = \mathscr{I}(\varphi_F)$, while the horizontal maps on the right-hand-side are induced from the multiplication map $\otimes s$. Each homomorphism of μ_σ's and ν_σ's is the composition of the maps on the corresponding row.

Through a simple diagram-chasing, one sees that, for each $\sigma \geq 1$, if the homomorphisms $\mu_{\sigma-1}$ and ν_σ satisfy $\ker \mu_{\sigma-1} = \ker \iota_{\sigma-1}$ and $\ker \nu_\sigma = \ker \tau_\sigma$ respectively, then it follows that $\ker \mu_\sigma = \ker \iota_\sigma$. One then obtains the following theorem via induction.

THEOREM 2.2. *If one has* $\ker \mu_0 = \ker \iota_0$ *and* $\ker \nu_\sigma = \ker \tau_\sigma$ *for* $\sigma = 1, \dots, \sigma_{\mathrm{mlc}}$, *then* $\mu_{\sigma_{\mathrm{mlc}}}$ *is injective (as* $\iota_{\sigma_{\mathrm{mlc}}}$ *is the identity map). In particular, when* $M := F^{\otimes m}$ *and when* $\varphi_M := m\varphi_F$ *is smooth, Fujino's conjecture holds true under these given assumptions.*

In [**7**], the assumption $\ker \mu_0 = \ker \iota_0$ is proved when φ_F and φ_M have neat analytic singularities with a suitable positivity assumption. More precisely,

THEOREM 2.3 ([**7**, Thm. 1.2.1]). *Suppose* (X, ϕ_D, φ_F) *is as described in Section* 0.1 *and let* φ_M *be a quasi-psh potential on* M *having the same regularity properties as* φ_F *described in Section* 0.1 *such that*

- $i\partial\bar{\partial}\varphi_F \geq 0$ *and* $i\partial\bar{\partial}\varphi_M \leq C i\partial\bar{\partial}\varphi_F$ *on* X *in the sense of currents for some constant* $C > 0$ *(where* $i := \frac{\sqrt{-1}}{2\pi}$ [2]*), and*
- $P_F \cup P_M \cup D$ *has only snc (where* $P_M := \varphi_M^{-1}(-\infty)$*).*

[2]The notation is chosen by mimicking the reduced Planck constant $\hbar = \frac{h}{2\pi}$. It can be typeset with the code `{\raisebox{-0.9ex}{$\mathchar'26$}\mkern-6.7mu i}`.

Suppose further that there exists a non-trivial holomorphic section $s \in H^0(X, M)$ *such that* $\sup_X |s|^2_{\varphi_M} < \infty$. *Then, given the commutative diagram*

$$\begin{array}{ccc} H^q(X, K_X \otimes D \otimes F \otimes \mathscr{I}(\phi_D + \varphi_F)) & & \\ \downarrow \iota_0 \quad \circlearrowleft & \searrow^{\mu_0} & \\ H^q(X, K_X \otimes D \otimes F \otimes \mathscr{I}(\varphi_F)) & \xrightarrow{\otimes s} & H^q(X, K_X \otimes D \otimes F \otimes M \otimes \mathscr{I}(\varphi_F + \varphi_M)) , \end{array}$$

in which ι_0 *is induced from the inclusion* $\mathscr{I}(\phi_D + \varphi_F) \subset \mathscr{I}(\varphi_F)$, *one has* $\ker \mu_0 = \ker \iota_0$ *for every* $q \geq 0$.

This theorem with smooth φ_F and φ_M is first proved by Matsumura by smoothing the canonical potential ϕ_D of D ([**20**, Thm. 3.9]) when applying the Bochner–Kodaira formula with a (smooth) Kähler form on X. Instead of smoothing out the lc singularities of ϕ_D, [**7**] considers the (twisted) Bochner–Kodaira formula (hence, in particular, the formal adjoint operator ϑ of $\overline{\partial}$) with respect to the *complete* Kähler form ω on $X^\circ = X \setminus (P_F \cup P_M)$ (or simply the Kähler form on X when φ_F and φ_M are smooth) and the potential $\phi_D + \varphi_F$ $(+\varphi_M)$ (which is singular along general points of D on X°). Such Bochner–Kodaira formula is shown to be applicable to the harmonic L^2 forms with respect to the data $(\phi_D, \varphi_F, \varphi_M, \omega)$ on X° by making use of a family of smooth cut-off functions $\{\theta_\varepsilon\}_{\varepsilon>0}$ supported away from D in X° such that $\theta_\varepsilon \nearrow 1$ almost everywhere in X° as $\varepsilon \searrow 0$. The proof (see [**7**, Prop. 3.2.5 and Cor. 3.2.6]) takes advantage of the residue computation in Proposition 1.3. A flavour of the computation can also be found in the following idea of proof of Theorem 2.3 (in the evaluation of $\langle\!\langle su, \overline{\partial} v_{(\infty)} \rangle\!\rangle_{\varphi+\varphi_M,\omega}$).

The outline of the proof of Theorem 2.3 is given without the treatment to the singularities on φ_F and φ_M, as the proof for the case when they are smooth already illustrates the use of the residue computation. For the full treatment for singular φ_F and φ_M, see [**7**, §3.3].

IDEA OF THE PROOF OF THEOREM 2.3. Note that any class in $H^q(X, K_X \otimes D \otimes F \otimes \mathscr{I}(\varphi))$ is represented by a unique element in the space of harmonic L^2 forms $\mathcal{H}^{n,q}_{\varphi,\omega}$ on X° with respect to $\varphi = \phi_D + \varphi_F$ and ω (see [**19**, Prop. 5.5] and [**20**, Prop. 2.8] for a proof). It is enough to show that, for any

$$[u] \in H^q(X, K_X \otimes D \otimes F \otimes \mathscr{I}(\varphi)) \quad \text{with} \quad u \in \mathcal{H}^{n,q}_{\varphi,\omega} ,$$

if $[u] \in (\ker \iota_0)^\perp \cap \ker \mu_0$ (in which the orthogonal complement $(\cdot)^\perp$ is taken with respect to the L^2 norm $\|\cdot\|_{\varphi,\omega}$ on $\mathcal{H}^{n,q}_{\varphi,\omega}$), then $u = 0$.

Since $0 = [su] \in H^q(X, K_X \otimes D \otimes F \otimes M \otimes \mathscr{I}(\varphi_F + \varphi_M))$, $su = \overline{\partial} v$ for some $v \in L^{n,q-1}_{(2)}(D \otimes F \otimes M)_{\varphi_D^{\mathrm{sm}}+\varphi_F+\varphi_M,\omega}$ which can be decomposed into $v = v_{(2)} + v_{(\infty)}$ such that

$$v_{(2)} \in L^{n,q-1}_{(2)}(D \otimes F \otimes M)_{\phi_D+\varphi_F+\varphi_M,\omega} \quad \text{and}$$
$$v_{(\infty)} \in \mathscr{A}^{n,q-1}_X(X; D \otimes F \otimes M \otimes \mathscr{I}(\varphi_F + \varphi_M))$$

via the L^2 Dolbeault isomorphism (see [**7**, Lemma 3.2.1]). Furthermore, the positivity $i\partial\overline{\partial}\varphi_F \geq 0$, together with the refined hard Lefschetz theorem (see [**20**, Thm. 3.3] or [**7**, Thm. 2.5.1]), implies that $*_\omega u$ is holomorphic on X, which then implies that $\frac{u}{\mathsf{s}_D}$ is smooth on X°. From

$$\|su\|^2_{\varphi+\varphi_M,\omega} = \langle\!\langle su, \overline{\partial} v \rangle\!\rangle_{\varphi+\varphi_M,\omega} = \langle\!\langle su, \overline{\partial} v_{(2)} \rangle\!\rangle_{\varphi+\varphi_M,\omega} + \langle\!\langle su, \overline{\partial} v_{(\infty)} \rangle\!\rangle_{\varphi+\varphi_M,\omega} ,$$

it suffices to show that both of the inner products vanish, which then results in $su = 0$, hence $u = 0$.

The 1st term: $\langle\!\langle su, \overline{\partial} v_{(2)} \rangle\!\rangle_{\varphi+\varphi_M,\omega} = \langle\!\langle \vartheta_{\varphi_M}(su), v_{(2)} \rangle\!\rangle = 0$, where ϑ_{φ_M} is the formal adjoint operator of $\overline{\partial}$ with respect to $\varphi + \varphi_M$ and ω.

From the fact that $\frac{u}{s_D}$ is smooth on X° and by making use of the family of cut-off functions $\{\theta_\varepsilon\}_{\varepsilon>0}$ supported away from D, the Bochner–Kodaira formula with respect to the potential φ and metric ω is valid for u (see [**7**, Prop. 3.2.5]). Together with the positivity $i\partial\overline{\partial}\varphi_F \geq 0$ on X, one obtains
(eq 2.2)

$$\nabla^{(0,1)} u = 0 \quad \text{and} \quad \left(i\partial\overline{\partial}\varphi_F\right)^{\omega}(u,u)_{\omega} = 0 \quad \text{on } X^\circ \qquad ([\mathbf{7}, \text{Prop. 3.2.5}]),$$

where $\nabla^{(0,1)}$ is the covariant differential operator of type $(0,1)$ induced from the Chern connection. Consequently, these results, together with the Bochner–Kodaira formula with respect to $\varphi + \varphi_M$ and ω, lead to

$$\vartheta_{\varphi_M}(su) = 0 \quad \text{and} \quad su \in \operatorname{Dom} \overline{\partial}^*_{\varphi_M} \qquad ([\mathbf{7}, \text{Cor. 3.2.6}]),$$

where ϑ_{φ_M} is the formal adjoint of $\overline{\partial}$ with respect to the L^2 norm $\|\cdot\|_{\varphi+\varphi_M,\omega}$ and $\overline{\partial}^*_{\varphi_M}$ is the corresponding Hilbert space adjoints on the L^2 space $L^{n,\bullet}_{(2)}(D \otimes F \otimes M)_{\varphi+\varphi_M,\omega}$. Since $v_{(2)}$ is L^2 with respect to $\|\cdot\|_{\varphi+\varphi_M,\omega}$ and lies inside $\operatorname{Dom}\overline{\partial}$, one can apply the integration by parts and the inner product under consideration vanishes as desired.

The 2nd term: $\langle\!\langle su, \overline{\partial} v_{(\infty)} \rangle\!\rangle_{\varphi+\varphi_M,\omega} = 0$.

Note that $v_{(\infty)}$ is L^2 with respect to the potential $\varphi_F + \varphi_M$ but not necessarily to $\phi_D + \varphi_F + \varphi_M$. To overcome this difficulty, the cut-off function θ_ε (as described before) is introduced (and defined explicitly here) to facilitate the integration by parts. Let $\theta\colon [0,\infty) \to [0,1]$ be a smooth non-decreasing cut-off function such that $\theta|_{[0,\frac{1}{2}]} \equiv 0$ and $\theta|_{[1,\infty)} \equiv 1$. For $\varepsilon \geq 0$, set $\theta_\varepsilon := \theta \circ \frac{1}{|\psi_D|^\varepsilon}$ and $\theta'_\varepsilon := \theta' \circ \frac{1}{|\psi_D|^\varepsilon}$ for convenience (where θ' is the derivative of θ). Note that both θ_ε and θ'_ε have compact supports inside $X \setminus D$ for $\varepsilon > 0$. One also has $\theta_\varepsilon \nearrow 1$ almost everywhere in X as $\varepsilon \searrow 0$. As $\varepsilon \to 0^+$, one has

$$\begin{aligned}
\langle\!\langle su, \overline{\partial} v_{(\infty)} \rangle\!\rangle_{\varphi+\varphi_M,\omega} &\leftarrow \langle\!\langle su, \theta_\varepsilon \overline{\partial} v_{(\infty)} \rangle\!\rangle_{\varphi+\varphi_M,\omega} \\
&= \langle\!\langle su, \overline{\partial}\big(\theta_\varepsilon v_{(\infty)}\big) \rangle\!\rangle_{\varphi+\varphi_M,\omega} - \left\langle\!\!\left\langle su, \frac{\varepsilon\theta'_\varepsilon}{|\psi_D|^{1+\varepsilon}} \overline{\partial}\psi_D \wedge v_{(\infty)} \right\rangle\!\!\right\rangle_{\varphi+\varphi_M,\omega} \\
&= \underbrace{\langle\!\langle \vartheta_{\varphi_M}(su), \theta_\varepsilon v_{(\infty)} \rangle\!\rangle_{\varphi+\varphi_M,\omega}}_{\to 0} - \varepsilon \left\langle\!\!\left\langle (\partial\psi_D)^{\omega} \lrcorner\, su, \frac{\theta'_\varepsilon v_{(\infty)}}{|\psi_D|^{1+\varepsilon}} \right\rangle\!\!\right\rangle_{\varphi+\varphi_M,\omega} .
\end{aligned}$$

Hinted by the argument $(\partial\psi_D)^\omega \lrcorner\, u$ in the inner product, one is led to consider the twisted Bochner–Kodaira formula with respect to φ, ω and the twisting factor $|\psi_D|^{1-\varepsilon}$ for $\varepsilon > 0$, given by

$$\begin{aligned}
&\int_{X^\circ} \left|\overline{\partial}\zeta\right|^2_{\varphi,\omega} |\psi_D|^{1-\varepsilon} + \int_{X^\circ} |\vartheta\zeta|^2_{\varphi,\omega} |\psi_D|^{1-\varepsilon} \\
&= \int_{X^\circ} \left|\nabla^{(0,1)}\zeta\right|^2_{\varphi,\omega} |\psi_D|^{1-\varepsilon} + \int_{X^\circ} \left(i\partial\overline{\partial}\varphi + \frac{1-\varepsilon}{|\psi_D|} i\partial\overline{\partial}\psi_D \right)^\omega (\zeta,\zeta)_{\varphi,\omega} |\psi_D|^{1-\varepsilon} \\
&\quad - 2(1-\varepsilon)\operatorname{Re}\int_{X^\circ} \left\langle \vartheta\zeta, \frac{(\partial\psi_D)^\omega \lrcorner\, \zeta}{|\psi_D|} \right\rangle_{\varphi,\omega} |\psi_D|^{1-\varepsilon} + \varepsilon\int_{X^\circ} \frac{1-\varepsilon}{|\psi_D|^2} |(\partial\psi_D)^\omega \lrcorner\, \zeta|^2_{\varphi,\omega} |\psi_D|^{1-\varepsilon}
\end{aligned}$$

for any compactly supported $\zeta \in \mathscr{A}^{n,q}_{X\,c}(X^\circ \setminus D; D \otimes F)$, where ϑ is the formal adjoint of $\overline{\partial}$ with respect to $\|\cdot\|_{\varphi,\omega}$ (see [**7**, §2.4] for more details). Putting $\theta_{\varepsilon'} u$ into ζ for some $\varepsilon' > 0$ (assuming that the singularities of φ_F has been dealt with or φ_F is smooth), together with the facts

- (eq 2.2) and $i\partial\overline{\partial}\phi_D = 0$ on $X \setminus D$,
- (u being harmonic) $\overline{\partial}(\theta_{\varepsilon'} u) = \overline{\partial}\theta_{\varepsilon'} \wedge u$, $\nabla^{(0,1)}(\theta_{\varepsilon'} u) = \overline{\partial}\theta_{\varepsilon'} \otimes u$ and
$$\vartheta(\theta_{\varepsilon'} u) = -(\partial\theta_{\varepsilon'})^\omega \lrcorner\, u = -\frac{\varepsilon'\theta'_{\varepsilon'}}{|\psi_D|^{1+\varepsilon'}} (\partial\psi_D)^\omega \lrcorner\, u \,,$$
- $\left|\overline{\partial}\theta_{\varepsilon'} \wedge u\right|^2_{\varphi,\omega} + |(\partial\theta_{\varepsilon'})^\omega \lrcorner\, u|^2_{\varphi,\omega} = \left|\overline{\partial}\theta_{\varepsilon'} \otimes u\right|^2_{\varphi,\omega}$ on X° (see [**7**, footnote 9]),

one obtains

$$\begin{aligned}
0 = &- \int_{X^\circ} \frac{1-\varepsilon}{|\psi_D|^\varepsilon} \left(i\partial\overline{\partial}\varphi_D^{\mathrm{sm}}\right)^\omega (\theta_{\varepsilon'} u, \theta_{\varepsilon'} u)_{\varphi,\omega} \\
&+ 2(1-\varepsilon)\operatorname{Re}\int_{X^\circ} \left\langle \frac{\varepsilon'\theta'_{\varepsilon'} (\partial\psi_D)^\omega \lrcorner\, u}{|\psi_D|^{1+\varepsilon'}}, \frac{(\partial\psi_D)^\omega \lrcorner\, \theta_{\varepsilon'} u}{|\psi_D|} \right\rangle_{\varphi,\omega} |\psi_D|^{1-\varepsilon} \\
&+ \varepsilon\int_{X^\circ} \frac{1-\varepsilon}{|\psi_D|^{1+\varepsilon}} |(\partial\psi_D)^\omega \lrcorner\, \theta_{\varepsilon'} u|^2_{\varphi,\omega} \,.
\end{aligned}$$

Dividing $(1-\varepsilon)$ from both sides and rearranging terms yield

$$\int_{X^\circ} \frac{\theta_{\varepsilon'}^2}{|\psi_D|^\varepsilon} \left(i\partial\overline{\partial}\varphi_D^{\mathrm{sm}}\right)^\omega (u,u)_{\varphi,\omega} = \int_{X^\circ} \left(\frac{\varepsilon\theta_{\varepsilon'}^2}{|\psi_D|^{1+\varepsilon}} + \frac{2\varepsilon'\theta'_{\varepsilon'}\theta_{\varepsilon'}}{|\psi_D|^{1+\varepsilon+\varepsilon'}} \right) |(\partial\psi_D)^\omega \lrcorner\, u|^2_{\varphi,\omega} \,.$$

Since $i\partial\overline{\partial}\varphi_D^{\mathrm{sm}}$ is a smooth $(1,1)$-form on X, the integral on the left-hand-side has finite limits as $\varepsilon' \to 0^+$ and $\varepsilon \to 0^+$ by the dominated convergence theorem. As both $\theta_{\varepsilon'}$ and $\theta'_{\varepsilon'}$ are non-negative for all $\varepsilon' > 0$, it follows (from letting $\varepsilon' \to 0^+$) that $\frac{|(\partial\psi_D)^\omega \lrcorner u|^2_{\varphi,\omega}}{|\psi_D|^{1+\varepsilon}}$ is integrable on X° for any $\varepsilon > 0$ and, as a result,

$$\text{(eq 2.3)} \qquad \int_X \left(i\partial\overline{\partial}\varphi_D^{\mathrm{sm}}\right)^\omega (u,u)_{\varphi,\omega} = \lim_{\varepsilon\to 0^+} \varepsilon \int_{X^\circ} \frac{|(\partial\psi_D)^\omega \lrcorner\, u|^2_{\varphi,\omega}}{|\psi_D|^{1+\varepsilon}}$$

(see [**7**, Prop. 3.2.8]; see also [**7**, Lemma 3.3.3] for the treatment when φ_F is singular).

Via Takegoshi's argument (see [**7**, §3.1, Step IV]) and the assumption $[u] \in (\ker \iota_0)^\perp$, one obtains

$$\text{(eq 2.4)} \qquad \int_X \left(i\partial\overline{\partial}\varphi_D^{\mathrm{sm}}\right)^\omega (u,u)_{\varphi,\omega} = \left\langle\!\left\langle i\partial\overline{\partial}\varphi_D^{\mathrm{sm}} \Lambda_\omega u,\, u \right\rangle\!\right\rangle_{\varphi,\omega} = 0 \,.$$

On the other hand, since $\frac{u}{\mathsf{s}_D}$ is smooth on X° and, on any admissible open set $V \Subset X$, one has

$$\partial\psi_D|_V = \sum_{j=1}^{\sigma_V} \frac{dz_j}{z_j} - \partial\varphi_D^{\mathrm{sm}} \quad (\text{and } \mathsf{s}_D|_V = z_1 \cdots z_{\sigma_V}) ,$$

it then follows from (eq 1.1) that

$$(\partial\psi_D)^\omega \lrcorner\, u \in \mathscr{A}_X^{n,q-1}(X^\circ; D \otimes F \otimes \mathscr{J}_1) .$$

Proposition 1.3 and (eq 2.4) then give

$$0 = \lim_{\varepsilon \to 0^+} \varepsilon \int_{X^\circ} \frac{|(\partial\psi_D)^\omega \lrcorner\, u|^2_{\phi_D+\varphi_F,\omega}}{|\psi_D|^{1+\varepsilon}} = \pi \sum_{i \in I_D} \int_{D_i} \left| \mathcal{R}_{D_i}\left((\partial\psi_D)^\omega \lrcorner\, \frac{u}{\mathsf{s}_D} \right) \right|^2_{\varphi_F,\omega} ,$$

which implies that

$$\mathcal{R}_{D_i}\left((\partial\psi_D)^\omega \lrcorner\, \frac{u}{\mathsf{s}_D} \right) \equiv 0 \quad \text{on } D_i \text{ for all } i \in I_D .$$

Via the argument in [**7**, §3.1, Step V, (eq 3.1.5′)], it follows that

$$\text{coef. of } (\partial\psi_D)^\omega \lrcorner\, \frac{u}{\mathsf{s}_D} \text{ are locally in } \mathscr{C}^\infty_{X*} \text{ on } X^\circ .$$

It follows that

$$\begin{aligned} &\left| \left\langle\!\!\left\langle (\partial\psi_D)^\omega \lrcorner\, su, \frac{\theta'_\varepsilon v_{(\infty)}}{|\psi_D|^{1+\varepsilon}} \right\rangle\!\!\right\rangle_{\varphi+\varphi_M,\omega} \right| \\ &\qquad \le \int_X \frac{\left| \left\langle (\partial\psi_D)^\omega \lrcorner\, \frac{u}{\mathsf{s}_D}, v_{(\infty)} e^{-\frac{1}{2}(\varphi_D^{\mathrm{sm}}+\varphi_M)} \right\rangle \right|_{\varphi_F,\omega} |s|_{\varphi_M} \, |\theta'_\varepsilon|}{|\mathsf{s}_D|_{\varphi_D^{\mathrm{sm}}} \, |\psi_D|^{1+\varepsilon}} , \end{aligned}$$

which is finite as $\frac{1}{|\mathsf{s}_D|_{\varphi_D^{\mathrm{sm}}}}$ is locally integrable while the other factors are bounded from above. This leads to the desired vanishing of the inner product as $\varepsilon \to 0^+$, thus completes the proof. □

When (X, D) is plt, i.e. all D_i's in $D = \sum_{i \in I_D^\sigma} D_i$ are the mutually disjoint, one has $\sigma_{\mathrm{mlc}} = 1$ and the homomorphism τ_1 in (eq 2.1) is the identity map. In order to prove the injectivity of μ_1 in this case, it is enough to show that $\ker \nu_1 = \ker \tau_1 = 0$ by Theorem 2.2 and Theorem 2.3. It follows from (eq 1.3) that the homomorphism ν_1 is reduced to

$$\begin{aligned} \nu_1 \colon &\bigoplus_{i \in I_D} H^q\big(D_i, K_{D_i} \otimes F|_{D_i} \otimes \mathscr{I}_{D_i}(\varphi_F|_{D_i})\big) \\ &\qquad \longrightarrow \bigoplus_{i \in I_D} H^q\big(D_i, K_{D_i} \otimes (F \otimes M)|_{D_i} \otimes \mathscr{I}_{D_i}\big((\varphi_F + \varphi_M)|_{D_i}\big)\big) , \end{aligned}$$

which maps the i-th summand to the i-th summand via the multiplication $\otimes\, s|_{D_i}$. The result in [**19**, Thm. 1.3] or the argument given in [**7**, §3.1] (with $D = 0$) concludes readily that each $\nu_1|_{D_i}$ is injective. This recovers the injectivity theorem for plt pairs in [**20**, Thm. 3.16].

In our recent preprint with Shin-ichi Matsumura [**8**], under the assumption that φ_F and φ_M are smooth, the claims $\ker \nu_\sigma = \ker \tau_\sigma$ for $\sigma = 1, \dots, \sigma_{\mathrm{mlc}}$ with general

$\sigma_{\mathrm{mlc}} \geq 1$ are proved. In fact, by considering a commutative diagram analogous to (eq 2.1) induced from the short exact sequences

$$\text{(eq 2.5)} \qquad 0 \longrightarrow \frac{\mathscr{I}_\sigma}{\mathscr{I}_\rho} \longrightarrow \frac{\mathscr{I}_\tau}{\mathscr{I}_\rho} \longrightarrow \frac{\mathscr{I}_\tau}{\mathscr{I}_\sigma} \longrightarrow 0 \qquad \text{for } 0 \leq \rho \leq \sigma \leq \tau ,$$

together with the isomorphisms $\mathscr{R}_\sigma \cong \frac{\mathscr{I}_\sigma}{\mathscr{I}_{\sigma-1}}$ for $\sigma = 1, \dots, \sigma_{\mathrm{mlc}}$ obtained from (eq 1.2), the following injectivity theorem for a special kind of singular spaces is obtained.

THEOREM 2.4 ([**8**, Thm. 1.2]). *Let (X, ω) be a compact Kähler manifold and D a reduced snc divisor on X. Let (F, φ_F) and (M, φ_M) be hermitian line bundles on X equipped with smooth hermitian metrics satisfying*

$$i\partial\overline{\partial}\varphi_F \geq 0 \quad \textit{and} \quad i\partial\overline{\partial}\varphi_M \leq C i\partial\overline{\partial}\varphi_F \qquad \textit{on } X \ \textit{ for some constant } C > 0 .$$

Let s be a section of M such that s does not vanish identically on any lc centres of (X, D). Then, the multiplication map

$$H^q(D, K_D \otimes F) \xrightarrow{\otimes s} H^q(D, K_D \otimes F \otimes M)$$

induced by the tensor product with s is injective for every $q \geq 0$.

3. An extension theorem

Note that the positivity assumption

$$\text{(eq 3.1a)} \qquad i\partial\overline{\partial}(\varphi_F + \phi_D + \lambda\psi_D) \geq 0 \quad \text{for all } \lambda \in [0, \delta] \text{ on } X$$

for some given constant $\delta > 0$ is the standard positivity assumption taken in the L^2 extension theorems (see, for example, [**9**] and [**6**]) and is stronger than the assumption in previous sections. In what follows, consider also the positivity assumption

$$\text{(eq 3.1b)} \quad i\partial\overline{\partial}\big(\varphi_F + \phi_{(p)} + \lambda\psi_{(p)}\big) \geq 0 \quad \text{for all } \lambda \in [0, \delta] \text{ on } \mathtt{D}_p^\sigma \text{ for each } p \in I_D^\sigma$$

and for each $\sigma = 1, 2, \dots, \sigma_{\mathrm{mlc}}$. Write (eq 3.1b)_σ to mean the assumption for the specific σ. It will become apparent that it is convenient to set

$$\text{(eq 3.1b)}_0 := \text{(eq 3.1a)}$$

and

$$\mathscr{I}_{-1} := 0 \quad \text{and} \quad \mathscr{R}_0 := \frac{\mathscr{I}_0}{\mathscr{I}_{-1}} = \mathscr{I}_0 .$$

Consider the long exact sequence

$$\cdots \longrightarrow \mathbb{H}^{q-1}\Big(\frac{\mathscr{I}_{\sigma'}}{\mathscr{I}_\sigma}\Big) \longrightarrow \mathbb{H}^q(\mathscr{R}_\sigma) \xrightarrow{\tau_\sigma^{\sigma'}} \mathbb{H}^q\Big(\frac{\mathscr{I}_{\sigma'}}{\mathscr{I}_{\sigma-1}}\Big) \longrightarrow \mathbb{H}^q\Big(\frac{\mathscr{I}_{\sigma'}}{\mathscr{I}_\sigma}\Big) \longrightarrow \cdots$$

for each $q \geq 1$ and each $\sigma = 0, 1, \dots, \sigma'$. The following theorem can then be obtained via the techniques developed in [**7**].

THEOREM 3.1. *Under the positivity assumption* (eq 3.1b)_σ *(for a given σ), one has $\ker \tau_\sigma^{\sigma'} = 0$ for all $q \geq 0$ and for any $\sigma' \geq \sigma$. The long exact sequence thus splits into short exact sequences*

$$0 \longrightarrow \mathbb{H}^q(\mathscr{R}_\sigma) \xrightarrow{\tau_\sigma^{\sigma'}} \mathbb{H}^q\Big(\frac{\mathscr{I}_{\sigma'}}{\mathscr{I}_{\sigma-1}}\Big) \longrightarrow \mathbb{H}^q\Big(\frac{\mathscr{I}_{\sigma'}}{\mathscr{I}_\sigma}\Big) \longrightarrow 0$$

for all $q \geq 0$ and any $\sigma' \geq \sigma$. In particular, if the assumption (eq 3.1b) *holds for all $\sigma = 0, 1, \ldots, \sigma'$ for some given $\sigma' \geq 1$, then, for every*

$$f \in \mathbb{H}^0\left(\frac{\mathscr{J}_{\sigma'}}{\mathscr{J}_{\sigma'-1}}\right) = H^0\left(\operatorname{lc}_X^{\sigma'}(D), K_X \otimes D \otimes F \otimes \frac{\mathscr{J}_{\sigma'}}{\mathscr{J}_{\sigma'-1}}\right),$$

there exists a global holomorphic section

$$F \in \mathbb{H}^0(\mathscr{J}_{\sigma'}) = H^0(X, K_X \otimes D \otimes F \otimes \mathscr{J}_{\sigma'})$$

such that $f \equiv F \mod \mathscr{J}_{\sigma'-1}$ on X, i.e. F is a holomorphic extension of f.

REMARK 3.2. The last claim in Theorem 3.1 is the claim suggested in [**5**, Conj. 2.2.3], but with a slightly different positivity assumption and without any estimates for the extensions.

PROOF. Note that $\mathcal{I}_{\operatorname{lc}_X^{\sigma+1}(D)}\Big|_{\mathsf{D}_p^\sigma} = \mathcal{I}_{\operatorname{Diff}_p^* D} = \left\langle \mathsf{s}_{(p)} \right\rangle = \mathscr{I}_{\mathsf{D}_p^\sigma}\left(\phi_{(p)}\right)$ and thus

$$\begin{aligned} K_{\mathsf{D}_p^\sigma} \otimes F|_{\mathsf{D}_p^\sigma} \otimes \mathscr{I}_{\mathsf{D}_p^\sigma}(\varphi_F) &\cong K_{\mathsf{D}_p^\sigma} \otimes \operatorname{Diff}_p^* D \otimes F|_{\mathsf{D}_p^\sigma} \otimes \mathscr{I}_{\mathsf{D}_p^\sigma}(\varphi_F) \cdot \mathcal{I}_{\operatorname{Diff}_p^* D} \\ &= K_{\mathsf{D}_p^\sigma} \otimes \operatorname{Diff}_p^* D \otimes F|_{\mathsf{D}_p^\sigma} \otimes \mathscr{J}_{\mathsf{D}_p^\sigma,0}\left(\varphi_F; \psi_{(p)}\right) \\ &= K_{\mathsf{D}_p^\sigma} \otimes \operatorname{Diff}_p^* D \otimes F|_{\mathsf{D}_p^\sigma} \otimes \mathscr{I}_{\mathsf{D}_p^\sigma}\left(\phi_{(p)} + \varphi_F\right) . \end{aligned}$$

It can be seen from this as well as the argument below that the computations on D_p^σ and X are formally the same, so only the proof of $\ker \tau_0^{\sigma'} = 0$ is given below.

Let ω be a complete Kähler form on $X^\circ := X \setminus \varphi_F^{-1}(-\infty)$ as described in [**7**, §2.2, item (4)]. Let $|\cdot|_{\phi_D} := |\cdot|_{\varphi_F + \phi_D, \omega}$ and $\|\cdot\|_{X,\phi_D} := \|\cdot\|_{X, \varphi_F + \phi_D, \omega}$ be respectively the pointwise and L^2 norms on X° induced from data shown in the subscript. From the L^2 Dolbeault isomorphism (see [**19**, Prop. 5.8] and [**20**, Prop. 2.8(1)]), one has

$$\begin{aligned} H^q(X, K_X \otimes F \otimes \mathscr{I}(\varphi_F)) &\cong H^q(X, K_X \otimes D \otimes F \otimes \mathscr{I}(\phi_D + \varphi_F)) \\ &\cong \mathcal{H}^{n,q}(X; D \otimes F)_{\varphi_F + \phi_D, \omega} =: \mathcal{H}^{n,q}_{\phi_D} , \end{aligned}$$

the space of harmonic forms with respect to $\|\cdot\|_{X,\phi_D}$. Using the Bochner–Kodaira formula and the residue computation shown in [**7**, Prop. 3.2.3, Remark 3.2.4, Prop. 3.2.5 and Prop. 3.2.8] (as well as [**7**, Prop. 3.3.2 and Lemma 3.3.3] to handle the singularities of φ_F; also cf. (eq 2.2) and (eq 2.3)), all $u \in \mathcal{H}^{n,q}_{\phi_D}$ should satisfy

$$\nabla^{(0,1)} u = 0 , \quad \left(i\partial\overline{\partial}\varphi_F\right)^\omega (u,u)_{\phi_D} = 0 \quad \text{on } X^\circ$$

$$\begin{aligned} \text{and} \quad \pi \sum_{i \in I_D} \int_{D_i} \left| \mathcal{R}_{D_i}\left((\partial\psi_D)^\omega \lrcorner \frac{u}{\mathsf{s}_D} \right) \right|^2_{\varphi_F, \omega} &= \lim_{\varepsilon \to 0^+} \varepsilon \int_{X^\circ} \frac{\left|(\partial\psi_D)^\omega \lrcorner u\right|^2_{\phi_D}}{|\psi_D|^{1+\varepsilon}} \\ &= \int_{X^\circ} \left(i\partial\overline{\partial}\varphi_D^{\mathrm{sm}}\right)^\omega (u,u)_{\phi_D} . \end{aligned}$$

However, the positivity assumption (eq 3.1b)$_0$ implies that $\left(i\partial\overline{\partial}\varphi_D^{\mathrm{sm}}\right)^\omega (u,u)_{\phi_D} \leq 0$, which then forces the equality

$$\mathcal{R}_{D_i}\left((\partial\psi_D)^\omega \lrcorner \frac{u}{\mathsf{s}_D} \right) \equiv 0 \quad \text{on } D_i \text{ for all } i \in I_D .$$

Via the argument in [**7**, §3.1, Step V, (eq 3.1.5′)], it follows that

$$\text{coef. of } (\partial\psi_D)^\omega \lrcorner \frac{u}{\mathsf{s}_D} \text{ are locally in } \mathscr{C}^\infty_{X^*} \text{ on } X^\circ .$$

Using the Čech–Dolbeault isomorphism, every element in $\ker \tau_0^{\sigma'} \subset \mathcal{H}^{n,q}_{\phi_D}$ can be written as $\overline{\partial} v_{(2)} + \overline{\partial} v_{(\infty)}$, where $v_{(2)}$ is a $K_X \otimes D \otimes F$-valued $(0, q-1)$-form on X° with L^2 coefficients with respect to $\|\cdot\|_{X,\phi_D}$, while $v_{(\infty)}$ is a smooth $K_X \otimes D \otimes F$-valued $(0, q-1)$-form on X° which is L^2 with respect to $\|\cdot\|_{X,\varphi_F+\varphi_D^{\mathrm{sm}}}$ (but need not be $\|\cdot\|_{X,\phi_D}$) (cf. [**7**, Lemma 3.2.1]). Let $\theta \colon [0,\infty) \to [0,1]$ be a smooth non-decreasing cut-off function and set $\theta_\varepsilon := \theta \circ \frac{1}{|\psi_D|^\varepsilon}$ and $\theta'_\varepsilon := \theta' \circ \frac{1}{|\psi_D|^\varepsilon}$ for $\varepsilon \geq 0$ as in [**7**, §3.1] (where θ' is the derivative of θ). Recall that u is harmonic. Therefore, a computation similar to [**7**, §3.1, Step III and Step V] yields

$$
\begin{aligned}
\left| \left\langle\!\!\left\langle \overline{\partial} v_{(2)} + \overline{\partial} v_{(\infty)}, u \right\rangle\!\!\right\rangle_{X,\phi_D} \right|
&= \left| \left\langle\!\!\left\langle \overline{\partial} v_{(\infty)}, u \right\rangle\!\!\right\rangle_{X,\phi_D} \right| \\
&\xleftarrow{\varepsilon \to 0^+} \left| \left\langle\!\!\left\langle \theta_\varepsilon \overline{\partial} v_{(\infty)}, u \right\rangle\!\!\right\rangle_{X,\phi_D} \right| \\
&= \left| \cancelto{0}{\left\langle\!\!\left\langle \overline{\partial}\left(\theta_\varepsilon v_{(\infty)}\right), u \right\rangle\!\!\right\rangle_{X,\phi_D}} - \varepsilon \left\langle\!\!\left\langle \frac{\theta'_\varepsilon v_{(\infty)}}{|\psi_D|^{1+\varepsilon}}, \left(\partial \psi_D\right)^\omega \lrcorner\, u \right\rangle\!\!\right\rangle_{X,\phi_D} \right| \\
&\leq \varepsilon \int_{X^\circ} \frac{|\theta'_\varepsilon| \left| \left\langle v_{(\infty)} e^{-\frac{1}{2}\varphi_D^{\mathrm{sm}}}, \left(\partial\psi_D\right)^\omega \lrcorner\, \frac{u}{\mathsf{s}_D} \right\rangle_{\varphi_F,\omega} \right|}{|\mathsf{s}_D|_{\varphi_D^{\mathrm{sm}}} \, |\psi_D|^{1+\varepsilon}} \\
&\xrightarrow{\varepsilon \to 0^+} 0 \, ,
\end{aligned}
$$

that is, $u \in \left(\ker \tau_0^{\sigma'}\right)^\perp$. However, $u \in \mathcal{H}^{n,q}_{\phi_D}$ is arbitrary, so $\ker \tau_0^{\sigma'} = 0$.

Once the statements $\ker \tau_\sigma^{\sigma'} = 0$ for $\sigma = 0, 1, \dots, \sigma'$, and hence the short exact sequences, are obtained, the last claim follows from a simple induction. □

References

[1] F. Ambro, *An injectivity theorem*, Compos. Math. **150** (2014), no. 6, 999–1023, DOI 10.1112/S0010437X13007768. MR3223880

[2] J. Cao and M. Păun, *$\partial\bar{\partial}$-lemmas and a conjecture of O. Fujino*, Preprint, `arXiv:2303.16239 [math.AG]`, 2023.

[3] T. O. M. Chan, *On an L^2 extension theorem from log-canonical centres with log-canonical measures*, Math. Z. **301** (2022), no. 2, 1695–1717, DOI 10.1007/s00209-021-02890-9. MR4418335

[4] T. O. M. Chan, *A new definition of analytic adjoint ideal sheaves via the residue functions of log-canonical measures I*, J. Geom. Anal. **33** (2023), no. 9, Paper No. 279, 68, DOI 10.1007/s12220-023-01314-w. MR4605571

[5] T. O. M. Chan, *Residue functions and extension problems*, `arXiv:2211.00885 [math.CV]` (2022). To appear in Proc. CCGA2022 and KSCV14.

[6] T. O. M. Chan and Y.-J. Choi, *Extension with log-canonical measures and an improvement to the plt extension of Demailly-Hacon-Păun*, Math. Ann. **383** (2022), no. 3-4, 943–997, DOI 10.1007/s00208-021-02152-3. MR4458394

[7] T. O. M. Chan and Y.-J. Choi, *On an injectivity theorem for log-canonical pairs with analytic adjoint ideal sheaves*, Trans. Amer. Math. Soc. **376** (2023), no. 12, 8337–8381, DOI 10.1090/tran/8935. MR4669299

[8] T. O. M. Chan, Y.-J. Choi, and S. Matsumura, *An injectivity theorem on snc compact Kähler spaces: an application of the theory of harmonic integrals on log-canonical centers via adjoint ideal sheaves*, Preprint, `arXiv:2307.12025 [math.CV]`, 2023.

[9] J.-P. Demailly, *Extension of holomorphic functions defined on non reduced analytic subvarieties*, The legacy of Bernhard Riemann after one hundred and fifty years. Vol. I, Adv. Lect. Math. (ALM), vol. 35, Int. Press, Somerville, MA, 2016, pp. 191–222. MR3525916

[10] I. Enoki, *Kawamata-Viehweg vanishing theorem for compact Kähler manifolds*, Einstein metrics and Yang-Mills connections (Sanda, 1990), Lecture Notes in Pure and Appl. Math., vol. 145, Dekker, New York, 1993, pp. 59–68. MR1215279

[11] H. Esnault and E. Viehweg, *Lectures on vanishing theorems*, DMV Seminar, vol. 20, Birkhäuser Verlag, Basel, 1992, DOI 10.1007/978-3-0348-8600-0. MR1193913

[12] O. Fujino, *Fundamental theorems for the log minimal model program*, Publ. Res. Inst. Math. Sci. **47** (2011), no. 3, 727–789, DOI 10.2977/PRIMS/50. MR2832805

[13] O. Fujino, *A transcendental approach to Kollár's injectivity theorem*, Osaka J. Math. **49** (2012), no. 3, 833–852. MR2993068

[14] O. Fujino, *On semipositivity, injectivity and vanishing theorems*, Hodge theory and L^2-analysis, Adv. Lect. Math. (ALM), vol. 39, Int. Press, Somerville, MA, 2017, pp. 245–282. MR3751293

[15] Y. Gongyo and S. Matsumura, *Versions of injectivity and extension theorems* (English, with English and French summaries), Ann. Sci. Éc. Norm. Supér. (4) **50** (2017), no. 2, 479–502, DOI 10.24033/asens.2325. MR3621435

[16] J. Kollár, *Higher direct images of dualizing sheaves. I*, Ann. of Math. (2) **123** (1986), no. 1, 11–42, DOI 10.2307/1971351. MR825838

[17] J. Kollár, *Singularities of the minimal model program*, Cambridge Tracts in Mathematics, vol. 200, Cambridge University Press, Cambridge, 2013. With a collaboration of Sándor Kovács, DOI 10.1017/CBO9781139547895. MR3057950

[18] R. Lazarsfeld, *Positivity in algebraic geometry. I*, Ergebnisse der Mathematik und ihrer Grenzgebiete. 3. Folge. A Series of Modern Surveys in Mathematics [Results in Mathematics and Related Areas. 3rd Series. A Series of Modern Surveys in Mathematics], vol. 48, Springer-Verlag, Berlin, 2004. Classical setting: line bundles and linear series, DOI 10.1007/978-3-642-18808-4. MR2095471

[19] S.-i. Matsumura, *An injectivity theorem with multiplier ideal sheaves of singular metrics with transcendental singularities*, J. Algebraic Geom. **27** (2018), no. 2, 305–337, DOI 10.1090/jag/687. MR3764278

[20] S.-i. Matsumura, *A transcendental approach to injectivity theorem for log canonical pairs*, Ann. Sc. Norm. Super. Pisa Cl. Sci. (5) **19** (2019), no. 1, 311–334. MR3923849

Email address: `mariochan@pusan.ac.kr`

Department of Mathematics, Pusan National University, 2, Busandaehak-ro 63beon-gil, Geumjeong-gu, Busan 46241, Republic of Korea

Email address: `youngjun.choi@pusan.ac.kr`

Department of Mathematics, Pusan National University, 2, Busandaehak-ro 63beon-gil, Geumjeong-gu, Busan 46241, Republic of Korea

Contemporary Mathematics
Volume **810**, 2025
https://doi.org/10.1090/conm/810/16210

A Hilbert bundle approach to the sharp strong openness theorem and the Ohsawa-Takegoshi extension theorem

Tai Terje Huu Nguyen and Xu Wang

ABSTRACT. The following paper is around parts of the first named author's thesis. We discuss (what we call) a Hilbert bundle approach to complex Brunn-Minkowski theory and obtain a general monotonicity theorem. As two applications, we prove a generalization of Guan's sharp strong openness theorem and a sharp Ohsawa-Takegoshi extension theorem. A second proof of Guan-Zhou's strong openness theorem using a Donnelly-Fefferman estimate is also given.

CONTENTS

1. Introduction

In this paper, we are interested in using techniques from Berndtsson's complex Brunn-Minkowski theory to discuss sharp estimates in the strong openness theorem and the Ohsawa-Takegoshi extension theorem. The Ohsawa-Takegoshi extension theorem, which first appeared in [**OT**], is a very powerful result, and much studied topic, in complex geometry. In [**Oh**], Ohsawa found that the sharp version of the theorem would directly give the Suita conjecture (see [**Su**]), on a precise Robin constant lower bound for the Bergman kernel (of a plane domain). The sharp version was finally proved by Blocki (for pseudoconvex domains in $\mathbb{C}^n$) and Guan-Zhou

2020 *Mathematics Subject Classification.* Primary 32A25; Secondary 32Q05.

Key words and phrases. Hilbert bundle, complex Brunn-Minkowski theory, sharp strong openness, sharp Ohsawa-Takegoshi estimates.

It is a pleasure to thank Bo Berndtsson, Bo-Yong Chen and Takeo Ohsawa for reading an early version of this paper and their valuable comments. Thanks are also given to the anonymous referees for valuable suggestions that greatly improved the paper, in particular, for finding a mistake in our first proof of Theorem 5.1 and the counterexample in the remark to Lemma 5.2.

(a very general form) in [**Bl**, **GZ1**]. A completely different proof was given by Berndtsson-Lempert in [**BL**] using complex Brunn-Minkowski theory, more specifically, using the convexity theorem of Berndtsson in [**B5**, **B6**]. Another application of Berndtsson's convexity theorem is a proof of the openness conjecture, given by Berndtsson in [**B7**]. In the same paper (section 5, in [**B7**]), Berndtsson also proposed a conjectural picture for the strong openness conjecture using the convexity theorem. Later, in [**GZ2**], Guan-Zhou used the Ohsawa-Takegoshi extenstion theorem to prove the strong openness conjecture. In [**B9**], Berndtsson proved a compact Kähler case of his conjecture from [**B7**], which can be viewed as a global case of the strong openness problem. The main difficulty in proving the conjecture of Berndtsson in the more general case is to obtain an infinite dimensional version of Lemma 2.3 in [**B9**]. This is still an open problem, as far as we know. In this paper, we shall instead get by by using an L^2 Hörmander-type of result that we also prove. We have chosen to name the latter "the Jet selection theorem" [1]. The Jet selection theorem (see Theorem 2.4 in section 2) allows us to generalize Berndtsson's method in [**B9**] to the non-compact case and prove a general monotonicity result, which is the main theorem (see Theorem A in section 2) in this paper. As two applications of it, we prove a generalization of Guan's sharp version [**Gu**] of the strong openness theorem (see Theorem B in section 2) and a sharp Ohsawa-Takegoshi theorem (see Theorem C in section 2).

We end our short introduction with the organization of the paper. In section 2, we present (the statements of) our Jet selection theorem and the main theorem. We also discuss in there applications of the main theorem to the sharp strong openness theorem and the sharp Ohsawa-Takegoshi theorem. In section 3, we introduce (what we call) a Hilbert bundle approach to the complex Brunn-Minkowski theory and prove our main theorem, assuming the statement of the Jet selection theorem. In section 4, we prove the Jet selection theorem in a simple setting and give the reference for the proof of the general case. The last section of this paper, section 5, is on further applications and results. We in there give a proof of the Demailly-Kollár semi-continuity theorem using the Jet selection theorem, and a second proof of Guan-Zhou's strong openness theorem using a Donnelly-Fefferman version of the Jet selection theorem.

2. Jet selection theorem, main theorem and applications

In this section, we give the statements of the Jet selection theorem (Theorem 2.4) and the main theorem (Theorem A), and discuss applications of the latter to sharp strong openness and sharp L^2-holomorphic extension. We commence by giving the set-up for the statements the theorems.

2.1. Quasi-complete Kähler manifolds and the Jet selection theorem.

Let X be a complex manifold, and denote by $C_0^\infty(X, [0,1])$ the space of smooth maps from X to $[0,1]$ with compact support in X. We will be dealing with the following class of Kähler manifolds, which comes from the main result in [**CWW**] (see also section 6.1 in [**W2**]); the support of a function χ is denoted $\operatorname{supp}\chi$:

DEFINITION 2.1. Let (X, ω) be a Kähler manifold. Then we say that it is (a) quasi-complete Kähler (manifold) if there exist a subfamily $\{\chi_j\}_{j\geq 1} \subset C_0^\infty(X, [0,1])$,

[1] Thanks to one of the referees for suggesting the name "Jet selection theorem" to us

and Kähler open neighborhoods (X_j, ω_j) of $\operatorname{supp} \chi_j$ such that:

(i) $X_j \subset X_{j+1}$ with $X = \bigcup_{j=1}^{\infty} X_j$,

(ii) $\omega_j \geq \omega$ on X_j,

(iii)

$$\lim_{j\to\infty} \sup_X |\overline{\partial}\chi_j|_{\omega_j} = 0,$$

for every compact subset K of X,

$$\lim_{j\to\infty} \sup_K |\omega_j - \omega|_\omega = 0,$$

and

(iv) for every compact subset K of X,

$$\chi_j|_K \equiv 1$$

for all large j.

In the case that (i) - (iv) hold, we also call (X_j, ω_j, χ_j) an approximation family for (X, ω).

REMARK. For simplicity, we shall use the abbreviation qck for "quasi-complete Kähler".

Recall that a Kähler manifold $(X, \hat{\omega})$ is said to be complete if there exists a real-valued smooth function ρ on X such that for every $c \in \mathbb{R}$, the sub-level set $\{\rho < c\}$ is relatively compact in X, and $|\overline{\partial}\rho|_{\hat{\omega}} \leq 1$ on X. One of our main motivations for introducing qck manifolds is the following fact:

PROPOSITION 2.2. Assume that $(X, \hat{\omega})$ is complete Kähler. Then (X, ω) is qck for every Kähler form ω on X.

PROOF. One may take $X_j := X$, $\omega_j := \omega + (1/j)\hat{\omega}$, $\chi_j := \chi(\rho/j^2)$, with $\chi : \mathbb{R} \to \mathbb{R}$ smooth such that $\chi = 1$ on $(-\infty, 1]$, $\chi = 0$ on $[3, \infty)$, $0 \leq \chi \leq 1$, and $|\chi'| \leq 1$. □

Another of our motivations is rooted in trying to find the right set-up for utilizing the Demailly approximation in Hörmander's L^2-estimate, in which case one needs to solve the $\overline{\partial}$-equation on the complement of an analytic subset of a complete Kähler manifold (see [**Ng**, page 89-91]). The point is then that while the complement might not be complete Kähler in general (in fact, it is a conjecture that there exists a counterexample), it is always qck by the above proposition and the following result (see [**Ng**, page 74, Lemma 4.1] for the proof):

PROPOSITION 2.3. Assume that (X, ω) is qck. Then $(X \setminus S, \omega)$ is qck for every analytic subset S of X.

Proposition 2.3 allows us to prove the Jet selection theorem mentioned in the introduction (Theorem 2.4), whose statement we are now ready to give. We use the following conventions: K_X denotes the canonical line bundle of X, and a line bundle $(L, e^{-\phi})$ (over X) being pseudoeffective is to mean that the local weight ϕ is plurisubharmonic. Then the Jet selection theorem is the following:

THEOREM 2.4 (The Jet selection theorem). *Let $(L, e^{-\phi})$ be a pseudoeffective line bundle over an n-dimensional qck manifold X, and introduce a family* *of norms* $\|\cdot\|_t$ *on the space* $H^0(X, \mathcal{O}(K_X+L))$ *indexed by* $t \in \mathbb{R}$, *by letting for* $u \in H^0(X, \mathcal{O}(K_X+L))$,

$$\|u\|_t^2 := \int_X i^{n^2} u \wedge \overline{u}\, e^{-\phi-\lambda \max\{G-t,0\}}, \tag{2.1}$$

where $G \leq 0$ *is a function on* X *such that* $\phi + \lambda G$ *is plurisubharmonic on* X *for some constant* $\lambda > 0$. *Fix* $t < 0$ *and* $0 \leq \alpha < \lambda$, *and let* $F \in H^0(X, \mathcal{O}(K_X+L))$ *with* $\|F\|_t^2 < \infty$. *Then there exists* $\tilde{F} \in H^0(X, \mathcal{O}(K_X+L))$ *such that*

$$\|\tilde{F} - F\|^2_{\phi+\alpha G} \leq \frac{9\lambda(1-t)^{\lambda-\alpha+1}e^{-2\alpha t}}{\lambda-\alpha}\|F\|_t^2 + 2e^{-\alpha t}\|F\|^2_\phi \tag{2.2}$$

and

$$\|\tilde{F}\|^2_\phi \leq \frac{4\lambda(1-t)^{\lambda-\alpha+1}e^{-\alpha t}}{\lambda-\alpha}\|F\|_t^2, \tag{2.3}$$

where $\|u\|^2_\psi := \int_X i^{n^2} u \wedge \bar{u}\, e^{-\psi}$.

Our main observation in this paper is the following: One may use the Jet selection theorem (Theorem 2.4) to replace the infinite dimensional version of Lemma 2.3 in [**B9**] and generalize the theory in [**B9**] to a large class of non-compact manifolds (see the main theorem, Theorem A, below).

2.2. The main theorem, a generalized sharp strong openness theorem, and a sharp Ohsawa-Takegoshi extension theorem. We next give the statement of the main theorem in this paper, and discuss applications of it to a generalization of the sharp strong openness theorem and a sharp Ohsawa-Takegoshi extension theorem. For a locally integrable function ψ, we will denote by $\mathcal{I}(\psi)$ the sheaf of germs of holomorphic functions f such that $|f|^2 e^{-\psi}$ is locally integrable. Our main theorem is the following:

THEOREM A (See page 88 of [**Ng**]). *Let* $(L, e^{-\phi})$ *be a pseudoeffective line bundle over a qck manifold* (X, ω), *and let* $G \leq 0$ *be a function on* X *such that* $\phi + \lambda G$ *is plurisubharmonic on* X *for some constant* $\lambda > 0$. *Fix* $F \in H^0(X, \mathcal{O}(K_X+L))$ *with* $\|F\|^2_\phi < \infty$. *Then for every* $0 \leq \alpha < \lambda$,

$$\mathbb{R} \ni t \mapsto e^{-\alpha t} \inf\left\{\|\tilde{F}\|_t^2 : \tilde{F} - F \in H^0(X, \mathcal{O}(K_X+L) \otimes \mathcal{I}(\phi+\alpha G))\right\}$$

is decreasing, where the norms $\|\cdot\|_t$ *are defined as in* (2.1), *by*

$$\|u\|_t^2 := \int_X i^{n^2} u \wedge \overline{u}\, e^{-\phi-\lambda \max\{G-t,0\}}.$$

In the case that X is compact, the above theorem follows from the main theorem in [**B9**]. Our motivation is to use the Jet selection theorem (Theorem 2.4) to prove the non-compact case. The proof of Theorem A is given in the next section, and the proof of the Jet selection theorem (Theorem 2.4) is given (in a simple setting) in section 4.

We now discuss applications, starting with strong openness. When X is the unit ball in $\mathbb{C}^n$, Theorem A gives the following effective strong openness theorem:

THEOREM B. *Let F be a holomorphic function on the unit ball $\mathbb{B}$ in $\mathbb{C}^n$, ϕ a plurisubharmonic function on $\mathbb{B}$, and $G \le 0$ a function on $\mathbb{B}$ such that $\phi + \lambda G$ is plurisubharmonic for some positive constant $\lambda > 0$. Suppose that*

$$\int_{\mathbb{B}} |F|^2 \, e^{-\phi-\beta G} < \infty$$

for some constant $0 < \beta < \lambda$, where integration is with respect to the Lebesgue measure. Then $|F|^2 \, e^{-\phi-\alpha G}$ is integrable near the origin for some $\alpha > \beta$. Moreover, letting

$$\alpha_0 := \sup\{\alpha \ge 0 : F_0 \in \mathcal{I}(\phi + \alpha G)_0\},$$

then if $\alpha_0 < \lambda$, we have

$$I_0 := \inf \left\{ \int_{\mathbb{B}} |\tilde{F}|^2 e^{-\phi} : (\tilde{F} - F)_0 \in \bigcup_{\alpha_0 < \alpha \le \lambda} \mathcal{I}(\phi + \alpha G)_0 \right\} > 0, \tag{2.4}$$

and

$$\alpha_0 - \beta \ge \frac{I_0}{\frac{\lambda}{(\lambda-\beta)\beta} \int_{\mathbb{B}} |F|^2 \, e^{-\phi-\beta G} - \frac{1}{\beta} \int_{\mathbb{B}} |F|^2 e^{-\phi}}. \tag{2.5}$$

PROOF. It suffices to prove (2.4) and (2.5).

STEP 1 (Proof of (2.4)). Nadel [**Na**] proved that each $\mathcal{I}(\phi + \alpha G)$ is a coherent ideal of the sheaf $\mathcal{O}_{\mathbb{B}}$ of germs of local holomorphic functions on $\mathbb{B}$. Since $G \le 0$, we get that $\mathcal{I}(\phi + \alpha G)$ is decreasing with respect to α. Thus the strong Noetherian property of coherent analytic sheaves (see [**De**, page 90]) gives

$$\bigcup_{\alpha_0 < \alpha \le \lambda} \mathcal{I}(\phi + \alpha G)_0 = \mathcal{I}(\phi + \boldsymbol{\alpha} G)_0 \text{ for some } \alpha_0 < \boldsymbol{\alpha} \le \lambda. \tag{2.6}$$

Now we have

$$I_0 = \inf \left\{ \int_{\mathbb{B}} |\tilde{F}|^2 e^{-\phi} : (\tilde{F} - F)_0 \in \mathcal{I}(\phi + \boldsymbol{\alpha} G)_0 \right\} \tag{2.7}$$

for some $\alpha_0 < \boldsymbol{\alpha} \le \lambda$. If $I_0 = 0$ then we can find holomorphic functions F_j on $\mathbb{B}$ with

$$\int_{\mathbb{B}} |F_j|^2 e^{-\phi} \le 2^{-j}, \ (F_j - F)_0 \in \mathcal{I}(\phi + \boldsymbol{\alpha} G)_0.$$

Thus $F_j \to 0$ locally uniformly on $\mathbb{B}$ and Cartan's closedness of analytic ideals gives $F \in \mathcal{I}(\phi + \boldsymbol{\alpha} G)_0$, which contradicts that $\alpha_0 < \boldsymbol{\alpha}$. Hence $I_0 > 0$.

STEP 2 (Proof of (2.5)). Using the following calculus identity from [**B7**] (which also follows by a direct computation):

$$\int_{-\infty}^{0} e^{-\lambda \max\{s-t,0\}} e^{-\beta t} dt = \frac{\lambda}{(\lambda - \beta)\beta} e^{-\beta s} - \frac{1}{\beta}, \ \ \forall\, \lambda > \beta, \ \ s \le 0, \tag{2.8}$$

we get

$$\int_{-\infty}^{0} \|F\|_t^2 e^{-\beta t} dt = \frac{\lambda}{(\lambda - \beta)\beta} \int_{\mathbb{B}} |F|^2 \, e^{-\phi-\beta G} - \frac{1}{\beta} \int_{\mathbb{B}} |F|^2 e^{-\phi}, \tag{2.9}$$

where

$$\|F\|_t^2 := \int_{\mathbb{B}} |F|^2 e^{-\phi - \lambda \max\{G-t,0\}}. \tag{2.10}$$

Our main theorem (Theorem A) with $X := \mathbb{B}$ and $\alpha_0 < \alpha < \boldsymbol{\alpha}$ gives

$$e^{-\alpha t} A(t) \geq A(0), \ \forall\, t < 0,$$

where

$$A(t) := \inf \left\{ \|\tilde{F}\|_t^2 : \tilde{F} - F \in H^0(X, \mathcal{O}(K_X + L) \otimes \mathcal{I}(\phi + \alpha G)) \right\}.$$

Thus we have

$$\|F\|_t^2 \geq A(t) \geq e^{\alpha t} A(0) \geq e^{\alpha t} I_0, \ \forall\, \alpha_0 < \alpha < \boldsymbol{\alpha}.$$

Letting $\alpha \to \alpha_0$, we get $\|F\|_t^2 \geq e^{\alpha_0 t} I_0$, which gives

$$\int_{-\infty}^{0} \|F\|_t^2 e^{-\beta t} dt \geq \left(\int_{-\infty}^{0} e^{\alpha_0 t} e^{-\beta t} dt \right) I_0 = \frac{I_0}{\alpha_0 - \beta}.$$

Hence we have

$$\alpha_0 - \beta \geq \frac{I_0}{\int_{-\infty}^{0} \|F\|_t^2 e^{-\beta t} dt}.$$

Equation (2.5) now follows from (2.9).

□

REMARK. In the case that G is plurisubharmonic and $\phi = 0$, letting $\lambda \to \infty$, we obtain

$$\alpha_0 - \beta \geq \frac{\beta\, I_0}{\int_{\mathbb{B}} |F|^2 e^{-\beta G} - \int_{\mathbb{B}} |F|^2} \geq \frac{\beta\, I_0}{\int_{\mathbb{B}} |F|^2 e^{-\beta G} - I_0},$$

which directly gives Guan's sharp estimate

$$\frac{\alpha_0 - \beta}{\alpha_0} \geq \frac{I_0}{\int_{\mathbb{B}} |F|^2 e^{-\beta G}}. \tag{2.11}$$

The non-effective version is called the strong openness conjecture of Demailly (see [**De1**, Remark 15.2.2]). It was first proved in general by Guan-Zhou in [**GZ2**], while the $n = 2$ case was first proved in [**FJ**, **JM**] (see also [**Hi**, **Le1**, **GZ3**] for other proofs). In the case that $F = 1$, the strong openness conjecture is also called the openness conjecture, and was first proved by Berndtsson in [**B7**]. Recent developments around the strong openness theory include [**BG**, **BG1**, **Gu1**, **GM**, **GY**]. In [**BG**], Bao-Guan improved (2.11) by replacing I_0 with a Bergman kernel term. The effective L^p version of the strong openness property is proved in [**BG1**, **GY**]. The proof in [**GY**] depends on the concavity property of minimal L^2 integrals over the sub-level sets of plurisubharmonic functions in [**Gu1**, **GM**]. In Theorem B, we look at a more general quasi-plurisubharmonic G. As far as we know, the concavity property associated with the sub-level sets of quasi-plurisubharmonic functions is still unknown. Our proof is also different from [**BG**, **GY**] in the sense that we only use the Berndtsson convexity theorem for product domains and the classical Hörmander L^2- estimate. That is, our arguments involve neither of the non-product version of Berndtsson's theorem, the Ohsawa-Takegoshi theorem, and the ODE-trick for twisted Bochner-Kodaira-Nakano formula.

Next, we turn to L^2-holomorphic extension and our second application of Theorem A. That is, the following sharp Ohsawa-Takegoshi estimate; the reader may refer to [**B1**, **B2**, **BCP**, **BL**, **Bl**, **Bl1**, **Ch**, **CC**, **CDM**, **C**, **CWW**, **De3**, **DHP**, **GZ1**, **Ho**, **K1**, **K2**, **KS**, **Le2**, **Ma**, **MV1**, **MV2**, **MV3**, **O2**, **O3**, **Oh**, **Siu**, **W1**, **ZZ**] for many other interesting generalizations and applications of the Ohsawa-Takegoshi theorem [**OT**]:

THEOREM C (See page 102 of [**Ng**]). *Let $(L, e^{-\phi})$ be a pseudoeffective line bundle over an n-dimensional qck manifold (X, ω), and let $G \leq 0$ be a function on X such that $\phi + \lambda G$ is plurisubharmonic on X for some constant $\lambda > 1$. Fix $F \in H^0(X, \mathcal{O}(K_X + L))$ with*

$$\|F\|_\phi^2 = \int_X i^{n^2} F \wedge \bar{F} e^{-\phi} < \infty.$$

Then there exists $\tilde{F}$ satisfying

$$\tilde{F} - F \in H^0(X, \mathcal{O}(K_X + L) \otimes \mathcal{I}(\phi + G))$$

and

$$\|\tilde{F}\|_\phi^2 \leq \frac{\lambda}{\lambda - 1} \limsup_{s \to -\infty} e^{-s} \int_{G<s} i^{n^2} F \wedge \bar{F} \, e^{-\phi}. \tag{2.12}$$

PROOF. By Theorem A with $\alpha = 1$, it suffices to show that

$$\limsup_{t \to -\infty} e^{-t} \|F\|_t^2 \leq \frac{\lambda}{\lambda - 1} \limsup_{s \to -\infty} e^{-s} \int_{G<s} i^{n^2} F \wedge \bar{F} \, e^{-\phi}. \tag{2.13}$$

Put $d\mu := i^{n^2} F \wedge \bar{F} \, e^{-\phi}$. Then

$$\mu(X) = \|F\|_\phi^2, \quad \mu(G < s) = \int_{G<s} i^{n^2} F \wedge \bar{F} \, e^{-\phi},$$

and we have

$$\|F\|_t^2 = \int_X e^{-\lambda \max\{G-t,0\}} \, d\mu = e^{\lambda t} \mu(X) - \int_{-\infty}^0 \mu(G < s) \, d e^{-\lambda \max\{s-t,0\}}.$$

Since $\mu(X)$ is finite and $\lambda > 1$, we have

$$\lim_{t \to -\infty} e^{-t} e^{\lambda t} \mu(X) = 0,$$

and hence

$$\begin{aligned}
\limsup_{t \to -\infty} e^{-t} \|F\|_t^2 &= \limsup_{t \to -\infty} \left(-e^{-t} \int_{-\infty}^0 \mu(G < s) \, d e^{-\lambda \max\{s-t,0\}} \right) \\
&= \limsup_{t \to -\infty} \left(-e^{-t} \int_t^0 \mu(G < s) \, d e^{-\lambda(s-t)} \right) \\
&= \limsup_{t \to -\infty} \left(\lambda \int_t^0 \mu(G < s) e^{-s} e^{-(\lambda-1)(s-t)} \, ds \right) \\
&= \limsup_{t \to -\infty} \left(\lambda \int_0^{-t} \mu(G < x + t) e^{-(x+t)} e^{-(\lambda-1)x} \, dx \right) \\
&\leq \left(\lambda \int_0^\infty e^{-(\lambda-1)x} \, dx \right) \limsup_{s \to -\infty} \left(\mu(G < s) e^{-s} \right) \\
&= \frac{\lambda}{\lambda - 1} \limsup_{s \to -\infty} e^{-s} \int_{G<s} i^{n^2} F \wedge \bar{F} \, e^{-\phi},
\end{aligned}$$

which gives the inequality. □

REMARK. In the case that X is a pseudoconvex domain in $\mathbb{C}^n$ and G is plurisubharmonic, letting $\lambda \to \infty$, Theorem C reduces to the main theorem of Berndtsson and Lempert in [**BL**]. See also [**Al, NW**] for other applications of the Berndtsson-Lempert method.

Let us denote by Z the non-integrable locus of $e^{-\phi-G}$. Then theorem C tells us that $\tilde{F}$ is a good L^2-extension of the restriction to Z, say $f := F|_Z$, of F. Of course, F is automatically an L^2-extension of f. Thus our theorem is only the "estimate" part of the sharp L^2-extension theorem. In the case that X is Stein or projective, Ohsawa-Takegoshi [**OT**] found that the "estimate" part also gives the "existence" part of the L^2-extension. For a more general X, one may use a partition of unity, $1 = \sum \chi_j$, to glue the local L^2-extensions f_j and obtain a global smooth L^2-extension

$$F := \sum \chi_j f_j.$$

In general, $\overline{\partial} F$ is not zero, but its coefficients are always locally square integrable with respect to $e^{-\phi-G}$. We will write for this $\overline{\partial} F \in L^2_{loc,\phi+G}$ and use a similar meaning for the notation $L^2_{loc,\phi}$. In the case that X is weakly pseudoconvex Kähler, the following sharp Ohsawa-Takegoshi theorem is proved in [**ZZ**]:

THEOREM 2.5. *Let $(L, e^{-\phi})$ be a holomorphic line bundle with quasi-plurisubharmonic metric (that is, ϕ is locally the sum of a smooth function and a plurisubharmonic function) over an n-dimensional weakly pseudoconvex Kähler manifold X, and let $G \leq 0$ be a function on X such that*

$$\phi + G \text{ and } \phi + \lambda G \text{ are plurisubharmonic}$$

for some constant $\lambda > 1$ on X. Then for every smooth section F of $K_X + L$ over X with $F \in L^2_{loc,\phi}$, $\overline{\partial} F \in L^2_{loc,\phi+G}$ and

$$\|F\|_G^2 := \sup_{K \text{ compact in } X} \limsup_{t\to-\infty} \int_{\{t<G<t+1\}\cap K} i^{n^2} F \wedge \overline{F}\, e^{-(\phi+G)} < \infty, \tag{2.14}$$

there exists $\tilde{F} \in H^0(X, \mathcal{O}(K_X + L) \otimes \mathcal{I}(\phi))$ such that $\tilde{F} - F \in L^2_{loc,\phi+G}$ and

$$\|\tilde{F}\|_\phi^2 \leq \frac{\lambda}{\lambda - 1} \|F\|_G^2. \tag{2.15}$$

REMARK. As above, one may think of F as a smooth extension of some $f \in H^0(X, \mathcal{O}(K_X + L) \otimes \mathcal{I}(\phi)/\mathcal{I}(\phi + G))$. Then (2.14) is known as the Ohsawa norm (see [**Oh**]) for f. It does not depend the choice of the smooth extension F. In fact, if F_1, F_2 are two such smooth extensions (from different partitions of unity and different local L^2-extensions) then one may verify that $F_1 - F_2 \in L^2_{loc,\phi+G}$. Hence we have

$$\limsup_{t\to-\infty} \int_{\{t<G<t+1\}\cap K} i^{n^2} F_1 \wedge \overline{F_1}\, e^{-(\phi+G)} = \limsup_{t\to-\infty} \int_{\{t<G<t+1\}\cap K} i^{n^2} F_2 \wedge \overline{F_2}\, e^{-(\phi+G)}$$

for every compact set K in X, from which we get that $\|F_1\|_G^2 = \|F_2\|_G^2$. Thus Theorem 2.5 is an equivalent form of [**ZZ**, Theorem 1.2, $R(t) = e^{-t}$]. Assume that

$$\limsup_{s\to-\infty} e^{-s} \int_{\{G<s\}\cap K} i^{n^2} F \wedge \overline{F}\, e^{-\phi} = \liminf_{s\to-\infty} e^{-s} \int_{\{G<s\}\cap K} i^{n^2} F \wedge \overline{F}\, e^{-\phi} \tag{2.16}$$

for every compact set K in X. Then we have

$$\|F\|_G^2 = \sup_{K \text{ compact in } X} \limsup_{s\to-\infty} e^{-s} \int_{\{G<s\}\cap K} i^{n^2} F \wedge \overline{F}\, e^{-\phi},$$

and hence

$$\|F\|_G^2 \leq \limsup_{s\to-\infty} e^{-s} \int_{G<s} i^{n^2} F \wedge \overline{F}\, e^{-\phi}. \tag{2.17}$$

Thus if (2.16) in the above remark holds and X is weakly pseudoconvex, then Theorem 2.5 implies our Theorem C (with weaker curvature assumptions). The proofs, however, of the two theorems even in this special case, are quite different. The qck case of Theorem 2.5 is open as far as we know.

We end this section with the following remark:

REMARK. Using the language of the above remarks, the following sharp Ohsawa-Takegoshi extension theorem may be viewed as a corollary of our Theorem C; note the existence assumption:

COROLLARY 2.6. *Under the same assumptions as in Theorem C, for any smooth section F of $K_X + L$ with $F \in L^2_{loc,\phi}, \bar\partial F \in L^2_{loc,\phi+G}$, if there exists an extension $F_1 \in H^0(X, \mathcal{O}(K_X + L) \otimes \mathcal{I}(\phi))$ in the sense that $F_1 - F \in L^2_{loc,\phi+G}$, with $\|F_1\|_\phi < \infty$, then we may choose an extension F_2 such that*

$$\|F_2\|^2_\phi \leq \frac{\lambda}{\lambda - 1} \limsup_{s \to -\infty} e^{-s} \int_{G<s} i^{n^2} F_1 \wedge \bar{F}_1 e^{-\phi}.$$

PROOF. Replace the F that appears in Theorem C by F_1. The assertion now follows by writing $F_2 - F = (F_2 - F_1) - (F - F_1)$ and noting that both of the terms in the parenthesis on the right-hand side lie in $L^2_{loc,\phi+G}$ by assumption. Hence, so does the left-hand side. □

3. Hilbert bundle approach to complex Brunn-Minkowski theory and proof of the main theorem

In this section, we prove the main theorem (Theorem A), assuming the statement of the Jet selection theorem 2.4 (the proof of which comes in section 4). Aside from the latter, the main ingredient in the proof is a version of Berndtsson's convexity theorem in the general setting of qck manifolds (see Theorem 3.9 below). That is, is the convexity of (the logarithm of) certain norms of continuous linear functionals on certain "Bergman type of spaces" (to be elaborated below). These norms depend on some (real) parameter t, and the idea is to compute the second order derivative of the norms with respect to t directly (and then estimate using L^2-methods). For this, we (first) need to know that the norms depend smoothly on t. In the classical approach to the complex Brunn-Minkowski theory, in the setting of pseudoconvex domains, this smoothness property, which follows from regularity of the Bergman projection, is established by means of the Hamilton-Kohn regularity theory for the $\bar\partial$-Neumann operator on smoothly bounded strictly pseudoconvex domains. To deal with the smoothness of the norms in the more general setting, we introduce (what we call) a Hilbert bundle approach to the complex Brunn-Minkowski theory, which allows us to deal with the question of regularity using much more elementary means. In fact, the key ingredient becomes the implicit mapping theorem (see Theorem 3.7 and its proof below). The argument involving the implicit mapping theorem was taught to us by Berndtsson. To introduce the smoothness result, let us first start by properly defining the corresponding Bergman type of spaces, and norms, that we are going to work with.

3.1. Positively curved Hilbert bundles. Let H be a complex Hilbert space. We shall call the product

$$\mathcal{H} := H \times \mathbb{C}$$

a (trivial) Hilbert bundle. By a sub-bundle of $\mathcal{H}$ we will simply mean a Hilbert bundle $\mathcal{H}_0 = H_0 \times \mathbb{C}$ where H_0 is a closed subspace of H. Our Bergman type of spaces will be such sub-bundles (of an ambient Hilbert bundle whose underlying Hilbert space is a particular L^2-space), and the norms that we shall be working with will be induced from what we call smooth hermitian metrics on Hilbert bundles as $\mathcal{H}$ above; see Definitions 3.2 and 3.4 below.

Now let $\mathcal{H}$ be as above. We follow [**Le3**, section 2] and use the following definition for the notion of holomorphic sections of $\mathcal{H}$:

DEFINITION 3.1. Let U be a domain in $\mathbb{C}$. A smooth mapping $U \to H$ is said to be holomorphic if its differential is $\mathbb{C}$-linear.

We then identify (local) holomorphic sections of $\mathcal{H}$ (over U) with holomorphic mappings $U \to H$ and denote by $H^0(U, \mathcal{O}(\mathcal{H}))$ the space of holomorphic sections of $\mathcal{H}$ over U; (local) smooth sections of $\mathcal{H}$ on U are of course identified with smooth maps $U \to H$. We may similarly define holomorphic sections for the dual bundle

$$\mathcal{H}^* := H^* \times \mathbb{C},$$

where H^* denotes the dual of H. Then $\sigma \in H^0(U, \mathcal{O}(\mathcal{H}^*))$ if and only if the mapping $U \times H \to \mathbb{C}$ defined by

$$(\tau, u) \to \sigma(\tau)(u)$$

is holomorphic and $\mathbb{C}$-linear in the second argument. In particular, each $f \in H^*$ defines a (constant) holomorphic section with the associated mapping $U \times H \to \mathbb{C}$ given by

$$(\tau, u) \to f(u).$$

DEFINITION 3.2. Let $\mathcal{H} = H \times \mathbb{C}$ be a Hilbert bundle as above. A smooth hermitian metric on $\mathcal{H}$ is a smooth assignment $h := \{h_\tau\}_{\tau \in \mathbb{C}}$ to each $\tau \in \mathbb{C}$ a complete inner product h_τ on H, where the smoothness means that for every pair of (local) smooth sections u and v of $\mathcal{H}$, the complex-valued function

$$\tau \mapsto h_\tau(u(\tau), v(\tau)) := h(u, v)(\tau)$$

is smooth.

Let $L(H)$ denote the space of bounded self-adjoint linear operators T on H such that T is bijective with positive spectrum, and let h be a smooth mapping $\mathbb{C} \to L(H)$. We identify h with the smooth (hermitian) metric on $\mathcal{H}$ (slightly abusing the notation) given by

$$h(u, v)(\tau) := (h(\tau)u, v), \quad u, v \in H, \tau \in \mathbb{C},$$

where $(\cdot, \cdot)$ denotes the inner product on H (and an element $x \in H$ is viewed as a constant section $\tau \mapsto x$ of $\mathcal{H}$).

REMARK. Given a smooth metric $h : \mathbb{C} \to L(H)$ on $\mathcal{H}$ as above, this defines a family of Hilbert norms $\|\cdot\|_{h,\tau} := \|\cdot\|_\tau$ on H where

$$\|u\|_\tau^2 := h(u, u)(\tau) = (h(\tau)u, u), \quad u \in H.$$

By the open mapping theorem, each new norm $\|\cdot\|_\tau$ is equivalent to the original norm on H.

If h is a smooth hermitian metric on $\mathcal{H}$, we denote the corresponding dual metric on the dual bundle $\mathcal{H}^*$ by h^*. The corresponding induced (family of) norm(s) $\|\cdot\|_{h^*,\tau} := \|\cdot\|_\tau$ is then given by

$$\|f\|_\tau = \sup\{|f(u)| : \|u\|_\tau = 1\}, \quad f \in H^*.$$

We now have the following definition for the the positivity of (the curvature of) smooth metrics $h : \mathbb{C} \to L(H)$ on $\mathcal{H}$:

DEFINITION 3.3. Let $h : \mathbb{C} \to L(H)$ be a smooth metric on $\mathcal{H}$. Then we say that h is positively curved if the dual metric h^* on $\mathcal{H}^*$ is negatively curved in the following sense: For every open set $U \subset \mathbb{C}$ and every $\sigma \in H^0(U, \mathcal{O}(\mathcal{H}^*))$, the norm function

$$\tau \mapsto \|\sigma(\tau)\|_\tau^2$$

is subharmonic on U, where $\|\cdot\|_\tau = \|\cdot\|_{h^*,\tau}$.

As explained by Berndtsson in page 2362 in [**B9**], the condition in the previous definition is equivalent to that $\tau \mapsto \log\|\sigma(\tau)\|_\tau$ is subharmonic (it can be identically equal to $-\infty$), and we shall usually use the latter directly for the positivity of smooth metrics on $\mathcal{H}$.

3.2. Complex Brunn-Minkowski theory. The complex Brunn-Minkowski theory established by Berndtsson (see [**B5, B6**], see also [**B8**] for a short survey) is around the positive curvature properties of certain vector bundles associated to holomorphic fibrations. In terms of our Hilbert bundle language, this corresponds to the positive curvature properties of the sub-bundle $\mathcal{H}_0$ of a fixed ambient Hilbert bundle $\mathcal{H}_1$. The underlying Hilbert spaces of these bundles, respectively H_0 and H_1, are defined as follows:

DEFINITION 3.4. Let $(L, e^{-\phi})$ be a pseudoeffective line bundle over a complex manifold X. We define H_1 to be the space of measurable sections u of $K_X + L$ with finite L^2-norm (squared)

$$\|u\|_\phi^2 := \int_X i^{n^2} u \wedge \bar{u}\, e^{-\phi} < \infty,$$

and H_0 to be the subspace $H_0 := H^0(X, \mathcal{O}(K_X + L)) \cap H_1$, called the Bergman space.

REMARK. Sometimes we will write $\|u\|_\phi$ simply as $\|u\|$.

To define a smooth Hermitian metric on our Hilbert bundles $\mathcal{H}_1$ and $\mathcal{H}_0$, we fix a non-positive measurable function $G : X \to [-\infty, 0]$, and a smooth function $\chi : \mathbb{R} \to [0, \infty)$ that vanishes identically on $(-\infty, 0]$. The following proposition is proved in [**Ng**, page 83-85].

PROPOSITION 3.5. The mapping $h : \mathbb{C} \to L(H_1)$ defined by

$$h(\tau)(u) := e^{-\chi(G - \operatorname{Re}\tau)} u, \quad \tau \in \mathbb{C}, \quad u \in H_1,$$

is smooth.

We take the induced smooth Hermitian metric from the map h in the previous proposition as a smooth metric on our Hilbert bundle $\mathcal{H}_1 = H_1 \times \mathbb{C}$. The associated family of new norms $\|\cdot\|_\tau$, on H_1 and H_0 is given by

$$\|u\|_\tau^2 := \int_X i^{n^2} u \wedge \bar{u}\, e^{-\phi - \chi(G - \operatorname{Re}\tau)}. \tag{3.1}$$

Thus our Hilbert bundles $\mathcal{H}_1$ and $\mathcal{H}_0$ are trivial families of vector spaces with non-trivial families of equivalent Hilbert norms. We are interested in the positivity of the curvature of the sub-bundle $\mathcal{H}_0$. Indeed, by Definition 3.3, this amounts to the convexity of the map $t \mapsto \log \|f\|_t$ for every holomorphic section f of the dual of $\mathcal{H}_0$. The following definition is of great relevance:

DEFINITION 3.6. Let H_1 and H_0 be as in Definition 3.4. We call the canonical orthogonal projection

$$P^\tau : H_1 \to H_0$$

with respect to the above $\|\cdot\|_\tau$-norm a Bergman projection.

The first step in establishing the positivity of the curvature of the sub-bundle $\mathcal{H}_0$, is to establish that $\tau \mapsto \|f(\tau)\|_\tau^2$ is smooth for every smooth section f of the dual of $\mathcal{H}_0$. To do this, we may prove that P^τ depends smoothly on τ. The Bergman projection P^τ naturally appears as we are now dealing with continuous linear functionals on the *sub*space H_0. This has to do with the fact that the ((1,0)-part of) the Chern connection of h restricted to $\mathcal{H}_0$ is related to that of h on $\mathcal{H}_1$ via the projection map P^τ, and that derivatives of the form $\frac{d}{d\tau}\|u\|_\tau^2, u \in H_i, i \in \{0,1\}$, can be computed from the ((1,0)-part of) Chern connection of h on $\mathcal{H}_i$. We omit here the details. They can be read in [**Ng**, chapter 1]. For smoothly bounded strongly pseudoconvex domains, the projection map P^τ amounts to the usual (weighted) Bergman projection and its smoothness follows as mentioned earlier from Kohn-Hamilton's regularity theory (see Lemma 2.1 in [**B5**] and Theorem 1.1 in [**B6**]). In our current setting, we have the following regularity theorem mentioned above using an implicit mapping theorem argument (see also [**Ng**] for generalizations):

THEOREM 3.7. *With the notation above,*

$$\begin{aligned} P : \mathbb{C} \times H_1 &\to H_0 \\ (\tau, u) &\mapsto P^\tau u, \end{aligned}$$

is a smooth map from $\mathbb{C} \times (H_1, \|\cdot\|)$ *to* $(H_0, \|\cdot\|)$.

PROOF. We give a short proof. Fix $(\tau_0, u_0) \in \mathbb{R} \times H$, and consider the smooth map

$$\begin{aligned} F : \mathbb{C} \times H_1 \times H_0 &\to H_0 \\ (\tau, u, v) &\mapsto P^{t_0}(e^{\chi(G - \operatorname{Re}\tau_0) - \chi(G - \operatorname{Re}\tau)}(v - u)) \end{aligned}$$

(smoothness of F follows from Proposition 3.5). One may show that $F(\tau, u, v) = 0$ if and only if $v = P^\tau u = P(\tau, u)$. Hence, since $F_{H_0}(\tau_0, u_0, v) \equiv id_{H_0}$ (that is, the partial derivative of F in the H_0-direction, at (t_0, u_0, v), is equal to the identity map on H_0), the smoothness of P follows directly from the implicit mapping theorem. □

Asssume that X is smoothly bounded strongly pseudoconvex. In [**B5**, Theorem 1.1], it is proved by Berndtsson, using Hörmander's theorem on L^2-estimates of solutions of the $\overline{\partial}$-operator, that if the total weight is (strictly) plurisubharmonic then $\mathcal{H}_0$ is positively curved. In our case, the total weight function is

$$\Phi : (z, \tau) \mapsto \phi(z) + \chi(G(z) - \operatorname{Re}\tau).$$

Note that if $\chi'' \geq 0$ then

$$i\partial\overline{\partial}\Phi \geq i\partial\overline{\partial}\phi + \chi' \, i\partial\overline{\partial}G.$$

Thus if ϕ and $\phi + \lambda G$ are plurisubharmonic for some constant $\lambda > 0$, and χ is convex with $0 \leq \chi' \leq \lambda$, then Φ is plurisubharmonic. With these assumptions, using Theorem 3.7 (and L^2-methods on qck manifolds; see [**Ng**, chapter 2]), the first named author gives in [**Ng**] the following version of [**B5**, Theorem 1.1] for all qck X.

THEOREM 3.8 (See Theorem 1.6 in page 59 of [**Ng**]). *Let $(L, e^{-\phi})$ be a pseudoeffective line bundle over an n-dimensional qck manifold X. Let $G \leq 0$ be a function on X such that $\phi + \lambda G$ is plurisubharmonic for some constant $\lambda > 0$. Let $\chi \geq 0$ be a smooth convex function on $\mathbb{R}$ that vanishes identically on $(-\infty, 0)$. Assume that $0 \leq \chi' \leq \lambda$. Then the metric defined in* (3.1) *on $\mathcal{H}_0$, by*

$$\|u\|_\tau^2 := \int_X i^{n^2} u \wedge \bar{u}\, e^{-\phi - \chi(G - \operatorname{Re}\tau)},$$

is positively curved in the sense of Definition 3.3.

Any fixed bounded $\mathbb{C}$-linear functional $f : H_0 \to \mathbb{C}$ defines a constant, thus holomorphic, section of $\mathcal{H}_0^*$. Hence, the above theorem implies that

$$\tau \mapsto \log \|f\|_\tau,$$

is subharmonic on $\mathbb{C}$, where

$$\|f\|_\tau := \sup\{|f(u)| : u \in H_0,\ \|u\|_\tau = 1\}.$$

Since $\|f\|_\tau$ depends only on the real part of τ, we get that $\log \|f\|_t$ is convex in $t \in \mathbb{R}$. Letting $\chi(s)$ converge to $\lambda \max\{s, 0\}$, we finally obtain the following Berndtsson's convexity theorem in our current setting:

THEOREM 3.9. *Let $(L, e^{-\phi})$ be a pseudoeffective line bundle over an n-dimensional qck manifold X. Let $G \leq 0$ be a function on X such that $\phi + \lambda G$ is plurisubharmonic for some constant $\lambda > 0$. Let f be a bounded $\mathbb{C}$-linear functional on*

$$H_0 := \left\{ u \in H^0(X, \mathcal{O}(K_X + L)) : \int_X i^{n^2} u \wedge \bar{u}\, e^{-\phi} < \infty \right\}.$$

Put

$$\|f\|_t := \sup \left\{ |f(u)| : u \in H_0,\ \int_X i^{n^2} u \wedge \bar{u}\, e^{-\phi - \lambda \max\{G - t, 0\}} = 1 \right\}, \quad t \in \mathbb{R}.$$

Then $t \mapsto \log \|f\|_t$ is convex on $\mathbb{R}$.

With this, we are now ready for the proof of the main theorem (Theorem A).

3.3. Proof of our main theorem. Assuming the (statement of the) Jet selection theorem (Theorem 2.4), we shall follow [**B9**] and prove our main theorem (Theorem A in section 2.2). Following [**B9**]. We look at quotient spaces of the form

$$Q := H_0/S.$$

Since $\mathcal{H}_0$ is positively curved (with respect to some smooth hermitian metric h), the corresponding Hilbert bundle $\mathcal{Q}$ is positively curved (this follows from a quotient bundle curvature formula, see [**Ng**, page 43-44]). Thus Theorem 3.9 applies and gives the following:

THEOREM 3.10. *Let $(L, e^{-\phi})$ be a pseudoeffective line bundle over an n-dimensional qck manifold X. Let $G \leq 0$ be a function on X such that $\phi + \lambda G$ is plurisubharmonic for some constant $\lambda > 0$. Let f be a bounded $\mathbb{C}$-linear functional on the quotient Hilbert space $Q := H_0/S$. Then $t \mapsto \log \|f\|_t$ is convex on $\mathbb{R}$, where*

$$\|f\|_t := \sup\{|f([u])| : [u] \in H_0/S,\ \|[u]\|_t = 1\},\quad t \in \mathbb{R},$$

and the quotient norm $\|[u]\|_t$ on H_0/S (of $[u] \in H_0/S$) is defined by

$$\|[u]\|_t^2 := \inf\left\{\int_X i^{n^2} v \wedge \bar{v}\, e^{-\phi-\lambda \max\{G-t,0\}} : v - u \in S\right\}. \tag{3.2}$$

REMARK. In order to prove our main theorem (Theorem A), we apply the above theorem to $S = S_\alpha$, $0 \leq \alpha < \lambda$, where

$$S_\alpha := H_0 \cap H^0(X, \mathcal{O}(K_X + L) \otimes \mathcal{I}(\phi + \alpha G)).$$

Then one may rephrase our main theorem (Theorem A) as the following:

THEOREM 3.11 (Equivalent to the main theorem). *Let $(L, e^{-\phi})$ be a pseudoeffective line bundle over an n-dimensional qck manifold X. Let $G \leq 0$ be a function on X such that $\phi + \lambda G$ is plurisubharmonic for some constant $\lambda > 0$. Then for every $0 \leq \alpha < \lambda$, $e^{-\alpha t}\|[u]\|_t^2$ is decreasing in $t \in \mathbb{R}$, where $\|[u]\|_t$ denotes the quotient norm on H_0/S_α.*

PROOF. Fix $t < 0$. For every $u \in H_0$, we may take $F \in H_0$ with $[F] = [u]$ (i.e. $F - u \in S_\alpha$) and $\|F\|_t = \|[u]\|_t$. Then the Jet selection theorem (Theorem 2.4) gives $\tilde{F} \in H_0$ with $[\tilde{F}] = [F]$ and

$$\|[u]\|_0^2 \leq \|\tilde{F}\|_\phi^2 \leq \frac{4\lambda(1-t)^{\lambda-\alpha+1}e^{-\alpha t}}{\lambda - \alpha}\|F\|_t^2 \leq C(\varepsilon, \lambda, \alpha)e^{-(\alpha+\varepsilon)t}\|[u]\|_t^2 \tag{3.3}$$

for every $\varepsilon > 0$. By (3.3), we have

$$\|[u]\|_0^2 \leq C(\varepsilon, \lambda, \alpha)e^{-(\alpha+\varepsilon)t}\|[u]\|_t^2,\quad \forall\, [u] \in H_0/S_\alpha,\quad t < 0,$$

which is equivalent to the following dual norm estimate:

$$\|f\|_t^2 \leq C(\varepsilon, \lambda, \alpha)e^{-(\alpha+\varepsilon)t}\|f\|_0^2,\quad \forall\, f \in (H_0/S_\alpha)^*,\quad t < 0.$$

In particular, we see that $\log(\|f\|_t^2) + (\alpha + \varepsilon)t$ is bounded from above at $t = -\infty$. By Theorem 3.10, $\log(\|f\|_t^2) + (\alpha + \varepsilon)t$ is convex in t. Hence it must be increasing. Letting ε go to zero, we find that $\alpha t + \log(\|f\|_t^2)$ is increasing for every $f \in (H_0/S_\alpha)^*$. Thus $e^{-\alpha t}\|[u]\|_t^2$ is decreasing for every $[u] \in H_0/S_\alpha$. □

REMARK. In the case that $\dim H_0 < \infty$, (3.3) is precisely [**B9**, Lemma 2.3].

4. Proof of the Jet selection theorem in a simple setting

In this section, we prove the Jet selection theorem (Theorem 2.4) in a special case, namely unit ball (local) case. The full proof, which depends heavily on the Demailly approximation theory [**De2**] and the $\overline{\partial}$-L^2 theory [**W2**] on qck manifolds, can be found in page 89-91 of [**Ng**]. In the unit ball case, the Jet selection theorem (Theorem 2.4) is the following:

THEOREM 4.1. *Let F be a holomorphic function on the unit ball $\mathbb{B}$ in $\mathbb{C}^n$, ϕ be a plurisubharmonic function on $\mathbb{B}$, and $G \le 0$ be a function on $\mathbb{B}$ such that $\phi + \lambda G$ is plurisubharmonic for some positive constant $\lambda > 0$. Then for every $t < 0$ and $0 \le \alpha < \lambda$ we can find a holomorphic function $\tilde{F}$ on $\mathbb{B}$ such that*

$$\int_{\mathbb{B}} |\tilde{F} - F|^2 e^{-\phi-\alpha G} \le \frac{9\lambda(1-t)^{\lambda-\alpha+1}e^{-2\alpha t}}{\lambda-\alpha}\|F\|_t^2 + 2e^{-\alpha t}\int_{\mathbb{B}} |F|^2 e^{-\phi} \tag{4.1}$$

and

$$\int_{\mathbb{B}} |\tilde{F}|^2 e^{-\phi} \le \frac{4\lambda(1-t)^{\lambda-\alpha+1}e^{-\alpha t}}{\lambda-\alpha}\|F\|_t^2, \tag{4.2}$$

where $\|F\|_t^2 := \int_{\mathbb{B}} |F|^2 e^{-\phi-\lambda\max\{G-t,0\}}$.

PROOF. We shall only prove the case when G and ϕ are smooth. For the general case, one may use the convolution with a smoothing kernel to replace ϕ and G with ϕ_ε and G_ε. Considering a slightly smaller ball, one may assume that ϕ_ε and G_ε are well defined on $\mathbb{B}$. Since we can choose a rotation invariant smoothing kernel, the submean inequality implies that for every $0 \le \alpha \le \lambda$, $\phi + \alpha G$ is the decreasing limit of $\phi_\varepsilon + \alpha G_\varepsilon$ as ε goes to zero. In particular,

$$\phi + \lambda\max\{G-t,0\} = \max\{\phi+\lambda G - \lambda t, \phi\}$$

is the decreasing limit of $\phi_\varepsilon + \lambda\max\{G_\varepsilon - t, 0\}$, and thus the general case follows by taking a weak limit of $\tilde{F}$. In the proof below, one may also replace max by the regularized max function. Put

$$\sigma(s) := \int_0^s 1 - e^{-\lambda\max\{x-t,0\}}\, dx. \tag{4.3}$$

Then

$$\sigma(0) = 0, \quad \sigma'(s) = 1 - e^{-\lambda\max\{s-t,0\}},$$

hence $0 \le \sigma' \le 1$ and $\sigma'(s) = 0$ when $s \le t$. Thus we have

$$t \le \sigma(t) = \sigma(-\infty) \le \sigma(G) \le 0 \tag{4.4}$$

and

$$\sigma''(G) = \lambda e^{-\lambda(G-t)} \quad \text{as } G > t. \tag{4.5}$$

Put

$$\psi := -\log(1-\sigma(G)).$$

We then have

$$i\partial\bar{\partial}\psi = \frac{i\partial\bar{\partial}\sigma(G)}{1-\sigma(G)} + i\partial\psi\wedge\bar{\partial}\psi \ge \frac{\sigma'(G)\, i\partial\bar{\partial}G + \sigma''(G)\, i\partial G\wedge\bar{\partial}G}{1-\sigma(G)}.$$

Hence our assumption gives

$$i\partial\bar{\partial}(\phi + \alpha G + (\lambda-\alpha)\psi) \ge \Theta := \frac{(\lambda-\alpha)\sigma''(G)\, i\partial G\wedge\bar{\partial}G}{1-\sigma(G)} \ge 0.$$

Applying the Hörmander L^2-estimate (see [**B4**, Theorem 6.2]), we get

$$\bar{\partial}u = \bar{\partial}((1-\sigma'(G))F)$$

with

$$\int_{\mathbb{B}} |u|^2 e^{-\phi-\alpha G-(\lambda-\alpha)\psi} \le \int_{\mathbb{B}} |\bar{\partial}((1-\sigma'(G))F)|^2_{\Theta} e^{-\phi-\alpha G-(\lambda-\alpha)\psi}. \tag{4.6}$$

Since $\sigma'(G) = 0$ when $G \leq t$, (4.5) and (4.4) imply

$$|\bar{\partial}((1-\sigma'(G))F)|^2_\Theta \, e^{-\alpha G-(\lambda-\alpha)\psi} \leq \frac{\lambda(1-t)^{\lambda-\alpha+1}e^{-\alpha t}}{\lambda-\alpha}\mathbf{1}_{G>t}\, e^{-\lambda(G-t)}|F|^2. \tag{4.7}$$

Hence (4.6) gives (note that $\psi \leq 0$)

$$\int_{\mathbb{B}} |u|^2 e^{-\phi-\alpha G} \leq \frac{\lambda(1-t)^{\lambda-\alpha+1}e^{-\alpha t}}{\lambda-\alpha}\|F\|_t^2. \tag{4.8}$$

Taking $\tilde{F} = (1-\sigma'(G))F - u$, we get

$$\int_{\mathbb{B}} |\tilde{F} - (1-\sigma'(G))F|^2 e^{-\phi-\alpha G} \leq \frac{\lambda(1-t)^{\lambda-\alpha+1}e^{-\alpha t}}{\lambda-\alpha}\|F\|_t^2. \tag{4.9}$$

Since $G \leq 0$, the above inequality gives

$$\begin{aligned}
\int_{\mathbb{B}} |\tilde{F}|^2 e^{-\phi} &\leq 2\int_{\mathbb{B}} |\tilde{F} - (1-\sigma'(G))F|^2 e^{-\phi} + 2\int_{\mathbb{B}} |(1-\sigma'(G))F|^2 e^{-\phi} \\
&\leq \frac{2\lambda(1-t)^{\lambda-\alpha+1}e^{-\alpha t}}{\lambda-\alpha}\|F\|_t^2 + 2\int_{\mathbb{B}} |(1-\sigma'(G))F|^2 e^{-\phi} \\
&= \frac{2\lambda(1-t)^{\lambda-\alpha+1}e^{-\alpha t}}{\lambda-\alpha}\|F\|_t^2 + 2\int_{\mathbb{B}} |F|^2 e^{-\phi-2\lambda\max\{G-t,0\}} \\
&\leq \left(\frac{2\lambda(1-t)^{\lambda-\alpha+1}e^{-\alpha t}}{\lambda-\alpha} + 2\right)\|F\|_t^2 \\
&\leq \frac{4\lambda(1-t)^{\lambda-\alpha+1}e^{-\alpha t}}{\lambda-\alpha}\|F\|_t^2.
\end{aligned}$$

Thus (4.2) follows. To prove (4.1), note that (4.9) gives

$$\int_{G<t} |\tilde{F} - F|^2 e^{-\phi-\alpha G} \leq \frac{\lambda(1-t)^{\lambda-\alpha+1}e^{-\alpha t}}{\lambda-\alpha}\|F\|_t^2. \tag{4.10}$$

Hence together with

$$\begin{aligned}
\int_{G\geq t} |\tilde{F} - F|^2 e^{-\phi-\alpha G} &\leq e^{-\alpha t}\int_{G\geq t} |\tilde{F} - F|^2 e^{-\phi} \\
&\leq 2e^{-\alpha t}\left(\int_{\mathbb{B}} |\tilde{F}|^2 e^{-\phi} + \int_{\mathbb{B}} |F|^2 e^{-\phi}\right),
\end{aligned}$$

we see that (4.1) follows from (4.10) and (4.2). □

5. Demailly-Kollár semi-continuity and a Donnelly-Fefferman proof of the strong openness theorem

This section is the final section of the paper, and on further applications and results. In here, we shall use Theorem 4.1 to give a proof of the Demailly-Kollár semi-continuity theorem (see Theorem 5.1), and a Donnelly-Fefferman version of Theorem 4.1 (see Corollary 5.6) to give a second proof of Guan-Zhou's strong openness theorem (see Theorem 5.7).

5.1. The semi-continuity theorem of Demailly-Kollár. In [**Na**] Nadel obtained a nice criterion for the existence of Kähler-Einstein metrics on certain Fano manifolds using his multiplier ideal sheaf. Later, Demailly-Kollár [**DK**] found a simpler proof of Nadel's criterion using the following semi-continuity result (see [**DK**, Theorem 0.2] for the $\phi = 0$ case). For the reader's convenience, we shall include a proof here using Theorem 4.1, but the idea is already implicitly included in [**DK**].

THEOREM 5.1. *Let ϕ be a psh function on the unit ball $\mathbb{B}$ in $\mathbb{C}^n$. Let $G, G_j \le 0$ such that*

$$\phi + \lambda G_j \text{ and } \phi + \lambda G \text{ are plurisubharmonic}$$

on $\mathbb{B}$ for some constant $\lambda > 1$, and assume that

$$\int_{\mathbb{B}} e^{-\phi-\beta G} < \infty, \quad \lim_{j\to\infty} \int_{\mathbb{B}} |G_j - G| = 0, \tag{5.1}$$

for some $1 < \beta < \lambda$. Then

$$\lim_{j\to\infty} \int_{|z|<r} |e^{-\phi-G_j} - e^{-\phi-G}| = 0, \quad \forall\, 0 < r < 1. \tag{5.2}$$

In the proof, we use Lemmas 5.2 and 5.3 below.

PROOF OF THEOREM 5.1. Assume for contradiction that (5.2) is not true. Then there exist $0 < r_0 < 1$, $\varepsilon_0 > 0$ and a sub-sequence, say $G_j^{(1)}$, of G_j, with

$$\int_{|z|<r_0} |e^{-\phi-G_j^{(1)}} - e^{-\phi-G}| \ge \varepsilon_0, \quad \forall\, j \ge 1. \tag{5.3}$$

By Lemma 5.2 below, we know that $G_j^{(1)}$ has a sub-sequence $G_j^{(2)}$ which converges to G almost everywhere. We shall use

(5.4)

$$\begin{aligned}\int |e^{-\phi-G_j^{(2)}} - e^{-\phi-G}| &= \int_{G_j^{(2)}\ge t} |e^{-\phi-G_j^{(2)}} - e^{-\phi-G}| + \int_{G_j^{(2)}<t} |e^{-\phi-G_j^{(2)}} - e^{-\phi-G}| \\ &\le \int_{G_j^{(2)}\ge t} |e^{-\phi-G_j^{(2)}} - e^{-\phi-G}| + \int_{G_j^{(2)}<t} e^{-\phi-G} + \int_{G_j^{(2)}<t} e^{-\phi-G_j^{(2)}}.\end{aligned}$$

The Lebesgue bounded convergence theorem implies that for every fixed $t \le -1$,

$$\limsup_{j\to\infty} \left(\int_{G_j^{(2)}\ge t} |e^{-\phi-G_j^{(2)}} - e^{-\phi-G}| + \int_{G_j^{(2)}<t} e^{-\phi-G} \right) \le \int_{G<t+1} e^{-\phi-G}. \tag{5.5}$$

The main difficulty is to control the last term $\int_{G_j^{(2)}<t} e^{-\phi-G_j^{(2)}}$ in (5.4), and the idea is to use Theorem 4.1 with $F = 1$, $G = G_j^{(2)}$ and $\alpha = 1$. Denote $\tilde{F}$ by $F_{j,t}$. Then (4.2) gives

$$\int_{\mathbb{B}} |F_{j,t}|^2 e^{-\phi} \le \frac{4\lambda(1-t)^\lambda e^{-t}}{\lambda - 1} \int_{\mathbb{B}} e^{-\phi-\lambda\max\{G_j^{(2)}-t,0\}} \tag{5.6}$$

and (4.10) gives

$$\int_{G_j^{(2)}<t} |F_{j,t} - 1|^2 e^{-\phi-G_j^{(2)}} \le \frac{\lambda(1-t)^\lambda e^{-t}}{\lambda - 1} \int_{\mathbb{B}} e^{-\phi-\lambda\max\{G_j^{(2)}-t,0\}}. \tag{5.7}$$

By the Lebesgue bounded convergence theorem we have

$$\lim_{j\to\infty}\int_{\mathbb{B}} e^{-\phi-\lambda\max\{G_j^{(2)}-t,0\}}=\int_{\mathbb{B}} e^{-\phi-\lambda\max\{G-t,0\}}. \tag{5.8}$$

Hence for every fixed $t\le -1$, we can choose $j(t)$ such that

$$\int_{\mathbb{B}} e^{-\phi-\lambda\max\{G_j^{(2)}-t,0\}}\le 2\int_{\mathbb{B}} e^{-\phi-\lambda\max\{G-t,0\}},\ \ \forall\, j\ge j(t).$$

Lemma 5.3 below further gives

$$e^{-\beta t}\int_{\mathbb{B}} e^{-\phi-\lambda\max\{G_j^{(2)}-t,0\}}\le \frac{4\lambda}{\lambda-\beta}\int_{\mathbb{B}} e^{-\phi-\beta G},\ \ \forall\, j\ge j(t),\ t\le -1.$$

Hence (5.6) implies that there exists $t_0\le -1$ such that

$$\sup_{|z|<r}|F_{j,t}(z)|\le\frac{1}{2},\ \ \forall\, j\ge j(t),\ t\le t_0.$$

Thus by (5.7) (when t_0 is small enough)

$$\int_{G_j^{(2)}<t,|z|<r} e^{-\phi-G_j^{(2)}}\le e^{(\beta-1)t/2},\ \ \forall\, j\ge j(t),\ t\le t_0.$$

By (5.5), we can choose $j(t)$ such that for all $j\ge j(t)$:

$$\int_{G_j^{(2)}\ge t,|z|<r}|e^{-\phi-G_j^{(2)}}-e^{-\phi-G}|+\int_{G_j^{(2)}<t,|z|<r} e^{-\phi-G}\le 2\int_{G<t+1,|z|<r} e^{-\phi-G}.$$

Thus (5.4) gives

$$\int_{|z|<r}|e^{-\phi-G_j^{(2)}}-e^{-\phi-G}|\le e^{(\beta-1)t/2}+2\int_{G<t+1,|z|<r} e^{-\phi-G},\ \ \forall\, j\ge j(t),\ t\le t_0,$$

which contradicts (5.3). This contradiction shows that (5.2) must be true. □

LEMMA 5.2. *Let (X,μ) be a measure space, let f_j be a sequence of measurable functions on X, and suppose that $f_j\to f$ in L^1. Then there exists a sub-sequence of f_j that converges pointwise to f almost everywhere on X.*

PROOF. Let us choose a sub-sequence $f_j^{(1)}$ such that

$$\sum_{j=1}^{\infty}\int_X |f_{j+1}^{(1)}-f_j^{(1)}|\,d\mu<\infty.$$

By the monotone convergence theorem, we know that

$$\int_X\sum_{j=1}^{\infty}|f_{j+1}^{(1)}-f_j^{(1)}|\,d\mu=\sum_{j=1}^{\infty}\int_X |f_{j+1}^{(1)}-f_j^{(1)}|\,d\mu<\infty.$$

Thus $E:=\left\{\sum_{j=1}^{\infty}|f_{j+1}^{(1)}-f_j^{(1)}|=\infty\right\}$ is a set of measure zero and $f_j^{(1)}$ converge pointwise to a measurable function g on $X\backslash E$. The dominated convergence theorem further implies that $f_j^{(1)}\to g$ in L^1. Thus $g=f$ almost everywhere and $f_j^{(1)}$ converge pointwise to f almost everywhere on X. □

REMARK. In general, L^1 convergence does not imply almost everywhere convergence, even for plurisubharmonic functions. The following is a counterexample found by the anonymous referee: Take a radial smooth function $0 \leq \chi \leq 1$ with compact support in the unit disc and $\chi(w) = 1$ for all $|w| < 1/2$. Then there exists a positive integer k such that for each $0 < r \leq 1/k$, $a \in \mathbb{C}$,

$$\phi_{a,r}(z) := \begin{cases} \max\{-1, |z|^2 + r^3\chi((z-a)/r)\log|z-a|\} & |z-a| < r; \\ |z|^2 & |z-a| \geq r. \end{cases}$$

is subharmonic on $\mathbb{C}$. Now for any integer $N \geq k$, there exist finite points a_j^N, $1 \leq j \leq j_N$ in the unit disc such that these balls $B(a_j^N, e^{-2N^3})$ cover the unit disc. Then the following sequence of subharmonic functions

$$\phi(a_1^k, 1/k),\ \phi(a_2^k, 1/k), \cdots, \phi(a_{j_k}^k, 1/k)$$
$$\phi(a_1^{k+1}, 1/(k+1)),\ \phi(a_2^{k+1}, 1/(k+1)), \cdots, \phi(a_{j_{k+1}}^{k+1}, 1/(k+1)), \cdots$$
$$\phi(a_1^N, 1/N),\ \phi(a_2^N, 1/N), \cdots, \phi(a_{j_N}^N, 1/N), \cdots$$

converges to $|z|^2$ in L^1 on a large disc, but does not converge at any point in the unit disc since $\phi(a_j^N, 1/N) = -1$ in the ball $B(a_j^N, e^{-2N^3})$.

LEMMA 5.3. For every $t \leq -1$ we have

$$e^{-\beta t} \int_{\mathbb{B}} e^{-\phi - \lambda \max\{G-t,0\}} \leq \frac{2\lambda}{\lambda - \beta} \int_{\mathbb{B}} e^{-\phi - \beta G}.$$

PROOF. Put

$$I(t) := \int_{\mathbb{B}} e^{-\phi - \lambda \max\{G-t,0\}}.$$

Notice that the $F = 1$ case of (2.9) gives

$$\int_{-\infty}^{0} I(t) e^{-\beta t}\, dt \leq \frac{\lambda}{(\lambda-\beta)\beta} \int_{\mathbb{B}} e^{-\phi - \beta G}. \tag{5.9}$$

Since $I(t)$ is increasing in t, we have

$$\int_{-\infty}^{0} I(s) e^{-\beta s}\, ds \geq \int_{t}^{0} I(s) e^{-\beta s}\, ds \geq I(t) \int_{t}^{0} e^{-\beta s}\, ds \geq \frac{I(t) e^{-\beta t}}{2\beta}$$

for every $t \leq -1$. Hence the lemma follows from (5.9). □

REMARK. Thanks to Berndtsson's solution [**B7**] of the openness conjecture, the following $\beta = 1$ case of Theorem 5.1 also holds:

THEOREM 5.4. *Let ϕ be a plurisubharmonic function on the unit ball $\mathbb{B}$ in $\mathbb{C}^n$, and let $G, G_j \leq 0$ be such that*

$$\phi + \lambda G_j \text{ and } \phi + \lambda G \text{ are plurisubharmonic}$$

on $\mathbb{B}$ for some constant $\lambda > 1$. Assume that

$$\int_{\mathbb{B}} e^{-\phi - G} < \infty, \quad \lim_{j \to \infty} \int_{\mathbb{B}} |G_j - G| = 0. \tag{5.10}$$

Then

$$\lim_{j \to \infty} \int_{|z|<r} |e^{-\phi - G_j} - e^{-\phi - G}| = 0, \quad \forall\, 0 < r < 1. \tag{5.11}$$

5.2. A Donnelly-Fefferman proof of the strong openness theorem. The following Donnelly-Fefferman theorem is a special ($r = 1$) case of Theorem 2.1 in [**B3**]:

THEOREM 5.5. *Let ϕ and ψ be plurisubharmonic functions of class C^2 in the unit ball $\mathbb{B}$ in $\mathbb{C}^n$, and assume that ϕ satisfies*

$$i\partial\phi \wedge \overline{\partial}\phi \leq i\partial\overline{\partial}\phi. \tag{5.12}$$

Then for any smooth $\overline{\partial}$-closed $(0,1)$-form f on $\mathbb{B}$, there exists a solution, u, to

$$\overline{\partial}u = f, \tag{5.13}$$

satisfying the estimate

$$\int_{\mathbb{B}} |u|^2 e^{-\psi} \leq 4 \int_{\mathbb{B}} |f|^2_{i\partial\overline{\partial}\phi} e^{-\psi}. \tag{5.14}$$

PROOF. We shall follow the proof in [**B3**]. Similar to the proof of Theorem 4.1, one may assume that ϕ and ψ are C^2 up to the boundary and strictly plurisubharmonic. Let u be the L^2-minimal solution of $\overline{\partial}(\cdot) = f$ with respect to the weight $\psi + \phi/2$, i.e.

$$\int_{\mathbb{B}} u\bar{h}\, e^{-\psi-\phi/2} = 0, \ \forall\, h \in A^2_{\psi+\phi/2}.$$

Since ϕ is assumed to be bounded, we see that $e^{\phi/2}u$ is the L^2-minimal solution of the equation $\overline{\partial}(\cdot) = \overline{\partial}(e^{\phi/2}u)$ with respect to the weight $\phi+\psi$. Thus the Hörmander estimate gives

$$\int_{\mathbb{B}} |e^{\phi/2}u|^2 e^{-\phi-\psi} \leq \int_{\mathbb{B}} |\overline{\partial}(e^{\phi/2}u)|^2_{i\partial\overline{\partial}(\phi+\psi)} e^{-\phi-\psi} \leq \int_{\mathbb{B}} |\overline{\partial}(e^{\phi/2}u)|^2_{i\partial\overline{\partial}\phi} e^{-\phi-\psi}. \tag{5.15}$$

Note that (5.12) implies

$$|\overline{\partial}(e^{\phi/2}u)|^2_{i\partial\overline{\partial}\phi} e^{-\phi} \leq 2|f|^2_{i\partial\overline{\partial}\phi} + \frac{1}{2}|u|^2,$$

and hence (5.15) gives (5.14). □

The following corollary is a Donnelly-Fefferman version of Theorem 4.1:

COROLLARY 5.6. Let F be a holomorphic function on the unit ball $\mathbb{B}$ in $\mathbb{C}^n$, and let $G \leq 0$ be a plurisubharmonic function on $\mathbb{B}$. Then for every $t < -1$ and $\alpha \geq 0$, we can find a holomorphic function $\tilde{F}$ on $\mathbb{B}$ such that

$$\int_{\mathbb{B}} |\tilde{F} - F|^2 e^{-\alpha G} \leq 24\, |t|\, e^{-2\alpha t} \|F\|^2_t + 2\, e^{-\alpha t} \int_{\mathbb{B}} |F|^2 \tag{5.16}$$

and

$$\int_{\mathbb{B}} |\tilde{F}|^2 \leq 10\, |t|\, e^{-\alpha t} \|F\|^2_t, \tag{5.17}$$

where $\|F\|^2_t := \int_{\mathbb{B}} |F|^2 e^{-\max\{G-t,0\}}$.

PROOF. Similar to the proof of Theorem 4.1, one may assume that G is smooth up to the boundary. Put $\sigma(s) := \int_0^s 1 - e^{-\max\{x-t,0\}}\, dx$. Applying the above Theorem 5.5 to

$$\phi := -\log(-\sigma(G)), \ \psi := \alpha G, \ \ v = \overline{\partial}((1 - \sigma'(G))F),$$

we get a solution u of $\bar\partial u = \bar\partial((1-\sigma'(G))F)$ with

$$\int_{\mathbb{B}} |u|^2 e^{-\alpha G} \le 4\int_{\mathbb{B}} |\sigma''(G)\bar\partial G|^2_{i\partial\bar\partial\phi}\cdot |F|^2 e^{-\alpha G}.$$

Notice that

$$i\partial\bar\partial\phi \ge \frac{i\partial\bar\partial(\sigma(G))}{-\sigma(G)} \ge \frac{\sigma''(G) i\partial G\wedge\bar\partial G}{-\sigma(G)}$$

gives

$$|\sigma''(G)\bar\partial G|^2_{i\partial\bar\partial\phi} \le \sigma''(G)|\sigma(G)| \le |t|\, e^{-\max\{G-t,0\}}.$$

Hence we have

$$\int_{\mathbb{B}} |u|^2 e^{-\alpha G} \le 4|t|\int_{G\ge t} |F|^2 e^{-\alpha G} e^{-\max\{G-t,0\}} \le 4|t|e^{-\alpha t}\|F\|_t^2.$$

Taking $\tilde F = (1-\sigma'(G))F - u$, similar to the proof of Theorem 4.1, one may verify that $\tilde F$ satisfies (5.16) and (5.17). □

We end by showing how to use the above corollary to prove the following Guan-Zhou's strong openness theorem (see [**GZ2**]):

THEOREM 5.7. *Let F be a holomorphic function on the unit ball $\mathbb{B}$ in $\mathbb{C}^n$, and let $G\le 0$ be a plurisubharmonic function on $\mathbb{B}$. If $\int_{\mathbb{B}} |F|^2\, e^{-\beta G} < \infty$ for some $\beta > 0$, then there exists $\alpha > \beta$ such that $|F|^2\, e^{-\alpha G}$ is integrable near the origin of $\mathbb{B}$.*

PROOF. Assume that

$$\alpha_0 := \sup\{\alpha\ge 0 : F_0 \in \mathcal{I}(\alpha G)_0\} < \infty,$$

and

$$I_0 := \inf\left\{\int_{\mathbb{B}} |\tilde F|^2 : (\tilde F - F)_0 \in \bigcup_{\alpha>\alpha_0} \mathcal{I}(\alpha G)_0\right\}.$$

Step 1 in the proof of Theorem B implies that $I_0 > 0$. By Corollary 5.6, we have

$$0 < I_0 \le 10\,|t|\, e^{-\alpha_0 t}\|F\|_t^2, \quad \forall\, t < -1. \tag{5.18}$$

Since $\int_{\mathbb{B}} |F|^2\, e^{-\beta G} < \infty$, we know that (2.9) (with $\lambda = \beta+1$) gives

$$\int_{-\infty}^0 \|F\|_t^2 e^{-\beta t} dt < \infty.$$

By (5.18), we obtain

$$\int_{-\infty}^0 \frac{e^{-\beta t}}{|t|e^{-\alpha_0 t}} dt < \infty,$$

which implies $\alpha_0 > \beta$, and completes the proof. □

References

[Al] Roberto Albesiano, *A degeneration approach to Skoda's division theorem*, Math. Z. **306** (2024), no. 2, Paper No. 34, 19, DOI 10.1007/s00209-024-03435-6. MR4696321 ↑105

[BG] Shi Jie Bao and Qi An Guan, *L^2 extension and effectiveness of strong openness property*, Acta Math. Sin. (Engl. Ser.) **38** (2022), no. 11, 1949–1964, DOI 10.1007/s10114-022-1220-5. MR4515127 ↑104

[BG1] Shi Jie Bao and Qi An Guan, *L^2 extension and effectiveness of L^p strong openness property*, Acta Math. Sin. (Engl. Ser.) **39** (2023), no. 5, 814–826, DOI 10.1007/s10114-023-1368-7. MR4598989 ↑104

[B1] Bo Berndtsson, *The extension theorem of Ohsawa-Takegoshi and the theorem of Donnelly-Fefferman* (English, with English and French summaries), Ann. Inst. Fourier (Grenoble) **46** (1996), no. 4, 1083–1094. MR1415958 ↑104

[B2] Bo Berndtsson, *Integral formulas and the Ohsawa-Takegoshi extension theorem*, Sci. China Ser. A **48** (2005), no. suppl., 61–73, DOI 10.1007/BF02884696. MR2156491 ↑104

[B3] Bo Berndtsson, *Weighted estimates for the $\overline{\partial}$-equation*, Complex analysis and geometry (Columbus, OH, 1999), Ohio State Univ. Math. Res. Inst. Publ., vol. 9, de Gruyter, Berlin, 2001, pp. 43–57. MR1912730 ↑118

[B4] Bo Berndtsson, *An introduction to things $\overline{\partial}$*, Analytic and algebraic geometry, IAS/Park City Math. Ser., vol. 17, Amer. Math. Soc., Providence, RI, 2010, pp. 7–76, DOI 10.1090/pcms/017/02. MR2743815 ↑113

[B5] Bo Berndtsson, *Subharmonicity properties of the Bergman kernel and some other functions associated to pseudoconvex domains* (English, with English and French summaries), Ann. Inst. Fourier (Grenoble) **56** (2006), no. 6, 1633–1662. MR2282671 ↑100, 109, 110, 111

[B6] Bo Berndtsson, *Curvature of vector bundles associated to holomorphic fibrations*, Ann. of Math. (2) **169** (2009), no. 2, 531–560, DOI 10.4007/annals.2009.169.531. MR2480611 ↑100, 109, 110

[B7] Bo Berndtsson, *The openness conjecture and complex Brunn-Minkowski inequalities*, Complex geometry and dynamics, Abel Symp., vol. 10, Springer, Cham, 2015, pp. 29–44. MR3587460 ↑100, 103, 104, 117

[B8] Bo Berndtsson, *Complex Brunn-Minkowski theory and positivity of vector bundles*, Proceedings of the International Congress of Mathematicians—Rio de Janeiro 2018. Vol. II. Invited lectures, World Sci. Publ., Hackensack, NJ, 2018, pp. 859–884. MR3966792 ↑109

[B9] Bo Berndtsson, *Lelong numbers and vector bundles*, J. Geom. Anal. **30** (2020), no. 3, 2361–2376, DOI 10.1007/s12220-017-9873-5. MR4105121 ↑100, 102, 109, 111, 112

[BL] Bo Berndtsson and László Lempert, *A proof of the Ohsawa-Takegoshi theorem with sharp estimates*, J. Math. Soc. Japan **68** (2016), no. 4, 1461–1472, DOI 10.2969/jmsj/06841461. MR3564439 ↑100, 104, 105

[BCP] Junyan Cao, Mihai Paun, and Bo Berndtsson, *On the Ohsawa-Takegoshi extension theorem*, J. Geom. Anal. **34** (2024), no. 1, Paper No. 25, 47, DOI 10.1007/s12220-023-01466-9. MR4668049 ↑104

[Bl] Zbigniew Błocki, *Suita conjecture and the Ohsawa-Takegoshi extension theorem*, Invent. Math. **193** (2013), no. 1, 149–158, DOI 10.1007/s00222-012-0423-2. MR3069114 ↑100, 104

[Bl1] Zbigniew Błocki, *Bergman kernel and pluripotential theory*, Analysis, complex geometry, and mathematical physics: in honor of Duong H. Phong, Contemp. Math., vol. 644, Amer. Math. Soc., Providence, RI, 2015, pp. 1–10, DOI 10.1090/conm/644/12785. MR3372456 ↑104

[CDM] JunYan Cao, Jean-Pierre Demailly, and Shin-ichi Matsumura, *A general extension theorem for cohomology classes on non reduced analytic subspaces*, Sci. China Math. **60** (2017), no. 6, 949–962, DOI 10.1007/s11425-017-9066-0. MR3647124 ↑104

[Ch] Tsz On Mario Chan, *On an L^2 extension theorem from log-canonical centres with log-canonical measures*, Math. Z. **301** (2022), no. 2, 1695–1717, DOI 10.1007/s00209-021-02890-9. MR4418335 ↑104

[CC] Tsz On Mario Chan and Young-Jun Choi, *Extension with log-canonical measures and an improvement to the plt extension of Demailly-Hacon-Păun*, Math. Ann. **383** (2022), no. 3-4, 943–997, DOI 10.1007/s00208-021-02152-3. MR4458394 ↑104

[C] B. Chen, *A simple proof of the Ohsawa-Takegoshi extension theorem*, arXiv:1105.2430, 2011. 104

[CWW] Bo-Yong Chen, Jujie Wu, and Xu Wang, *Ohsawa-Takegoshi type theorem and extension of plurisubharmonic functions*, Math. Ann. **362** (2015), no. 1-2, 305–319, DOI 10.1007/s00208-014-1117-4. MR3343879 ↑100, 104

[De] J.-P. Demailly, *Complex analytic and differential geometry*, book in https://www-fourier.ujf-grenoble.fr/~demailly/manuscripts/agbook.pdf. 103

[De1] Jean-Pierre Demailly, *Multiplier ideal sheaves and analytic methods in algebraic geometry*, School on Vanishing Theorems and Effective Results in Algebraic Geometry (Trieste, 2000), ICTP Lect. Notes, vol. 6, Abdus Salam Int. Cent. Theoret. Phys., Trieste, 2001, pp. 1–148. MR1919457 ↑104

[De2] Jean-Pierre Demailly, *Analytic methods in algebraic geometry*, Surveys of Modern Mathematics, vol. 1, International Press, Somerville, MA; Higher Education Press, Beijing, 2012. MR2978333 ↑112

[De3] Jean-Pierre Demailly, *Extension of holomorphic functions and cohomology classes from non reduced analytic subvarieties*, Geometric complex analysis, Springer Proc. Math. Stat., vol. 246, Springer, Singapore, 2018, pp. 97–113, DOI 10.1007/978-981-13-1672-2_8. MR3923220 ↑104

[DHP] Jean-Pierre Demailly, Christopher D. Hacon, and Mihai Păun, *Extension theorems, non-vanishing and the existence of good minimal models*, Acta Math. **210** (2013), no. 2, 203–259, DOI 10.1007/s11511-013-0094-x. MR3070567 ↑104

[DK] Jean-Pierre Demailly and János Kollár, *Semi-continuity of complex singularity exponents and Kähler-Einstein metrics on Fano orbifolds* (English, with English and French summaries), Ann. Sci. École Norm. Sup. (4) **34** (2001), no. 4, 525–556, DOI 10.1016/S0012-9593(01)01069-2. MR1852009 ↑115

[FJ] Charles Favre and Mattias Jonsson, *Valuations and multiplier ideals*, J. Amer. Math. Soc. **18** (2005), no. 3, 655–684, DOI 10.1090/S0894-0347-05-00481-9. MR2138140 ↑104

[Gu] Qi'an Guan, *A sharp effectiveness result of Demailly's strong openness conjecture*, Adv. Math. **348** (2019), 51–80, DOI 10.1016/j.aim.2019.03.017. MR3924390 ↑100

[Gu1] Q. Guan, *General concavity of minimal L^2 integrals related to multiplier sheaves*, arXiv:1811.03261v4, 2018. 104

[GM] Qi'an Guan and Zhitong Mi, *Concavity of minimal L^2 integrals related to multiplier ideal sheaves*, Peking Math. J. **6** (2023), no. 2, 393–457, DOI 10.1007/s42543-021-00047-5. MR4619598 ↑104

[GY] Qi'an Guan and Zheng Yuan, *Effectiveness of strong openness property in L^p*, Proc. Amer. Math. Soc. **151** (2023), no. 10, 4331–4339, DOI 10.1090/proc/16408. MR4643322 ↑104

[GZ1] Qi'an Guan and Xiangyu Zhou, *A solution of an L^2 extension problem with an optimal estimate and applications*, Ann. of Math. (2) **181** (2015), no. 3, 1139–1208, DOI 10.4007/annals.2015.181.3.6. MR3296822 ↑100, 104

[GZ2] Qi'an Guan and Xiangyu Zhou, *A proof of Demailly's strong openness conjecture*, Ann. of Math. (2) **182** (2015), no. 2, 605–616, DOI 10.4007/annals.2015.182.2.5. MR3418526 ↑100, 104, 119

[GZ3] Qi'an Guan and Xiangyu Zhou, *Effectiveness of Demailly's strong openness conjecture and related problems*, Invent. Math. **202** (2015), no. 2, 635–676, DOI 10.1007/s00222-014-0575-3. MR3418242 ↑104

[Hi] Pham Hoang Hiep, *The weighted log canonical threshold* (English, with English and French summaries), C. R. Math. Acad. Sci. Paris **352** (2014), no. 4, 283–288, DOI 10.1016/j.crma.2014.02.010. MR3186914 ↑104

[Ho] Genki Hosono, *On sharper estimates of Ohsawa-Takegoshi L^2-extension theorem*, J. Math. Soc. Japan **71** (2019), no. 3, 909–914, DOI 10.2969/jmsj/80018001. MR3984247 ↑104

[JM] Mattias Jonsson and Mircea Mustaţă, *Valuations and asymptotic invariants for sequences of ideals* (English, with English and French summaries), Ann. Inst. Fourier (Grenoble) **62** (2012), no. 6, 2145–2209 (2013), DOI 10.5802/aif.2746. MR3060755 ↑104

[K1] Dano Kim, *L^2 extension of adjoint line bundle sections* (English, with English and French summaries), Ann. Inst. Fourier (Grenoble) **60** (2010), no. 4, 1435–1477. MR2722247 ↑104

[K2] Dano Kim, *A remark on L^2 extension theorems*, Bull. Korean Math. Soc. **58** (2021), no. 5, 1235–1245, DOI 10.4134/BKMS.b200911. MR4319455 ↑104

[KS] Dano Kim and Hoseob Seo, *On L^2 extension from singular hypersurfaces*, Math. Z. **303** (2023), no. 4, Paper No. 89, 21, DOI 10.1007/s00209-023-03248-z. MR4562223 ↑104

[Le1] László Lempert, *Modules of square integrable holomorphic germs*, Analysis meets geometry, Trends Math., Birkhäuser/Springer, Cham, 2017, pp. 311–333, DOI 10.1007/978-3-319-52471-9_19. MR3773623 ↑104

[Le2] László Lempert, *Extrapolation, a technique to estimate*, Functional analysis, harmonic analysis, and image processing: a collection of papers in honor of Björn Jawerth, Contemp. Math., vol. 693, Amer. Math. Soc., Providence, RI, 2017, pp. 271–281. MR3682614 ↑104

[Le3] László Lempert, *A maximum principle for Hermitian (and other) metrics*, Proc. Amer. Math. Soc. **143** (2015), no. 5, 2193–2200, DOI 10.1090/S0002-9939-2015-12472-0. MR3314125 ↑108

[Ma] Shin-ichi Matsumura, *An injectivity theorem with multiplier ideal sheaves of singular metrics with transcendental singularities*, J. Algebraic Geom. **27** (2018), no. 2, 305–337, DOI 10.1090/jag/687. MR3764278 ↑104

[MV1] Jeffery D. McNeal and Dror Varolin, *Analytic inversion of adjunction: L^2 extension theorems with gain* (English, with English and French summaries), Ann. Inst. Fourier (Grenoble) **57** (2007), no. 3, 703–718. MR2336826 ↑104

[MV2] Jeffery D. McNeal and Dror Varolin, *L^2 extension of $\overline{\partial}$-closed forms from a hypersurface*, J. Anal. Math. **139** (2019), no. 2, 421–451, DOI 10.1007/s11854-019-0062-2. MR4041108 ↑104

[MV3] J. D. McNeal and D. Varolin,*Extension of Jets With L^2 estimates, and an application*, `arXiv:1707.04483`, 2017. 104

[Na] Alan Michael Nadel, *Multiplier ideal sheaves and existence of Kähler-Einstein metrics of positive scalar curvature*, Proc. Nat. Acad. Sci. U.S.A. **86** (1989), no. 19, 7299–7300, DOI 10.1073/pnas.86.19.7299. MR1015491 ↑103, 115

[Ng] T. Nguyen, *A Hilbert bundles description of complex Brunn-Minkowski theory*, `arXiv:2305.07476v1`, 2023. 101, 102, 105, 109, 110, 111, 112

[NW] Tai Terje Huu Nguyen and Xu Wang, *On a remark by Ohsawa related to the Berndtsson-Lempert method for L^2-holomorphic extension*, Ark. Mat. **60** (2022), no. 1, 173–182, DOI 10.4310/arkiv.2022.v60.n1.a7. MR4423274 ↑105

[O2] Takeo Ohsawa, *On the extension of L^2 holomorphic functions. II*, Publ. Res. Inst. Math. Sci. **24** (1988), no. 2, 265–275, DOI 10.2977/prims/1195175200. MR944862 ↑104

[O3] Takeo Ohsawa, *On the extension of L^2 holomorphic functions. III. Negligible weights*, Math. Z. **219** (1995), no. 2, 215–225, DOI 10.1007/BF02572360. MR1337216 ↑104

[Oh] Takeo Ohsawa, *On the extension of L^2 holomorphic functions. V. Effects of generalization*, Nagoya Math. J. **161** (2001), 1–21, DOI 10.1017/S0027763000022108. MR1820210 ↑99, 104, 106

[OT] Takeo Ohsawa and Kenshō Takegoshi, *On the extension of L^2 holomorphic functions*, Math. Z. **195** (1987), no. 2, 197–204, DOI 10.1007/BF01166457. MR892051 ↑99, 104, 106

[Siu] Yum-Tong Siu, *Invariance of plurigenera*, Invent. Math. **134** (1998), no. 3, 661–673, DOI 10.1007/s002220050276. MR1660941 ↑104

[Su] Nobuyuki Suita, *Capacities and kernels on Riemann surfaces*, Arch. Rational Mech. Anal. **46** (1972), 212–217, DOI 10.1007/BF00252460. MR367181 ↑99

[W1] X. Wang, *An explicit estimate of the Bergman kernel for positive line bundles*, to appear in Ann. Fac. Sci. Toulouse Math. 104

[W2] X. Wang, *Hörmander $\overline{\partial}$ theory on quasi-complete Kähler manifold*, lecture notes in `https://folk.ntnu.no/xuwa/res/course_all.pdf`. 100, 112

[ZZ] Xiangyu Zhou and Langfeng Zhu, *Extension of cohomology classes and holomorphic sections defined on subvarieties*, J. Algebraic Geom. **31** (2022), no. 1, 137–179. MR4372411 ↑104, 106

Department of Mathematical Sciences, Norwegian University of Science and Technology, NO-7491 Trondheim, Norway

Current address: Department of Business, Strategy and Political Sciences, University of South Eastern Norway Hasbergsvei 36, 3616 Kongsberg, Norway

Email address: taiterje@hotmail.com

Department of Mathematical Sciences, Norwegian University of Science and Technology, NO-7491 Trondheim, Norway

Email address: xu.wang@ntnu.no

Contemporary Mathematics
Volume **810**, 2025
https://doi.org/10.1090/conm/810/16215

Curvature of smooth proper direct images by way of a holomorphic Gauss formula

Dror Varolin

Dedicated to Bo Berndtsson on the occasion of his 70th birthday

Abstract. The direct image of the relative canonical bundle of a smooth proper holomorphic family, twisted by a holomorphic vector bundle E with smooth Hermitian metric $\mathfrak{h}$ with relatively Nakano-positive curvature, is locally trivial. The underlying vector bundle has a natural Hermitian metric called the L^2 metric. In his 2009 Annals paper Berndtsson computed the curvature of this metric when E has rank 1 and $\mathfrak{h}$ has semi-positive curvature form. Later, Liu and Yang computed the curvature of the relative canonical bundle twisted by any Hermitian holomorphic vector bundle, under the assumption that this direct image is locally trivial. We give a new proof of the results of Berndtsson-Liu-Yang. Our proof uses a generalization of the holomorphic Gauss formula to the setting of BLS fields, introduced by the author.

Introduction

Let X and B be complex manifolds, let $p : X \to B$ be a surjective holomorphic submersion, and let $E \to X$ be a holomorphic vector bundle with Hermitian metric $\mathfrak{h}$. We call

$$(E, \mathfrak{h}) \to X \overset{p}{\to} B \tag{1}$$

a *holomorphic family of Hermitian vector bundles*, and we say that this family is *proper* if $p : X \to B$ is a proper map, i.e., $p^{-1}(K) \subset\subset X$ whenever $K \subset\subset B$.

To the holomorphic family (1) we attach a family of Hilbert spaces, $\mathscr{H} \to B$ whose fiber over $t \in B$ is

$$\mathscr{H}_t := \left\{ s \in H^0(X_t, \mathscr{O}_{X_t}(K_{X_t} \otimes E_t)) \ ; \ \int_{X_t} |s|^2_{\mathfrak{h}} < +\infty \right\},$$

where $X_t := p^{-1}(t)$ and

$$|s|^2_{\mathfrak{h}} := \sqrt{\text{-}1}^{n^2} \langle s \wedge \bar{s}, \mathfrak{h} \rangle$$

denotes the smooth positive measure obtained by pairing the $E \otimes E^{\dagger}$-valued (n, n)-form $\sqrt{\text{-}1}^{n^2} s \wedge \bar{s}$ with the metric $\mathfrak{h}$, viewed as a section of $E^* \otimes (E^*)^{\dagger}$. The vector space $\mathscr{H}_t$ is a Hilbert space with respect to the inner product

$$(s_1, s_2)_t := \int_{X_t} \sqrt{\text{-}1}^{n^2} \langle s_1 \wedge \bar{s}_2, \mathfrak{h} \rangle . \tag{2}$$

2020 *Mathematics Subject Classification.* Primary 32L05.

We adopt the following terminology introduced by Lempert and Szőke in [**LS-2014**].

DEFINITION 0.1. A *complex Hilbert field* over a set B is a map $\mathfrak{H} \to B$ whose fibers are complex Hilbert spaces. ◇

We shall omit the adjective 'complex' from here on; all Hilbert spaces considered in the present article are complex Hilbert spaces.

The Hilbert field $\mathscr{H} \to B$ defined above is given additional structure by defining smooth (resp. holomorphic) sections of $\mathscr{H} \to B$ to be smooth (resp. holomorphic) sections of $E \otimes K_{X/B} \to X$ whose pullbacks to the fibers of $X \to B$ are holomorphic and square-integrable, where $K_{X/B} := K_X \otimes p^*K_B^*$ denotes the relative canonical bundle. One can then ask whether $\mathscr{H} \to B$ can be given the structure of a holomorphic vector bundle in such a way that the smooth and holomorphic sections just defined are precisely the smooth and holomorphic sections in the usual sense of holomorphic vector bundles. Without further assumptions on the family of Hermitian holomorphic vector bundles (1), the answer is *no*.

EXAMPLE 0.2. If one further assumes that (1) is a proper holomorphic family then in particular the fibers $X_t := p^{-1}(\{t\})$ are compact. Hence the fibers of $\mathscr{H} \to B$ are finite-dimensional. If the base B is connected then by a theorem of Ehresmann the fibers X_t are pairwise diffeomorphic. Thus by a theorem due to Kodaira-Spencer $\mathscr{H} \to B$ is a holomorphic vector bundle if and only if the function $B \ni t \mapsto \dim \mathscr{H}_t \in \mathbb{Z}$ is locally constant. ◇

In [**B-2009**] Bo Berndtsson studies the Hilbert field $\mathscr{H}$ associated to a proper holomorphic family of semipositive Hermitian line bundles. Berndtsson proves the following theorem.

THEOREM 0.3 (Berndtsson, [**B-2009**]). *Let X be a Kähler manifold and let B be a complex manifold. Let $(E, e^{-\varphi}) \to X \overset{p}{\to} B$ be a proper holomorphic family of Hermitian line bundles. Assume that the metric $e^{-\varphi}$ has non-negative (resp. positive) curvature. Then the Hilbert field $\mathscr{H} \to B$ is a holomorphic vector bundle whose curvature, with respect to the L^2 metric, is non-negative (resp. positive) in the sense of Nakano.*

As Berndtsson notes, the positive curvature hypothesis $\sqrt{-1}\partial\bar\partial\varphi \ge 0$ implies the local triviality of $\mathscr{H}$ via the L^2 Extension Theorem of Ohsawa-Takegoshi. Berndtsson then computes the curvature of $\mathscr{H}$ directly, and uses Hodge Theory to prove that the curvature is positive.

The purpose of the present article is to obtain Berndtsson's curvature formula in a somewhat different manner. We consider the case of a proper holomorphic family of Hermitian vector bundles. But rather than including an assumption on the family (1) that yields a holomorphic vector bundle structure for $\mathscr{H} \to B$, we simply (and somewhat less naturally) *assume* this to be the case.

Standing Assumption: *The holomorphic family* (1) *is proper, and the associated Hilbert field $\mathscr{H} \to B$ is a holomorphic vector bundle.*

REMARK. The Hilbert field $\mathscr{H} \to B$ associated to a proper holomorphic family (1) is a holomorphic vector bundle 'almost everywhere': Given any proper holomorphic family (1), there exists an analytic subvariety $Z \subset B$ such that $\mathscr{H}|_{B-Z}$ is a holomorphic vector bundle. This is the case because by a famous theorem of

Grauert the 0th direct image sheaf $p_*\mathcal{O}_X(K_{X/B}\otimes E)\to B$ is a coherent sheaf (as are the higher direct images, though we will not consider those here). Hence the function

$$\rho : B \ni b \mapsto \text{Rank}\left(p_*\mathcal{O}_X(K_{X/B}\otimes E)\right)_b$$

is upper semi-continuous in the Zariski topology. Because the function ρ is $\mathbb{N}$-valued, its minimum value r is achieved on the Zariski open subset $B - Z := \rho^{-1}([0, r+1))$. Another theorem of Grauert then yields

$$\left(p_*\mathcal{O}_X(K_{X/B}\otimes E)\right)_b = \mathscr{C}^{\infty}_{B,b}\otimes H^0(X_b, \mathcal{O}_{X_b}(K_{X_b}\otimes E)).$$

There are also some known hypotheses that force local triviality 'everywhere'.

(i) If E has rank 1 and $\mathfrak{h}$ has non-negative curvature form then $\mathscr{H}$ is locally trivial.

(ii) Similarly, if $(E,\mathfrak{h})|_{X_t}\to X_t$ is Nakano-positive for every t, or if $(E,\mathfrak{h})\to X$ is Griffiths nonnegative and $(E,\mathfrak{h})|_{X_t}\to X_t$ is Nakano-nonnegative for every t then $\mathscr{H}$ is again a holomorphic vector bundle.

Thus the standing assumption holds in many important situations. ◇

REMARK. A large part of the article [**Va-2022**] concerns the case in which the Hilbert field $\mathscr{H}\to B$ associated to the proper holomorphic family (1) need not be locally trivial. To study this more general setting, a much more involved theory was developed. ◇

REMARK. There is also a lot of work on the much more complicated non-proper case, due to Lempert and Szőke [**LS-2014**], Wang [**Wang-2017**] and Upadrashta [**U-2023**]. ◇

Let us recall Berndtsson's definition of the complex structure for the Hilbert fields $\mathscr{H}\to B$. As already mentioned, a section $\mathfrak{f}$ of $\mathscr{H}\to B$ is identified with a section f of the twisted relative canonical bundle $K_{X/B}\otimes E\to X$ such that for each $t\in B$

$$\iota^*_{X_t} f \in H^0(X_t, \mathcal{O}_{X_t}(K_{X_t}\otimes E|_{X_t})).$$

(Because of our standing assumption of properness, fiberwise square-integrability follows.)

Notational convention: *We use gothic letters (e.g.* $\mathfrak{f}$*) to denote sections of* $\mathscr{H}\to B$ *and their latin versions (e.g.* f*) to denote the corresponding sections of* $K_{X/B}\otimes E\to X$.

A section $\mathfrak{f}$ of $\mathscr{H}\to B$ is said to be smooth (resp. holomorphic) if the corresponding section f of $K_{X/B}\otimes E\to X$ is smooth (resp. holomorphic). We denote by $\Gamma(B,\mathscr{C}^\infty(\mathscr{H}))$ and $\Gamma(B,\mathcal{O}(\mathscr{H}))$ the set of smooth and holomorphic sections of $\mathscr{H}\to B$ respectively. The standing assumption can be equivalently stated as follows:

(i) The holomorphic family (1) is proper, and

(ii) for each $t\in B$ there exists a neighborhood U of t in B with the following property:

For each $s\in\mathscr{H}_t$ there is a section $\mathfrak{f}\in\Gamma(U,\mathcal{O}(\mathscr{H}))$ such that $\iota^*_{X_t} f = s$.

(Indeed, Condition (ii) implies that the function $t\mapsto \dim\mathscr{H}_t$ is lower semicontinuous. As for upper semicontinuity, one could again apply Grauert's Theorem on the coherence of direct images of proper maps, but in our present setting of proper surjective submersions, upper semicontinuity is a simple consequence of Montel's Theorem.)

Given a smooth section $\mathfrak{f} \in \Gamma(B, \mathscr{C}^\infty(\mathscr{H}))$, we define

$$\bar\partial^{\mathscr{H}}_{\bar\tau} \mathfrak{f} := \bar\partial^X_{\bar\xi} f \quad \text{for some } \xi \in H^0(X_t, \mathscr{C}^\infty(T^{1,0}_X|_{X_t})) \text{ such that } dp\xi = \tau. \tag{3}$$

The operator $\bar\partial^{\mathscr{H}}$ is well-defined independent of the choice of lift ξ. Indeed, if two vector fields $\xi_1, \xi_2 \in H^0(X_t, \mathscr{C}^\infty(T^{1,0}_X|_{X_t}))$ are lifts of a vector $\tau \in T^{1,0}_{B,t}$ then

$$\bar\partial^X_{\bar\xi_1} f - \bar\partial^X_{\bar\xi_2} f = \bar\partial^X_{\bar\xi_1 - \bar\xi_2} f = 0,$$

since $\xi_1 - \xi_2$ is tangent to X_t and f is holomorphic along fibers. (Actually, for the definition of $\bar\partial^{\mathscr{H}}$ the lifts ξ, ξ_1, ξ_2 need not be smooth sections, but they certainly can be, and in most calculations one needs smooth lifts.) Note also that

$$\mathfrak{f} \in \Gamma(B, \mathcal{O}(\mathscr{H})) \iff \bar\partial^{\mathscr{H}} \mathfrak{f} = 0,$$

and that

$$\bar\partial^{\mathscr{H}} \bar\partial^{\mathscr{H}} = 0.$$

One can define another Hilbert field $\mathscr{L} \to B$ by 'removing the fiberwise-holomorphic requirement', i.e., setting

$$\mathscr{L}_t := \left\{ s \in H^0(X_t, L^2_{\ell oc}(K_{X_t} \otimes E|_{X_t})) \ ; \ \int_{X_t} |s|^2_{\mathfrak{h}} < +\infty \right\}.$$

A section $\mathfrak{f}$ of $\mathscr{L} \to B$ is then by definition identified with a fiberwise-square-integrable section of $K_{X/B} \otimes E \to X$. However, a *smooth* section of $\mathscr{L} \to B$ is by definition a smooth, hence automatically fiberwise-square-integrable, section of $K_{X/B} \otimes E \to X$. (We present the general theory of such Hilbert fields in Sections 2 and 3.) The set of such smooth sections over open $U \subset B$ forms a sheaf $\mathscr{C}^\infty(\mathscr{L})$ of $\mathscr{C}^\infty_B$-modules over B.

Remark. This notion of smoothness is not the same notion one would have if $\mathscr{L} \to B$ somehow had the structure of a smooth vector bundle.[1] Indeed, the vector bundle notion of smoothness would not require smoothness along the fibers. ◇

The critical feature of the Hilbert field $\mathscr{L} \to B$ is that for each $t \in B$ the stalk

$$\mathscr{C}^\infty(\mathscr{L})_t = \left\{ s \in \mathscr{L}_t \ ; \ \text{there exists } \mathfrak{f} \in \Gamma(B, \mathscr{C}^\infty(\mathscr{L})) \text{ such that } \iota^*_{X_t} f = s \right\} \subset \mathscr{L}_t$$

is *dense* in the L^2-norm of $\mathscr{L}_t$.

The idea of considering a Hilbert field $\mathfrak{H}$ with a sheaf of *smooth* sections $\mathscr{C}^\infty(\mathfrak{H})$ (as well as a few other properties allowing differentiation of these sections) is due to Lempert and Szőke [**LS-2014**], who call such objects *smooth Hilbert fields.*

In [**Va-2022**] the author noticed that one can equip the smooth Hilbert field $(\mathscr{L}, \mathscr{C}^\infty(\mathscr{L}))$ with a large collection of $\bar\partial$-operators. These $\bar\partial$-operators of the Hilbert field $\mathscr{L} \to B$ are parametrized by smooth subbundles $\theta \subset T^{1,0}_X$ such that $dp|_\theta : \theta \to p^* T^{1,0}_B$ is a vector bundle isomorphism; such θ are called *horizontal distributions.* A horizontal distribution θ defines a unique smooth lift $\xi^\theta_\tau \in \Gamma(X_t, \mathscr{C}^\infty(T^{1,0}_X|_{X_t}))$ of $\tau \in T^{1,0}_{B,t}$. We then define

$$\tau \lrcorner \bar\partial^\theta \mathfrak{f} = \bar\xi^\theta_\tau \lrcorner \bar\partial f,$$

[1] In fact, $\mathscr{L} \to B$ does indeed admit a smooth vector bundle structure because of Ehresmann's Theorem, but this smooth vector bundle structure in general cannot be extended to a holomorphic vector bundle structure.

where $\bar{\xi}^\theta_\tau \in H^0(X_t, \mathscr{C}^\infty(T^{0,1}_X|_{X_t}))$ is the complex conjugate of the lift ξ^θ_τ of $\tau \in T^{1,0}_{X_t}$ and the $\bar\partial$-operator on the right hand side is the $\bar\partial$-operator for the holomorphic vector bundle $K_{X/B} \otimes E \to X$. We note that

(i) $\mathscr{C}^\infty(\mathscr{H}) \subset \mathscr{C}^\infty(\mathscr{L})$, and if $\mathfrak{f} \in \Gamma(U, \mathscr{C}^\infty(\mathscr{H}))$ then $\bar\partial^\theta \mathfrak{f} = \bar\partial^{\mathscr{H}} \mathfrak{f}$ for every horizontal distribution θ, and

(ii) $\bar\partial^\theta \bar\partial^\theta = 0$ if and only if $[\theta, \theta] \subset \theta$, i.e., the 'complex structure' of $(\mathscr{L}, \mathscr{C}^\infty(\mathscr{L}), \bar\partial^\theta)$ is integrable if and only if θ is integrable in the sense of Frobenius (Proposition 3.6).

The merging of the idea of 'complex structure' introduced by Berndtsson in the case of $\mathscr{H}$, and the notion of 'smooth Hilbert field' introduced by Lempert and Szőke, led the author to name Hilbert fields that combine these two concepts *Berndtsson-Lempert-Szőke* fields, or BLS fields for short.

BLS fields enjoy certain properties of Hermitian holomorphic vector bundles. In particular, BLS fields have a unique connection that generalizes the Chern connection of a holomorphic vector bundle.

In [**Va-2022**] we computed the curvature of the BLS field $\mathscr{L}^\theta := (\mathscr{L}, \mathscr{C}^\infty(\mathscr{L}), \bar\partial^\theta)$. More precisely, we obtained the following theorem, whose proof is outlined in Section 4.

THEOREM 0.4 ([**Va-2022**, Theorem 8.5]). *Let $(E, \mathfrak{h}) \to X \stackrel{p}{\to} B$ be a proper holomorphic family of Hermitian vector bundles and let θ be any horizontal distribution. Assume there is a closed $(1,1)$-form ω on X whose pullback to each fiber of $p : X \to B$ is Kähler. Then the $(1,1)$-component of the curvature $\Theta^{\mathscr{L}^\theta}$ of $\mathscr{L}^\theta \to B$ is given by*

$$\begin{aligned}(4)\qquad \left(\Theta^{\mathscr{L}^\theta}_{\sigma\bar\tau} s_1, s_2\right)_t &= \int_{X_t} \left\langle \Theta(\mathfrak{h})_{\xi^\theta_\sigma \bar\xi^\theta_\tau} s_1 \wedge \bar{s}_2, \mathfrak{h}\right\rangle - \int_{X_t} \left\langle \kappa^\theta_\sigma s_1 \wedge \overline{\kappa^\theta_\tau s_2}, \mathfrak{h}\right\rangle \\ &\quad + \int_{X_t} \left\langle \bar\partial_{[\xi^\theta_\sigma, \bar\xi^\theta_\tau]^{0,1}} s_1 \wedge \bar{s}_2, \mathfrak{h}\right\rangle + \int_{X_t} \left\langle s_1 \wedge \overline{\bar\partial_{[\xi^\theta_\tau, \bar\xi^\theta_\sigma]^{0,1}} s_2}, \mathfrak{h}\right\rangle,\end{aligned}$$

where $t \in B$, $\sigma, \tau \in T^{1,0}_{B,t}$, $s_1, s_2 \in \mathscr{L}_t$ are the evaluations of smooth sections at t, $\Theta(\mathfrak{h})$ is the curvature of $(E, \mathfrak{h}) \to X$, $\sigma, \tau \in T^{1,0}_{B,t}$ are extended to holomorphic vector fields near t, and

$$\kappa^\theta_\sigma s := (\bar\partial \xi^\theta_\sigma) \lrcorner s.$$

In particular, if $s_1, s_2 \in \mathscr{H}_t$ then

$$(5)\qquad \left(\Theta^{\mathscr{L}^\theta}_{\sigma\bar\tau} s_1, s_2\right) = \left(\Theta(\mathfrak{h})_{\xi^\theta_\sigma \bar\xi^\theta_\tau} s_1, s_2\right) - \int_{X_t} \sqrt{-1}^{n^2} \left\langle \kappa^\theta_\sigma s_1 \wedge \overline{\kappa^\theta_\tau s_2}, \mathfrak{h}\right\rangle.$$

REMARK. We show (Proposition 3.6) that the $(0,2)$-part of the curvature of $\mathscr{L}^\theta$ is

$$\left(\Theta^{\mathscr{L}^\theta}_{\bar\sigma\bar\tau} \mathfrak{f}\right)(t) := \left\langle \bar\partial^\theta \bar\partial^\theta \mathfrak{f}, \sigma \wedge \bar\tau \right\rangle (t) = \iota^*_{X_t}\left(\left(\bar\partial_{[\bar\xi^\theta_\sigma, \bar\xi^\theta_\tau]} - \bar\partial_{\bar\xi^\theta_{[\sigma,\tau]}}\right) u\right);$$

the section $u \in H^0(X, \mathscr{C}^\infty(\Lambda^{n,0}_X \otimes E))$ on the right hand side is any representative, in the sense explained in Paragraph 1.1, of the section $f \in H^0(X, \mathscr{C}^\infty(K_{X/B} \otimes E))$ corresponding to $\mathfrak{f}$. Since the BLS Chern connection is in particular Hermitian, this formula also determines the $(2,0)$-part of the curvature. ◇

REMARK. Note that if $s \in \mathscr{L}_t$ is the evaluation at t of a smooth section of $\mathscr{L}$ near t then $\kappa^\theta_\sigma s$ is an E-valued $(n-1,1)$-form. Therefore in general the term

$$-\int_{X_t} \sqrt{-1}^{n^2} \left\langle \kappa^\theta_\sigma s_1 \wedge \overline{\kappa^\theta_\tau s_2}, \mathfrak{h} \right\rangle$$

on the right hand side of (4) has no sign. If the sections s_1, s_2 are holomorphic, as is the case in (5), and if θ is chosen to be the ω-horizontal distribution, then this term is indeed positive. We shall not need this fact in the present article. ◇

As just suggested, Theorem 0.4 is proved not so much for its own sake, but rather in the service of the computation of the curvature of $\mathscr{H}$. Thus, since $\bar\partial^{\mathscr{H}} \bar\partial^{\mathscr{H}} = 0$, we really only need the $(1,1)$-component of the curvature of $\mathscr{L}^\theta$, i.e., Formula (5). To obtain the curvature of $\mathscr{H}$ from the curvature of $\mathscr{L}^\theta$, we use the holomorphic Gauss Formula introduced by Griffiths in the case of holomorphic subbundles of holomorphic vector bundles. The holomorphic Gauss formula generalizes to the BLS setting (c.f. Proposition 2.21).

To compute the curvature of $\mathscr{H}$ using the BLS version of the holomorphic Gauss formula we require, in addition to Theorem 0.4, a formula for the squared-norm of the second fundamental form. This formula is one of the main results of the present article. To state it, we denote by $\mathbf{I\!I}^\theta$ the *second fundamental form* of $\mathscr{H}$ in $\mathscr{L}^\theta$. As we will see, $\mathbf{I\!I}^\theta$ is defined by the formula

$$\mathbf{I\!I}^\theta_\tau s := P^\perp_{t_o} \iota^*_{X_{t_o}} \left(\nabla^{1,0}(\xi^\theta_\tau \lrcorner u) + \xi^\theta_\tau \lrcorner (\nabla^{1,0} u) \right), \quad s \in \mathscr{H}_{t_o}, \ \tau \in T^{1,0}_{B,t_o},$$

where $P_{t_o} : \mathscr{L}_{t_o} \to \mathscr{H}_{t_o}$ is the Bergman projection and $u \in H^0(X, \mathscr{C}^\infty(\Lambda^{n,0}_X \otimes E))$ is a section satisfying $\iota^*_{X_{t_o}} u = s$ and $\iota^*_{X_t} \bar\partial u = 0$ for all t near t_o, i.e., the section of $K_{X/B} \otimes E$ represented by u is holomorphic in a neighborhood of X_{t_o}. (For each $s \in \mathscr{H}_{t_o}$ one has such a u because $\mathscr{H} \to B$ is a holomorphic vector bundle by our standing assumption.)

THEOREM 1. *Let $(E, \mathfrak{h}) \to X \xrightarrow{p} B$ be a proper family of holomorphic Hermitian vector bundles such that the complex manifold X is Kähler, let θ be any smooth horizontal distribution, let $t \in B$ and let ω_t be a Kähler form on X_t. Then for every $s_1, s_2 \in \mathscr{H}_t$ there exist holomorphic sections $\mathfrak{f}_1, \mathfrak{f}_2 \in \Gamma(U, \mathcal{O}(\mathscr{H}))$, where $U \subset B$ is a neighborhood of t, satisfying $\iota^*_{X_t} \mathfrak{f}_i = s_i$, and smooth representatives $u_1, u_2 \in H^0(p^{-1}(U), \mathscr{C}^\infty(\Lambda^{n,0}_X \otimes E))$ of $f_1, f_2 \in H^0(p^{-1}(U), \mathcal{O}(K_{X/B} \otimes E))$, depending on θ and ω_t, such that*

$$\begin{aligned} \int_{X_t} \sqrt{-1}^{n^2} \left\langle \mathbf{I\!I}^\theta_\sigma s_1 \wedge \overline{\mathbf{I\!I}^\theta_\tau s_2}, \mathfrak{h} \right\rangle &= \left(\Theta(\mathfrak{h})_{\xi^\theta_\sigma \bar\xi^\theta_\tau} f_1, f_2 \right)_t - \int_{X_t} \sqrt{-1}^{n^2} \left\langle \kappa^\theta_\sigma s_1 \wedge \overline{\kappa^\theta_\tau s_2}, \mathfrak{h} \right\rangle \\ &\quad - \left(\xi^\theta_\sigma \lrcorner \bar\partial u_1, \xi^\theta_\tau \lrcorner \bar\partial u_2 \right) - \left\langle p_* \left(\sqrt{-1}^{n^2} \langle (\Theta(\mathfrak{h}) u_1 \wedge \overline{u_2}, h \rangle \right), \sigma \wedge \bar\tau \right\rangle (t) \end{aligned} \tag{6}$$

for all $\sigma, \tau \in T^{1,0}_{B,t}$.

REMARK. As we show in [**Va-2022**], Theorem 1 also holds in the setting in which $\mathscr{H} \to B$ is not necessarily a holomorphic vector bundle, but in that case the proof is significantly more complicated. ◇

By combining Theorems 0.4 and 1, one obtains the following Theorem.

THEOREM 2. *Let $(E, \mathfrak{h}) \to X \xrightarrow{p} B$ be a proper Kähler family of holomorphic Hermitian vector bundles, let θ be a smooth horizontal distribution, let $t \in B$ and* *let ω_t be a Kähler form on X_t. Then for every $s_1, s_2 \in \mathscr{H}_t$ there exist smooth E-valued $(n,0)$-forms u_1, u_2 such that $\iota^*_{X_t} u_i = s_i$ and*

$$(7)\quad (\Theta^{\mathscr{H}}_{\sigma\bar\tau} s_1, s_2) = \left\langle p_* \left(\sqrt{-1}^{n^2} \langle (\Theta(\mathfrak{h}) u_1) \wedge \overline{u_2}, \mathfrak{h} \rangle \right), \sigma \wedge \bar\tau \right\rangle + \left(\xi^\theta_\sigma \lrcorner \bar\partial u_1, \xi^\theta_\tau \lrcorner \bar\partial u_2 \right).$$

Formula (7) is not new. In the rank 1 case it is the aforementioned formula for the curvature of $\mathscr{H}$ computed in [**B-2009**, Section 4], and in the higher rank case it was proved by Liu and Yang [**LY-2014**, Theorem 1.6]. The proofs of Berndtsson and Liu-Yang use the same method mentioned earlier: the curvature is obtained from the formula

$$\partial\bar\partial (f,g)_{\mathscr{H}} = (\Theta^{\mathscr{H}} f, g) + \left(D^{1,0} f, D^{1,0} g \right)$$

by explicitly computing the left hand side and then using Hodge Theory to prove the positivity of the right-most term of (7). Our proof is different, using Theorems 0.4 and 1 together with the holomorphic holomorphic Gauss Formula. The Gauss Formula approach sheds a different light on the geometric origins of Formula (7). In particular, our proof shows that from a certain geometric perspective Formula (7) is a consequence of some cancellation. This perspective is compatible with the observation that the two terms on the right hand side of (7) are not canonically determined by the s_i; they depend on the representatives u_i, though their sum does not.

At the end of the introduction of [**B-2009**] Berndtsson suggests that perhaps (7) can be interpreted as a formula for the curvature of $\mathscr{H}$ that realizes the latter as a quotient bundle. We have not been able to find a geometric construction that realizes this intriguing possibility.

Theorem 2 immediately yields the following result, which generalizes Theorem 0.3.

THEOREM 3. *Let $(E, \mathfrak{h}) \to X \xrightarrow{p} B$ be a proper Kähler family of Hermitian holomorphic vector bundles such that $\mathscr{H} \to B$ is a holomorphic vector bundle and let $k \ge 1$ be an integer. If the metric $\mathfrak{h}$ is k-non-negatively curved (resp. k-positively curved) then $\mathscr{H} \to B$ is k-non-negatively curved (resp. k-positively curved).*

The notion of k-positivity was introduced by Demailly in [**D-1982**, Définition 2.2]. Metrics are 1-positively curved if and only if they are positively curved in the sense of Griffiths, while metrics are k-positively curved for every k if and only if they are positively curved in the sense of Nakano. Notions of positivity are recalled in Paragraph 2.13.

The proof of Theorem 2 (and hence Theorem 3) presented here makes heavy use of our standing assumption. We do not know if $\mathscr{H} \to B$ is a holomorphic vector bundle under the assumption that $(E, \mathfrak{h})$ is k-positive but not Nakano positive.

1. Calculus requirements and other preliminaries

We recall some basic notions from complex differential calculus, mostly for the purpose of setting up notation. We also recall some notation introduced in [**Va-2022**], and state some lemmas that will be useful in the article.

1.1. The relative canonical bundle. The relative canonical bundle of a holomorphic family $p : X \to B$ is the line bundle

$$K_{X/B} := K_X \otimes p^* K_B^*.$$

Since $p : X \to B$ is a submersion, there is an equivalent way of defining the relative canonical bundle as the set of equivalence classes of E-valued $(n,0)$-forms, where n is the fiber dimension. Consider the holomorphic vector bundle $\Lambda^{n,0}_X \to X$ whose local sections are $(n,0)$-forms on X, and the subbundle $Z(p)^{n,0} \to X$ whose fiber at $x \in X$ is defined by

$$Z(p)^{n,0}_x = \Big\{ f \in \Lambda^{n,0}_{X,x} \ ; \ \langle f, \xi_1 \wedge \cdots \wedge \xi_n \rangle = 0 \\ \text{for all } \xi_1 \wedge \cdots \wedge \xi_n \in \Lambda^n(\mathrm{Kernel}(dp(x))) \Big\}.$$

Thus $Z(p)^{n,0} \to X$ is the annihilator of $\bigwedge^n T^{1,0}_{X/B}$, where $T^{1,0}_{X/B} := \mathrm{kernel}(dp)$ is the so-called *vertical* $(1,0)$*-tangent bundle* of $p : X \to B$. Then one has the exact sequence of vector bundles

$$0 \to Z(p)^{n,0} \to \Lambda^{n,0}_X \to K_{X/B} \to 0.$$

That is to say, a section of $K_{X/B} \otimes E \to X$ can be identified with equivalence class of E-valued $(n,0)$-forms on X, where two sections are equivalent if their difference is a section of $Z(p)^{n,0} \otimes E \to X$. We note also that a section u of $\Lambda^{n,0}_X \otimes E \to X$ is a section of $Z(p)^{n,0} \otimes E \to X$ if and only if $\iota^*_{X_t} u = 0$ for every $t \in B$.

REMARK 1.1. In computation one often works locally on the base B, in some coordinate chart. In terms of such coordinates the projection may be written $p = (p^1, \ldots, p^m)$, where m is the dimension of B. The functions p^i are constant on the fibers of p, and their differentials dp^i and $d\bar{p}^i$ are global forms on $p^{-1}(U)$. Every section of $Z(p)^{n,0}|_U \to U$ is of the form $dp^1 \wedge v_1 + \cdots + dp^m \wedge v_m$ for some $v_1, \ldots, v_m \in \Gamma(p^{-1}(U), \mathscr{C}^\infty_X(\Lambda^{n-1,0}_X \otimes E))$. Thus two sections u_1, u_2 of $\Lambda^{n,0}_X \otimes E|_U$ define the same section of $K_{X/B} \otimes E$ over U if and only if

$$u_2 - u_1 = dp^1 \wedge v_1 + \cdots + dp^m \wedge v_m$$

for some $v_1, \ldots, v_m \in \Gamma(p^{-1}(U), \mathscr{C}^\infty_X(\Lambda^{n-1,0}_X \otimes E))$, or equivalently

$$u_1 \wedge dp^1 \wedge \cdots \wedge dp^m = u_2 \wedge dp^1 \wedge \cdots \wedge dp^m$$

on U. ◇

1.2. Lie derivatives of twisted forms. For a real differential form α and a smooth real vector field ξ the Lie derivative $L_\xi \alpha$ is by definition

$$L_\xi \alpha := \left. \frac{d}{d\lambda} \right|_{\lambda=0} (\Phi^\lambda_\xi)^* \alpha,$$

where Φ_ξ denotes the flow of ξ. One has E. Cartan's formula

$$L_\xi \alpha = d(\xi \lrcorner \alpha) + \xi \lrcorner (du).$$

One can extend the Lie derivative to complex vector fields by setting

$$L_\xi := L_{\mathrm{Re}\ \xi} + \sqrt{\text{-}1} L_{\mathrm{Im}\ \xi}.$$

Evidently Cartan's formula still holds.

We shall also need a twisted version of the Lie derivative.

DEFINITION 1.2. Let X be a complex manifold, let $E \to X$ be a holomorphic vector bundle with connection ∇, let u be an E-valued (p,q)-form and let ξ be a complex vector field on X, i.e., a section of $T_X \otimes \mathbb{C} \to X$. Then

$$L^\nabla_\xi u := \nabla(\xi \lrcorner u) + \xi \lrcorner \nabla u$$

is called the *twisted Lie derivative of the E-valued (p,q)-form u along ξ*. ◇

We shall usually write L_ξ in place of L^∇_ξ, since typically the connection is clear from the context, and there is no confusion which Lie derivative is being used since the form u is twisted.

Via the splitting $\nabla = \nabla^{1,0} + \nabla^{0,1}$ one can define the $(1,0)$- and $(0,1)$-Lie derivative of a (p,q)-form u with values in a holomorphic vector bundle along a complex vector field ξ by

$$L^{\nabla 1,0}_\xi u := \nabla^{1,0}(\xi \lrcorner u) + \xi \lrcorner \nabla^{1,0} u \quad \text{and} \quad L^{\nabla 0,1}_\xi u := \nabla^{0,1}(\xi \lrcorner u) + \xi \lrcorner \nabla^{0,1} u.$$

As with the total Lie derivative, we often omit reference to the connection in our notation, writing $L^{1,0}_\xi$ and $L^{0,1}_\xi$ instead of $L^{\nabla 1,0}_\xi$ and $L^{\nabla 0,1}_\xi$.

REMARK. Note that $L^{1,0}_\xi$ (resp. $L^{0,1}_\xi$) preserves the bi-degree (p,q) if and only if ξ is a $(1,0)$ (resp. $(0,1)$) vector field or $q = 0$ (resp. $p = 0$). ◇

REMARK. One can also define $L^{0,1}_\xi u := \xi \lrcorner \bar\partial u + \bar\partial(\xi \lrcorner u)$. Since we only make use of connections ∇ satisfying $\nabla^{0,1} = \bar\partial$, there should be no confusion. ◇

We quite some useful results from [**Va-2022**].

PROPOSITION 1.3 ([**Va-2022**, Proposition 3.5]). *Let X be a complex manifold, let $Z \subset X$ be an n-dimensional submanifold, let $E \to X$ be a holomorphic vector bundle equipped with Hermitian metric h and its Chern connection, and let u, u' be E-valued $(n,0)$-forms on X. If ξ is a $(1,0)$-vector field on X then*

$$\iota_Z^* L_\xi \langle u \wedge \bar u', h\rangle = \iota_Z^* \left\langle \left(L^{1,0}_\xi u\right) \wedge \bar u', h\right\rangle + \iota_Z^* \left\langle u \wedge \overline{L^{0,1}_{\bar\xi} u'}, h\right\rangle$$

and

$$\iota_Z^* L_{\bar\xi} \langle u \wedge \bar u', h\rangle = \iota_Z^* \left\langle \left(L^{0,1}_{\bar\xi} u\right) \wedge \bar u', h\right\rangle + \iota_Z^* \left\langle u \wedge \overline{L^{1,0}_{\xi} u'}, h\right\rangle.$$

LEMMA 1.4 ([**Va-2022**, Lemma 3.6]). *Let $p : X \to B$ be a proper holomorphic family with fiber-dimension n, let $E \to X$ be a holomorphic vector bundle equipped with a Hermitian metric h its Chern connection, and let u and u' be E-valued $(n,0)$-forms on X. If τ is a $(1,0)$-vector field on B and ξ_τ is a $(1,0)$-vector field on X such that $dp\xi_\tau = \tau$ then*

$$\tau \int_{X_t} \langle u \wedge \bar u', h\rangle = \int_{X_t} \left\langle L^{1,0}_{\xi_\tau} u \wedge \bar u', h\right\rangle + \int_{X_t} \left\langle u \wedge \overline{L^{0,1}_{\bar\xi_\tau} u'}, h\right\rangle,$$

where $X_t := p^{-1}(t)$.

LEMMA 1.5 ([**Va-2022**, Lemma 3.7]). *Let X, Z, E and h be as in Proposition 1.3, let α and β be E-valued forms of bidegree $(n,1)$ and $(n-1,0)$ respectively. If ξ is a $(1,0)$-vector field on X then*

$$L^{1,0}_\xi \langle \alpha \wedge \bar\beta, h\rangle = \left\langle (L^{1,0}_\xi \alpha) \wedge \bar\beta, h\right\rangle + \left\langle \alpha \wedge \overline{L^{0,1}_{\bar\xi} \beta}, h\right\rangle,$$

and hence

$$\int_Z L^{1,0}_\xi \langle \alpha \wedge \bar\beta, h\rangle = \int_Z \left\langle (L^{1,0}_\xi \alpha) \wedge \bar\beta, h\right\rangle + \int_Z \left\langle \alpha \wedge \overline{L^{0,1}_{\bar\xi} \beta}, h\right\rangle.$$

PROPOSITION 1.6 ([**Va-2022**, Proposition 3.8]). *Let X, Z, E and h be as in Proposition 1.3, let u and u' be E-valued $(n,0)$-forms on X and let γ be an E-valued $(n-1,1)$-form on X. If ξ is a $(1,0)$-vector field on X then*

$$L^{1,0}_{\bar\xi} \langle \gamma \wedge \bar u, h\rangle = \left\langle L^{1,0}_{\bar\xi} \gamma \wedge \bar u, h\right\rangle + \left\langle \gamma \wedge \overline{L^{0,1}_{\xi} u}, h\right\rangle \tag{8}$$

and

$$L^{0,1}_{\xi} \langle u \wedge \bar{u}', h \rangle = \left\langle (L^{0,1}_{\xi} u) \wedge \bar{u}', h \right\rangle . \tag{9}$$

In particular,

$$\iota_Z^* L_{\bar{\xi}} \langle \gamma \wedge \bar{u}, h \rangle = \iota_Z^* \left\langle L^{1,0}_{\bar{\xi}} \gamma \wedge \bar{u}, h \right\rangle + \iota_Z^* \left\langle \gamma \wedge \overline{L^{0,1}_{\xi} u}, h \right\rangle \tag{10}$$

LEMMA 1.7 ([**Va-2022**, Lemma 3.9]). *Let $(E,h) \to X \xrightarrow{p} B$ be a holomorphic family of Hermitian holomorphic vector bundles, let τ be a holomorphic vector field on B, let ξ_τ be a lift of τ to X, and let u be an E-valued $(n,0)$-form on X, where n is the fiber dimension of $p : X \to B$. Then with respect to the Chern connection of (E,h)*

$$\bar{\partial} \iota^*_{X_t} \left(L^{1,0}_{\xi_\tau} u \right) = \iota^*_{X_t} \left(-(\xi_\tau \lrcorner \Theta(h)) u - \nabla^{1,0}((\bar{\partial}\xi_\tau) \lrcorner u) + L^{1,0}_{\xi_\tau} \bar{\partial} u \right),$$

where the $(n,1)$-form $\iota^*_{X_t}((\xi_\tau \lrcorner \Theta(h))u)$ is defined by

$$\bar{\zeta} \lrcorner \left(\iota^*_{X_t} ((\xi_\tau \lrcorner \Theta(h))u) \right) := \iota^*_{X_t} \left(\Theta(h)_{\xi_\tau \bar{\zeta}} u \right).$$

PROPOSITION 1.8 ([**Va-2022**, Proposition 3.10]). *Let $(E,h) \to X \xrightarrow{p} B$ be a holomorphic family of holomorphic Hermitian vector bundles, let n be the dimension of the fibers of $p : X \to B$, let $t \in B$ and let $\tau \in H^0(B, \mathcal{O}(T^{1,0}_B))$. Let ξ_τ be a lift of τ, i.e., a $(1,0)$-vector field ξ_τ on X such that $dp\xi_\tau = \tau$. Then with respect to the Chern connection for (E,h), one has the following statements.*

(a) *If $u, \tilde{u} \in H^0(X, \mathscr{C}^\infty(\Lambda^{n,0}_X \otimes E))$ agree on the fibers of $X \to B$, i.e., represent the same smooth section $[u] = [\tilde{u}]$ of $K_{X/B} \otimes E \to X$, then*

$$\iota^*_{X_t} L^{1,0}_{\xi_\tau} u = \iota^*_{X_t} L^{1,0}_{\xi_\tau} \tilde{u} \quad \text{and} \quad \iota^*_{X_t} L^{1,0}_{\xi_\tau} \bar{\partial} u = \iota^*_{X_t} L^{1,0}_{\xi_\tau} \bar{\partial} \tilde{u}.$$

(b) *If moreover $\iota^*_{X_t} u \in \mathscr{H}_t$ then*

$$\iota^*_{X_t} L^{1,0}_{\tilde{\xi}_\tau} \bar{\partial} u = \iota^*_{X_t} L^{1,0}_{\xi_\tau} \bar{\partial} u$$

for any other lift $\tilde{\xi}_\tau$ of τ.

(c) *Finally, if $\iota^*_{X_t} u = 0$ and $\iota^*_{X_t} L^{1,0}_{\xi_\tau} \bar{\partial} u = 0$ then $\iota^*_{X_t} L^{1,0}_{\xi_\tau} u \in \mathscr{H}_t$.*

1.3. Horizontal lifts.

DEFINITION 1.9. A horizontal distribution for a proper holomorphic submersion $p : X \to B$ is a distribution $H \subset T_X$ such that $dp(x)H_x = T_{B,p(x)}$ for all $x \in X$, i.e., $dp|_H : H \to p^*T_B$ is an isomorphism of vector bundles. Equivalently, there is a morphism of vector bundles $\theta : p^*T_B \to T_X$ such that $dp \circ \theta = \mathrm{Id}_{p^*T_B}$. ◇

We abuse notation and identify θ with its image H; thus θ is both a subbundle of T_X and a right inverse to $dp : T_X \to p^*T_B$.

Since the total space X and base manifold B are complex manifolds and $p : X \to B$ is holomorphic, $dpJ = Jdp$, where J denotes the almost complex structures. The map

$$s^{1,0} : T_X \ni \xi \mapsto \tfrac{1}{2}(\xi - \sqrt{-1}J\xi) \in T^{1,0}_X.$$

(whose inverse is $2\,\mathrm{Re}$) assigns to any horizontal distribution $\theta \subset T_X$ a unique horizontal $(1,0)$-distribution $\theta^{1,0} = s^{1,0} \circ \theta \subset T^{1,0}_X$. The correspondence $\theta \leftrightarrow \theta^{1,0}$ is bijective, and since $p : X \to B$ is holomorphic

$$dp\theta^{1,0} = p^*T^{1,0}_B \quad \text{and} \quad dp\theta^{0,1} = p^*T^{0,1}_B,$$

where $\theta^{0,1} := \overline{\theta^{1,0}}$. To make the notation slightly more readable, we shall write

$$\xi^\theta_\tau := \theta^{1,0}(\tau) \quad \text{and} \quad \bar\xi^{\bar\theta}_\tau := \theta^{0,1}(\bar\tau) = \overline{\theta^{1,0}(\tau)}, \qquad \tau \in T^{1,0}_B.$$

PROPOSITION 1.10 ([**Va-2022**, Proposition 3.12]). *Let $p : X \to B$ be a holomorphic submersion and let θ be a horizontal distribution for p. Then θ is integrable if and only if*

$$[\xi^\theta_\sigma, \xi^\theta_\tau] = \xi^\theta_{[\sigma,\tau]} \tag{11}$$

for all local $(1,0)$-vector fields σ, τ on B.

1.4. Selections from twisted Hodge theory. A thorough discussion of Hodge Theory may be found in several texts. See, for example, [**Vo-2007**]. Here we present a summary of just the few results that we use.

Let Z be a compact Hermitian manifold equipped with a Hermitian metric g and a Hermitian holomorphic vector bundle $(E, \mathfrak{h})$. We denote by

$$\Box := \bar\partial\bar\partial^* + \bar\partial^*\bar\partial$$

the $\bar\partial$-Laplacian, and by

$$\Box' := \nabla^{1,0}\nabla^{1,0*} + \nabla^{1,0*}\nabla^{1,0}$$

the Laplacian associated to the Chern connection of $(E, \mathfrak{h})$. These operators act on the bigraded module of smooth E-valued differential forms, and preserve the bigrading.

1.4.1. *The Hodge theorem.* We denote by $\mathrm{Harm}_{p,q}(g, \mathfrak{h}) := \mathrm{Ker}(\Box)$ the space of E-valued harmonic (p,q)-forms.

THEOREM 1.11 (Hodge Theorem for vector bundle-valued forms). *Let (Z, g) be a compact Hermitian manifold and let $(E, \mathfrak{h}) \to Z$ be a holomorphic Hermitian vector bundle. Then the vector spaces $\mathrm{Harm}_{p,q}(g, \mathfrak{h})$ are finite dimensional and consist of smooth E-valued (p,q)-forms. Moreover, with*

$$\wp_{p,q} : L^2_{p,q}(g, \mathfrak{h}) \to \mathrm{Harm}_{p,q}(g, \mathfrak{h})$$

denoting the orthogonal projection, there exists a compact self-adjoint operator

$$G_{p,q} : L^2(g, \mathfrak{h}) \to L^2(g, \mathfrak{h}),$$

called the Green operator, which preserves the bigrading and satisfies

(a) *$G_{p,q} W^{s,2}_{p,q}(Z) \subset W^{s+2,2}_{p,q}(Z)$ for all $s \ge 0$,*
(b) *$G_{p,q}\,\mathrm{Harm}_{p,q}(g, \mathfrak{h}) = 0$, $[G, \bar\partial] = [G, \bar\partial^*] = 0$ and $\|G_{p,q}\| \le \frac{1}{\lambda^1_{p,q}}$, where $\lambda^1_{p,q}$ is the smallest positive eigenvalue of $\Box$ in $L^2_{p,q}(g, \mathfrak{h})$, and*
(c) *one has the orthogonal decomposition $\mathrm{Id} = \wp_{p,q} + \Box G_{p,q}$ on $L^2_{p,q}(g, \mathfrak{h})$.*

REMARK. The analogous result for the operator $\Box'$ is left to the reader. ◇

1.4.2. *Bergman theory on compact Hermitian manifolds.* We return to the Hilbert fields $\mathscr{L}$ and $\mathscr{H}$ associated to the proper holomorphic family (1). Let

$$P_t : \mathscr{L}_t \to \mathscr{H}_t$$

denote the *Bergman projection.* From a section $f \in \Gamma(X, K_{X/B} \otimes E)$ that is smooth (or even just L^2) on the fibers of p one has a section $\mathcal{P}f \in \Gamma(X, K_{X/B} \otimes E)$ defined by

$$\iota^*_{X_t}\mathcal{P}f := P_t \iota^*_{X_t} f.$$

The orthogonal projection $P_t^\perp : \mathscr{L}_t \to \mathscr{H}_t^\perp$ onto the orthogonal complement of $\mathscr{H}_t$ in $\mathscr{L}_t$ is the Neumann operator. Given a $K_{X_t} \otimes E|_{X_t}$-valued, $\bar\partial$-exact $(0,1)$-form α, any two solutions $u_o, u_1 \in \mathscr{L}_t$ of the equation $\bar\partial u = \alpha$ differ by an element of $\mathscr{H}_t$. Thus if $u \in \mathscr{L}_t$ satisfies $\bar\partial u = \alpha$ then so does $P_t^\perp u$, and the latter has minimal $\mathscr{L}_t$-norm.

On a compact Hermitian manifold the Hodge Theorem (for the vector bundle $K_{X/B} \otimes E$ with Hermitian metric $\mathfrak{h}/\det g$) computes P_t and $P_t^\perp$. Indeed, by the Hodge Theorem every $\bar\partial_{X_t}$-exact form is orthogonal to the harmonic forms, and if $\alpha = \Box_{X_t} u$ is $\bar\partial_{X_t}$-closed then $\alpha = \bar\partial_{X_t}\bar\partial^*_{X_t} u$. Thus if $f \in \mathscr{L}_t$ then $\bar\partial_{X_t} f$ has no harmonic component, and hence $\bar\partial_{X_t} f = \bar\partial_{X_t}\bar\partial^*_{X_t} G_t \bar\partial_{X_t} f$. Since $f - \bar\partial^*_{X_t} G_t \bar\partial_{X_t} f$ is holomorphic and $(\bar\partial^*_{X_t} G_t \bar\partial_{X_t} f, f_o) = (G_t \bar\partial_{X_t} f, \bar\partial_{X_t} f_o) = 0$ for any $f_o \in \mathscr{H}_t$, we see that

$$P_t f = f - \bar\partial^*_{X_t} G_t \bar\partial_{X_t} f \quad \text{and thus} \quad P_t^\perp f = \bar\partial^*_{X_t} G_t \bar\partial_{X_t} f. \tag{12}$$

REMARK 1.12. It is natural to ask whether $\mathcal{P}$ (and hence $\mathcal{P}^\perp = \mathrm{Id} - \mathcal{P}$), is a smooth operator, i.e., maps smooth sections to smooth sections. In view of (12) $\mathcal{P}$ is smooth if and only if the operator

$$G : H^0(X, \mathscr{C}^\infty(K_{X/B} \otimes E)) \to \mathscr{L}$$

defined by $(Gf)_t := G_t f_t$ is smooth. Kodaira and Spencer proved that if the dimensions of the harmonic spaces are independent of t then G is smooth [**K-1986**, Theorem 7.6, page 344]. The converse, though not explicitly stated in [**K-1986**], is also true. Thus, in view of Example 0.2, $\mathcal{P}$ is smooth if and only if $\mathscr{H}$ is a holomorphic vector bundle. ◇

1.4.3. *Twisted Hodge identities.* Now suppose Z is a Kähler manifold with Kähler form ω. The Hodge Identities (often also called the Kähler Identities) in the twisted setting are

$$[\bar\partial^*, L_\omega] = \sqrt{\text{-}1}\nabla^{1,0} \quad \text{and} \quad [\nabla^{1,0*}, L_\omega] = -\sqrt{\text{-}1}\bar\partial,$$

and their adjoints are

$$[\Lambda_\omega, \bar\partial] = -\sqrt{\text{-}1}\nabla^{1,0*} \quad \text{and} \quad [\Lambda_\omega, \nabla^{1,0}] = \sqrt{\text{-}1}\bar\partial^*.$$

Here L_ω is the *Lefschetz operator* of wedging with ω, and Λ_ω is its pointwise adjoint. From these identities one derives

$$\Box - \Box' = [\sqrt{\text{-}1}\Theta(\mathfrak{h}), \Lambda_\omega]. \tag{13}$$

1.4.4. *Twisted primitive forms.*

DEFINITION 1.13. An E-valued form α of degree $k \le n$ is *primitive* if $\alpha \wedge \omega^{n+1-k} = 0$, or equivalently if $\Lambda_\omega \alpha = 0$. ◇

We require the following simple but crucial result, which may be seen as a generalization, to the twisted case, of the only part of the Hodge-Riemann Bilinear Relations for which such a generalization makes any sense.

PROPOSITION 1.14. *Let $(E, \mathfrak{h})$ be a Hermitian holomorphic vector bundle over a compact Kähler manifold (X, ω). If an E-valued (p,q)-form α is primitive then*

$$\frac{\sqrt{\text{-}1}^{(p+q)^2}}{(n-(p+q))!}\left\langle \alpha \wedge \bar\alpha \wedge \omega^{n-(p+q)}, \mathfrak{h}\right\rangle = \sqrt{\text{-}1}^{p-q} |\alpha|^2_{g,\mathfrak{h}} dV_\omega.$$

See [**Va-2022**, Proposition 6.9] for a proof in when $(p,q)=(n-1,1)$; the only case we use.

1.5. The Kodaira-Spencer map. The Kodaira-Spencer map of a germ at $o \in B$ of a proper holomorphic family $p : X \to B$ is a map

$$K_o : T^{1,0}_{B,o} \to H^{0,1}_{\bar\partial}\left(X_o, T^{1,0}_{X_o}\right),$$

whose definition we now recall. There are three $\bar\partial$-operators in play: the operators $\bar\partial^X$, $\bar\partial^{X/B}$ and $\bar\partial^B$ for the holomorphic vector bundles $T^{1,0}_X$, $T^{1,0}_{X/B}$, and $T^{1,0}_B$ respectively.

Let $\tau \in T^{1,0}_{B,o}$ be a vector, which we extend to a holomorphic vector field in a neighborhood of o (the extension still denoted τ), and let ξ_τ be any smooth lift of τ to a neighborhood of X_o in X. The $T^{1,0}_X$-valued $(0,1)$-form $\bar\partial^X\xi_\tau$ takes values in the vertical bundle $T^{1,0}_{X/B} \to X$. Indeed, since p is holomorphic,

$$dp\left(\bar\partial^X\xi_\tau\right) = \bar\partial^B dp(\xi_\tau) = \bar\partial^B\tau = 0.$$

While $\bar\partial^X\xi_\tau$ is obviously $\bar\partial$-exact as a $T^{1,0}_X$-valued $(0,1)$-form, it is in general not $\bar\partial$-exact as a $T^{1,0}_{X/B}$-valued $(0,1)$-form. However, $\bar\partial^X\xi_\tau$ is $\bar\partial^{X/B}$-closed. Indeed, if η is a $T^{(1,0)}_{X/B}$-valued form then $\bar\partial^{X/B}\eta = \bar\partial^X\eta$, so $\bar\partial^{X/B}\bar\partial^X\xi_\tau = \bar\partial^X\bar\partial^X\xi_\tau = 0$.

It is not hard to show that the cohomology class $[\bar\partial^X\xi_\tau] \in H^{0,1}_{\bar\partial}\left(X_o, T^{1,0}_{X_o}\right)$ is independent of the choice of lift ξ_τ or the holomorphic extension of $\tau \in T^{1,0}_{B,o}$, and that the map $\tau \mapsto [\bar\partial\xi_\tau]$ is linear. (See, for example, [**Va-2022**, Section 7].)

DEFINITION 1.15. The map $K_o : T^{1,0}_{B,o} \to H^{0,1}_{\bar\partial}\left(X_o, T^{1,0}_{X_o}\right)$ defined by

$$K_o(\tau) := \left[\bar\partial^X\xi_\tau\right]$$

is called the Kodaira-Spencer Map of the germ at o of the holomorphic family $p : X \to B$. ◇

REMARK. To simplify notation, we write $\bar\partial\xi_\tau$ for the $\bar\partial^{X/B}$-closed $T^{1,0}_{X/B}$-valued $(0,1)$-form obtained by applying $\bar\partial^X$ to the lift ξ_τ of the holomorphic $(1,0)$-vector field τ. This notation can be confusing at first, but it provides a slight improvement in removing some clutter from our already-too-cluttered formulas. ◇

The Kodaira-Spencer Map appears in the curvature formula for $\mathscr{L}^\theta$ (and hence, by the holomorphic Gauss formula, in the curvature formula for $\mathscr{H}$) as follows. Let τ be a holomorphic vector field on B. One can contract the $T^{1,0}_{X/B}$-valued $(0,1)$-form $\bar\partial\xi_\tau$ with a section u of $\Lambda^{n,0}_X \otimes E$ to produce an E-valued $(n-1,1)$-form

$$\bar\partial\xi_\tau \lrcorner u,$$

which we think of as a family of E-valued $(n-1,1)$-forms on the fibers of $X \to B$. If u is a representative of a section f of $K_{X/B} \otimes E$ then $\bar\partial\xi_\tau \lrcorner u$ does not depend on the choice of representative u. Indeed, since τ is holomorphic, $\bar\partial\xi_\tau$ is a $T^{1,0}_{X/B}$-valued $(0,1)$-form, so if v is an E-valued $(n,0)$-form that vanishes on the fibers of $X \to B$ then

$$\bar\partial\xi_\tau \lrcorner v = 0 \quad \text{as } (n-1,1)\text{-forms.}$$

Thus $\bar{\partial}\xi_\tau \lrcorner (u+v) = \bar{\partial}\xi_\tau \lrcorner u$, so we have a map

$$\kappa_\tau : \Gamma(X, \mathscr{C}^\infty(K_{X/B}\otimes E)) \to \Gamma(X, \mathscr{C}^\infty(\Lambda_X^{n-1,1}\otimes E))$$

defined by

$$\kappa_\tau f := \bar{\partial}\xi_\tau \lrcorner u.$$

Such a term appears in the formulas (4) and (5) of Theorem 0.4. Hence by the holomorphic Gauss formula, we find these terms in formulas (6) and (7).

REMARK. Note that if $f \in H^0(X, \mathscr{C}^\infty(K_{X/B}\otimes E)$ is holomorphic on a fiber of X_t (which means that for any representative u of f the E-valued $(n,1)$-form $\bar{\partial}u$ satisfies $\iota_{X_t}^*\bar{\partial}u = 0$) then $\iota_{X_t}^*\kappa_\tau f$ is a $\bar{\partial}$-closed $E|_{X_t}$-valued $(n,0)$-form. Indeed, one has

$$\bar{\partial}\iota_{X_t}^*(\bar{\partial}\xi_\tau \lrcorner u) = \iota_{X_t}^*\bar{\partial}(\bar{\partial}\xi_\tau \lrcorner u) = \iota_{X_t}^*(\bar{\partial}\xi_\tau \lrcorner \bar{\partial}u) = (\bar{\partial}\xi_\tau|_{X_t}) \lrcorner \iota_{X_t}^*\bar{\partial}u = 0,$$

and the second-to-last equality holds because $\bar{\partial}\xi_\tau$ is a $T^{1,0}_{X/B}$-valued $(0,1)$-form. ◇

Formulas (6) and (7) also contain terms of the form

$$\xi_\tau \lrcorner \bar{\partial}u.$$

The latter form is cohomologous to the form[2] $\kappa_\tau f$, since

$$\bar{\partial}(\xi_\tau \lrcorner u) = \bar{\partial}\xi_\tau \lrcorner u - \xi_\tau \lrcorner \bar{\partial}u.$$

Unlike $\bar{\partial}\xi_\tau \lrcorner u$, the form $\xi_\tau \lrcorner \bar{\partial}u$ does depend on the choice of representative u. But far from being a problem, this dependence allows one to choose representatives that demonstrate the positivity of the rightmost term in (7).

2. The abstract theory of Berndtsson-Lempert-Szőke fields

In this section we define Hilbert fields with two special features: a smooth structure and a complex structure. The notion of smooth structures was introduced by Lempert and Szőke [**LS-2014**, Definition 2.2.2]. The complex structure is a generalization of the work of Berndtsson, who defined the complex structure for the Hilbert field $\mathscr{H}$. It thus seems to call such Hilbert fields Berndtsson-Lempert-Szőke fields, or *BLS* fields for short.

2.1. Quasi-Hilbert fields.

Let B be a set.

DEFINITION 2.1. A *quasi-Hilbert field* is a map $\pi : \mathfrak{H} \to B$ whose fibers are topological vector spaces each of which is isomorphic to a complex Hilbert space. ◇

Each fiber of a quasi-Hilbert field has a collection of equivalent norms as part of its data. In general there is no relation between different fibers. Especially, different fibers need not be isomorphic as topological vector spaces.

---

[2]The form $\kappa_\tau f$ is $\bar{\partial}$-closed if $f \in H^0(X, \mathcal{O}(K_{X/B}\otimes E))$, but need not be $\bar{\partial}$-closed in general.

2.2. Smooth structures. Let $\Gamma(U, \mathfrak{H})$ denote the set of sections of $\pi : \mathfrak{H} \to B$ over $U \subset B$.

DEFINITION 2.2. Let B be a smooth manifold and let $\mathfrak{H} \to B$ be a quasi-Hilbert field. A *smooth structure* for a quasi-Hilbert field $\mathfrak{H} \to B$ is a sheaf $\mathscr{C}^\infty(\mathfrak{H}) \to B$ of $\mathscr{C}^\infty_B$-modules, whose elements are germs of sections of $\pi : \mathfrak{H} \to B$, such that for each $t \in B$ the stalk $\mathscr{C}^\infty(\mathfrak{H})_t$ is mapped by the evaluation-at-t map to a dense subspace of the Hilbert space $\mathfrak{H}_t$. The associated presheaf assigns to an open $U \subset B$ a set of sections of $\pi|_{\pi^{-1}(U)} : \pi^{-1}(U) \to U$ denoted $\Gamma(U, \mathscr{C}^\infty(\mathfrak{H}))$, whose members are called *smooth sections over* U. ◇

For each $t \in B$ one has an *evaluation-at-t* map. We shall write

$$\mathscr{C}^\infty(\mathfrak{H})(t) := \{\mathfrak{f}(t) \ ; \ \mathfrak{f} \in \mathscr{C}^\infty(\mathfrak{H})_t\}.$$

2.3. Maps. One must of course deal with maps between quasi-Hilbert fields. It is convenient to have some nomenclature that draws attention to the sorts of maps that might arise. We follow the convention that a linear map A from a topological vector spaces V_1 to a topological vector space V_2 is defined on some (often not explicitly referred to) subspace $D \subset V_1$, called the domain of A. If the domain D of A is dense then the map A is said to be densely defined. We shall use the notation $A : V_1 \to V_2$ which, though not uncommon, unfortunately can be misleading.

DEFINITION 2.3. Let $\mathfrak{H}^1, \mathfrak{H}^2$ be quasi-Hilbert fields over a smooth manifold B.

(a) A map $F : \mathfrak{H}^1 \to \mathfrak{H}^2$ is said to be a *map of quasi-Hilbert fields* if it is fiber-preserving, i.e., one has the commutative diagram

$$\begin{array}{ccc} \mathfrak{H}^1 & \xrightarrow{F} & \mathfrak{H}^2 \\ \downarrow & & \downarrow \\ B & \xrightarrow{\mathrm{Id}} & B \end{array},$$

and linear on fibers. We write $F \in \mathrm{Lin}(\mathfrak{H}^1, \mathfrak{H}^2)$.
If the fiber restrictions are defined on dense subsets of the fibers we will say that F is *densely-defined.*

(b) A map of quasi-Hilbert fields $F : \mathfrak{H}^1 \to \mathfrak{H}^2$ is called a *morphism* if its fiber restrictions are densely-defined and continuous. In this case we write $F \in \mathrm{Hom}(\mathfrak{H}^1, \mathfrak{H}^2)$.

(c) Suppose moreover that $\mathscr{C}^\infty(\mathfrak{H}_1)$ and $\mathscr{C}^\infty(\mathfrak{H}_2)$ are smooth structures for $\mathfrak{H}_1$ and $\mathfrak{H}_2$ respectively. A map $F \in \mathrm{Lin}(\mathfrak{H}^1, \mathfrak{H}^2)$ is said to be *smooth* if it maps $\mathscr{C}^\infty(\mathfrak{H}^1)$ to $\mathscr{C}^\infty(\mathfrak{H}^2)$, i.e., if for each $t \in B$ the subspace $\mathscr{C}^\infty(\mathfrak{H}^1)(t)$ lies in the domain of $F|_{\mathfrak{H}^1_t}$, and one has $F\mathfrak{f} \in \mathscr{C}^\infty(\mathfrak{H}^2)_t$ for every $\mathfrak{f} \in \mathscr{C}^\infty(\mathfrak{H}^1)_t$. ◇

DEFINITION 2.4. Let B be a smooth manifold and let $\mathfrak{H}^1, \mathfrak{H}^2$ be quasi-Hilbert fields over B, equipped with smooth structures $\mathscr{C}^\infty(\mathfrak{H}^1)$ and $\mathscr{C}^\infty(\mathfrak{H}^2)$.

(a) A map of sheaves $\mathbf{L} : \mathscr{C}^\infty(\mathfrak{H}^1) \to \mathscr{C}^\infty(\mathfrak{H}^2)$ is said to be $\mathscr{C}^\infty$*-linear* if

$$\mathbf{L}(f\mathfrak{f}) = f\mathbf{L}\mathfrak{f}$$

for each $t \in B$, all $\mathfrak{f} \in \mathscr{C}^\infty(\mathfrak{H}^1)_t$ and all $f \in \mathscr{C}^\infty_{B,t}$.

(b) A $\mathscr{C}^\infty$-linear $\mathbf{L}$ is said to be *tensorial at* $t \in B$ if

$$\mathfrak{f} \in \mathscr{C}^\infty(\mathfrak{H}^1)_t \text{ and } \mathfrak{f}(t) = 0 \quad \Rightarrow \quad (\mathbf{L}\mathfrak{f})(t) = 0.$$

We say $\mathbf{L}$ is *tensorial* if $\mathbf{L}$ is tensorial at every point of B. ◇

Tensorial $\mathscr{C}^\infty$-linear maps determine smooth maps of quasi-Hilbert fields in the sense of Definition 2.3 (c). (The converse is obvious.) Indeed, if $\mathbf{L}$ is a tensorial $\mathscr{C}^\infty$-linear map then one can define the densely-defined map of quasi-Hilbert fields L by

$$L\mathfrak{f}(t) := (\mathbf{L}\mathfrak{f})(t), \quad \mathfrak{f} \in \mathscr{C}^\infty(\mathfrak{H})_t,$$

and by the tensorial condition L is well-defined.

REMARK. If $\mathscr{C}^\infty(\mathfrak{H}^1)_t$ is generated, as a $\mathscr{C}^\infty_{B,t}$-module, by germs that do not vanish at p then every $\mathscr{C}^\infty$-linear map is tensorial at t. This is the situation for vector bundles with their usual smooth structure consisting of the sheaf of germs of smooth sections. ◇

2.4. Smooth metrics.

DEFINITION 2.5. Let $\mathfrak{H}$ be a quasi-Hilbert field over a smooth manifold B.

(a) A *metric* $\mathfrak{h}$ for $\mathfrak{H}$ is an assignment to each fiber $\mathfrak{H}_t$ of a complete inner product $\mathfrak{h}_t$ so that $(\mathfrak{H}_t, \mathfrak{h}_t)$ is linearly isomorphic to $\mathfrak{H}_t$ for every $t \in B$.

(b) If $\mathscr{C}^\infty(\mathfrak{H})$ is a smooth structure for $\mathfrak{H}$ then the metric $\mathfrak{h}$ is *smooth* if

$$\mathfrak{h}(\mathfrak{f}_1, \mathfrak{f}_2) \in \mathscr{C}^\infty(U)$$

for all smooth sections $\mathfrak{f}_1, \mathfrak{f}_2 \in \Gamma(U, \mathscr{C}^\infty(\mathfrak{H}))$. ◇

2.5. Connections and curvature.

DEFINITION 2.6. Let B be a smooth manifold and let $\mathfrak{H}$ be a quasi-Hilbert field over B, equipped with a smooth structure $\mathscr{C}^\infty(\mathfrak{H})$.

(a) A *connection* for $\mathfrak{H}$ is a map

$$\nabla : \Gamma(B, \mathscr{C}^\infty(T_B)) \times \Gamma(B, \mathscr{C}^\infty(\mathfrak{H})) \in (\xi, \mathfrak{f}) \mapsto \nabla_\xi \mathfrak{f} \in \Gamma(B, \mathscr{C}^\infty(\mathfrak{H}))$$

such that

$$\nabla_{f\xi+\eta} = f\nabla_\xi + \nabla_\eta \quad \text{ and } \nabla_\xi(f\mathfrak{f}) = (\xi f)\mathfrak{f} + f\nabla_\xi \mathfrak{f}$$

for all real-valued $f \in \mathscr{C}^\infty(B)$, $\xi \in \Gamma(B, \mathscr{C}^\infty(T_B))$ and $\mathfrak{f} \in \Gamma(B, \mathscr{C}^\infty(\mathfrak{H}))$.

(b) The map $\Theta(\nabla) : \Gamma(B, \mathscr{C}^\infty(\Lambda^2(T_B))) \times \Gamma(B, \mathscr{C}^\infty(\mathfrak{H})) \to \Gamma(B, \mathscr{C}^\infty(\mathfrak{H}))$ defined by

$$\langle \Theta(\nabla), (\xi \wedge \eta, \mathfrak{f}) \rangle := \nabla_\xi \nabla_\eta \mathfrak{f} - \nabla_\eta \nabla_\xi \mathfrak{f} - \nabla_{[\xi,\eta]} \mathfrak{f}$$

is called the *curvature* of ∇. ◇

Definition 2.6 extends naturally to allow complex valued $f \in \mathscr{C}^\infty(B)$ and complex vector fields $\xi \in \Gamma(B, \mathscr{C}^\infty(T_B \otimes \mathbb{C}))$ provided that we set

$$\nabla_\xi := \nabla_{\mathrm{Re}\ \xi} + \sqrt{-1}\nabla_{\mathrm{Im}\ \xi}.$$

DEFINITION 2.7. Let B be a smooth manifold, let $\mathfrak{H}$ be a quasi-Hilbert field over B equipped with a smooth structure $\mathscr{C}^\infty(\mathfrak{H})$, and let $\mathfrak{h}$ be a smooth metric for $\mathfrak{H}$. A connection ∇ for $\mathfrak{H}$ is said to be *compatible with the metric* $\mathfrak{h}$ if

$$\langle d\left(\mathfrak{h}(\mathfrak{f}_1, \mathfrak{f}_2)\right), \xi \rangle = \mathfrak{h}(\nabla_\xi \mathfrak{f}_1, \mathfrak{f}_2) + \mathfrak{h}(\mathfrak{f}_1, \nabla_\xi \mathfrak{f}_2)$$

for all $t \in B$, all $\mathfrak{f}_1, \mathfrak{f}_2 \in \mathscr{C}^\infty(\mathfrak{H})_t$ and all $\xi \in T_{B,t}$. ◇

One defines the covariant derivative of a smooth metric $\mathfrak{h}$ by

$$\nabla\mathfrak{h}(\mathfrak{f}_1,\mathfrak{f}_2) := d(\mathfrak{h}(\mathfrak{f}_1,\mathfrak{f}_2)) - \mathfrak{h}(\nabla\mathfrak{f}_1,\mathfrak{f}_2) - \mathfrak{h}(\mathfrak{f}_1,\nabla\mathfrak{f}_2). \tag{14}$$

In terms of this derivative, a connection ∇ is compatible with a smooth metric $\mathfrak{h}$ if and only if $\nabla\mathfrak{h} = 0$.

The difference of two connections and the curvature of a connection are $\mathscr{C}^\infty$-linear. If the connections are metric compatible then one can say more.

PROPOSITION 2.8 ([**Va-2022**, Proposition 2.11]). *Let $\mathfrak{H} \to B$ be a quasi-Hilbert field with smooth structure, let $\mathfrak{h}$ be a smooth metric for $\mathfrak{H}$ and let $\nabla, \nabla_1, \nabla_2$ be $\mathfrak{h}$-compatible connections. Then*

$$\nabla_2 - \nabla_1 \quad \textit{and} \quad \Theta(\nabla)$$

are tensorial.

2.6. Smooth (quasi-)Hilbert fields.

DEFINITION 2.9. Let B be a smooth manifold.

(a) A *smooth quasi-Hilbert field* is a quasi-Hilbert field with a smooth structure that admits a connection.

(b) A smooth quasi-Hilbert field $(\mathfrak{H}, \mathscr{C}^\infty(\mathfrak{H}))$ over B is said to be a *smooth Hilbert field* if it has a smooth metric $\mathfrak{h}$ and a connection $\mathring{\nabla}$, called a *reference connection*, with the following continuity property. For each $t \in B$, $\xi \in T_{B,t}$ and $\mathfrak{f} \in \mathscr{C}^\infty(\mathfrak{H})_t$

$$\left|(\mathring{\nabla}_\xi\mathfrak{h})(\mathfrak{f},\mathfrak{g}))(t)\right|^2 \le C\mathfrak{h}(\mathfrak{g},\mathfrak{g})(t) \quad \text{for all } \tilde{\mathfrak{f}} \in \mathscr{C}^\infty(\mathfrak{H})_t \tag{15}$$

for some constant $C = C(t,\xi,\mathfrak{f})$. ◇

PROPOSITION 2.10 ([**Va-2022**, Proposition 2.17]). *A smooth Hilbert field $(\mathfrak{H}, \mathfrak{h})$ has a metric compatible connection.*

2.7. Formulation with differential forms. One might sometimes prefer to think of the connection as a total differential, rather than to talk about directional derivatives. To do so it is convenient to introduce differential forms with values in $\mathfrak{H}$.

DEFINITION 2.11. Let $\mathfrak{H} \to B$ be a smooth quasi-Hilbert field. An $\mathfrak{H}$-valued differential k-form is a section $\mathfrak{b}$ of $\Lambda^k_B \otimes \mathfrak{H} \to B$. If $\mathfrak{H} \to B$ is a smooth quasi-Hilbert field then we say that $\mathfrak{b}$ is *smooth* if for each k-tuple of vector fields $\tau_1, \dots, \tau_k \in \Gamma(B, \mathscr{C}^\infty(T_B))$ the section $\langle \alpha, \tau_1 \wedge \cdots \wedge \tau_k \rangle$ of $\mathfrak{H} \to B$ obtained by contracting in the factor Λ^k_B lies in $\Gamma(B, \mathscr{C}^\infty(\mathfrak{H}))$. We denote by $\Gamma(B, \mathscr{C}^\infty(\Lambda^k_B \otimes \mathfrak{H}))$ the collection of smooth $\mathfrak{H}$-valued k-forms. ◇

From the 'forms' point of view

$$\nabla : \Gamma(B, \mathscr{C}^\infty(\mathfrak{H})) \to \Gamma(B, \mathscr{C}^\infty(\Lambda^1_B \otimes \mathfrak{H})),$$

and we can extend ∇ to higher twisted exterior derivatives

$$\nabla : \Gamma(B, \mathscr{C}^\infty(\Lambda^k_B \otimes \mathfrak{H})) \to \Gamma(B, \mathscr{C}^\infty(\Lambda^{k+1}_B \otimes \mathfrak{H}))$$

in the usual way. The difference $\nabla_1 - \nabla_2$ of two connections, or the curvature $\nabla\nabla$ of a connection, are $\mathscr{C}^\infty(B)$-linear, and if they are also tensorial then

$$\nabla_2 - \nabla_1 \in \Gamma(B, \mathscr{C}^\infty(\Lambda^1_B \otimes \mathrm{Lin}(\mathfrak{H}, \mathfrak{H})))$$

and

$$\nabla\nabla =: \Theta(\nabla) \in \Gamma(B, \mathscr{C}^\infty(\Lambda^2_B \otimes \mathrm{Lin}(\mathfrak{H}, \mathfrak{H}))).$$

As noted in Proposition 2.8, this is the case for metric connections.

2.8. Holomorphic Quasi-Hilbert fields and BLS fields. Assume now that B is a complex manifold. Then the splitting $T_B \otimes \mathbb{C} = T^{1,0}_B \oplus T^{0,1}_B$ yields a decomposition

$$\nabla = \nabla^{1,0} + \nabla^{0,1},$$

and one can talk about $\mathfrak{H}$-valued (p,q)-forms as sections of $\Lambda^{p,q}_B \otimes \mathfrak{H} \to B$. Such forms are said to be smooth if they are smooth as $\mathfrak{H}$-valued $(p+q)$-forms. The set of smooth $\mathfrak{H}$-valued (p,q)-forms is denoted $\Gamma(B, \mathscr{C}^\infty(\Lambda^{p,q}_B \otimes \mathfrak{H}))$.

DEFINITION 2.12. Let $\mathfrak{H}$ be a smooth quasi-Hilbert field over a complex manifold B.

(a) An almost complex structure for $\mathfrak{H} \to B$ is a family of linear maps of sheaves

$$\bar\partial : \mathscr{C}^\infty(\Lambda^{p,q}_B \otimes \mathfrak{H}) \to \mathscr{C}^\infty(\Lambda^{p,q+1}_B \otimes \mathfrak{H}), \quad 0 \le p, q \le n,$$

compatible with the $\bar\partial$-operator on B in the sense that

$$\bar\partial(f \wedge \mathfrak{b}) = \bar\partial f \wedge \mathfrak{b} + (-1)^{r+s} f \wedge \bar\partial \mathfrak{b}, \quad f \in \mathscr{C}^\infty(\Lambda^{r,s}_B).$$

The almost complex structure is *involutive* or *integrable*, or simply a *complex structure*, if $\bar\partial\bar\partial = 0$.

(b) A smooth section $\mathfrak{f}$ of the smooth quasi-Hilbert field $\mathfrak{H} \to B$ with almost complex structure $\bar\partial$ is said to be *holomorphic* on an open subset $U \subset B$ if $\bar\partial \mathfrak{f} = 0$ on U. The set of holomorphic sections on U is denoted $\Gamma(U, \mathscr{O}(\mathfrak{H}))$.

(c) An *(almost) holomorphic quasi-Hilbert field* is a smooth quasi-Hilbert field with an (almost) complex structure.

(d) A *Berndtsson-Lempert-Szőke* (BLS) field is an almost holomorphic smooth Hilbert field admitting a reference connection $\mathring{\nabla}$ such that $\mathring{\nabla}^{0,1} = \bar\partial$.

(e) A BLS field is said to be *integrable* if its almost complex structure is integrable. ◇

PROPOSITION 2.13 ([**Va-2022**, Proposition 2.21]). *On every BLS field $(\mathfrak{H}, \mathfrak{h}, \bar\partial)$ there exists a unique connection ∇, called the* BLS Chern connection*, or simply the* Chern connection*, such that*

$$\nabla \mathfrak{h} = 0 \quad \textit{and} \quad \nabla^{0,1} = \bar\partial.$$

We often refer to the BLS field $(\mathfrak{H}, \mathfrak{h})$ and omit explicit reference to the connection and the almost complex structure. We even omit reference to the metric when the latter is clear from the context. Moreover, as with holomorphic Hermitian vector bundles, the curvature of the Chern connection for $(\mathfrak{H}, \mathfrak{h})$ is often just called the curvature of $\mathfrak{h}$.

PROPOSITION 2.14 ([**Va-2022**, Proposition 2.23]). *The curvature $\Theta(\mathfrak{h})$ of the BLS field $\mathfrak{H}$ is a smooth* $\mathrm{Lin}(\mathfrak{H}, \mathfrak{H})$*-valued 2-form that is $\mathfrak{h}$-anti-Hermitian. Moreover, if $\bar\partial\bar\partial = 0$ then the curvature is a* $\mathrm{Lin}(\mathfrak{H}, \mathfrak{H})$*-valued* $(1,1)$*-form.*

2.9. Smooth (Quasi-)Hilbert subfields.

DEFINITION 2.15. Let $(\mathfrak{L}, \mathfrak{h}^{\mathfrak{L}})$ be a smooth Hilbert field over a smooth manifold B.

(a) A smooth Hilbert subfield of $(\mathfrak{L}, \mathfrak{h}^{\mathfrak{L}})$ is a smooth Hilbert field $(\mathfrak{H}, \mathfrak{h}^{\mathfrak{H}})$ over B such that
  i) $\mathfrak{H}_t$ is a closed subspace of $\mathfrak{L}_t$ for every $t \in B$,
  ii) $\mathscr{C}^\infty(\mathfrak{H}) = \{\mathfrak{f} \in \mathscr{C}^\infty(\mathfrak{L})\ ;\ \mathfrak{f}(t) \in \mathfrak{H}_t \text{ for all } t \in \text{Domain}(\mathfrak{f})\}$, and
  iii) $\mathfrak{h}^{\mathfrak{L}}|_{\mathfrak{H}} = \mathfrak{h}^{\mathfrak{H}}$.

(b) To any smooth Hilbert subfield $\mathfrak{H} \subset \mathfrak{L}$ there is associated the *orthogonal projector*: the unique morphism $P = P_{\mathfrak{H}} : \mathfrak{L} \to \mathfrak{H}$ defined by fiberwise orthogonal projection.

(c) We say that the smooth Hilbert subfield $\mathfrak{H} \subset \mathfrak{L}$ is *regular* if the projection morphism $P_{\mathfrak{H}} : \mathfrak{L} \to \mathfrak{H}$ is smooth. ◇

REMARK. A smooth quasi-Hilbert subfield $\mathfrak{H} \subset \mathfrak{L}$ is a smooth quasi-Hilbert field that satisfies (a.i) and (a.ii) of Definition 2.15. ◇

REMARK. To any smooth Hilbert subfield $\mathfrak{H} \subset \mathfrak{L}$ one can associate its orthogonal complement $\mathfrak{H}^\perp$. In general $\mathfrak{H}^\perp$ need not be a smooth Hilbert subfield of $\mathfrak{L}$, but this is certainly the case if $\mathfrak{H}$ is regular. ◇

2.10. The second fundamental map. Let $\mathfrak{L}$ be a smooth (quasi-)Hilbert field over a smooth manifold B and $\mathfrak{H} \subset \mathfrak{L}$ a smooth (quasi-)Hilbert subfield. Suppose both fields are equipped with connections $\nabla^{\mathfrak{L}}$ and $\nabla^{\mathfrak{H}}$. Consider the map

$$\mathbf{N}^{\mathfrak{L}/\mathfrak{H}} := \nabla^{\mathfrak{L}}|_{\mathfrak{H}} - \nabla^{\mathfrak{H}} : \mathscr{C}^\infty(\mathfrak{H}) \to \mathscr{C}^\infty(\Lambda^1_B \otimes \mathfrak{L}).$$

It is easy to see that $\mathbf{N}^{\mathfrak{L}/\mathfrak{H}}$ is a $\mathscr{C}^\infty$-linear map (c.f. Definition 2.4 (a).

DEFINITION 2.16. The linear map of sheaves $\mathbf{N}^{\mathfrak{L}/\mathfrak{H}}$ is called the *second fundamental map* of the (quasi-)Hilbert fields with connections. ◇

If $\mathfrak{L}$ is a Hilbert field, $\mathfrak{H} \subset \mathfrak{L}$ is a Hilbert subfield and the connections $\nabla^{\mathfrak{L}}$ and $\nabla^{\mathfrak{H}}$ are metric-compatible then the map $\mathbf{N}^{\mathfrak{L}/\mathfrak{H}}$ is tensorial. Therefore one has a densely defined map of Hilbert fields $\mathrm{I\!I}^{\mathfrak{L}/\mathfrak{H}} : \mathfrak{H} \to \Lambda^1_B \otimes \mathfrak{L}$ such that

$$\mathbf{N}^{\mathfrak{L}/\mathfrak{H}}\mathfrak{f} = \mathrm{I\!I}^{\mathfrak{L}/\mathfrak{H}}\mathfrak{f} \quad \text{for all } \mathfrak{f} \in \mathscr{C}^\infty(\mathfrak{H}).$$

DEFINITION 2.17. The map of Hilbert fields $\mathrm{I\!I}^{\mathfrak{L}/\mathfrak{H}}$ is called the *second fundamental form* of the Hilbert fields with metric-compatible connections $(\mathfrak{L}, \nabla^{\mathfrak{L}})$ and $(\mathfrak{H}, \nabla^{\mathfrak{H}})$. ◇

If $\mathfrak{H} \subset \mathfrak{L}$ is a regular Hilbert subfield then any connection $\nabla^{\mathfrak{L}}$ for $\mathfrak{L}$ defines a connection $\nabla^{\mathfrak{H}} := P_{\mathfrak{H}}\nabla^{\mathfrak{L}}$. In this case, $\mathbf{N}^{\mathfrak{L}/\mathfrak{H}} = P^\perp_{\mathfrak{H}}\nabla^{\mathfrak{L}}$. And if $\nabla^{\mathfrak{L}}$ is metric-compatible then for any $\mathfrak{f}, \mathfrak{g} \in \mathscr{C}^\infty(\mathfrak{H})$ one has

$$\mathfrak{h}(\nabla^{\mathfrak{H}}\mathfrak{f}, \mathfrak{g}) = \mathfrak{h}(\nabla^{\mathfrak{L}}\mathfrak{f}, \mathfrak{g}) = d(\mathfrak{h}(\mathfrak{f}, \mathfrak{g})) - \mathfrak{h}(\mathfrak{f}, \nabla^{\mathfrak{L}}\mathfrak{g}) = d(\mathfrak{h}(\mathfrak{f}, \mathfrak{g})) - \mathfrak{h}(\mathfrak{f}, \nabla^{\mathfrak{H}}\mathfrak{g}),$$

so that $\nabla^{\mathfrak{H}}$ is also metric compatible. Therefore $\mathbf{N}^{\mathfrak{L}/\mathfrak{H}}$ is tensorial, and one has a second fundamental form $\mathrm{I\!I}^{\mathfrak{L}/\mathfrak{H}}$, which moreover takes its values in $\Lambda^1_B \otimes \mathfrak{H}^\perp$.

2.11. BLS subfields.

DEFINITION 2.18. Let $(\mathfrak{L}, \bar\partial^{\mathfrak{L}})$ be an almost holomorphic quasi-Hilbert field over a complex manifold B.

(a) An almost holomorphic quasi-Hilbert field $(\mathfrak{H}, \bar\partial^{\mathfrak{H}})$ over B is said to be a *subfield* of $\mathfrak{L}$ if
  (i) $\mathfrak{H}_t$ is a closed subspace of $\mathfrak{L}_t$ for every $t \in B$,
  (ii) $\mathscr{C}^\infty(\mathfrak{H}) = \{\mathfrak{f} \in \mathscr{C}^\infty(\mathfrak{L}))\ ;\ \mathfrak{f}(t) \in \mathfrak{H}_t$ for all $t \in \mathrm{Domain}(\mathfrak{f})\}$, and
  (iii) $\bar\partial^{\mathfrak{L}}|_{\mathscr{C}^\infty(\mathfrak{H})} = \bar\partial^{\mathfrak{H}}$.

(b) If $\mathfrak{L}$ is a BLS field, a subfield $\mathfrak{H}$ of $\mathfrak{L}$ is a *BLS subfield* of $\mathfrak{L}$ if $P_{\mathfrak{H}}\mathscr{C}^\infty(\mathfrak{L}) \subset \mathscr{C}^\infty(\mathfrak{H})$. ◇

REMARK 2.19. If $\mathfrak{H} \subset \mathfrak{L}$ are BLS then the inclusions $P_{\mathfrak{H}}\mathscr{C}^\infty(\mathfrak{L}) \subset \mathscr{C}^\infty(\mathfrak{H}) \subset \mathscr{C}^\infty(\mathfrak{L})$, which follow from ii and i(b) respectively, imply that $\mathfrak{H}$ is a regular subfield of $\mathfrak{L}$ and that $P_{\mathfrak{H}}\mathscr{C}^\infty(\mathfrak{L}) = \mathscr{C}^\infty(\mathfrak{H})$. In particular, a smooth section of $\mathfrak{L}$ that takes all its values in $\mathfrak{H}$ is a smooth section of $\mathfrak{H}$.

Conversely, if $\mathfrak{L}$ is a BLS field and $\mathfrak{H} \subset \mathfrak{L}$ is a regular subfield such that the subspace $\{(P_{\mathfrak{H}}\mathfrak{f})(p)\ ;\ \mathfrak{f} \in \mathscr{C}^\infty(\mathfrak{L})_p\} \subset \mathfrak{H}_p$ is dense for every $p \in B$ then $\mathfrak{H} \subset \mathfrak{L}$ is a BLS subfield. ◇

PROPOSITION 2.20 ([**Va-2022**, Proposition 2.32]). *If $\mathfrak{H} \subset \mathfrak{L}$ is a BLS subfield then the BLS Chern connection for $\mathfrak{H} \to B$ is $\nabla^{\mathfrak{H}} = P_{\mathfrak{H}}\nabla^{\mathfrak{L}}|_{\mathfrak{H}}$, where $\nabla^{\mathfrak{L}}|_{\mathfrak{H}}$ denotes the restriction of the connection $\nabla^{\mathfrak{L}}$ to $\mathscr{C}^\infty(\mathfrak{H})$.*

By Definition 2.18 one also sees that

$$\mathbf{N}^{\mathfrak{L}/\mathfrak{H}} = P_{\mathfrak{H}}^{\perp}\nabla^{\mathfrak{L}1,0}|_{\mathfrak{H}}.$$

Consequently the second fundamental map $\mathbf{N}^{\mathfrak{L}/\mathfrak{H}}$ of a BLS subfield $\mathfrak{H} \subset \mathfrak{L}$ is tensorial, and its associated second fundamental form $\mathbf{I\!I}^{\mathfrak{L}/\mathfrak{H}}$ takes values in $\Lambda^{1,0}_B \otimes \mathfrak{H}^{\perp}$.

2.12. Holomorphic Gauss formula.

PROPOSITION 2.21 (Holomorphic Gauss Formula). *Let $(\mathfrak{L}, \mathfrak{h}^{\mathfrak{L}})$ be a BLS field over a complex manifold B and let $\mathfrak{H} \subset \mathfrak{L}$ be a BLS subfield.*

(a) *For any smooth sections $\mathfrak{f}_1, \mathfrak{f}_2 \in \Gamma(B, \mathscr{C}^\infty(\mathfrak{H}))$ and any $(1,0)$-vectors $\xi, \eta \in T^{1,0}_{B,t}$, $t \in B$, one has*

$$\mathfrak{h}(\Theta(\mathfrak{h}^{\mathfrak{L}})(\xi,\bar\eta)\mathfrak{f}_1, \mathfrak{f}_2) = \mathfrak{h}(\Theta(\mathfrak{h}^{\mathfrak{H}})(\xi,\bar\eta)\mathfrak{f}_1, \mathfrak{f}_2) + \mathfrak{h}(\mathbf{I\!I}^{\mathfrak{L}/\mathfrak{H}}(\xi)\mathfrak{f}_1, \mathbf{I\!I}^{\mathfrak{L}/\mathfrak{H}}(\eta)\mathfrak{f}_2) \tag{16}$$

(b) *If moreover the quadratic form*

$$\mathbb{G}_{\xi\bar\eta}(f_1, f_2) := \mathfrak{h}(\mathbf{I\!I}^{\mathfrak{L}/\mathfrak{H}}(\xi)f_1, \mathbf{I\!I}^{\mathfrak{L}/\mathfrak{H}}(\eta)f_2)$$

defines a bounded conjugate-linear functional $\mathbb{G}_{\xi,\bar\eta}(f,\cdot)$ for every $t \in B$, every $f \in \mathfrak{H}_t$ and every pair of $(1,0)$-vectors $\xi, \eta \in T^{1,0}_{B,t}$, then

$$\Theta(\mathfrak{h}^{\mathfrak{L}})|_{\mathfrak{H}} = \Theta(\mathfrak{h}^{\mathfrak{H}}) - (\mathbf{I\!I}^{\mathfrak{L}/\mathfrak{H}})^{\dagger} \wedge \mathbf{I\!I}^{\mathfrak{L}/\mathfrak{H}}. \tag{17}$$

For the proof, see [**Va-2022**, Proposition 2.33], in which one also finds the following definition.

DEFINITION 2.22. Let $\mathfrak{L}$ be a BLS field and let $\mathfrak{H} \subset \mathfrak{L}$ be a BLS subfield.

(a) The quadratic form $\mathbb{G}$ defined in Proposition 2.21.b is called the *Gauss form* of the BLS subfield $\mathfrak{H} \subset \mathfrak{L}$.

(b) If the Gauss form satisfies the continuity condition of Proposition 2.21.b then we say that $\mathfrak{H} \subset \mathfrak{L}$ is a *strongly BLS subfield.* ◇

2.13. Positivity. The usual notions of positivity in complex geometry have direct analogues in the setting of BLS fields, with one important change. Since one wants curvature to measure an infinitesimal quantity related to local variation, it is natural to evaluate the curvature only on vectors that can be continued to a local section. Thus for a BLS field $(\mathfrak{H}, \mathfrak{h}) \to B$ the curvature $\Theta(\nabla^{\mathfrak{H}})$ is evaluated at the fiber $\mathfrak{H}_t$ only on vectors in $\mathscr{C}^\infty(\mathfrak{H})_t(t)$.

Let $(\mathfrak{H}, \mathfrak{h})$ be a BLS field over a complex manifold B. Using the metric $\mathfrak{h}$, one defines Hermitian forms $\{\ ,\ \}_{\mathfrak{h},\Theta(\mathfrak{h})}$ on the fibers $T^{1,0}_{B,t} \otimes \mathscr{C}^\infty(\mathfrak{H})_t(t)$ by

$$\{\tau_1 \otimes \mathfrak{f}_1, \tau_2 \otimes \mathfrak{f}_2\}_{\mathfrak{h},\Theta(\mathfrak{h})} := \mathfrak{h}(\Theta(\mathfrak{h})_{\tau_1,\bar{\tau}_2}\mathfrak{f}_1, \mathfrak{f}_2)$$

for rank-1 tensors in $T^{1,0}_{B,t} \otimes \mathscr{C}^\infty(\mathfrak{H})_t(t)$, and extended sesqui-linearly on each fiber.

DEFINITION 2.23. Let $\mathfrak{H} \to B$ be a BLS field with smooth Hermitian metric $\mathfrak{h}$, and fix a smooth Hermitian metric g on B.

(a) We say that $\mathfrak{h}$ has Griffiths-positive curvature if at $t \in B$ if for some $c > 0$

$$\{\tau \otimes \mathfrak{f}, \tau \otimes \mathfrak{f}\}_{\mathfrak{h},\Theta(\mathfrak{h})} \geq c\mathfrak{h}(\mathfrak{f}, \mathfrak{f})g(\tau, \tau)$$

for all $\tau \otimes \mathfrak{f} \in T^{1,0}_{B,t} \otimes \mathscr{C}^\infty(\mathfrak{H})_t(t)$.

(b) We say that $\mathfrak{h}$ has Nakano-positive curvature at $t \in B$ if for some $c > 0$

$$\left\{\sum_{j=1}^{N} \tau_j \otimes \mathfrak{f}_j, \sum_{k=1}^{N} \tau_k \otimes \mathfrak{f}_k\right\}_{\mathfrak{h},\Theta(\mathfrak{h})} \geq c\sum_{j=1}^{N} \mathfrak{h}(\mathfrak{f}_j, \mathfrak{f}_j)g(\tau_j, \tau_j)$$

for all $N \in \mathbb{N}$ and all $\tau_1 \otimes \mathfrak{f}_1, \ldots, \tau_N \otimes \mathfrak{f}_N \in T^{1,0}_{B,t} \otimes \mathscr{C}^\infty(\mathfrak{H})_t(t)$.

(c) Let $k \in \mathbb{N}$ be a positive integer. We say that $\mathfrak{h}$ has k-positive curvature at $t \in B$ if for some $c > 0$

$$\left\{\sum_{j=1}^{k} \tau_j \otimes \mathfrak{f}_j, \sum_{\ell=1}^{k} \tau_\ell \otimes \mathfrak{f}_\ell\right\}_{\mathfrak{h},\Theta(\mathfrak{h})} \geq c\sum_{j=1}^{k} \mathfrak{h}(\mathfrak{f}_j, \mathfrak{f}_j)g(\tau_j, \tau_j)$$

for all $\tau_1 \otimes \mathfrak{f}_1, \ldots, \tau_k \otimes \mathfrak{f}_k \in T^{1,0}_{B,t} \otimes \mathscr{C}^\infty(\mathfrak{H})_t(t)$ ◇

Clearly the definitions are independent of the choice of metric g. Note also that 1-positivity is just Griffiths positivity, that if $k \geq \min(\dim_{\mathbb{C}} B, \mathrm{Rank}(\mathscr{C}^\infty(\mathfrak{H})_t(t)))$ then k-positivity is just Nakano positivity, and that m-positivity implies k-positivity if $m \geq k$.

3. Hilbert fields of sections

We fix a proper holomorphic family

$$(E, \mathfrak{h}) \to X \xrightarrow{p} B$$

of Hermitian vector bundles, and describe the associated Hilbert fields $\mathscr{H}$ and $\mathscr{L}$ in the language of BLS fields.

3.1. Berndtsson's Hilbert field.

DEFINITION 3.1. The Hilbert field $\mathscr{H}$ with its L^2-inner product (2) is called the *Berndtsson Hilbert field of the family* $(E, \mathfrak{h}) \to X \xrightarrow{p} B$. ◇

Smooth and holomorphic sections of $\mathscr{H}$. The following definition was put forth by Berndtsson [**B-2009**], perhaps in slightly different language.

DEFINITION 3.2. Let $(E, \mathfrak{h}) \to X \overset{p}{\to} B$ be a proper holomorphic family.

(a) A section $\mathfrak{f}$ of $\mathscr{H}$ over $U \subset B$ corresponds to a section
$$f \in \Gamma(p^{-1}(U), K_{X/B} \otimes E)$$
such that for each $t \in U$ $\iota^*_{X_t} f \in \mathscr{H}_t = H^0(X_t, \mathcal{O}(K_{X_t} \otimes E|_{X_t}))$. We write
$$\mathfrak{f} \in \Gamma(U, \mathscr{H}).$$

(b) We say $\mathfrak{f} : U \to \mathscr{H}$ is holomorphic (resp. $\mathscr{C}^r$, measurable , etc.) if f is holomorphic (resp. $\mathscr{C}^r$, measurable , etc.). We denote by $\Gamma(U, \mathscr{C}^\infty(\mathscr{H}))$ and $\Gamma(U, \mathcal{O}(\mathscr{H}))$ the set of smooth and holomorphic sections over an open set $U \subset B$. Finally, we let $\mathscr{C}^\infty(\mathscr{H})$ and $\mathcal{O}(\mathscr{H})$ denote the sheaves of germs of smooth and holomorphic sections of $\mathscr{H} \to B$ respectively. ◇

It is sometimes convenient to employ the notation
$$\mathfrak{f} = \mathfrak{i}(f) \quad \text{and} \quad f = \mathfrak{a}(\mathfrak{f}),$$
and say that $\mathfrak{f} \in \Gamma(B, \mathscr{H})$ is induced by f, and that $f \in \Gamma(X, K_{X/B} \otimes E)$ is associated to $\mathfrak{f}$.

Our standing assumption that $\mathscr{H} \to B$ is a holomorphic vector bundle means that $(E, \mathfrak{h}) \to X \overset{p}{\to} B$ is proper and for each $t \in B$ and $s \in \mathscr{H}_t$ there exists $\mathfrak{f} \in \mathscr{C}^\infty(\mathscr{H})_t$ such that
$$\iota^*_{X_t} f = s.$$
In particular, by our standing assumption $\mathscr{C}^\infty(\mathscr{H})$ defines a smooth structure on $\mathscr{H}$.

Berndtsson's $\bar\partial^{\mathscr{H}}$ operator. As explained in the introduction, if u is an E-valued $(n, 0)$-form on the total space of the family $X \to B$, $t \in B$ and $\iota^*_{X_t} u \in \mathscr{H}_t$ then for any lifts ξ^1_τ and ξ^2_τ of the $(1, 0)$-vector $\tau \in T^{1,0}_{B,t}$ one has
$$\bar\xi^1_\tau \lrcorner \bar\partial u = \bar\xi^2_\tau \lrcorner \bar\partial u$$
as E-valued $(n, 0)$-forms on X along X_t. This fact allows us to define the operator $\bar\partial^{\mathscr{H}}$ on $\mathscr{C}^\infty(\mathscr{H})$ by Formula (3), i.e., for any lift ξ_τ of any $(1, 0)$-vector τ one has
$$\langle \bar\partial^{\mathscr{H}} \mathfrak{f}, \bar\tau \rangle := \langle \bar\partial u, \bar\xi_\tau \rangle \quad \text{where } u \text{ represents } f = \mathfrak{a}(\mathfrak{f}).$$
Equivalently, if $\mathfrak{f} \in \Gamma(B, \mathscr{C}^\infty(\mathscr{H}))$ then the section
$$f := \mathfrak{a}(\mathfrak{f}) \in H^0(X, \mathscr{C}^\infty(K_{X/B} \otimes E))$$
satisfies
$$\bar\partial f(\bar\zeta) = 0 \quad \text{for all } \zeta \in T^{1,0}_{X/B}.$$
Thus, in view of the exact sequence of holomorphic vector bundles
$$0 \to T^{1,0}_{X/B} \to T^{1,0}_X \to p^* T^{1,0}_B \to 0,$$
for any lift ξ_τ of the $(1, 0)$-vector $\tau \in T^{1,0}_{B,t}$ one has
$$\langle \bar\partial^{K_{X/B} \otimes E} f, \bar\xi_\tau \rangle = \langle \bar\partial^E u, \bar\xi_\tau \rangle \quad \text{for any representative } u \text{ of } f.$$
Note that

(i) $\mathfrak{f} \in \Gamma(B, \mathcal{O}(\mathscr{H}))$ in the sense of Definition 3.2 if and only if $\bar\partial^{\mathscr{H}} \mathfrak{f} = 0$, i.e., if and only if $\mathfrak{f} \in \Gamma(B, \mathcal{O}(\mathscr{H}))$ in the sense of Definition 2.12, and

(ii) for any $\mathscr{H}$-valued (p,q)-form $\mathfrak{b}$, expressed in terms of a $\bar{\partial}$-closed local frame $\alpha^1, \ldots, \alpha^N$ for $\Lambda_B^{p,q}$ as $\mathfrak{b} = \alpha^i \otimes \mathfrak{b}_i$ for some $\mathfrak{b}_i \in \mathscr{C}^\infty(\mathscr{H})$, the operator $\bar{\partial}^{\mathscr{H}}$ is given via Definition 2.12 by

$$\bar{\partial}^{\mathscr{H}} \mathfrak{b} = \alpha^i \otimes \bar{\partial}^{\mathscr{H}} \mathfrak{b}_i,$$

and therefore $\bar{\partial}^{\mathscr{H}} \bar{\partial}^{\mathscr{H}} = 0$ provided $\bar{\partial}^{\mathscr{H}} \bar{\partial}^{\mathscr{H}} \mathfrak{f} = 0$ for all smooth sections $\mathfrak{f}$ of $\mathscr{H}$.

PROPOSITION 3.3. *For every* $\mathfrak{f} \in \Gamma(U, \mathscr{C}^\infty(\mathscr{H}))$ *one has* $\bar{\partial}^{\mathscr{H}} \bar{\partial}^{\mathscr{H}} \mathfrak{f} = 0$.

PROOF. It suffices to assume that U is a coordinate chart in B with coordinates $s^1, \ldots, s^m$, say. We define the functions $t^i := p^* s^i \in \mathcal{O}(p^{-1}(U))$. Let $f \in H^0(p^{-1}(U), \mathscr{C}^\infty(K_{X/B} \otimes E))$ be the section associated to $\mathfrak{f}$, and let $u \in H^0(p^{-1}(U), \mathscr{C}^\infty(\Lambda_B^{n,0} \otimes E))$ be a representative of f.

In general, the functions $t^1, \ldots, t^m$ can be completed to a holomorphic coordinate system on X only locally. However, one can define the operators $\frac{\partial}{\partial \bar{t}^j}$ on representatives u of sections f associated to $\mathfrak{f}$ globally by setting

$$\frac{\partial u}{\partial \bar{t}^j} := \bar{\partial}_{\bar{\xi}_j} u,$$

where ξ_j is any smooth lift of $\frac{\partial}{\partial s^j}$. If $(t^1, \ldots, t^m)$ is completed to a local coordinate system on some neighborhood in X then the action of the operators $\frac{\partial}{\partial \bar{t}^j}$ on u is given by the usual partial derivatives, hence the notation. (In particular, the operators $\frac{\partial}{\partial \bar{t}^1}, \ldots, \frac{\partial}{\partial \bar{t}^m}$ pairwise-commute.) Therefore

$$\bar{\partial} u = \frac{\partial u}{\partial \bar{t}^i} d\bar{t}^i.$$

The sections $\frac{\partial u}{\partial \bar{t}^i} \in H^0(U, \mathscr{C}^\infty(\Lambda_B^{n,0} \otimes E))$ are again representatives of sections that induce smooth sections of $\mathscr{H}$. Hence $\mathfrak{i}(\bar{\partial}^{\mathscr{H}} \bar{\partial}^{\mathscr{H}} \mathfrak{f})$ is represented by

$$\frac{\partial^2 u}{d\bar{t}^i d\bar{t}^j} d\bar{t}^i \wedge d\bar{t}^j = 0,$$

and the proof is complete. □

Thus $\mathscr{H} \to B$ is a holomorphic Hilbert field.

3.2. The $\mathbf{L^2}$ BLS fields. We now construct a large collection of BLS fields, each of which, in view of our standing assumption, contains the Berndtsson Hilbert field $\mathscr{H} \to B$ as a BLS subfield. The underlying smooth structure of all of these BLS fields is the same, and it is the complex structure that changes.

The underlying Hilbert field.

The fibers of the underlying Hilbert field are the L^2-spaces

$$\mathscr{L}_t = \left\{ \text{measurable sections } f \in \Gamma(X_t, K_{X_t} \otimes E|_{X_t}) \ ; \ (-1)^{\frac{n^2}{2}} \int_{X_t} \langle f \wedge \bar{f}, \mathfrak{h} \rangle < +\infty \right\}.$$

Since we want $\mathscr{H}$ to be a Hilbert subfield, we are forced to take the L^2-metric for $\mathscr{L}$.

The smooth structure.

DEFINITION 3.4. A smooth section $\mathfrak{f}$ of $\mathscr{L} \to B$ is a section

$$f \in H^0(X, \mathscr{C}^\infty(K_{X/B} \otimes E)).$$

We write $\mathfrak{f} \in \Gamma(B, \mathscr{C}^\infty(\mathscr{L}))$. ◇

As with $\mathscr{H} \to B$, we might write $f = \mathfrak{a}(\mathfrak{f})$ and $\mathfrak{f} = \mathfrak{i}(f)$ to relate sections of $\mathscr{L} \to B$ with sections of $K_{X/B} \otimes E \to X$.

As in the case of $\mathscr{H}$, differential k-forms and (p,q)-forms with values in $\mathscr{L}$ are sections of $\Lambda^k_B \otimes \mathscr{L} \to B$ and $\Lambda^{p,q}_B \otimes \mathscr{L} \to B$ respectively, and correspond to sections of the vector bundles $p^*\Lambda^k_B \otimes K_{X/B} \otimes E \to X$ and $p^*\Lambda^{p,q}_B \otimes K_{X/B} \otimes E \to X$. It is also standard that

(i) the collection of smooth members of $\mathscr{L}_t$ is dense,
(ii) every smooth $s \in \mathscr{L}_t$ has a smooth extension to a section of $K_{X/B} \otimes E \to X$, and
(iii) the L^2 metric is smooth with respect to $\mathscr{C}^\infty(\mathscr{L})$.

Thus $\mathscr{L}$ is a smooth Hilbert field.

Reference connections.

Given any connection D for $E \to X$, one defines the connection $\mathring{\nabla}$ for $\mathscr{L} \to B$ by

$$\mathring{\nabla}_{\sigma+\bar{\tau}} \mathfrak{f} = \mathfrak{i}\left(L^{D1,0}_{\xi^\theta_\sigma} \mathfrak{a}(\mathfrak{f})\right) + \mathfrak{i}\left(L^{D0,1}_{\bar{\xi}^\theta_\tau} \mathfrak{a}(\mathfrak{f})\right), \quad \sigma, \tau = T^{1,0}_{B,t}. \tag{18}$$

A direct computation shows that

$$\mathring{\nabla}_{\sigma+\bar{\tau}} \mathfrak{h}(\mathfrak{f}_1, \mathfrak{f}_2) = \int_X \langle u_1 \wedge \bar{u}_2, Dh \rangle,$$

where u_i is a representative of the section $f_i := \mathfrak{a}(\mathfrak{f}_i)$, so that the continuity condition (15) holds. Thus $\mathring{\nabla}$ is a reference connection in the sense of Definition 2.9(b).

The almost complex structures.

DEFINITION 3.5. Let $(E, \mathfrak{h}) \to X \xrightarrow{p} B$ be a proper holomorphic family. For a horizontal distribution $\theta \subset T_X$ and a smooth section $\mathfrak{f} \in \Gamma(B, \mathscr{C}^\infty(\mathscr{L}))$

$$\bar{\partial}^\theta_{\bar{\tau}} \mathfrak{f} := \mathfrak{i}\left(\bar{\partial}_{\bar{\xi}^\theta_\tau} \mathfrak{a}(\mathfrak{f})\right), \quad \tau \in T^{1,0}_B$$

is the almost complex structure (i.e., $\bar{\partial}$-operator) for $\mathscr{L}$ associated to θ. ◇

The operator $\bar{\partial}^\theta$ is extended to smooth $\mathscr{L}$-valued differential forms in the usual way.

PROPOSITION 3.6 ([**Va-2022**, Proposition 4.13]). *Let $p : X \to B$ be a proper holomorphic submersion and let $\theta \subset T_X$ be a horizontal distribution. Then*

$$\left\langle \bar{\partial}^\theta \bar{\partial}^\theta \mathfrak{f}, \sigma \wedge \bar{\tau} \right\rangle = \mathfrak{i}\left(\left(\bar{\partial}_{[\bar{\xi}^\theta_\sigma, \bar{\xi}^\theta_\tau]} - \bar{\partial}_{\bar{\xi}^\theta_{[\sigma,\tau]}}\right) \mathfrak{a}(\mathfrak{f})\right). \tag{19}$$

In particular, $\bar{\partial}^\theta \bar{\partial}^\theta = 0$ if and only if the distribution θ is integrable in the sense of Frobenius.

The L^2 BLS field of $(E, \mathfrak{h}) \to X \xrightarrow{p} B$.

Note that if in Formula (18) we take the connection D to be complex, i.e., such that $D^{0,1} = \bar\partial$, then the resulting reference connection $\mathring{\nabla}$ for $(\mathscr{L}, \bar\partial^\theta)$ satisfies $\mathring{\nabla}^{0,1} = \bar\partial^\theta$. Therefore all of the properties of a BLS field required by Definition 2.12(d) are satisfied for $(\mathscr{L}, \bar\partial^\theta)$.

DEFINITION 3.7. The BLS field $(\mathscr{L}, \mathfrak{h}, \bar\partial^\theta) \to B$ is called the *L^2 BLS field of the family* $(E, \mathfrak{h}) \to X \xrightarrow{p} B$ associated to the horizontal distribution $\theta \subset T_X$. ◇

To simplify some notation we write $\mathscr{L}^\theta$ in place of the more cumbersome $(\mathscr{L}, \mathfrak{h}, \bar\partial^\theta)$.

The BLS subfield $\mathscr{H}$. In Paragraph 3.1 we defined the $\bar\partial$-operator for $\mathscr{H} \to B$ by

$$\bar\partial_{\bar\tau}\mathfrak{f} := \mathfrak{i}(\bar\xi^\theta_\tau \lrcorner \bar\partial \mathfrak{a}(\mathfrak{f})), \quad \tau \in T^{1,0}_B,$$

and observed that the right hand side is independent of the choice of lift. Hence for every distribution $\theta \subset T_X$ the operator $\bar\partial^\theta|_{\mathscr{H}}$ agrees with the $\bar\partial$-operator for $\mathscr{H}$ given in Paragraph 3.1. In particular, since we have assumed that $\mathscr{H} \to B$ is a holomorphic vector bundle, we see that for any horizontal distribution θ the holomorphic Hilbert field $(\mathscr{H}, \bar\partial)$ is a BLS subfield of the BLS field $\mathscr{L}^\theta$.

REMARK. In fact, $\mathscr{H}$ is a strongly BLS subfield of $\mathscr{L}^\theta$. That is to say, the Gauss form

$$\mathbb{G}(\mathfrak{f}, \mathfrak{g}) := \left(\mathbf{I\!I}^{\mathfrak{L}^\theta/\mathscr{H}}\mathfrak{f}, \mathbf{I\!I}^{\mathfrak{L}^\theta/\mathscr{H}}\mathfrak{g}\right)$$

has the continuity property stated in Definition 2.21.b. This fact is a consequence of the identity (7), whose proof is completed in Section 4.2. ◇

3.3. Semi-local descriptions of sections and representatives. For later computation it is useful to describe the actions of $\nabla^{1,0}$ and $\bar\partial^\theta$ on sections of $\mathscr{L}$ and $\mathscr{H}$ somewhat more explicitly when working locally over the base.

If we fix a local coordinate chart $U \subset B$ and write $p = (p^1, \dots, p^m)$ as in Remark 1.1 then a section of $K_{X/B} \otimes E$ over U is an equivalence class of E-valued $(n, 0)$-forms. Fix a representative E-valued $(n, 0)$-form u of the twisted relative canonical form f associated to a smooth section $\mathfrak{f}$ of $\mathscr{L} \to B$. Then $D^{1,0}u$ is an E-valued $(n+1, 0)$-form, so there exist global $(n, 0)$-forms $\phi_1, \dots, \phi_m$ in $p^{-1}(U)$ such that

$$D^{1,0}u = dp^1 \wedge \phi_1 + \dots + dp^\mu \wedge \phi_\mu. \tag{20}$$

The forms ϕ_μ depend on the representative u, but their restrictions to the fibers, i.e.,

$$\iota^*_{X_t}\phi_\mu, \quad \mu = 1, \dots, m, \ t \in U,$$

are uniquely determined by $\mathfrak{f}$, i.e., they are independent of the representative u of $\mathfrak{f}$.

For $\bar\partial$ the situation is just slightly more subtle. One has

$$\bar\partial u = \psi + dp^1 \wedge \alpha_1 + \dots + dp^m \wedge \alpha_m + d\bar p^1 \wedge \beta_{\bar 1} + \dots + d\bar p^m \wedge \beta_{\bar m} \tag{21}$$

for some E-valued $(n-1, 1)$-forms $\alpha_1, \dots, \alpha_m$ and E-valued $(n, 0)$-forms $\beta_{\bar 1}, \dots, \beta_{\bar m}$ (that depend on the choice of local coordinates in U, but are global on $p^{-1}(U)$) and some $(n, 1)$-form ψ. The forms $\alpha_1, \dots, \alpha_m, \beta_{\bar 1}, \dots \beta_{\bar m}$ and ψ are not at all unique. In fact, obviously one can take $\psi = u$ and $\alpha_\mu = 0, \beta_{\bar\nu} = 0$ for all $\mu, \nu = 1, \dots, m$.

However, though there can be different ways to express u in the form (21), for any such representation, and for each $t \in U$, the forms

$$\iota^*_{X_t}\left(\alpha_\mu + \xi^\theta_{\frac{\partial}{\partial t^\mu}} \lrcorner \psi\right) \quad \text{and} \quad \iota^*_{X_t}\left(\beta_{\bar\mu} + \bar\xi^\theta_{\frac{\partial}{\partial t^\mu}} \lrcorner \psi\right), \qquad \mu = 1, \dots, m,$$

are uniquely determined.

The expression for $\bar\partial$ gets slightly simpler if one is looking at sections of $\mathscr{H} \to B$.

PROPOSITION 3.8 ([**Va-2022**, Proposition 4.23]). *Let $E \to X \xrightarrow{p} B$ be a holomorphic family of holomorphic vector bundles and let $\mathfrak{f} \in \Gamma(B, \mathscr{L})$. Let $u \in H^0(X, \mathscr{C}^\infty(\Lambda^{n,0}_X \otimes E))$ be a representative of $\mathfrak{f}$. Fix a coordinate neighborhood $U \subset B$ and write $p = (p^1, \dots, p^m)$ in these coordinates.*

(a) *If $t \in U$ then $\mathfrak{f}(t) \in \mathscr{H}_t$ if and only if whenever u is a representative of $f := \mathfrak{a}(\mathfrak{f})$ and $\bar\partial u$ is expressed in the form (21), one has $\iota^*_{X_t}\psi = 0$.*
(b) *The section $\mathfrak{f}$ of $\mathscr{L}$ restricts to a section of $\mathscr{H}$ over U if and only if for any (and hence every) representative u of $f := \mathfrak{a}(\mathfrak{f})$ there exist forms $\alpha_1, \dots, \alpha_m, \beta_{\bar 1}, \dots, \beta_{\bar m}$ such that $\bar\partial u$ is given by (21) with $\psi \equiv 0$.*
(c) *The section $\mathfrak{f}$ of $\mathscr{L}$ is a holomorphic section of $\mathscr{H}|_U$ if and only if for some (and hence every) representative u of $f = \mathfrak{a}(\mathfrak{f})$ the section $\bar\partial u$ is given by (21) with $\psi \equiv 0$, $\beta_{\bar\nu} \equiv 0$, $1 \le \nu \le m$.*

4. The proof of Theorem 1

The following abuse of notation simplifies the presentation somewhat. If f is a smooth section of the twisted relative canonical bundle $K_{X/B} \otimes E$ and u is an E-valued $(n,0)$-form representing f then the twisted Lie derivative $L^{1,0}_{\xi^\theta_\tau} u$ depends only on f (c.f. Proposition 1.8.a). We shall therefore sometimes write $L^{1,0}_{\xi^\theta_\tau} f$ instead of $L^{1,0}_{\xi^\theta_\tau} u$.

4.1. Outline of the proof of Theorem 0.4.

4.1.1. *Formula for the Chern connection of $\mathscr{L}^\theta$.* Let $E \to X$ be a holomorphic vector bundle with Hermitian metric h. As was proved in [**Va-2022**, Proposition 8.1], the BLS-Chern connection $\nabla^{\mathscr{L}^\theta}$ for $\mathscr{L}^\theta$ is given by the formula

$$\nabla^{\mathscr{L}^\theta 1,0}_\tau \mathfrak{f} = \mathfrak{i}\left(L^{\nabla 1,0}_{\xi^\theta_\tau} \mathfrak{a}(\mathfrak{f})\right), \quad \tau \in T^{1,0}_{B,t}, \tag{22}$$

where ∇ is the Chern connection for $(E, \mathfrak{h})$.

REMARK. Although we will not use the Chern connection for $\mathscr{H}$ directly, one can now compute it because by our standing assumption $\mathscr{H}$ is BLS. Indeed, by Proposition 2.20, Remark 1.12 and the fact that $\bar\partial^\theta|_{\mathscr{H}} = \bar\partial^{\mathscr{H}}$, the Chern connection for $\mathscr{H}$ is given by the formula

$$\nabla^{\mathscr{H}}_\tau \mathfrak{f} = \nabla^{\mathscr{H} 1,0}_\tau \mathfrak{f} = \mathfrak{i}\left(\mathcal{P}\left(L^{1,0}_{\xi_\tau} \mathfrak{a}(\mathfrak{f})\right)\right) \quad \text{for any } \mathfrak{f} \in \Gamma(B, \mathscr{C}^\infty(\mathscr{H})) \text{ and } \tau \in T^{1,0}_{B,t},$$

where $\mathcal{P}$ is the fiberwise Bergman projection operator defined in Paragraph 1.4.2 and ξ_τ is any lift of τ to a $(1,0)$-vector field on X_t. ◇

4.1.2. *Main points in the proof of Theorem* 0.4. A detailed proof of Theorem 0.4 can be found in [**Va-2022**]. We include a sketch here for the reader's convenience.

We work in local coordinates (z,t) on X such that $p(z,t)=t$. We fix $\mathfrak{f}\in\Gamma(b,\mathscr{C}^\infty(\mathscr{L}))$, and we denote by $f=\mathfrak{a}(\mathfrak{f})$ the associated section of $K_{X/B}\otimes E\to X$. Furthermore, we choose a representative $u\in H^0(X,\mathscr{C}^\infty(\Lambda_X^{n,0}\otimes E))$ for f, and write

$$u=u_o d\vec{z}+dt^\mu\wedge\alpha_\mu$$

in the above local coordinates, where $d\vec{z}:=dz^1\wedge\cdots\wedge dz^n$.

1. One first proves ([**Va-2022**, Lemma 8.6]) that if $v=v_o d\vec{z}+dt^\mu\wedge\beta_\mu$ is an E-valued $(n,0)$-form, where v_o is a local section of E and the β_μ are local E-valued $(n-1,0)$-forms, then

$$L^{1,0}_{\xi^\theta_\sigma}v=(\nabla^{1,0}_{\xi^\theta_\sigma}v_o)d\vec{z}+v_o\partial\left(\xi^\theta_\sigma\lrcorner d\vec{z}\right)+\partial\sigma^\mu\wedge\beta_\mu+dt^\mu\wedge L^{1,0}_{\xi^\theta_\sigma}\beta_\mu. \tag{23}$$

2. Using (23) for u and then for $L^{0,1}_{\bar{\xi}^\theta_\tau}u$, one calculates that

$$\begin{aligned}L^{0,1}_{\bar{\xi}^\theta_\tau}L^{1,0}_{\xi^\theta_\sigma}u&=\left(\bar\partial_{\bar{\xi}^\theta_\tau}\nabla^{1,0}_{\xi^\theta_\sigma}u_o\right)d\vec{z}+(\bar\partial u_o(\bar{\xi}^\theta_\tau))\partial\left(\xi^\theta_\sigma\lrcorner d\vec{z}\right)-u_o\bar{\xi}^\theta_\tau\lrcorner\left(\partial\left(\bar\partial\xi^\theta_\sigma\lrcorner d\vec{z}\right)\right)\\&\quad+\partial\sigma^\mu\wedge\left(\bar{\xi}^\theta_\tau\lrcorner\bar\partial\alpha_\mu\right)+dt^\mu\wedge L^{0,1}_{\bar{\xi}^\theta_\tau}L^{1,0}_{\xi^\theta_\sigma}\alpha_\mu,\end{aligned}$$

and

$$L^{1,0}_{\xi^\theta_\sigma}L^{0,1}_{\bar{\xi}^\theta_\tau}u=(\nabla^{1,0}_{\xi^\theta_\sigma}\bar\partial_{\bar{\xi}^\theta_\tau}u_o)d\vec{z}+(\bar\partial u_o(\bar{\xi}^\theta_\tau))\partial\left(\xi^\theta_\sigma\lrcorner d\vec{z}\right)+\partial\sigma^\mu\wedge L^{0,1}_{\bar{\xi}^\theta_\tau}\alpha_\mu+dt^\mu\wedge L^{1,0}_{\xi^\theta_\sigma}L^{0,1}_{\bar{\xi}^\theta_\tau}\alpha_\mu$$

respectively.

3. If σ and τ are holomorphic vector fields then $[\xi^\theta_\sigma,\bar{\xi}^\theta_\tau]$ is a vertical vector field. Therefore

$$L_{[\xi^\theta_\sigma,\bar{\xi}^\theta_\tau]}(u_o d\vec{z}+dt^\mu\wedge\alpha_\mu)=\nabla_{[\xi^\theta_\sigma,\bar{\xi}^\theta_\tau]}u_o d\vec{z}-u_o\partial(\bar{\xi}^\theta_\tau\lrcorner(\bar\partial\xi^\theta_\sigma\lrcorner d\vec{z}))+dt^\mu\wedge L_{[\xi^\theta_\sigma,\bar{\xi}^\theta_\tau]}\alpha_\mu.$$

4. Using Definition 2.6 (b), Formula (22) and the computations in 2 and 3, one has

$$\begin{aligned}\mathfrak{a}\left(\Theta^{\mathscr{L}^\theta}_{\sigma\bar\tau}\mathfrak{f}\right)&=\Theta(\mathfrak{h})_{\xi^\theta_\sigma\bar{\xi}^\theta_\tau}u_o d\vec{z}+L_{[\xi^\theta_\sigma,\bar{\xi}^\theta_\tau]}\mathfrak{a}(\mathfrak{f})+u_o\left(\bar{\xi}^\theta_\tau\lrcorner\partial(\bar\partial\xi^\theta_\sigma\lrcorner d\vec{z})+\partial(\bar{\xi}^\theta_\tau\lrcorner(\bar\partial\xi^\theta_\sigma\lrcorner d\vec{z}))\right)\\&\quad+dt^\mu\wedge\left(L^{1,0}_{\xi^\theta_\sigma}L^{0,1}_{\bar{\xi}^\theta_\tau}-L^{0,1}_{\bar{\xi}^\theta_\tau}L^{1,0}_{\xi^\theta_\sigma}-L_{[\xi^\theta_\sigma,\bar{\xi}^\theta_\tau]}\right)\alpha_\mu.\end{aligned}$$

5. One computes directly that

$$L^{1,0}_{\bar{\xi}^\theta_\tau}\left(u_o(\bar\partial\xi^\theta_\sigma\lrcorner d\vec{z})+dt^\mu\wedge\alpha_\mu\right)=u_o\partial\left(\bar{\xi}^\theta_\tau\lrcorner(\bar\partial\xi^\theta_\sigma\lrcorner d\vec{z})\right)+u_o\partial(\bar\partial\xi^\theta_\sigma\lrcorner dz)+dt^\mu\wedge L^{1,0}_{\bar{\xi}^\theta_\tau}\alpha_\mu$$

and that

$$L^{0,1}_{\xi^\theta_\sigma}f=(\bar\partial\xi^\theta_\sigma)\lrcorner f.$$

Since $L^{1,0}_{\xi^\theta_\sigma}f=0$ for reasons of type, one concludes that

$$\iota^*_{X_t}\Theta^{\mathscr{L}^\theta}_{\sigma\bar\tau}\mathfrak{f}=\iota^*_{X_t}\left(\Theta(\mathfrak{h})_{\xi^\theta_\sigma\bar{\xi}^\theta_\tau}f+L^{1,0}_{\bar{\xi}^\theta_\tau}L^{0,1}_{\xi^\theta_\sigma}f\right)+\iota^*_{X_t}\left(L_{[\xi^\theta_\sigma,\bar{\xi}^\theta_\tau]}f\right),$$

and integration yields

$$\begin{aligned}(\Theta^{\mathscr{L}^\theta}_{\sigma\bar\tau}\mathfrak{f}_1,\mathfrak{f}_2)_t&=\int_{X_t}\left\langle\left(\Theta(\mathfrak{h})_{\xi^\theta_\sigma\bar{\xi}^\theta_\tau}f_1\right)\wedge\bar{f}_2,\mathfrak{h}\right\rangle\\&\quad+\int_{X_t}\left\langle L^{1,0}_{\bar{\xi}^\theta_\tau}L^{0,1}_{\xi^\theta_\sigma}f_1\wedge\bar{f}_2,\mathfrak{h}\right\rangle+\int_{X_t}\left\langle L_{[\xi^\theta_\sigma,\bar{\xi}^\theta_\tau]}f_1\wedge\bar{f}_2,\mathfrak{h}\right\rangle.\end{aligned}$$

6. Since $L^{0,1}_{\xi^\theta_\sigma}\mathfrak{a}(\mathfrak{f}_1)$ is an $(n-1,1)$-form, a computation using Proposition 1.6 yields

$$\int_{X_t}\left\langle L^{1,0}_{\bar{\xi}^\theta_\tau}L^{0,1}_{\xi^\theta_\sigma}f_1\wedge\bar{f}_2,\mathfrak{h}\right\rangle=-\int_{X_t}\left\langle L^{0,1}_{\xi^\theta_\sigma}f_1\wedge\overline{L^{0,1}_{\xi^\theta_\tau}f_2},\mathfrak{h}\right\rangle.$$

Finally, for reasons of bi-degree

$$\int_{X_t}\left\langle L_{[\xi^\theta_\sigma,\bar{\xi}^\theta_\tau]}f_1)\wedge\bar{f}_2,\mathfrak{h}\right\rangle=\int_{X_t}\left\langle L^{0,1}_{[\xi^\theta_\sigma,\bar{\xi}^\theta_\tau]^{0,1}}f_1\wedge\bar{f}_2,\mathfrak{h}\right\rangle+\int_{X_t}\left\langle L^{1,0}_{[\xi^\theta_\sigma,\bar{\xi}^\theta_\tau]^{1,0}}f_1\wedge\bar{f}_2,\mathfrak{h}\right\rangle,$$

and by Lemma 1.4 and the fact that $[\xi^\theta_\sigma,\bar{\xi}^\theta_\tau]$ is vertical,

$$\int_{X_t}\left\langle L^{1,0}_{[\xi^\theta_\sigma,\bar{\xi}^\theta_\tau]^{1,0}}\mathfrak{a}(\mathfrak{f}_1)\wedge\overline{\mathfrak{a}(\mathfrak{f}_2)},\mathfrak{h}\right\rangle=\int_{X_t}\left\langle\mathfrak{a}(\mathfrak{f}_1)\wedge\overline{L^{0,1}_{[\xi^\theta_\tau,\bar{\xi}^\theta_\sigma]^{0,1}}\mathfrak{a}(\mathfrak{f}_2)},\mathfrak{h}\right\rangle.$$

This completes the sketch of proof. □

4.2. The Second fundamental form of $\mathscr{H}\subset\mathscr{L}^\theta$. Let $\theta\subset T_X$ be a smooth horizontal distribution. By our standing assumption $\mathscr{H}\subset\mathscr{L}^\theta$ is regular in the sense of Definition 2.15. Thus the second fundamental form

$$\mathbf{II}^{\mathscr{L}^\theta/\mathscr{H}}=P^\perp_{\mathscr{H}}\nabla^{\mathscr{L}^\theta 1,0},$$

depends smoothly on $t\in B$.

Since $\iota_t\mathbf{II}^{\mathscr{L}^\theta/\mathscr{H}}(\tau)\mathfrak{f}=P^\perp_t\iota^*_{X_t}L^{1,0}_{\xi^\theta_\tau}\alpha$, we have

$$\mathbf{II}^{\mathscr{L}^\theta/\mathscr{H}}(\sigma)\mathfrak{f}_1=P^\perp_t\iota^*_{X_t}L^{1,0}_{\xi^\theta_\sigma}u_1=P^\perp_t\iota^*_{X_t}\left(\nabla^{1,0}(\xi^\theta_\sigma\lrcorner u_1)+\xi^\theta_\sigma\lrcorner(\nabla^{1,0}u_1)\right).\tag{24}$$

4.3. Berndtsson's method of representatives. Since $\mathscr{H}$ is locally trivial by assumption, we need only the following special case of [**Va-2022**, Theorem 9.2], which is a slight generalization of (the essential part of) [**B-2009**, Proposition 4.2]. The proof we give here is different from Berndtsson's, but only in that it uses the Hodge Theorem directly, rather than the proof of the Hodge Theorem.

PROPOSITION 4.1 (Berndtsson's Method). *Let $(E,\mathfrak{h})\to X\xrightarrow{p}B$ be a proper holomorphic family whose associated Berndtsson field $\mathscr{H}\to B$ is a holomorphic vector bundle. Fix a horizontal lift θ, a point $t\in B$ and a Kähler form ω_t on X_t. Then for any $\mathfrak{f}\in\mathscr{C}^\infty(\mathscr{H})_t$ and any $\ell\in\mathscr{C}^\infty(\Lambda^{n,0}_X\otimes E)_t$ such that $\iota^*_{X_t}\ell=0$ and $\iota^*_{X_t}(\xi^\theta_\tau\lrcorner\nabla^{1,0}\ell)\perp\mathscr{H}_t$ for all $\tau\in\mathcal{O}(T^{1,0}_B)_t$ there exists a representative $u\in H^0(X,\mathscr{C}^\infty(\Lambda^{n,0}_X\otimes E))$ of $f=\mathfrak{a}(\mathfrak{f})$ with the following properties.*

(a) *The E-valued $(n-1,1)$-forms $\iota^*_{X_t}(\xi^\theta_\tau\lrcorner\bar\partial u)$ are primitive for all $\tau\in\mathcal{O}(T^{1,0}_B)_t$.*
(b) $P^\perp_t\iota^*_{X_t}(\xi^\theta_\tau\lrcorner\nabla^{1,0}u)=\iota^*_{X_t}(\xi^\theta_\tau\lrcorner\nabla^{1,0}\ell)$.

PROOF. Let $u^o\in H^0(X,\mathscr{C}^\infty(\Lambda^{n,0}_X\otimes E))$ be an initial representative of f. Writing $p=(t^1,\dots,t^m)$ near X_t, Proposition 3.8 implies that

$$\bar\partial u^o=dt^\mu\wedge\alpha^o_\mu+d\bar{t}^\nu\wedge\beta_{\bar\nu}\quad\text{and}\quad\nabla^{1,0}u^o=dt^\mu\wedge\phi^o_\mu\quad\text{near }X_t,$$

where α^o_μ are E-valued $(n-1,1)$-forms and $\phi^o_\mu,\beta_{\bar\nu}$ are E-valued $(n,0)$-forms.

If u is another representative of f then $u=u^o+dt^\mu\wedge v_\mu$ for some E-valued $(n-1,0)$-forms v_μ, and thus

$$\bar\partial u=dt^\mu\wedge(\alpha^o_\mu-\bar\partial v_\mu)+d\bar{t}^\nu\wedge\beta_{\bar\nu}\quad\text{and}\quad\nabla^{1,0}u=dt^\mu\wedge(\phi^o_\mu-\nabla^{1,0}v_\mu).$$

Therefore we seek v_μ satisfying

$$(25)\quad \omega_t \wedge (\iota^*_{X_t}\alpha^o_\mu - \iota^*_{X_t}\bar\partial v_\mu) = 0 \quad \text{and} \quad \nabla^{1,0} v_\mu = P^\perp_t \iota^*_{X_t}\phi^o_\mu - \iota^*_{X_t}(\xi^\theta_{\frac{\partial}{\partial t^\mu}} \lrcorner \nabla^{1,0}\ell).$$

Note that $\bar\partial(\omega \wedge u^o) = \omega \wedge \bar\partial u^o = dt^\mu \wedge \alpha^o_\mu \wedge \omega$, and hence

$$(26)\qquad \iota^*_{X_t}\left(\xi^\theta_\tau \lrcorner \bar\partial(\omega \wedge u^o)\right) = \iota^*_{X_t}(\tau^\mu \iota^*_{X_t}\alpha^o_\mu) = \tau^\mu \omega_t \wedge \iota^*_{X_t}\alpha^o_\mu.$$

On the other hand, since $\omega \wedge u^o$ has bidegree $(n+1,1)$, we can write $\omega \wedge u^o = dt^\mu \wedge c_\mu$. Then $\xi^\theta_\tau \lrcorner(\omega \wedge u^o) = \tau^\mu c_\mu + dt^\mu \wedge (\xi^\theta_\tau \lrcorner c_\mu)$.

As $\bar\partial \xi^\theta_\tau$ is vertical, we have $\iota^*_{X_t}\left(\bar\partial \xi^\theta_\tau \lrcorner(\omega \wedge u^o)\right) = 0$. Hence

$$\begin{aligned}\iota^*_{X_t}\left(\xi^\theta_\tau \lrcorner \bar\partial(\omega \wedge u^o)\right) &= \iota^*_{X_t}\left((\bar\partial \xi^\theta_\tau)\lrcorner(\omega \wedge u^o) - \bar\partial(\xi^\theta_\tau \lrcorner(\omega \wedge u^o))\right)\\ &= -\iota^*_{X_t}\left(\bar\partial(\xi^\theta_\tau \lrcorner(\omega \wedge u^o))\right),\end{aligned}$$

and as τ is holomorphic and defined locally on the base B,

$$(27)\qquad \iota^*_{X_t}\left(\bar\partial(\xi^\theta_\tau \lrcorner(\omega \wedge u^o))\right) = \iota^*_{X_t}\left(\bar\partial\left(\tau^\mu c_\mu + dt^\mu \wedge (\xi^\theta_\tau \lrcorner c_\mu)\right)\right) = \tau^\mu \bar\partial(\iota^*_{X_t} c_\mu).$$

By (26) and (27) $\omega_t \wedge \iota^*_{X_t}\alpha^o_\mu$ is $\bar\partial$-exact on X_t. Hence the $E|_{X_t}$-valued $(n-1,0)$-form

$$V^1_\mu := \Lambda_{\omega_t}\bar\partial^* G_t\left(\omega_t \wedge \iota^*_{X_t}\alpha^o_\mu\right)$$

satisfies

$$\omega_t \wedge (\bar\partial V^1_\mu - \iota^*_{X_t}\alpha^o_\mu) = \bar\partial(\omega_t \wedge V^1_\mu) - \omega_t \wedge \iota^*_{X_t}\alpha^o_\mu = \bar\partial\bar\partial^* G_t(\omega_t \wedge \iota^*_{X_t}\alpha^o) - \omega_t \wedge \iota^*_{X_t}\alpha^o = 0.$$

The second equality follows because the Lefschetz operator $L_{\omega_t} : \alpha \mapsto \omega_t \wedge \alpha$ and its adjoint Λ_ω satisfy $[L_{\omega_t}, \Lambda_{\omega_t}] = L_{\omega_t}\Lambda_{\omega_t} = \mathrm{Id}$ on (E-valued) $(n,1)$-forms, and in the last equality we have used that $\Box = \bar\partial\bar\partial^*$ on $\bar\partial$-closed forms. Note also that since $\sqrt{-1}[\nabla^{1,0}, \Lambda_{\omega_t}] = \bar\partial^*$,

$$\nabla^{1,0} V^1_\mu = \nabla^{1,0}\Lambda_{\omega_t}\bar\partial^* G_t(\omega_t \wedge \iota^*_{X_t}\alpha^o_\mu) = [\nabla^{1,0}, \Lambda_{\omega_t}]\bar\partial^* G_t(\omega_t \wedge \iota^*_{X_t}\alpha^o_\mu) = 0.$$

The second equality follows because $\nabla^{1,0}$ annihilates forms of degree $(n,*)$.

Next, $P^\perp_t \iota^*_{X_t}\phi^o_\mu$ is trivially orthogonal to the holomorphic E-valued n-forms, which are precisely the harmonic E-valued $(n,0)$-forms. By our assumption on ℓ,

$$\iota^*_{X_t}(\xi^\theta_{\frac{\partial}{\partial t^\mu}} \lrcorner \nabla^{1,0}\ell) \perp \mathscr{H}_t \quad \text{for every } \mu \in \{1, \dots, m\}.$$

Since on E-valued $(n,0)$-forms $\Box = \Box' = \nabla^{1,0}\nabla^{1,0*}$ and hence $G' = G$,

$$\begin{aligned}V^2_\mu &:= G'_t \nabla^{1,0*}\left(P^\perp_t \iota^*_{X_t}\phi^o_\mu - \iota^*_{X_t}(\xi^\theta_{\frac{\partial}{\partial t^\mu}} \lrcorner \nabla^{1,0}\ell)\right)\\ &= \nabla^{1,0*} G'_t\left(P^\perp_t \iota^*_{X_t}\phi^o_\mu - \iota^*_{X_t}(\xi^\theta_{\frac{\partial}{\partial t^\mu}} \lrcorner \nabla^{1,0}\ell)\right)\end{aligned}$$

solves the equation

$$\nabla^{1,0} V^2_\mu = P^\perp_t \iota^*_{X_t}\phi^o_\mu - \iota^*_{X_t}(\xi^\theta_{\frac{\partial}{\partial t^\mu}} \lrcorner \nabla^{1,0}\ell).$$

And since $L_{\omega_t}\alpha = 0$ for any $(n,0)$-form α, and $[L_{\omega_t}, \nabla^{1,0*}] = -\sqrt{-1}\bar\partial$,

$$\begin{aligned}\omega_t \wedge \bar\partial V^2_\mu &= \bar\partial L_{\omega_t}\nabla^{1,0*} G_t\left(P^\perp_t \iota^*_{X_t}\phi^o_\mu - \iota^*_{X_t}(\xi^\theta_{\frac{\partial}{\partial t^\mu}} \lrcorner \nabla^{1,0}\ell)\right)\\ &= \bar\partial[L_{\omega_t}, \nabla^{1,0*}] G_t\left(P^\perp_t \iota^*_{X_t}\phi^o_\mu - \iota^*_{X_t}(\xi^\theta_{\frac{\partial}{\partial t^\mu}} \lrcorner \nabla^{1,0}\ell)\right) = 0.\end{aligned}$$

Therefore any smooth extension v_μ of $V_\mu := V^1_\mu + V^2_\mu$ to X satisfies the two equations (25), and the proof is complete. □

4.4. The squared-norm of the second fundamental form. Let us fix a neighborhood U of t in B, and sections $\mathfrak{f}_1, \mathfrak{f}_2 \in \Gamma(U, \mathcal{O}(\mathscr{H}))$. We denote by f_1, f_2 the associated twisted relative canonical sections, and we choose representatives u_i of f_i, $i = 1, 2$, as in Proposition 4.1 with $\ell = 0$. That is to say, u_1 and u_2 have the following properties.

(i) The $E|_{X_t}$-valued $(n-1,1)$-forms $\iota^*_{X_t}(\xi^\theta_\tau \lrcorner \bar\partial u_i)$ are primitive.
(ii) The $E|_{X_t}$-valued $(n,0)$-forms $\iota^*_{X_t}(\xi^\theta_\tau \lrcorner \nabla^{1,0} u_i)$ are holomorphic.

From (24) and Property b we have

$$\mathbf{II}^{\mathscr{L}^0/\mathscr{H}}(\sigma)\mathfrak{f}_1 = P_t^\perp \nabla^{1,0}(\iota^*_{X_t}(\xi^\theta_\sigma \lrcorner u_1)) = \nabla^{1,0}(\iota^*_{X_t}(\xi^\theta_\sigma \lrcorner u_1)).$$

The last equality holds because the image of $\nabla^{1,0}$ in $\Gamma(X, \mathscr{C}^\infty(\Lambda^{n,0}_X \otimes E))$ is orthogonal to the kernel of $\Box' = \Box$, and the latter kernel consists precisely of holomorphic forms. Similarly $\mathbf{II}^{\mathscr{L}^0/\mathscr{H}}(\tau)\mathfrak{f}_2 = \nabla^{1,0}(\iota^*_{X_t}(\xi^\theta_\tau \lrcorner u_2))$. Therefore

$$\begin{aligned}
&\left(\mathbf{II}^{\mathscr{L}^0/\mathscr{H}}(\sigma)\mathfrak{f}_1, \mathbf{II}^{\mathscr{L}^0/\mathscr{H}}(\tau)\mathfrak{f}_2\right) \\
&= \sqrt{-1}^{n^2} \int_{X_t} \left\langle \nabla^{1,0}(\iota^*_{X_t}(\xi^\theta_\sigma \lrcorner u_1)) \wedge \overline{\nabla^{1,0}(\iota^*_{X_t}(\xi^\theta_\tau \lrcorner u_2))}, \mathfrak{h}\right\rangle \\
&= (-1)^{n+1}\sqrt{-1}^{n^2} \int_{X_t} \left\langle \bar\partial\nabla^{1,0}(\iota^*_{X_t}(\xi^\theta_\sigma \lrcorner u_1)) \wedge \overline{\iota^*_{X_t}(\xi^\theta_\tau \lrcorner u_2)}, \mathfrak{h}\right\rangle \\
&= \sqrt{-1}^{n^2-2n+1} \int_{X_t} \left\langle (\sqrt{-1}\Theta(\mathfrak{h}_t)\iota^*_{X_t}(\xi^\theta_\sigma \lrcorner u_1)) \wedge \overline{\iota^*_{X_t}(\xi^\theta_\tau \lrcorner u_2)}, \mathfrak{h}\right\rangle \\
&\quad - (-1)^{n+1}\sqrt{-1}^{n^2} \int_{X_t} \left\langle \nabla^{1,0}\bar\partial(\iota^*_{X_t}(\xi^\theta_\sigma \lrcorner u_1)) \wedge \overline{\iota^*_{X_t}(\xi^\theta_\tau \lrcorner u_2)}, \mathfrak{h}\right\rangle \\
&= \sqrt{-1}^{(n-1)^2} \int_{X_t} \left\langle (\sqrt{-1}\Theta(\mathfrak{h}_t)(\xi^\theta_\sigma \lrcorner u_1)) \wedge \overline{(\xi^\theta_\tau \lrcorner u_2)}, \mathfrak{h}\right\rangle \\
&\quad - \sqrt{-1}^{n^2} \int_{X_t} \left\langle (\bar\partial(\xi^\theta_\sigma \lrcorner u_1)) \wedge \overline{\bar\partial(\xi^\theta_\tau \lrcorner u_2)}, \mathfrak{h}\right\rangle.
\end{aligned}$$

Since

$$L_{\omega_t}\Lambda_{\omega_t}\alpha = \alpha \quad \text{for any } (n,1)\text{-form } \alpha, \tag{28}$$

$$\begin{aligned}
\sqrt{-1}\Theta(\mathfrak{h}_t)\iota^*_{X_t}(\xi^\theta_\sigma \lrcorner u_1) &= L_{\omega_t}\Lambda_{\omega_t}\sqrt{-1}\Theta(\mathfrak{h}_t)\iota^*_{X_t}(\xi^\theta_\sigma \lrcorner u_1)) \\
&= -L_{\omega_t}[\sqrt{-1}\Theta(\mathfrak{h}_t), \Lambda_{\omega_t}]\iota^*_{X_t}(\xi^\theta_\sigma \lrcorner u_1),
\end{aligned} \tag{29}$$

and thus

$$\begin{aligned}
&\sqrt{-1}^{(n-1)^2} \int_{X_t} \left\langle (\sqrt{-1}\Theta(\mathfrak{h}_t)(\xi^\theta_\sigma \lrcorner u_1)) \wedge \overline{(\xi^\theta_\tau \lrcorner u_2)}, \mathfrak{h}\right\rangle \\
&\qquad = -\left(\left[\sqrt{-1}\Theta(\mathfrak{h}_t), \Lambda_{\omega_t}\right] \xi^\theta_\sigma \lrcorner u_1, \xi^\theta_\tau \lrcorner u_2\right).
\end{aligned}$$

From the algebraic identity

$$(a-b, a-b) = (a,a) - (b,b) - (a-b, b) - (b, a-b)$$

for any bilinear form $(\cdot,\cdot)$ we obtain

$$\begin{aligned}
&\int_{X_t} \left\langle (\bar\partial(\xi^\theta_\sigma \lrcorner u_1)) \wedge \overline{\bar\partial(\xi^\theta_\tau \lrcorner u_2)}, \mathfrak{h} \right\rangle \\
&= \int_{X_t} \left\langle ((\bar\partial\xi^\theta_\sigma)\lrcorner u_1) \wedge \overline{((\bar\partial\xi^\theta_\tau)\lrcorner u_2)}, \mathfrak{h} \right\rangle - \int_{X_t} \left\langle \xi^\theta_\sigma \lrcorner(\bar\partial u_1)) \wedge \overline{\xi^\theta_\tau \lrcorner(\bar\partial u_2))}, \mathfrak{h} \right\rangle \\
&- \int_{X_t} \left\langle (\bar\partial(\xi^\theta_\sigma \lrcorner u_1)) \wedge \overline{(\xi^\theta_\tau \lrcorner(\bar\partial u_2))}, \mathfrak{h} \right\rangle - \int_{X_t} \left\langle (\xi^\theta_\sigma \lrcorner(\bar\partial u_1)) \wedge \overline{(\bar\partial(\xi^\theta_\tau \lrcorner u_2)}, \mathfrak{h} \right\rangle .
\end{aligned}$$

Now,

$$\begin{aligned}
&\int_{X_t} \left\langle (\xi^\theta_\sigma \lrcorner(\bar\partial u_1)) \wedge \overline{(\bar\partial(\xi^\theta_\tau \lrcorner u_2)}, \mathfrak{h} \right\rangle = (-1)^{n-1} \int_{X_t} \left\langle \nabla^{1,0}(\xi^\theta_\sigma \lrcorner(\bar\partial u_1)) \wedge \overline{\xi^\theta_\tau \lrcorner u_2}, \mathfrak{h} \right\rangle \\
&= (-1)^n \int_{X_t} \left\langle (\xi^\theta_\sigma \lrcorner(\nabla^{1,0}\bar\partial u_1)) \wedge \overline{\xi^\theta_\tau \lrcorner u_2}, \mathfrak{h} \right\rangle - (-1)^n \int_{X_t} \left\langle (L^{1,0}_{\xi^\theta_\sigma}\bar\partial u_1) \wedge \overline{\xi^\theta_\tau \lrcorner u_2}, \mathfrak{h} \right\rangle \\
&= (-1)^n \int_{X_t} \left\langle (\xi^\theta_\sigma \lrcorner(\nabla^{1,0}\bar\partial u_1)) \wedge \overline{\xi^\theta_\tau \lrcorner u_2}, \mathfrak{h} \right\rangle .
\end{aligned}$$

We claim that $\iota^*_{X_t}(\xi^\theta_\sigma \lrcorner(\nabla^{1,0}\bar\partial u_1)) = \iota^*_{X_t}(\xi^\theta_\sigma \lrcorner(\Theta(\mathfrak{h})u_1))$. Indeed, since $\nabla^{1,0}u_1 = dt^\mu \wedge \phi_\mu$ with ϕ_μ holomorphic on X_t,

$$\begin{aligned}
\xi^\theta_\sigma \lrcorner(\bar\partial\nabla^{1,0}u_1) &= (\bar\partial\xi^\theta_\sigma)\lrcorner\nabla^{1,0}u_1 - \bar\partial(\xi^\theta_\sigma \lrcorner\nabla^{1,0}u_1) \\
&= dt^\mu \wedge ((\bar\partial\xi^\theta_\sigma)\lrcorner\phi_\mu) - \bar\partial(\sigma^\mu\phi_\mu - dt^\mu \wedge (\xi^\theta_\sigma \lrcorner\phi_\mu)) \\
&= -\sigma^\mu\bar\partial\phi_\mu + dt^\mu \wedge \left((\bar\partial\xi^\theta_\sigma)\lrcorner\phi_\mu - \bar\partial(\xi^\theta_\sigma \lrcorner\phi_\mu)\right) .
\end{aligned}$$

Both of these terms vanish on X_t, which proves our claim. Therefore

$$\int_{X_t} \left\langle (\xi^\theta_\sigma \lrcorner(\bar\partial u_1)) \wedge \overline{(\bar\partial(\xi^\theta_\tau \lrcorner u_2)}, \mathfrak{h} \right\rangle = (-1)^n \int_{X_t} \left\langle (\xi^\theta_\sigma \lrcorner(\Theta(\mathfrak{h})u_1)) \wedge \overline{\xi^\theta_\tau \lrcorner u_2}, \mathfrak{h} \right\rangle .$$

By interchanging $\sigma \leftrightarrow \tau$ and $u_1 \leftrightarrow u_2$ and then taking complex conjugates,

$$\int_{X_t} \left\langle (\bar\partial(\xi^\theta_\sigma \lrcorner u_1)) \wedge \overline{(\xi^\theta_\tau \lrcorner(\bar\partial u_2))}, \mathfrak{h} \right\rangle = (-1)^n \int_{X_t} \left\langle (\xi^\theta_\sigma \lrcorner u_1) \wedge \overline{\xi^\theta_\tau \lrcorner(\Theta(\mathfrak{h})u_2)}, \mathfrak{h} \right\rangle$$

Using (28) again, we have

$$\begin{aligned}
&(-1)^n \int_{X_t} \sqrt{\text{-}1}^{n^2} \left\langle (\xi^\theta_\sigma \lrcorner(\Theta(\mathfrak{h})u_1)) \wedge \overline{\xi^\theta_\tau \lrcorner u_2}, \mathfrak{h} \right\rangle \\
&= -\sqrt{\text{-}1}^{n^2-2n+1} \int_{X_t} \left\langle \omega_t \wedge \Lambda_{\omega_t}\iota^*_{X_t}\left(\xi^\theta_\sigma \lrcorner(\sqrt{\text{-}1}\Theta(\mathfrak{h})u_1)\right) \wedge \overline{\xi^\theta_\tau \lrcorner u_2}, \mathfrak{h} \right\rangle \\
&= -(\Lambda_{\omega_t}(\xi^\theta_\sigma \lrcorner(\sqrt{\text{-}1}\Theta(\mathfrak{h})u_1)), \xi^\theta_\tau \lrcorner u_2) \\
&= -\left(\Lambda_{\omega_t}\iota^*_{X_t}\left(\xi^\theta_\sigma \lrcorner\sqrt{\text{-}1}\Theta(\mathfrak{h})\right)u_1, \xi^\theta_\tau \lrcorner u_2\right) - \left(\Lambda_{\omega_t}\sqrt{\text{-}1}\Theta(\mathfrak{h}_t)(\xi^\theta_\sigma \lrcorner u_1), \xi^\theta_\tau \lrcorner u_2\right) \\
&= -\left(\Lambda_{\omega_t}\iota^*_{X_t}\left(\xi^\theta_\sigma \lrcorner\sqrt{\text{-}1}\Theta(\mathfrak{h})\right)u_1, \xi^\theta_\tau \lrcorner u_2\right) + \left(\left[\sqrt{\text{-}1}\Theta(\mathfrak{h}_t), \Lambda_{\omega_t}\right](\xi^\theta_\sigma \lrcorner u_1), \xi^\theta_\tau \lrcorner u_2\right),
\end{aligned}$$

Since $[\sqrt{\text{-}1}\Theta(\mathfrak{h}_t), \Lambda_{\omega_t}] = \Box - \Box'$ is Hermitian symmetric for the L^2-norm, we find that

$$\sqrt{\text{-}1}^{n^2} \int_{X_t} \left\langle (\xi^\theta_\sigma \lrcorner (\bar\partial u_1)) \wedge \overline{(\bar\partial(\xi^\theta_\tau \lrcorner u_2)}, \mathfrak{h} \right\rangle + \sqrt{\text{-}1}^{n^2} \int_{X_t} \left\langle (\bar\partial(\xi^\theta_\sigma \lrcorner u_1)) \wedge \overline{(\xi^\theta_\tau \lrcorner (\bar\partial u_2))}, \mathfrak{h} \right\rangle$$
$$= 2\left(\left[\sqrt{\text{-}1}\Theta(\mathfrak{h}_t), \Lambda_{\omega_t}\right](\xi^\theta_\sigma \lrcorner u_1), \xi^\theta_\tau \lrcorner u_2\right)$$
$$- \left(\Lambda_{\omega_t}\iota^*_{X_t}\left(\xi^\theta_\sigma \lrcorner \sqrt{\text{-}1}\Theta(\mathfrak{h})\right)u_1, \xi^\theta_\tau \lrcorner u_2\right) - \left(\xi^\theta_\sigma \lrcorner u_1, \Lambda_{\omega_t}\iota^*_{X_t}\left(\xi^\theta_\tau \lrcorner \sqrt{\text{-}1}\Theta(\mathfrak{h})\right)u_2\right).$$

Adding everything up yields

$$\left(\mathbf{II}^{\mathscr{L}^\theta/\mathscr{H}}(\sigma)\mathfrak{f}_1, \mathbf{II}^{\mathscr{L}^\theta/\mathscr{H}}(\tau)\mathfrak{f}_2\right)$$
$$= -\left(\xi^\theta_\sigma \lrcorner (\bar\partial u_1), \xi^\theta_\tau \lrcorner (\bar\partial u_2)\right) - \int_{X_t} \sqrt{\text{-}1}^{n^2} \left\langle \left((\bar\partial \xi^\theta_\sigma) \lrcorner u_1\right) \wedge \overline{\left((\bar\partial \xi^\theta_\tau) \lrcorner u_2\right)}, \mathfrak{h} \right\rangle$$
$$+ \left([\sqrt{\text{-}1}\Theta(\mathfrak{h}_t), \Lambda_{\omega_t}](\xi^\theta_\sigma \lrcorner u_1), \xi^\theta_\tau \lrcorner u_2\right)$$
$$- \left([\Lambda_{\omega_t}, \iota^*_{X_t}((\xi^\theta_\sigma \lrcorner \sqrt{\text{-}1}\Theta(\mathfrak{h}))]u_1, \xi^\theta_\tau \lrcorner u_2\right) - \left(\xi^\theta_\sigma \lrcorner u_1, [\Lambda_{\omega_t}, \iota^*_{X_t}((\xi^\theta_\tau \lrcorner \sqrt{\text{-}1}\Theta(\mathfrak{h}))]u_2\right).$$

Finally, to obtain Formula (6) one uses the following *integration-by-parts* computation [**Va-2022**, Proposition 9.3]

$$\left\langle p_*\left(\sqrt{\text{-}1}^{n^2}\langle(\Theta(\mathfrak{h})u_1 \wedge \overline{u_2}, \mathfrak{h}\rangle\right), \sigma \wedge \bar\tau \right\rangle$$
$$= \left(\Theta(\mathfrak{h})_{\xi^\theta_\sigma \bar\xi^\theta_\tau} f_1, f_2\right) + \left(\Lambda_{\omega_t}\iota^*_{X_t}\left(\xi^\theta_\sigma \lrcorner \sqrt{\text{-}1}\Theta(\mathfrak{h})\right) f_1, \xi^\theta_\tau \lrcorner u_2\right)$$
$$+ \left(\xi^\theta_\sigma \lrcorner u_1, \Lambda_{\omega_t}\iota^*_{X_t}\left(\xi^\theta_\tau \lrcorner \sqrt{\text{-}1}\Theta(\mathfrak{h})\right) f_2\right)$$
$$- \left(\left[\sqrt{\text{-}1}\Theta(\mathfrak{h}_t), \Lambda_{\omega_t}\right](\xi^\theta_\sigma \lrcorner u_1), (\xi^\theta_\tau \lrcorner u_2)\right).$$

Thus Theorem 1 is proved. □

Acknowledgments

I am grateful to Roberto Albesiano, Randall Kayser, Luke Kiernan and Pranav Upadrashta for many delightful and enlightening conversations. I thank Robert Berman and Yanir Rubinstein for their encouragement to write the article. I thank the anonymous referee for a very careful reading and large number of suggestions for improvement, for which the reader should be thankful as well. Finally, I am grateful to Bo Berndtsson for all the inspiration that his work and, equally, his kindness and generosity, have provided me throughout my career.

References

[B-2009] Bo Berndtsson, *Curvature of vector bundles associated to holomorphic fibrations*, Ann. of Math. (2) **169** (2009), no. 2, 531–560, DOI 10.4007/annals.2009.169.531. MR2480611

[D-1982] Jean-Pierre Demailly, *Estimations L^2 pour l'opérateur $\bar\partial$ d'un fibré vectoriel holomorphe semi-positif au-dessus d'une variété kählérienne complète* (French), Ann. Sci. École Norm. Sup. (4) **15** (1982), no. 3, 457–511. MR690650

[K-1986] Kunihiko Kodaira, *Complex manifolds and deformation of complex structures*, Grundlehren der mathematischen Wissenschaften [Fundamental Principles of Mathematical Sciences], vol. 283, Springer-Verlag, New York, 1986. Translated from the Japanese by Kazuo Akao; With an appendix by Daisuke Fujiwara, DOI 10.1007/978-1-4613-8590-5. MR815922

[LS-2014] László Lempert and Róbert Szőke, *Direct images, fields of Hilbert spaces, and geometric quantization*, Comm. Math. Phys. **327** (2014), no. 1, 49–99, DOI 10.1007/s00220-014-1899-y. MR3177932

[LY-2014] Kefeng Liu and Xiaokui Yang, *Curvatures of direct image sheaves of vector bundles and applications*, J. Differential Geom. **98** (2014), no. 1, 117–145. MR3263516

[U-2023] Upadrashta, P., *Direct Images and Hilbert Fields*, `arXiv:/abs/2302.04826`, 2023.

[Va-2022] Dror Varolin, *Berndtsson-Lempert-Szőke fields associated to proper holomorphic families of vector bundles*, Adv. Math. **446** (2024), 109675, DOI 10.1016/j.aim.2024.109675. MR4736590

[Vo-2007] Voisin, C., *Hodge theory and complex algebraic geometry. I and II*, Cambridge Studies in Advanced Mathematics, 76 and 77, Cambridge University Press, Cambridge, 2007.

[Wang-2017] Xu Wang, *A curvature formula associated to a family of pseudoconvex domains* (English, with English and French summaries), Ann. Inst. Fourier (Grenoble) **67** (2017), no. 1, 269–313. MR3612332

DEPARTMENT OF MATHEMATICS, STONY BROOK UNIVERSITY, STONY BROOK, NEW YORK 11794-3651

Contemporary Mathematics
Volume **810**, 2025
https://doi.org/10.1090/conm/810/16207

Measures of finite energy in pluripotential theory: A synthetic approach

Sébastien Boucksom and Mattias Jonsson

To Bo Berndtsson, on the occasion of his 70ε-th birthday

Abstract. We introduce a synthetic approach to global pluripotential theory, covering in particular the case of a compact Kähler manifold and that of a projective Berkovich space over a non-Archimedean field. We define and study the space of measures of finite energy, introduce twisted energy and free energy functionals thereon, and show that coercivity of these functionals is an open condition with respect to the Kähler class.

Contents

Introduction

Pluripotential theory on compact Kähler spaces is by now a very well developed subject, with key applications to Kähler geometry. Generalizing classical concepts from potential theory, measures and potentials of finite energy play a central role in this theory, see for instance [**GZ07, BBGZ13, Berm13, DN15, DGL21**]. In parallel, a non-Archimedean version of pluripotential theory has also emerged, taking place on projective Berkovich spaces [**Berk90**] over a (complete) non-Archimedean field. Initially motivated by Arakelov geometry [**Zha95, Cha06**], it has found various other applications, including degenerations of Calabi–Yau manifolds [**Y.Li20**] and the Yau–Tian–Donaldson conjecture [**BBJ21, C.Li22, BoJ23**].

These two versions of pluripotential theory bear many similarities, and can be formulated in a quite parallel way. The main purpose of the present article is to introduce a synthetic formalism covering in particular these two cases, and use it to extend some of the main results of [**BoJ22**] and [**BoJ23**] (that were taking place

2020 *Mathematics Subject Classification.* Primary 31D05; Secondary 32Q26.

on projective Berkovich spaces over a trivially valued field, and applied to the study of K-stability). More specifically, we

- define and study the space of measures of finite energy;
- introduce the *twisted energy* and *free energy* functionals on the latter space; the composition of these functionals with the Monge–Ampère operator recover, respectively, the Donaldson J-functional and the Mabuchi K-energy in the Kähler case, and their analogues in the non-Archimedean case;
- show that coercivity of the free energy is an open condition with respect to the Kähler class.

The emphasis in this paper is on *measures* of finite energy, as opposed to *potentials* of finite energy, that we do not seek to investigate here (see for instance [**BFJ15, BoJ22, Reb21, DXZ23**] in the non-Archimedean context).

The general setup. Throughout this paper, we work with a compact topological space X equipped with the following data:

- a dense linear subspace $\mathcal{D} \subset \mathrm{C}^0(X)$ of *test functions*, containing all constants;
- a vector space $\mathcal{Z}$ of *admissible* $(1,1)$*-forms on* X, endowed with a nice[1] partial order, and a linear map $\mathrm{dd^c}\colon \mathcal{D} \to \mathcal{Z}$ vanishing on constants;
- an integer $n \geq 1$ (seen as the 'dimension' of X), and a nonzero n-linear symmetric map taking a tuple $(\theta_1, \dots, \theta_n)$ in $\mathcal{Z}$ to a signed Radon measure $\theta_1 \wedge \cdots \wedge \theta_n$ on X, assumed to be positive when all $\theta_i \geq 0$, and such that each bilinear form

$$\mathcal{D} \times \mathcal{D} \to \mathbb{R} \quad (\varphi, \psi) \mapsto \int \varphi\, \mathrm{dd^c}\psi \wedge \theta_1 \wedge \cdots \wedge \theta_{n-1} \tag{0.1}$$

  with $\theta_i \in \mathcal{Z}$ is symmetric, and seminegative for $\theta_i \geq 0$.

We then introduce the *Bott–Chern cohomology space*

$$\mathrm{H_{BC}}(X) := \mathcal{Z}/\mathrm{dd^c}\mathcal{D},$$

and define the *positive cone* $\mathrm{Pos}(X) \subset \mathrm{H_{BC}}(X)$ as the interior[2] of the image of the convex cone

$$\mathcal{Z}_+ := \{\theta \in \mathcal{Z} \mid \theta \geq 0\}.$$

This setup is primarily inspired by the case of a compact Kähler manifold X, where $\mathcal{D}$ is the space of smooth functions, and $\mathcal{Z}$ the space of closed $(1,1)$-forms. It also covers the case of a projective Berkovich space X over a complete non-Archimedean field k, where $\mathcal{D}$ is generated by *piecewise linear* (or *model*) functions, and elements of $\mathcal{Z}$ are represented by numerical classes on models over the valuation ring (or test configurations, in the trivially valued case) [**BFJ16a, GM16, BoJ22**]. At least under reasonable assumptions on X and k, we then have $\mathrm{H_{BC}}(X) = \mathrm{N}^1(X)$, and $\mathrm{Pos}(X)$ coincides with the ample cone, see §1.3.2.

[1] See §1.1 for the terminology.

[2] Here we use the finest vector space topology of $\mathrm{H_{BC}}(X)$, see §1.1.

Measures of finite energy. Fix a form $\omega \in \mathcal{Z}_+$ such that $[\omega] \in \operatorname{Pos}(X)$, with volume $V_\omega = \int \omega^n > 0$. The space of *$\omega$-plurisubharmonic* test functions is defined as

$$\mathcal{D}_\omega := \{\varphi \in \mathcal{D} \mid \omega_\varphi := \omega + dd^c\varphi \geq 0\},$$

and the *Monge–Ampère operator* takes $\varphi \in \mathcal{D}_\omega$ to the probability measure

$$\mathrm{MA}_\omega(\varphi) := V_\omega^{-1}\omega_\varphi^n.$$

It admits a primitive, the *Monge–Ampère energy* $\mathrm{E}_\omega \colon \mathcal{D}_\omega \to \mathbb{R}$, explicitly given by

$$\mathrm{E}_\omega(\varphi) = \frac{1}{n+1}\sum_{j=0}^{n} V_\omega^{-1}\int \varphi\,\omega_\varphi^j \wedge \omega^{n-j}.$$

Denote by $\mathcal{M}$ the space of (Radon) probability measures on X, and define the *energy* of a measure $\mu \in \mathcal{M}$ as the Legendre transform[3]

$$\mathrm{J}_\omega(\mu) := \sup_{\varphi\in\mathcal{D}_\omega}\{\mathrm{E}_\omega(\varphi) - \int \varphi\,\mu\} \in [0, +\infty].$$

Then $\mathrm{J}_\omega \colon \mathcal{M} \to [0, +\infty]$ is convex, and lsc in the weak topology; the space of *measures of finite energy* is defined as

$$\mathcal{M}_\omega^1 := \{\mu \in \mathcal{M} \mid \mathrm{J}_\omega(\mu) < \infty\},$$

equipped with the *strong topology*, i.e. the coarsest refinement of the weak topology in which J_ω becomes continuous.

As a consequence of the seminegativity of (0.1), the functional E_ω is concave. This is equivalent to the non-negativity of the *Dirichlet functional*

$$\mathrm{J}_\omega(\varphi, \psi) := \mathrm{E}_\omega(\varphi) - \mathrm{E}_\omega(\psi) + \int (\psi - \varphi)\,\mathrm{MA}_\omega(\varphi),$$

which is more explicitly given by the familiar expression

$$\mathrm{J}_\omega(\varphi, \psi) = \frac{1}{2}\int (\varphi - \psi) dd^c(\psi - \varphi)$$

when $n = 1$, and a positive linear combination of integrals of the form

$$\int (\varphi - \psi) dd^c(\psi - \varphi) \wedge \omega_\varphi^j \wedge \omega_\psi^{n-j-1} \quad (0 \leq j < n)$$

in general, see (1.31). Our first main result shows that the Dirichlet functional induces, via the Monge–Ampère operator, a complete quasi-metric[4] space structure on $\mathcal{M}_\omega^1$.

THEOREM A. *Assume that ω has the* orthogonality property. *Then:*

(i) *the image of the Monge–Ampère operator $\mathrm{MA}_\omega \colon \mathcal{D}_\omega \to \mathcal{M}$ is a dense subspace of $\mathcal{M}_\omega^1$;*

(ii) *there exists a unique quasi-metric δ_ω on $\mathcal{M}_\omega^1$ that defines the strong topology of $\mathcal{M}_\omega^1$ and such that*

$$\delta_\omega(\mathrm{MA}_\omega(\varphi), \mathrm{MA}_\omega(\psi)) = \mathrm{J}_\omega(\varphi, \psi)$$

for all $\varphi, \psi \in \mathcal{D}_\omega$;

(iii) *the quasi-metric space $(\mathcal{M}_\omega^1, \delta_\omega)$ is complete.*

[3]This corresponds to $\mathrm{E}_\omega^\vee(\mu)$ in the notation of [**BBGZ13, BoJ22**], and to $\|\mu\|_\omega$ in that of [**BoJ23**].

[4]See §1.1 for the notion of quasi-metric used in this paper.

The energy can be expressed in terms of the quasi-metric as

$$\mathrm{J}_\omega(\mu) = \delta_\omega(\mu, \mu_\omega) \quad \text{where} \quad \mu_\omega := V_\omega^{-1}\omega^n = \mathrm{MA}_\omega(0).$$

We refer to Definition 2.15 for the precise definition of the orthogonality property. Suffice it to say here that it only depends on $[\omega] \in \mathrm{Pos}(X)$, and holds when X is a compact Kähler manifold, or any projective Berkovich space X over a non-Archimedean field (as a consequence of [**BE21, BGM22**]). In the former case, Theorem A can be deduced from [**BBGZ13, BBEGZ19**]; in the latter, it was established in the trivially valued case in [**BoJ22**], and is thus extended here to the case of an arbitrary non-Archimedean ground field.

The strategy of proof of Theorem A follows the same lines as in the trivially case treated in [**BoJ22**]. The first key ingredient is a uniform differentiability property for the Legendre transform of the convex functional $\mu \mapsto \mathrm{J}_\omega(\mu)$, which is shown to be equivalent to the orthogonality property. This is used to prove that if (φ_i) is a maximizing sequence for a given $\mu \in \mathcal{M}^1_\omega$ (i.e. a sequence in $\mathcal{D}_\omega$ computing the supremum that defines $\mathrm{J}_\omega(\mu)$), then $\mathrm{MA}_\omega(\varphi_i)$ converges to μ. We emphasize that this 'asymptotic Calabi–Yau theorem' is sufficient for our purposes, and that we do not need to characterize the image of the Monge–Ampère operator (which is a delicate issue in the non-Archimedean context).

The rest of the proof relies on an extensive use of Hölder estimates for mixed Monge–Ampère integrals, obtained from repeated applications of the Cauchy–Schwarz inequality to the seminegative form (0.1). This approach, which goes back to [**Blo03**] and was further exploited in [**BBGZ13, BBEGZ19, BoJ22**], is put in a simple general setting in Appendix A.

Twisted energy, free energy, and coercivity. Assuming from now on the orthogonality property, we next investigate the dependence on ω of the space $\mathcal{M}^1_\omega$ and the energy functional $\mathrm{J}_\omega \colon \mathcal{M}^1_\omega \to \mathbb{R}_{\geq 0}$. To this end, we require the *submean value property*, i.e. the existence of $C \in \mathbb{R}_{\geq 0}$ such that

$$\sup \varphi \leq \int \varphi \, \mu_\omega + C$$

for all ω-psh test functions $\varphi \in \mathcal{D}_\omega$. We show that this condition is independent of ω, and that it is equivalent to the irreducibility of X when the latter is a compact Kähler or projective Berkovich space (see §1.5).

THEOREM B. *Assume that the submean value property holds. Then:*

- *the topological space $\mathcal{M}^1 = \mathcal{M}^1_\omega$ is independent of ω;*
- *for any $\theta \in \mathcal{Z}$, there exists a unique continuous functional $\mathrm{J}^\theta_\omega \colon \mathcal{M}^1 \to \mathbb{R}$ such that*

$$\mathrm{J}^\theta_\omega(\mu) = \frac{d}{dt}\Big|_{t=0} \mathrm{J}_{\omega+t\theta}(\mu)$$

for any $\mu \in \mathcal{M}^1$; furthermore, $\mathrm{J}^\theta_\omega(\mu)$ satisfies Hölder estimates with respect to ω.

The strategy of proof of Theorem B again globally follows the same lines as the trivially valued case treated in [**BoJ22, BoJ23**]. However, in the latter case the submean value inequality is actually an equality, i.e. one can take the constant C above to be 0, while an extra layer of complication arises in the general case to handle this constant, which forces us to take a slightly different route.

We call $J^\theta_\omega(\mu)$ the *θ-twisted energy* of $\mu \in \mathcal{M}^1$. It provides an analogue of the *Donaldson J-functional* on the level of measures, and its relevance comes from its relation to the *Mabuchi K-energy*, when X is a compact Kähler manifold or a smooth projective Berkovich space. In these two cases, the choice of a (smooth or PL) metric ρ on the canonical bundle K_X defines an *entropy* functional $\mathrm{Ent}\colon \mathcal{M} \to \mathbb{R} \cup \{+\infty\}$. In line with [**Berm13**], we then define the *free energy* $\mathrm{F}_\omega\colon \mathcal{M}^1 \to \mathbb{R} \cup \{+\infty\}$ by setting

$$\mathrm{F}_\omega(\mu) := \mathrm{Ent}(\mu) - \mathrm{Ent}(\mu_\omega) + \mathrm{J}^\theta_\omega(\mu),$$

where $\theta \in \mathcal{Z}$ denotes the curvature of ρ. The free energy so defined is independent of the choice of ρ, and its composition with the Monge–Ampère operator coincides with the *Mabuchi K-energy* $\mathrm{M}_\omega\colon \mathcal{D}_\omega \to \mathbb{R}$. As a consequence of Theorem B, we then show:

THEOREM C. *Assume X is a compact Kähler manifold or a smooth projective Berkovich space over a non-Archimedean field. Then the* coercivity threshold

$$\sigma(X,\omega) := \sup\{\sigma \in \mathbb{R} \mid \mathrm{F}_\omega \ge \sigma \mathrm{J}_\omega + A \text{ for some } A \in \mathbb{R}\}.$$

defines a continuous function of $[\omega] \in \mathrm{Pos}(X)$.

This result actually holds in much greater generality, for the twisted coercivity threshold of an arbitrary given functional on $\mathcal{M}^1$ with no a priori regularity whatsoever (see Theorem 5.5).

In the trivially valued case, the free energy $\mathrm{F}_\omega(\mu)$ coincides with the invariant $\beta_\omega(\mu)$ defined and studied in [**BoJ23**], and $\sigma(X,\omega)$ with the *divisorial stability threshold* of (X,ω), which is positive iff (X,ω) is *divisorially stable* (a strengthening of uniform K-stability, conjecturally equivalent to it, cf. §5.2.2). Specializing to the case of Dirac measures μ recovers the notion of *valuative stability* [**DL23, Liu23**], which in the Fano case is equivalent to K-stability [**Fuj19a, Li17**].

In the case of a compact Kähler manifold, we have $\sigma(X,\omega) > 0$ iff $[\omega]$ contains a unique constant scalar curvature Kähler (cscK) metric, as a consequence of [**CC21**] and [**DaR17, BDL20**]. Theorem C thus recovers the fact, originally due to LeBrun-Simanca [**LS94**], that the existence of a unique cscK metric in a Kähler class is an open condition on that class.

Structure of the paper. The article is organized as follows.

- Section 1 introduces the synthetic pluripotential theoretic formalism, including the energy pairing and the submean value property, and establishes basic properties of the Dirichlet functional.
- Section 2 studies the space of measures of finite energy. It introduces the orthogonality property, and proves Theorem A (cf. §2.6).
- Assuming the submean value property, Section 3 establishes the first part of Theorem B, along with various further estimates for the energy.
- Section 4 is devoted to the twisted energy, which is proved to compute the directional derivatives of the energy, concluding the proof of Theorem B.
- Section 5 studies the (twisted) coercivity threshold of a functional, and proves that it depends continuously on the cohomology classes. This is then applied to the free energy, yielding Theorem C.
- Finally, Appendix A establishes the relevant estimates needed for the Dirichlet functional in a simple general setting, while Appendix B studies the orthogonality property on compact Kähler spaces.

ACKNOWLEDGMENTS. The authors would like to thank Robert Berman, Ruadhaí Dervan, Philippe Eyssidieux, Charles Favre, Vincent Guedj, Léonard Pille-Schneider, Rémi Reboulet and Ahmed Zeriahi for helpful discussions related to the contents of this paper. We also thank the referees for many corrections and useful comments.

It is a great pleasure to dedicate this article to Bo Berndtsson. His many fundamental contributions to complex analysis, pluripotential theory and convex analysis have been a constant source of inspiration for us. We would especially like to express our admiration for his fantastic results on positivity of direct images that have played a key role in our own works (among many others!), as well as for his laid-back approach to mathematics in general, which makes Bo such a pleasant person to interact with.

The second author was supported by the grants DMS-1900025 and DMS-2154380 from the NSF, and award 1034361 from the Simons Foundation.

1. A synthetic pluripotential theoretic formalism

In this section, we introduce the general setup considered in this paper. This is designed to cover in a synthetic manner the case of a compact Kähler space and that of a projective Berkovich space over a non-Archimedean field.

1.1. Notation and terminology.

- For $x, y \in \mathbb{R}$, $x \lesssim y$ or $x = O(y)$ mean in this paper $x \le C_n y$ for a constant $C_n > 0$ only depending on a given integer n fixed in the setup, and $x \approx y$ if $x \lesssim y$ and $y \lesssim x$.
- Recall that any real vector space V admits a *finest vector space topology*, generated by its finite dimensional subspaces, i.e. a subset $A \subset V$ is open (or closed) iff, for each finite dimensional subspace $W \subset V$, $A \cap W$ is open (resp. closed) in the canonical vector space topology of W. This topology is not locally convex as soon as the dimension of V is uncountable.
- Consider a partially ordered $\mathbb{R}$-vector space $(V, \ge)$. We shall say for brevity that the partial order is *nice* if $V_+ := \{x \in V \mid x \ge 0\}$ spans V, and is closed in the finest vector space topology of V.
- In this paper, a *quasi-metric* on a set Z is a function $\delta\colon Z \times Z \to \mathbb{R}_{\ge 0}$ that is
  - *quasi-symmetric*, i.e. there exists $C > 0$ such that
  $$C^{-1}\delta(x, y) \le \delta(y, x) \le C\delta(x, y);$$
  - satisfies the *quasi-triangle inequality*, i.e. there exists $C > 0$ such that
  $$\delta(x, y) \le C(\delta(x, z) + \delta(z, y))$$
  - *separates points*, i.e. $\delta(x, y) = 0 \Leftrightarrow x = y$.

  A quasi-metric space (Z, δ) comes with a Hausdorff topology, and Cauchy sequences and completeness further make sense for (Z, δ).

1.2. Test functions and admissible $(1,1)$-forms. Throughout this paper, we work with a compact Hausdorff topological space X. We denote by $\mathrm{C}^0(X)^\vee$ the space of signed Radon measures on X, and by

$$\mathcal{M} \subset \mathrm{C}^0(X)^\vee$$

the subset of probability measures, which is convex and compact in the weak topology. Recall from the introduction that we assume X equipped with the following data:

- a dense linear subspace $\mathcal{D} \subset \mathrm{C}^0(X)$ of *test functions*, containing all constants;
- a vector space $\mathcal{Z}$ of *admissible* $(1,1)$*-forms on* X, endowed with a nice partial order, and a linear map $\mathrm{dd^c}\colon \mathcal{D} \to \mathcal{Z}$ vanishing on constants;
- an integer $n \geq 1$ (viewed as the 'dimension' of X), and a nonzero n-linear symmetric map taking a tuple $(\theta_1, \dots, \theta_n)$ in $\mathcal{Z}$ to a signed Radon measure $\theta_1 \wedge \dots \wedge \theta_n$ on X, assumed to be positive when all $\theta_i \geq 0$, and such that each bilinear form
$$\mathcal{D} \times \mathcal{D} \to \mathbb{R} \quad (\varphi, \psi) \mapsto \int \varphi \, \mathrm{dd^c}\psi \wedge \theta_1 \wedge \dots \wedge \theta_{n-1}$$
with $\theta_i \in \mathcal{Z}$ is symmetric, and seminegative when $\theta_i \geq 0$.

Symmetry in the last item amounts to the *integration-by-parts* formula

$$\int \varphi \, \mathrm{dd^c}\psi \wedge \theta_1 \wedge \dots \wedge \theta_{n-1} = \int \psi \, \mathrm{dd^c}\varphi \wedge \theta_1 \wedge \dots \wedge \theta_{n-1} \tag{1.1}$$

for all $\varphi, \psi \in \mathcal{D}$ and $\theta_i \in \mathcal{Z}$, while seminegativity requires

$$\int \varphi \, \mathrm{dd^c}\varphi \wedge \theta_1 \wedge \dots \wedge \theta_{n-1} \leq 0 \tag{1.2}$$

when $\theta_i \geq 0$ for all i.

REMARK 1.1. *The above setup induces a similar one by viewing $\theta_1 \wedge \dots \wedge \theta_n$ as a p-linear symmetric function of $(\theta_1, \dots, \theta_p)$ for $1 \leq p \leq n$ and $\theta_{p+1}, \dots, \theta_n \in \mathcal{Z}_+$ fixed, or by replacing $\mathcal{Z}$ with any linear subspace $\mathcal{Z}'$ containing* $\mathrm{dd^c}\mathcal{D}$.

DEFINITION 1.2. *For any $\theta \in \mathcal{Z}$ and $\varphi \in \mathcal{D}$, we set $\theta_\varphi := \theta + \mathrm{dd^c}\varphi$. We say that the test function φ is* θ-plurisubharmonic *(θ-psh for short) if $\theta_\varphi \geq 0$, and denote by*
$$\mathcal{D}_\theta := \{\varphi \in \mathcal{D} \mid \theta_\varphi \geq 0\}$$
the space so defined.

Note that if $\theta \in \mathcal{Z}$ and $\varphi \in \mathcal{D}$, we have

$$\varphi \in \mathcal{D}_{\theta_\tau} \iff \varphi + \tau \in \mathcal{D}_\theta. \tag{1.3}$$

Moreover, for all $\theta, \theta' \in \mathcal{Z}$ and $t \in \mathbb{R}_{>0}$ we have
$$\mathcal{D}_\theta + \mathcal{D}_{\theta'} \subset \mathcal{D}_{\theta+\theta'}, \quad \mathcal{D}_{t\theta} = t\mathcal{D}_\theta.$$
In particular, $\mathcal{D}_\theta$ is a convex subset of $\mathcal{D}$, and

$$\theta \leq \theta' \Longrightarrow \mathcal{D}_\theta \subset \mathcal{D}_{\theta'}. \tag{1.4}$$

Since $\mathrm{dd^c}$ vanishes on constants, we have:

EXAMPLE 1.3. *Constant functions on X are θ-psh iff $\theta \geq 0$.*

The two main instances of the above setup considered in this paper are as follows.

1.2.1. *The Kähler case.* The above formalism is primarily inspired by the case of a compact Kähler complex analytic space X. Here $n = \dim X$, $\mathcal{D} = \mathcal{C}^\infty(X)$ is the space of smooth functions on X, and $\mathcal{Z}$ is the space of closed $(1,1)$-forms θ on X that are locally $\mathrm{dd^c}$-exact (i.e. global sections of the image $\mathcal{Z}_X^{1,1}$ of the sheaf morphism $\mathrm{dd^c}\colon \mathcal{C}_X^\infty \to \Omega_X^{1,1}$), and $\theta \geq 0$ means that θ is semipositive as a smooth $(1,1)$-form (see for instance [**Dem85**] for the definition of smooth forms in this context).

When X is nonsingular, i.e. a compact Kähler manifold, $\mathcal{Z}$ coincides with the space of closed $(1,1)$-forms on X, but the inclusion might be strict in the singular case.

For all $\varphi, \psi \in \mathcal{D}$ and $\theta_i \in \mathcal{Z}$, the Stokes formula implies

$$\int \varphi \, \mathrm{dd^c}\psi \wedge \Theta = \int \psi \, \mathrm{dd^c}\varphi \wedge \Theta = -\int \mathrm{d}\varphi \wedge \mathrm{d^c}\psi \wedge \Theta$$

with $\Theta := \theta_1 \wedge \cdots \wedge \theta_{n-1}$, which yields (1.1) and (1.2), since the $(1,1)$-form $\mathrm{d}\varphi \wedge \mathrm{d^c}\varphi$ is semipositive.

1.2.2. *The non-Archimedean case.* Assume now that X is a projective Berkovich space over a non-Archimedean field k, i.e. the Berkovich analytification of a projective k-scheme, of dimension $n = \dim X$. We then take $\mathcal{D}$ to be the $\mathbb{R}$-vector space generated by *PL functions*, see [**BE21**, §5.4].

When k is nontrivially valued, $\mathcal{D}$ can be described in terms of *vertical divisors* on (projective, flat) models $\mathcal{X}$ of X over the spectrum S of the valuation ring. More precisely, we have

$$\mathcal{D} \simeq \varinjlim_{\mathcal{X}} \mathrm{VCar}(\mathcal{X})_{\mathbb{R}},$$

where $\mathrm{VCar}(\mathcal{X})_{\mathbb{R}}$ denotes the $\mathbb{R}$-vector space generated by Cartier divisors on $\mathcal{X}$ that are vertical, i.e. supported on the special fiber. The same description applies in the trivially valued case as well, if a model is now understood as a *test configuration* $\mathcal{X} \to S := \mathbb{A}^1$ (see [**BHJ17**, §6.1], [**BoJ22**, §2.2]).

In the nontrivially valued case, the space $\mathcal{Z}$ is defined by setting

$$\mathcal{Z} := \varinjlim_{\mathcal{X}} \mathrm{N}^1(\mathcal{X}/S), \tag{1.5}$$

where $\mathrm{N}^1(\mathcal{X}/S)$ denotes the (finite dimensional) vector space of relative numerical classes (see [**BFJ16a**, §4.2], [**GM16**, §4], the definition being inspired by [**BGS95**]). A form $\theta \in \mathcal{Z}$ is thus represented by a numerical class $\theta_{\mathcal{X}} \in \mathrm{N}^1(\mathcal{X}/S)$ for some model $\mathcal{X}$, called a *determination* of θ, two such classes being identified if they coincide after pulling back to some higher model, and we write $\theta \geq 0$ if $\theta_{\mathcal{X}}$ is (relatively) nef for some (hence any) determination $\mathcal{X}$ of θ.

The measure $\theta_1 \wedge \cdots \wedge \theta_n$ associated to a tuple of forms $\theta_i \in \mathcal{Z}$ is a finite linear combination of Dirac masses at divisorial points, whose coefficients can be described in terms of intersection numbers computed on models.

The linear map $\mathrm{dd^c}\colon \mathcal{D} \to \mathcal{Z}$ takes a vertical divisor $D \in \mathrm{VCar}(\mathcal{X})_{\mathbb{R}}$ to its numerical class in $\mathrm{N}^1(\mathcal{X}/S)$, and the seminegativity condition (1.2) follows the local Hodge index theorem of Yuan–Zhang [**YZ17**, Theorem 2.1].

Again, the same discussion applies to the trivially valued case as well, using test configurations instead of models. In that case, pulling back numerical classes on X to the product test configuration further yields an injection

$$\mathrm{N}^1(X) \hookrightarrow \mathcal{Z}. \tag{1.6}$$

Note that only forms lying in $\mathrm{N}^1(X)$ were considered in [**BoJ22**].

REMARK 1.4. *In the non-Archimedean case, one could also work with smooth functions and* $(1,1)$*-forms in the sense of* [**CD12**] (*see also* [**GK15, GK17, GJR21**]), *but this will not be considered in the present paper.*

REMARK 1.5. *More generally, test configurations of any compact Kähler manifold (as in* [**SD18, DeR17**]*) can also be approached using the above formalism. This is the topic of recent work of Pietro Mesquita-Piccione* [**MP24**].

REMARK 1.6. *Another setting where the above formalism applies is that of tropical toric pluripotential theory as in* [**BGJK21**].

1.3. Bott–Chern cohomology and positive classes.

DEFINITION 1.7. *We define the* Bott–Chern cohomology space *as*

$$\mathrm{H}_{\mathrm{BC}}(X) := \mathcal{Z}/\operatorname{Im} \mathrm{dd}^{\mathrm{c}},$$

and denote by $\theta \mapsto [\theta]$ *the quotient map* $\mathcal{Z} \to \mathrm{H}_{\mathrm{BC}}(X)$. *The* positive cone

$$\mathrm{Pos}(X) \subset \mathrm{H}_{\mathrm{BC}}(X)$$

is defined as the interior of the image of $\mathcal{Z}_+$.

Here the interior is taken with respect to the finest vector space topology (see §1.1). Concretely, a class $\alpha \in \mathrm{H}_{\mathrm{BC}}(X)$ belongs to $\mathrm{Pos}(X)$ if, for any $\beta \in \mathrm{H}_{\mathrm{BC}}(X)$, $\alpha + t\beta$ lies in the image of $\mathcal{Z}_+$ for all $t \in \mathbb{R}$ small enough.

Since it is assumed that $\mathcal{Z}_+$ spans $\mathcal{Z}$, its image in $\mathrm{H}_{\mathrm{BC}}(X)$ is a convex cone that generates $\mathrm{H}_{\mathrm{BC}}(X)$. As a consequence, the positive cone $\mathrm{Pos}(X)$ is non-empty as soon as $\mathrm{H}_{\mathrm{BC}}(X)$ is finite dimensional.

REMARK 1.8. *The image of* $\mathcal{Z}_+$ *in* $\mathrm{H}_{\mathrm{BC}}(X)$ *is not closed in general. Indeed, in the compact Kähler case, this means that a nef* $(1,1)$*-class on* X *does not always admit a smooth semipositive representative (see Example* 1.3.1 *below, and* [**DPS94**, Example 1.7] *for an explicit example).*

Since dd^{c} vanishes on $\mathbb{R} \subset \mathcal{D}$, (1.1) yields

$$\int \mathrm{dd}^{\mathrm{c}}\varphi \wedge \theta_1 \wedge \cdots \wedge \theta_{n-1} = \int \varphi \, \mathrm{dd}^{\mathrm{c}} 1 \wedge \theta_1 \wedge \cdots \wedge \theta_{n-1} = 0$$

for all $\varphi \in \mathcal{D}$ and $\theta_i \in \mathcal{Z}$. As a result, $(\theta_1, \ldots, \theta_n) \mapsto \int \theta_1 \wedge \cdots \wedge \theta_n$ descends to a symmetric n-linear pairing

$$\mathrm{H}_{\mathrm{BC}}(X)^n \to \mathbb{R} \quad (\alpha_1, \ldots, \alpha_n) \mapsto \alpha_1 \cdot \ldots \cdot \alpha_n,$$

which we call the *intersection pairing.*

LEMMA 1.9. *For all classes* $\alpha_1, \ldots, \alpha_n \in \mathrm{Pos}(X)$ *we have* $\alpha_1 \cdot \ldots \cdot \alpha_n > 0$.

PROOF. By assumption, the measure $\theta_1 \wedge \cdots \wedge \theta_n$ is nonzero for some tuple $\theta_1, \ldots, \theta_n \in \mathcal{Z}$. Since $\mathcal{Z}_+$ generates $\mathcal{Z}$, we can further assume $\theta_i \in \mathcal{Z}_+$ for all i. Then $[\theta_1] \cdot \ldots \cdot [\theta_n] = \int \theta_1 \wedge \cdots \wedge \theta_n > 0$. Now we can find $0 < \varepsilon \ll 1$ such that $\alpha_i - \varepsilon[\theta_i] \in \mathrm{Pos}(X)$ for all i, and hence $\alpha_1 \cdot \ldots \cdot \alpha_n \geq \varepsilon^n [\theta_1] \cdot \ldots \cdot [\theta_n] > 0$. □

DEFINITION 1.10. *For each* $\omega \in \mathcal{Z}_+$ *and* $\theta \in \mathcal{Z}$ *we set*

$$\|\theta\|_\omega := \inf\{C \geq 0 \mid \pm\theta \leq C\omega\} \in [0, +\infty],$$

and we say that θ *is* ω-bounded *if* $\|\theta\|_\omega < \infty$.

The set

$$\mathcal{Z}_\omega := \{\theta \in \mathcal{Z} \mid \|\theta\|_\omega < \infty\}$$

of ω-bounded forms is a linear subspace of $\mathcal{Z}$, on which $\|\cdot\|_\omega$ defines a norm.

REMARK 1.11. *In general, $\mathcal{Z}_\omega$ is a strict subspace of $\mathcal{Z}$. More precisely, $\mathcal{Z}_\omega = \mathcal{Z}$ iff ω lies in the interior of $\mathcal{Z}_+$ in the finest vector space topology of $\mathcal{Z}$, and this interior is empty in the non-Archimedean case (see §1.3.2 below).*

We can now characterize positive classes as follows:

PROPOSITION 1.12. *A class $\alpha \in \mathrm{H}_{\mathrm{BC}}(X)$ lies in $\mathrm{Pos}(X)$ iff, for each finite dimensional subspace $V \subset \mathcal{Z}$, α admits a representative $\omega \in \mathcal{Z}_+$ such that $V \subset \mathcal{Z}_\omega$.*

PROOF. Assume $\alpha \in \mathrm{Pos}(X)$, and pick a basis $(\theta_i)_{1\le i\le r}$ of a finite dimensional vector space $V \subset \mathcal{Z}$. Since $\mathcal{Z}_+$ spans $\mathcal{Z}$, for each i we can write $\theta_i = \theta_i^+ - \theta_i^-$ with $\theta_i^\pm \ge 0$. Since $\alpha \in \mathrm{Pos}(X)$ we have $\alpha - \varepsilon[\theta_i^\pm] \in \mathrm{Pos}(X)$ for all $0 < \varepsilon \ll 1$, and we can thus find $\varepsilon > 0$ and $\omega_i^\pm \in \alpha$ such that $\omega_i^\pm - \varepsilon\theta_i^\pm \ge 0$ for all i. Now set

$$\omega := \tfrac{1}{2r}\sum_{i=1}^r (\omega_i^+ + \omega_i^-) \in \mathcal{Z}_+.$$

Then $[\omega] = \alpha$, and for each i we have $\omega \ge \frac{\varepsilon}{2r}\theta_i^\pm \ge \pm\frac{\varepsilon}{2r}\theta_i$, and hence $\theta_i \in \mathcal{Z}_\omega$, i.e. $V \subset \mathcal{Z}_\omega$. This proves the 'only if' part, and the converse is clear. □

COROLLARY 1.13. *Pick $\theta \in \mathcal{Z}$ such that $[\theta] \in \mathrm{Pos}(X)$. Then any $f \in \mathcal{D}$ can be written as $f = f^+ - f^-$ with $f^\pm \in \mathcal{D}_{t\theta} = t\mathcal{D}_\theta$ for some $t > 0$. In particular, $\mathcal{D}_\theta$ spans $\mathcal{D}$.*

PROOF. By Proposition 1.12, we can find $\psi \in \mathcal{D}_\theta$ and $t > 0$ such that $-\mathrm{dd}^{\mathrm{c}} f \le t(\theta + \mathrm{dd}^{\mathrm{c}}\psi)$. Thus $f^+ := f + t\psi$ lies in $\mathcal{D}_{t\theta}$, and the result follows with $f^- := t\psi \in \mathcal{D}_{t\theta}$. □

Following [**Tho63**], we define the *Thompson distance* between $\omega, \omega' \in \mathcal{Z}_+$ as

$$\mathrm{d}_{\mathrm{T}}(\omega, \omega') := \inf\{\delta \ge 0 \mid e^{-\delta}\omega \le \omega' \le e^{\delta}\omega\} \in [0, +\infty]. \tag{1.7}$$

We say that ω and ω' are *commensurable* if $\mathrm{d}_{\mathrm{T}}(\omega, \omega') < \infty$. Note that this holds iff $\omega' \in \mathcal{Z}_\omega$ and $\omega \in \mathcal{Z}_{\omega'}$. Commensurability is an equivalence relation on $\mathcal{Z}_+$. The linear subspace $\mathcal{Z}_\omega$ only depends on the commensurability class of $\omega \in \mathcal{Z}_+$, and so does the equivalence class of the norm $\|\cdot\|_\omega$.

The next result is readily checked, and left to the reader.

LEMMA 1.14. *The commensurability class of any $\omega \in \mathcal{Z}_+$ forms an open convex cone in the normed vector space $(\mathcal{Z}_\omega, \|\cdot\|_\omega)$, whose topology is defined by the restriction of the Thompson metric.*

We conclude this section with the following fact, that we will put to use in §3.

PROPOSITION 1.15. *Each finite subset of $\mathrm{Pos}(X)$ can be represented by commensurable forms in $\mathcal{Z}_+$.*

PROOF. Consider $\alpha_1, \dots, \alpha_r \in \mathrm{Pos}(X)$ and set $\beta := \frac{1}{r}\sum_1^r \alpha_i$. If $0 < t \ll 1$, then the classes α_i' defined by $\alpha_i = (1-t)\alpha_i' + t\beta$ lie in $\mathrm{Pos}(X)$. Pick any representatives $\theta_i' \in \mathcal{Z}_+$ of α_i', $1 \le i \le r$. By Proposition 1.12, we can find a representative $\omega \in \mathcal{Z}_+$ of β such that $\theta_i' \in \mathcal{Z}_\omega$ for all i. Let $\omega_i := (1-t)\theta_i' + t\omega$ for $1 \le i \le r$. Then $\omega_i \in \mathcal{Z}_+$ is a representative of α_i and ω, ω_i are commensurable for all i. □

1.3.1. *The Kähler case.* As in §1.2.1, assume that X is a compact Kähler space. When X is nonsingular, $\mathrm{H}_{\mathrm{BC}}(X)$ coincides with the $(1,1)$-part of the Hodge decomposition of $\mathrm{H}^2(X,\mathbb{R})$. In general, denote by $\mathcal{PH}_X$ the sheaf of germs of *pluriharmonic functions*, i.e. the kernel of the sheaf morphism $\mathrm{dd}^{\mathrm{c}}\colon \mathcal{C}^\infty_X \to \Omega^{1,1}_X$. The existence of partitions of unity implies that the sheaf $\mathcal{C}^\infty_X$ is soft, and the cohomology long exact sequence associated to the short exact sequence of sheaves

$$0 \to \mathcal{PH}_X \to \mathcal{C}^\infty_X \to \mathcal{Z}^{1,1}_X \to 0$$

thus yields

$$\mathrm{H}_{\mathrm{BC}}(X) \simeq \mathrm{H}^1(X, \mathcal{PH}_X).$$

We do not know whether the right-hand side is always finite dimensional, but this holds at least when X is normal, as a consequence of the fact that $\mathcal{PH}_X$ then coincides with the sheaf $\Re\mathcal{O}_X$ of real parts of holomorphic functions, see [**BG13**, §4.6.1].

The positive cone $\mathrm{Pos}(X)$ is compatible with the usual definition, i.e. $\alpha \in \mathrm{H}_{\mathrm{BC}}(X)$ lies in $\mathrm{Pos}(X)$ iff α can be represented by a Kähler form. Indeed, given any Kähler form ω, a class $\alpha \in \mathrm{Pos}(X)$ can be represented by a form $\theta \in \mathcal{Z}$ such that $\theta - \varepsilon\omega \in \mathcal{Z}_+$ for some $0 < \varepsilon \ll 1$, so that θ is a Kähler form. In particular, Kähler forms constitute a single commensurability class that maps onto $\mathrm{Pos}(X)$.

1.3.2. *The non-Archimedean case.* Assume, as in §1.2.2, that X is a projective Berkovich space over a non-Archimedean field. By definition, any $\theta \in \mathcal{Z}$ is represented by a numerical class on some model $\mathcal{X}$ of X, and the restriction of θ to the generic fiber defines a numerical class $\{\theta\} \in \mathrm{N}^1(X)$ on (the projective variety underlying) X. This induces a surjective map

$$\mathrm{H}_{\mathrm{BC}}(X) \twoheadrightarrow \mathrm{N}^1(X), \tag{1.8}$$

which is an isomorphism when X is smooth and k is discretely valued of residue characteristic 0 [**BFJ16a**, Theorem 4.3], or when X is normal and k is algebraically closed [**Jell16**, Theorem 4.2.7]). It is also an isomorphism when k is trivially valued, with inverse provided by pulling back classes in $\mathrm{N}^1(X)$ to the trivial test configuration $\mathcal{X}_{\mathrm{triv}} = X \times \mathbb{A}^1$.

We claim that $\mathrm{Pos}(X)$ coincides with the preimage of the ample cone of X under (1.8), and that $\mathcal{Z}_+$ has empty interior in the finest vector space topology of $\mathcal{Z}$ (so that Kähler forms do not admit an analogue in the non-Archimedean case).

To see this, we say as in [**BFJ16a**, §5.1] that a form $\omega \in \mathcal{Z}_+$ is *$\mathcal{X}$-positive* for a given model/test configuration $\mathcal{X}$ if it is represented by a (relatively) ample class in $\mathrm{N}^1(\mathcal{X}/S)$. Note that all $\mathcal{X}$-positive forms are commensurable. As observed in [**BFJ16a**, Proposition 5.2] (see also [**GM16**, Proposition 4.14], [**BoJ22**, Lemma 3.11]), any $\alpha \in \mathrm{H}_{\mathrm{BC}}(X)$ whose image in $\mathrm{N}^1(X)$ under (1.8) is ample can be represented by an $\mathcal{X}$-positive form ω for any sufficiently high model $\mathcal{X}$. Now $\mathcal{Z}_\omega$ contains all forms determined on $\mathcal{X}$, but does not contain any $\mathcal{X}'$-positive form for a model $\mathcal{X}'$ strictly dominating $\mathcal{X}$. The claim easily follows.

1.4. The energy pairing. Mimicking the properties of induced metrics on Deligne pairings (see for instance [**BE21**, Theorem 8.16]), and generalizing [**SD18**, §2.2] (for a compact Kähler manifold) and [**BoJ22**, §3.2] (for a projective Berkovich space over a trivially valued field), we introduce:

PROPOSITION-DEFINITION 1.16. *The* energy pairing

$$(\mathcal{Z}\times\mathcal{D})^{n+1}\to\mathbb{R}\quad((\theta_0,\varphi_0),\dots,(\theta_n,\varphi_n))\mapsto(\theta_0,\varphi_0)\cdot\ldots\cdot(\theta_n,\varphi_n)$$

is defined as the unique $(n+1)$-linear symmetric map such that

$$(0,\varphi_0)\cdot(\theta_1,\varphi_1)\cdot\ldots\cdot(\theta_n,\varphi_n)=\int\varphi_0\,(\theta_1+\mathrm{dd^c}\varphi_1)\wedge\cdots\wedge(\theta_n+\mathrm{dd^c}\varphi_n)\tag{1.9}$$

and

$$(\theta_0,0)\cdot\ldots\cdot(\theta_n,0)=0.\tag{1.10}$$

It further satisfies

(1.11)

$$(\theta_0,\varphi_0+c_0)\cdot\ldots\cdot(\theta_n,\varphi_n+c_n)=(\theta_0,\varphi_0)\cdot\ldots\cdot(\theta_n,\varphi_n)+\sum_{i=0}^n c_i[\theta_0]\cdot\ldots\cdot\widehat{[\theta_i]}\cdot\ldots\cdot[\theta_n]$$

for all $c_i\in\mathbb{R}$, and

$$\begin{aligned}(\theta_0+\mathrm{dd^c}\tau_0,\varphi_0)\cdot\ldots\cdot(\theta_n+\mathrm{dd^c}\tau_n,\varphi_n)\\=(\theta_0,\tau_0+\varphi_0)\cdot\ldots\cdot(\theta_n,\tau_n+\varphi_n)-(\theta_0,\tau_0)\cdot\ldots\cdot(\theta_n,\tau_n).\end{aligned}\tag{1.12}$$

for all $\tau_i\in\mathcal{D}$.

PROOF. Using multilinearity, symmetry and (1.9), (1.10), we necessarily have

(1.13)

$$(\theta_0,\varphi_0)\cdot\ldots\cdot(\theta_n,\varphi_n)=\sum_{i=0}^n\int\varphi_i\,\theta_0\wedge\cdots\wedge\theta_{i-1}\wedge(\theta_{i+1}+\mathrm{dd^c}\varphi_{i+1})\wedge\cdots\wedge(\theta_n+\mathrm{dd^c}\varphi_n).$$

This proves uniqueness. To show existence, the only nontrivial part is to check that the right-hand side of (1.13) is a symmetric function of the (θ_i,φ_i). It suffices to see invariance under transpositions, which is an easy consequence of the integration-by-parts formula (1.1) (compare [**SD18**, Proposition 2.3]).

To see (1.11), we may assume $c_i=0$ for $i>0$, and the result is then a direct consequence of (1.9).

Finally, pick $\tau_i\in\mathcal{D}$, and set

$$F(\varphi_0,\dots,\varphi_n):=(\theta_0,\tau_0+\varphi_0)\cdot\ldots\cdot(\theta_n,\tau_n+\varphi_n)-(\theta_0+\mathrm{dd^c}\tau_0,\varphi_0)\cdot\ldots\cdot(\theta_n+\mathrm{dd^c}\tau_n,\varphi_n).$$

By (1.9), we have

$$(0,\varphi_0)\cdot(\theta_1,\tau_1+\varphi_1)\cdot\ldots\cdot(\theta_n,\tau_n+\varphi_n)-(0,\varphi_0)\cdot(\theta_1+\mathrm{dd^c}\tau_1,\varphi_1)\cdot\ldots\cdot(\theta_n+\mathrm{dd^c}\tau_n,\varphi_n)=0.$$

This implies that $F(\varphi_0,\dots,\varphi_n)$ is independent of φ_0, and hence equal to $F(0,\varphi_1,\dots,\varphi_n)$. Applying the same argument successively to $\varphi_1,\dots,\varphi_n$, we end up with $F(\varphi_0,\dots,\varphi_0)=F(0,\dots,0)=(\theta_0,\tau_0)\cdot\ldots\cdot(\theta_n,\tau_n)$, which proves (1.12). □

By (1.9), the seminegativity property (1.2) translates into

$$(0,\varphi)^2\cdot(\theta_1,\varphi_1)\cdot\ldots\cdot(\theta_{n-1},\varphi_{n-1})\le 0\tag{1.14}$$

for all $\varphi\in\mathcal{D}$ and $\varphi_i\in\mathcal{D}_{\theta_i}$.

As in [**BoJ22**, §3.2], we further note the following straightforward monotonicity properties:

PROPOSITION 1.17. *For all $\theta_i, \theta_i' \in \mathcal{Z}$ and $\varphi_i, \varphi_i' \in \mathcal{D}_{\theta_i}$, we have*

$$\varphi_i \le \varphi_i' \text{ for all } i \Longrightarrow (\theta_0, \varphi_0) \cdot \ldots \cdot (\theta_n, \varphi_n) \le (\theta_0, \varphi_0') \cdot \ldots \cdot (\theta_n, \varphi_n'); \tag{1.15}$$

$$\varphi_i \le 0 \text{ and } 0 \le \theta_i \le \theta_i' \text{ for all } i \Longrightarrow (\theta_0', \varphi_0) \cdot \ldots \cdot (\theta_n', \varphi_n) \le (\theta_0, \varphi_0) \cdot \ldots \cdot (\theta_n, \varphi_n). \tag{1.16}$$

Combining (1.15) and (1.11), we infer the Lipschitz property

$$|(\theta_0, \varphi_0) \cdot \ldots \cdot (\theta_n, \varphi_n) - (\theta_0, \varphi_0') \cdot \ldots \cdot (\theta_n, \varphi_n')| \le \sum_{i=0}^{n} \sup |\varphi_i - \varphi_i'| [\theta_0] \cdot \ldots \cdot \widehat{[\theta_i]} \cdot \ldots \cdot [\theta_n] \tag{1.17}$$

for all $\varphi_i, \varphi_i' \in \mathcal{D}_{\theta_i}$.

Following [**BoJ22**, §7.2], we next establish a lower bound for the energy pairing, which will play a crucial role in §3.2 below.

THEOREM 1.18. *Assume $\omega_0, \ldots, \omega_n \in \mathcal{Z}_+$ are commensurable, and set $\delta := \max_{i,j} \mathrm{d_T}(\omega_i, \omega_j)$. If $0 \ge \varphi_i \in \mathcal{D}_{\omega_i}$ for $i = 1, \ldots, n$, then*

$$0 \ge (\omega_0, \varphi_0) \cdot \ldots \cdot (\omega_n, \varphi_n) \gtrsim e^{O(\delta)} \min_i (\omega_i, \varphi_i)^{n+1}.$$

Here $\mathrm{d_T}$ denotes the Thompson metric (1.7), and the implicit constant in $O(\delta)$ only depends on n (see §1.1).

LEMMA 1.19. *For any $\theta \in \mathcal{Z}$, $\varphi \mapsto (\theta, \varphi)^{n+1}$ is concave on $\mathcal{D}_\theta$.*

PROOF. This is a formal consequence of (1.14), see Appendix A. □

LEMMA 1.20. *Pick $\omega, \omega' \in \mathcal{Z}_+$ and $t \ge 1$ such that $\omega \le \omega' \le t\omega$. For any $\varphi \in \mathcal{D}_\omega \subset \mathcal{D}_{\omega'}$ such that $\varphi \le 0$, we then have*

$$0 \ge (\omega', \varphi)^{n+1} \ge t^n (\omega, \varphi)^{n+1}.$$

PROOF. By (1.16) we have

$$0 \ge (\omega', \varphi)^{n+1} \ge (t\omega, \varphi)^{n+1} = t^{n+1} (\omega, t^{-1}\varphi)^{n+1}.$$

Since $t^{-1} \in [0, 1]$, concavity of the energy (Lemma 1.19) yields $(\omega, t^{-1}\varphi)^{n+1} \ge t^{-1}(\omega, \varphi)^{n+1}$, and the result follows. □

LEMMA 1.21. *Pick $\omega_0, \ldots, \omega_r \in \mathcal{Z}_+$ and $0 \ge \varphi_i \in \mathcal{D}_{\omega_i}$ for $i = 0, \ldots, r$. Assume also given $t \ge 1$ such that $\omega_i \le t\omega_j$ for all i, j. Then*

$$\left(\sum_i \omega_i, \sum_i \varphi_i \right)^{n+1} \ge C_{r,n} t^n \sum_i (\omega_i, \varphi_i)^{n+1} \tag{1.18}$$

with $C_{r,n} := (r(r+1))^n$.

PROOF. Set $\overline{\omega} := \frac{1}{r+1}\sum_0^r \omega_i$ and $\overline{\omega}_t := \frac{t}{r+t}\sum_0^r \omega_i$. Then $\overline{\omega} \le \overline{\omega}_t \le \frac{t(1+rt)}{t+r}\omega_i$ for $0 \le i \le r$. If we set $\overline{\varphi} := \frac{1}{r+1}\sum_i \varphi_i$, then

$$\begin{aligned}\left(\sum_i \omega_i, \sum_i \varphi_i\right)^{n+1} &= (r+1)^{n+1}(\overline{\omega},\overline{\varphi})^{n+1}\\ &\ge (r+1)^{n+1}(\overline{\omega}_t,\overline{\varphi})^{n+1}\\ &\ge (r+1)^n \sum_i (\overline{\omega}_t, \varphi_i)^{n+1}\\ &\ge (r+1)^n \left(\frac{t(1+rt)}{t+r}\right)^n \sum_i (\omega_i,\varphi_i)^{n+1}\\ &\ge C_{r,n} t^n \sum_i (\omega_i,\varphi_i)^{n+1},\end{aligned}$$

where the first inequality holds by (1.16), the second one by Lemma 1.19, and the last two inequalities from Lemma 1.20 and the estimate $\frac{1+rt}{t+r} \le r$. □

PROOF OF THEOREM 1.18. Expanding out $(\omega_0 + \cdots + \omega_n, \varphi_0 + \cdots + \varphi_n)^{n+1}$ yields

$$(n+1)!(\omega_0,\varphi_0)\cdot\ldots\cdot(\omega_n,\varphi_n) \ge (\omega_0+\cdots+\omega_n, \varphi_0+\cdots+\varphi_n)^{n+1},$$

and we conclude by Lemma 1.21 with $r = n$. □

1.5. The Monge–Ampère operator and the submean value property. Pick $\omega \in \mathcal{Z}_+$ with $[\omega] \in \mathrm{Pos}(X)$. We define its *volume* as

$$V_\omega := \int \omega^n = [\omega]^n,$$

which is positive by Lemma 1.9, and introduce the probability measure

$$\mu_\omega := V_\omega^{-1}\omega^n.$$

DEFINITION 1.22. *The* Monge–Ampère operator $\mathrm{MA}_\omega \colon \mathcal{D}_\omega \to \mathcal{M}$ *is defined by setting*

$$\mathrm{MA}_\omega(\varphi) := V_\omega^{-1}\omega_\varphi^n.$$

Equivalently,

$$\mathrm{MA}_\omega(\varphi) = \mu_{\omega_\varphi}. \tag{1.19}$$

As an illustration of the energy pairing formalism, we recover the following version of the classical *Chern–Levine–Nirenberg inequality.*

LEMMA 1.23. *For all $\varphi, \psi, \tau \in \mathcal{D}_\omega$ we have*

$$\left|\int \tau\,(\mathrm{MA}_\omega(\varphi) - \mathrm{MA}_\omega(\psi))\right| \le 2n \sup|\varphi-\psi|.$$

PROOF. By (1.9) we have

$$\begin{aligned}V_\omega \int \tau\,(\mathrm{MA}_\omega(\varphi) - \mathrm{MA}_\omega(\psi)) &= (0,\tau)\cdot(\omega,\varphi)^n - (0,\tau)\cdot(\omega,\psi)^n\\ &= (\omega,\tau)\cdot(\omega,\varphi)^n - (\omega,0)\cdot(\omega,\varphi)^n - (\omega,\tau)\cdot(\omega,\psi)^n + (\omega,0)\cdot(\omega,\psi)^n.\end{aligned}$$

Now (1.17) yields

$$|(\omega,\tau)\cdot(\omega,\varphi)^n-(\omega,\tau)\cdot(\omega,\psi)^n|\le n\sup|\varphi-\psi|V_\omega,$$
$$|(\omega,0)\cdot(\omega,\varphi)^n-(\omega,0)\cdot(\omega,\psi)^n|\le n\sup|\varphi-\psi|V_\omega,$$

and the result follows. □

We next consider the quantity

$$T_\omega:=\sup\left\{\sup\varphi-\int\varphi\,\mu_\omega\mid\varphi\in\mathcal{D}_\omega\right\}\in[0,+\infty]. \tag{1.20}$$

Thus $T_\omega<\infty$ iff there exists $C>0$ such that the submean value inequality

$$\sup\varphi\le\int\varphi\,\mu_\omega+C \tag{1.21}$$

holds for all $\varphi\in\mathcal{D}_\omega$.

Proposition 1.24. *If $T_\omega<\infty$, then $T_{\omega'}<\infty$ for any $\omega'\in\mathcal{Z}_+$ with $[\omega']\in$ $\mathrm{Pos}(X)$.*

Definition 1.25. *We say that the* submean value property *holds if $T_\omega<\infty$ for some (hence any) $\omega\in\mathcal{Z}_+$ with $[\omega]\in\mathrm{Pos}(X)$.*

When X is a compact Kähler or projective Berkovich space, this property holds iff X is irreducible (see Theorem 1.27 below).

Lemma 1.26. *Pick $\tau\in\mathcal{D}_\omega$, and $\omega'\in\mathcal{Z}_+$ commensurable to ω. Then*

$$T_{\omega_\tau}\le T_\omega+(2n+2)\sup|\tau| \tag{1.22}$$

and

$$T_{\omega'}\le e^{O(\delta)}T_\omega \tag{1.23}$$

with $\delta:=\mathrm{d_T}(\omega,\omega')$.

Recall that, in this paper, the implicit constant in O only depends on n (see §1.1).

Proof. Pick $\varphi\in\mathcal{D}_{\omega_\tau}$. Then $\varphi+\tau\in\mathcal{D}_\omega$, and the Chern–Levine–Nirenberg inequality (see Lemma 1.23) yields

$$\varphi+\tau\le\int(\varphi+\tau)\,\mathrm{MA}_\omega(0)+T_\omega\le\int(\varphi+\tau)\,\mathrm{MA}_\omega(\tau)+2n\sup|\tau|+T_\omega.$$

This proves (1.22), in view of (1.19).

For each $t>0$ we have $\mathcal{D}_{t\omega}=t\mathcal{D}_\omega$, and hence $T_{t\omega}=tT_\omega$. Further, $\omega'\le\omega$ implies $\mathcal{D}_{\omega'}\subset\mathcal{D}_\omega$ and

$$V_{\omega'}\mu_{\omega'}=(\omega')^n\le\omega^n=V_\omega\mu_\omega,$$

and hence $V_{\omega'}T_{\omega'}\le V_\omega T_\omega$. These properties imply (1.23). □

Proof of Proposition 1.24. The condition $T_\omega<\infty$ only depends on the commensurability class of ω, by (1.23), and it only depends on $[\omega]\in\mathrm{Pos}(X)$, by (1.22). This condition is thus independent of ω, since any two classes in $\mathrm{Pos}(X)$ admit commensurable representatives (see Proposition 1.15). □

Theorem 1.27. *Assume X is either a compact Kähler space or a projective Berkovich space over a non-Archimedean field k. Then the submean value property holds iff X is irreducible.*

LEMMA 1.28. *Let $\mathfrak{a} \subset \mathcal{O}_X$ be a coherent ideal sheaf. Then we can find $\omega \in \mathcal{Z}_+$ with $[\omega] \in \operatorname{Pos}(X)$, and a decreasing sequence $(\varphi_j)_j$ in $\mathcal{D}_\omega$ such that, for any given choice of local generators (f_ν) of $\mathfrak{a}$ near a point of X, $\varphi_j - \max\{\log \max_\nu |f_\nu|, -j\}$ is locally bounded uniformly with respect to j.*

In the non-Archimedean case, $\mathfrak{a}$ can be viewed as a coherent ideal sheaf on the underlying algebraic variety, by the GAGA principle.

PROOF. Assume first X is compact Kähler. Fix a Kähler form ω, and pick a finite open cover (U_α) of X such that $\mathfrak{a}$ is generated on each U_α by a finite subset $(f_{\alpha\nu})_\nu$ of $\mathcal{O}(U_\alpha)$. Introduce the psh function on U_α

$$\varphi_{\alpha j} := \tfrac{1}{2}\log\left(\sum_\nu |f_{\alpha\nu}|^2 + e^{-2j}\right) = \max\left\{\log\max_\nu |f_{\alpha\nu}|, -j\right\} + O(1),$$

and observe that, for all α, β, $|\varphi_{\alpha j} - \varphi_{\beta j}|$ is uniformly bounded with respect to j on each compact subset of $U_\alpha \cap U_\beta$. Now choose relatively compact open subsets $U''_\alpha \Subset U'_\alpha \Subset U_\alpha$ such that the U''_α still cover X. Since $\overline{U'_\alpha} \cap \overline{U'_\beta}$ is compact in $U_\alpha \cap U_\beta$, we have

$$C := \sup_{\alpha,\beta,j} \sup_{U'_\alpha \cap U'_\beta} |\varphi_{\alpha j} - \varphi_{\beta j}| < \infty.$$

Pick also cut-off functions $\chi_\alpha \in C_c^\infty(U'_\alpha)$ such that $\chi_\alpha \geq 0$ and $\chi_\alpha \equiv 1$ on U''_α. We claim that the smooth functions

$$\varphi_j := \log \sum_\alpha \chi_\alpha^2 e^{\varphi_{\alpha j}}$$

satisfy $\mathrm{dd^c}\varphi_j \geq -A\omega$ for a uniform constant $A > 0$, which will yield the desired conclusion. The result follows indeed from [**Dem92**, Lemma 3.5], whose proof we now briefly recall. A direct computation using $\mathrm{dd^c}\varphi_{\alpha j} \geq 0$ yields

$$\mathrm{dd^c}\varphi_j \geq -\frac{\sum_\alpha \theta_\alpha e^{\varphi_{\alpha j}}}{\sum_\alpha \chi_\alpha^2 e^{\varphi_{\alpha j}}}$$

with $\theta_\alpha := 2\mathrm{d}\chi_\alpha \wedge \mathrm{d^c}\chi_\alpha - 2\chi_\alpha \mathrm{dd^c}\chi_\alpha$. Since the latter are smooth $(1,1)$-forms, we have $\theta_\alpha \leq A\omega$ for some constant $A > 0$. Now each $x \in X$ in the support of some θ_α lies in $U'_\alpha \setminus U''_\alpha$, and also in U''_β for some β, and hence $\sum_\beta \chi_\beta^2 e^{\varphi_{\beta j}} \geq e^C e^{\varphi_{\alpha j}}$ at x. This shows

$$\frac{\sum_\alpha \theta_\alpha e^{\varphi_{\alpha j}}}{\sum_\alpha \chi_\alpha^2 e^{\varphi_{\alpha j}}} \leq e^C A\omega,$$

which yields the result.

Assume now X is a projective Berkovich space. Pick an ample line bundle L such that $L \otimes \mathfrak{a}$ is generated by global sections (s_ν), and choose a PL metric on L with curvature form $\omega \in \mathcal{Z}_+$. Then setting $\varphi_j := \max\{\log\max_\nu |s_\nu|, -j\}$ yields the result. □

PROOF OF THEOREM 1.27. Assume first that the submean value property holds, and pick an n-dimensional irreducible component Y of X. Applying Lemma 1.28 to the ideal sheaf of Y in X yields $\omega \in \mathcal{Z}_+$ with $[\omega] \in \operatorname{Pos}(X)$ and a decreasing sequence (φ_j) in $\mathcal{D}_\omega$ such that $\sup_Y \varphi_j \to -\infty$ while (φ_j) is uniformly bounded on compact subsets of $X \setminus Y$. Since $\int_Y \omega^n = [\omega]^n \cdot [Y] > 0$, we get $\int \varphi_j \, \mu_\omega \to -\infty$, and hence $\sup \varphi_j \to -\infty$, by the submean value property. It follows that $X = Y$ is irreducible.

Conversely, assume X is irreducible. In the Kähler case, pick a Kähler form ω on X, and choose a resolution of singularities $\pi\colon X' \to X$ with X' a connected compact Kähler manifold, and pick a Kähler form ω' on X' such that $\omega' \geq \pi^\star\omega$. Since $\mu := V_\omega^{-1}\pi^\star\omega^n = f\omega'^n$ satisfies $\mathrm{PSH}(X',\omega') \subset L^1(\mu)$, [**GZ05**, Proposition 2.7] yields $C > 0$ such that $\sup_{X'}\psi \leq \int \psi\,\mu + C$ for all $\psi \in \mathrm{PSH}(X',\omega')$. Applying this to $\psi = \pi^\star\varphi$ with $\varphi \in \mathcal{D}_\omega$ yields the submean value property.

In the non-Archimedean case, pick an ample line bundle L and a PL metric on L determined by an ample model/test configuration $(\mathcal{X},\mathcal{L})$ of (X,L), with curvature form $\omega \in \mathcal{Z}_+$. Each $\varphi \in \mathcal{D}_\omega$ satisfies $\sup\varphi = \max_\Gamma \varphi$, where $\Gamma \subset X$ is the (finite) set of Shilov points attached to $\mathcal{X}$ (see [**GM16**, Proposition 4.22] or [**BE21**, Lemma 6.3]). On the other hand, Γ is also the support of μ_ω, by definition of the measure ω^n in terms of intersection numbers. Now [**BoJ18a**, Theorem 2.21] yields $C > 0$ such that $|\varphi(x) - \varphi(y)| \leq C$ for all $x, y \in \Gamma$ and all $\varphi \in \mathcal{D}_\omega$, and we infer, as desired, $\sup\varphi \leq \int \varphi\,\mu_\omega + C$ for all $\varphi \in \mathcal{D}_\omega$. □

1.6. Monge–Ampère energy and the Dirichlet functional. From now on we fix $\omega \in \mathcal{Z}_+$ with $[\omega] \in \mathrm{Pos}(X)$.

DEFINITION 1.29. *The* Monge–Ampère energy $\mathrm{E}_\omega\colon \mathcal{D} \to \mathbb{R}$ *is defined by*

$$\mathrm{E}_\omega(\varphi) := \frac{(\omega,\varphi)^{n+1}}{(n+1)V_\omega}. \tag{1.24}$$

For all $\varphi, \psi \in \mathcal{D}$ we have

$$\mathrm{E}_\omega(\varphi) - \mathrm{E}_\omega(\psi) = \frac{1}{n+1}\sum_{j=0}^{n} V_\omega^{-1}\int (\varphi - \psi)\,\omega_\varphi^j \wedge \omega_\psi^{n-j}. \tag{1.25}$$

Indeed, by (1.9), this amounts to the basic identity

$$(\omega,\varphi)^{n+1} - (\omega,\psi)^{n+1} = (0,\varphi-\psi)\cdot\sum_{j=0}^{n}(\omega,\varphi)^j\cdot(\omega,\psi)^{n-j}.$$

Assume now $\varphi, \psi \in \mathcal{D}_\omega$. By (1.25), we then have

$$\varphi \leq \psi \Longrightarrow \mathrm{E}_\omega(\varphi) \leq \mathrm{E}_\omega(\psi). \tag{1.26}$$

Further, $\mathrm{E}_\omega((1-t)\varphi + t\psi)$ is a polynomial function of $t \in [0,1]$, with

$$\left.\frac{d}{dt}\right|_{t=0} \mathrm{E}_\omega((1-t)\varphi + t\psi) = \int (\psi - \varphi)\,\mathrm{MA}_\omega(\varphi). \tag{1.27}$$

This characterizes E_ω as the unique primitive of the Monge–Ampère operator that vanishes at $0 \in \mathcal{D}_\omega$. By Lemma 1.19, E_ω is concave on $\mathcal{D}_\omega$, which translates into

$$\mathrm{E}_\omega(\psi) \leq \mathrm{E}_\omega(\varphi) + \int (\psi - \varphi)\,\mathrm{MA}_\omega(\varphi) \tag{1.28}$$

for all $\varphi, \psi \in \mathcal{D}_\omega$.

As a direct consequence of (1.12), we have

$$\mathrm{E}_{\omega_\tau}(\varphi) = \mathrm{E}_\omega(\varphi + \tau) - \mathrm{E}_\omega(\tau). \tag{1.29}$$

for any $\varphi, \tau \in \mathcal{D}$.

DEFINITION 1.30. *We define the* Dirichlet functional $\mathrm{J}_\omega\colon \mathcal{D}_\omega \times \mathcal{D}_\omega \to \mathbb{R}_{\geq 0}$ *by setting*

$$\mathrm{J}_\omega(\varphi,\psi) := \mathrm{E}_\omega(\varphi) - \mathrm{E}_\omega(\psi) + \int (\psi - \varphi)\,\mathrm{MA}_\omega(\varphi).$$

By concavity of E_ω, $\mathrm{J}_\omega(\varphi,\psi)$ is a convex function of $\psi\in\mathcal{D}_\omega$. Note also that

$$\mathrm{J}_\omega(\varphi,\psi)+\mathrm{J}_\omega(\psi,\varphi)=\int(\varphi-\psi)(\mathrm{MA}_\omega(\psi)-\mathrm{MA}_\omega(\varphi)). \tag{1.30}$$

By (A.4), we have the formula

$$\mathrm{J}_\omega(\varphi,\psi)=V_\omega^{-1}\sum_{j=0}^{n-1}\frac{j+1}{n+1}\int(\varphi-\psi)\,\mathrm{dd^c}(\psi-\varphi)\wedge\omega_\varphi^j\wedge\omega_\psi^{n-1-j}. \tag{1.31}$$

We simply write

$$\mathrm{J}_\omega(\varphi):=\mathrm{J}_\omega(0,\varphi)=\int\varphi\,\mu_\omega-\mathrm{E}_\omega(\varphi). \tag{1.32}$$

EXAMPLE 1.31. *When $n=1$, we have*

$$\mathrm{J}_\omega(\varphi,\psi)=\mathrm{J}_\omega(\psi,\varphi)=\tfrac{1}{2}\int(\varphi-\psi)\mathrm{dd^c}(\psi-\varphi),$$

which recovers the usual expression for the Dirichlet functional on a Riemann surface.

For each $R>0$ we set

$$\mathcal{D}_{\omega,R}:=\{\varphi\in\mathcal{D}_\omega\mid \mathrm{J}_\omega(\varphi)\le R\}. \tag{1.33}$$

LEMMA 1.32. *For each $R>0$, $\mathcal{D}_{\omega,R}$ is a convex subset of $\mathcal{D}$ that generates it.*

PROOF. By convexity of $\mathrm{J}_\omega=\mathrm{J}_\omega(0,\cdot)$, $\mathcal{D}_{\omega,R}$ is convex. Since $\mathcal{D}_\omega$ spans $\mathcal{D}$ (see Corollary 1.13), to see that $\mathcal{D}_{\omega,R}$ spans it suffices to show that any $\varphi\in\mathcal{D}_\omega$ satisfies $\mathrm{J}_\omega(t\varphi)\le R$ for $0<t\ll 1$, which holds since $\mathrm{J}_\omega(t\varphi)$ is a polynomial function of t with $\mathrm{J}_\omega(0)=0$. □

We may now collect the fundamental properties of the Dirichlet functional in the next result.

THEOREM 1.33. *For all $\varphi,\varphi',\psi,\psi',\tau\in\mathcal{D}_\omega$ and $t\in[0,1]$, the following holds:*

- quasi-symmetry:

$$\mathrm{J}_\omega(\varphi,\psi)\approx\mathrm{J}_\omega(\psi,\varphi); \tag{1.34}$$

- quasi-triangle inequality:

$$\mathrm{J}_\omega(\varphi,\psi)\lesssim\mathrm{J}_\omega(\varphi,\tau)+\mathrm{J}_\omega(\tau,\psi); \tag{1.35}$$

- quadratic estimate:

$$\mathrm{J}_\omega(\varphi,(1-t)\varphi+t\psi)\lesssim t^2\mathrm{J}_\omega(\varphi,\psi); \tag{1.36}$$

- uniform concavity:

$$\mathrm{E}_\omega((1-t)\varphi+t\psi)-[(1-t)\,\mathrm{E}_\omega(\varphi)+t\,\mathrm{E}_\omega(\psi)]\gtrsim t(1-t)\mathrm{J}_\omega(\varphi,\psi).$$

For all $\varphi,\varphi',\psi,\psi'\in\mathcal{D}_{\omega,R}$, we further have the following Hölder estimates:

$$\left|\int(\varphi-\varphi')\,(\mathrm{MA}_\omega(\psi)-\mathrm{MA}_\omega(\psi'))\right|\lesssim\mathrm{J}_\omega(\varphi,\varphi')^\alpha\mathrm{J}_\omega(\psi,\psi')^{1/2}R^{1/2-\alpha}; \tag{1.37}$$

and

$$|\mathrm{J}_\omega(\varphi,\psi)-\mathrm{J}_\omega(\varphi',\psi')|\lesssim\max\{\mathrm{J}_\omega(\varphi,\varphi'),\mathrm{J}_\omega(\psi,\psi')\}^\alpha R^{1-\alpha}, \tag{1.38}$$

where $\alpha:=2^{-n}$.

PROOF. In view of (1.14), this is a direct consequence of Theorem A.3 applied to

- the vector space $V := \mathcal{Z} \times \mathcal{D}$ with projection $\pi\colon V \to \mathcal{Z}$ onto the first factor;
- the convex cone $P := \{(\theta, \varphi) \in V \mid \varphi \in \mathcal{D}_\theta\}$;
- the homogeneous polynomial $F\colon V \to \mathbb{R}$ defined by $F(\theta, \varphi) := -\frac{(\theta,\varphi)^{n+1}}{(n+1)V_\omega}$.

□

As a simple consequence of (1.29), we finally note:

LEMMA 1.34. *For each $\tau \in \mathcal{D}_\omega$ and $\varphi, \psi \in \mathcal{D}_{\omega_\tau}$ we have $\varphi + \tau, \psi + \tau \in \mathcal{D}_\omega$ and*

$$\mathrm{J}_{\omega_\tau}(\varphi, \psi) = \mathrm{J}_\omega(\varphi + \tau, \psi + \tau). \tag{1.39}$$

REMARK 1.35. *The above formalism recovers that of* [**BBGZ13, BBEGZ19, BFJ15, BoJ22**] *in the Kähler and non-Archimedean settings. However, in contrast to those works, we do not explicitly introduce the functional $\mathrm{I}_\omega(\varphi, \psi)$, which corresponds to the right-hand side of* (1.30).

2. Measures of finite energy

In what follows, we pick $\omega \in \mathcal{Z}_+$ with $[\omega] \in \mathrm{Pos}(X)$. We define the space $\mathcal{M}^1_\omega$ of measures of finite energy with respect to ω, and show, assuming a certain orthogonality property, that it is complete with respect to a quasi-metric δ_ω induced by the Dirichlet functional.

2.1. The energy of a measure.

DEFINITION 2.1. *We define the* energy of $\mu \in \mathcal{M}$ relative to $\psi \in \mathcal{D}_\omega$ *as*

$$\mathrm{J}_\omega(\mu, \psi) := \sup_{\varphi \in \mathcal{D}_\omega} \left\{ \mathrm{E}_\omega(\varphi) - \mathrm{E}_\omega(\psi) + \int (\psi - \varphi)\mu \right\} \in [0, +\infty]. \tag{2.1}$$

The choice of notation is justified by (2.5) below. When $\psi = 0$, we simply write

$$\mathrm{J}_\omega(\mu) := \mathrm{J}_\omega(\mu, 0) = \sup_{\varphi \in \mathcal{D}_\omega} \left\{ \mathrm{E}_\omega(\varphi) - \int \varphi\, \mu \right\}, \tag{2.2}$$

and call it the *energy*[5] *of* μ (with respect to ω). Note that

$$\mathrm{J}_\omega(\mu) = \sup_{\varphi \in \mathcal{D}_\omega} \left\{ \int \varphi(\mu_\omega - \mu) - \mathrm{J}_\omega(\varphi) \right\}, \tag{2.3}$$

by (1.32).

PROPOSITION 2.2. *For each $\psi \in \mathcal{D}_\omega$, the functional $\mathrm{J}_\omega(\cdot, \psi)\colon \mathcal{M} \to [0, +\infty]$ is convex and weakly lsc, and satisfies, for all $\mu \in \mathcal{M}$ and $\varphi \in \mathcal{D}_\omega$,*

$$\mathrm{J}_\omega(\mu, \psi) = \mathrm{J}_\omega(\mu) + \int \psi\, \mu - \mathrm{E}_\omega(\psi); \tag{2.4}$$

$$\mathrm{J}_\omega(\mathrm{MA}_\omega(\varphi), \psi) = \mathrm{J}_\omega(\varphi, \psi); \tag{2.5}$$

$$\mathrm{J}_\omega(\mathrm{MA}_\omega(\varphi)) \approx \mathrm{J}_\omega(\varphi); \tag{2.6}$$

$$\mathrm{J}_\omega(\varphi, \psi) \lesssim \mathrm{J}_\omega(\mu, \varphi) + \mathrm{J}_\omega(\mu, \psi). \tag{2.7}$$

[5]This corresponds to $\mathrm{E}^\vee_\omega(\mu)$ in the notation of [**BBGZ13, BoJ22**], and to $\|\mu\|_\omega$ in that of [**BoJ23**].

PROOF. Convexity and lower semicontinuity are clear from (2.1), which also directly yields (2.4). By (1.28), any $\varphi \in \mathcal{D}_\omega$ further computes the supremum defining $\mathrm{J}_\omega(\mathrm{MA}_\omega(\varphi), \psi)$, which is thus equal to

$$\mathrm{E}_\omega(\varphi) - \mathrm{E}_\omega(\psi) + \int (\psi - \varphi)\,\mathrm{MA}_\omega(\varphi) = \mathrm{J}_\omega(\varphi, \psi).$$

This proves (2.5), which implies

$$\mathrm{J}_\omega(\mathrm{MA}_\omega(\varphi)) = \mathrm{J}_\omega(\mathrm{MA}_\omega(\varphi), 0) = \mathrm{J}_\omega(\varphi, 0) \approx \mathrm{J}_\omega(0, \varphi) = \mathrm{J}_\omega(\varphi),$$

see (1.34). Finally, pick $\mu \in \mathcal{M}$, and set $\tau = \frac{1}{2}(\varphi + \psi) \in \mathcal{D}_\omega$. By (2.1), we have

$$\mathrm{J}_\omega(\mu, \varphi) \geq \mathrm{E}_\omega(\tau) - \mathrm{E}_\omega(\varphi) + \int (\varphi - \tau)\,\mu, \quad \mathrm{J}_\omega(\mu, \psi) \geq \mathrm{E}_\omega(\tau) - \mathrm{E}_\omega(\psi) + \int (\psi - \tau)\,\mu,$$

and hence

$$\mathrm{J}_\omega(\mu, \varphi) + \mathrm{J}_\omega(\mu, \psi) \geq 2\,\mathrm{E}_\omega(\tau) - (\mathrm{E}_\omega(\varphi) + \mathrm{E}_\omega(\psi)).$$

On the other hand,

$$2\,\mathrm{E}_\omega(\tau) - (\mathrm{E}_\omega(\varphi) + \mathrm{E}_\omega(\psi)) \gtrsim \mathrm{J}_\omega(\varphi, \psi)$$

by uniform concavity of E_ω (see Theorem 1.33), and (2.7) follows. □

Generalizing Lemma 1.34, we note:

LEMMA 2.3. *For all $\tau \in \mathcal{D}_\omega$, $\psi \in \mathcal{D}_{\omega_\tau}$ and $\mu \in \mathcal{M}$ we have*

$$\mathrm{J}_{\omega_\tau}(\mu, \psi) = \mathrm{J}_\omega(\mu, \psi + \tau). \tag{2.8}$$

In particular,

$$\mathrm{J}_{\omega_\tau}(\mu) = \mathrm{J}_\omega(\mu) + \int \tau\,\mu - \mathrm{E}_\omega(\tau). \tag{2.9}$$

PROOF. By (1.3) and (1.29) we have $\varphi \in \mathcal{D}_{\omega_\tau} \Leftrightarrow \varphi + \tau \in \mathcal{D}_\omega$, and

$$\mathrm{E}_{\omega_\tau}(\varphi) - \mathrm{E}_{\omega_\tau}(\psi) + \int (\varphi - \psi)\mu = \mathrm{E}_\omega(\varphi + \tau) - \mathrm{E}_\omega(\psi + \tau) + \int ((\varphi + \tau) - (\psi + \tau))\mu.$$

Taking the sup over φ yields (2.8), and (2.9) follows, by (2.4). □

REMARK 2.4. *If we drop the assumption that $\omega \geq 0$, but still require $[\omega] \in$ $\mathrm{Pos}(X)$, then $\mathrm{E}_\omega(\varphi)$ and $\mathrm{J}_\omega(\mu)$ can still be defined by (1.24) and (2.2), respectively. Then (1.29), and hence (2.9), remain valid for any $\tau \in \mathcal{D}$. This will only get used in the context of Theorem 4.8 below.*

2.2. Measures of finite energy.

DEFINITION 2.5. *The space of* measures of finite energy *(with respect to ω) is defined as*

$$\mathcal{M}^1_\omega := \{\mu \in \mathcal{M} \mid \mathrm{J}_\omega(\mu) < \infty\}.$$

It is endowed with the strong topology, *defined as the coarsest refinement of the weak topology in which $\mathrm{J}_\omega \colon \mathcal{M}^1_\omega \to \mathbb{R}_{\geq 0}$ becomes continuous.*

In other words, a net (μ_i) converges strongly to μ in $\mathcal{M}^1_\omega$ iff $\mu_i \to \mu$ weakly in $\mathcal{M}$ and $\mathrm{J}_\omega(\mu_i) \to \mathrm{J}_\omega(\mu)$. For any $R > 0$ we also set

$$\mathcal{M}^1_{\omega,R} := \{\mu \in \mathcal{M}^1_\omega \mid \mathrm{J}_\omega(\mu) \leq R\}. \tag{2.10}$$

By Proposition 2.2, this set is convex and weakly compact. By (2.4),

$$\mathrm{J}_\omega(\cdot, \psi) \colon \mathcal{M}^1_\omega \to \mathbb{R}_{\geq 0}$$

is continuous in the strong topology for any $\psi \in \mathcal{D}_\omega$. By Lemma 2.3, this yields:

PROPOSITION 2.6. *The topological space $\mathcal{M}^1_\omega$ only depends on the positive class $[\omega] \in \mathrm{Pos}(X)$.*

One should be wary of the fact that, in the present generality, even a 'nice' probability measure of the form $\mu = \theta_1 \wedge \cdots \wedge \theta_n$ with $\theta_i \in \mathcal{Z}_+$ need not be of finite energy with respect to ω in general (see however Theorem 3.4 below):

EXAMPLE 2.7. *Let X be either a compact Kähler or projective Berkovich space. For each irreducible component Y of X and each $\mu \in \mathcal{M}^1_\omega$, we then have*

$$\mu(Y) = \frac{[\omega|_Y]^n}{[\omega]^n}. \tag{2.11}$$

Indeed, this is proved in [**BoJ22**, Corollary 9.13] *in the trivially valued case, and the proof can be adapted to the general case. Now* (2.11) *fails in general for $\mu = \mu_{\omega'}$ with $\omega' \in \mathcal{Z}_+$ such that $[\omega] \neq [\omega'] \in \mathrm{Pos}(X)$, and hence $\mathcal{M}^1_{\omega'} \neq \mathcal{M}^1_\omega$.*

By (2.3), we have, for all $\varphi \in \mathcal{D}_\omega$ and $\mu \in \mathcal{M}$,

$$\int \varphi\,(\mu_\omega - \mu) \le \mathrm{J}_\omega(\varphi) + \mathrm{J}_\omega(\mu).$$

The following converse will come in handy.

LEMMA 2.8. *Assume that $\mu \in \mathcal{M}$ satisfies*

$$S := \sup_{\varphi \in \mathcal{D}_{\omega,R}} \int \varphi(\mu_\omega - \mu) < \infty$$

for some $R > 0$. Then μ has finite energy, and

$$\mathrm{J}_\omega(\mu) \lesssim S(1 + R^{-1} S). \tag{2.12}$$

PROOF. Pick $\varphi \in \mathcal{D}_\omega$ and set $J := \mathrm{J}_\omega(\varphi)$. By (1.36), we have $\mathrm{J}_\omega(t\varphi) \lesssim t^2 J$ for any $t \in [0, 1]$, and we can thus choose

$$1 \le a \lesssim 1 + (R^{-1} J)^{1/2}$$

such that $\mathrm{J}_\omega(a^{-1}\varphi) \le R$. By assumption, we then have $\int a^{-1}\varphi\,(\mu_\omega - \mu) \le S$, and hence

$$\begin{aligned} \mathrm{E}_\omega(\varphi) - \int \varphi\,\mu = \int \varphi\,(\mu_\omega - \mu) - \mathrm{J}_\omega(\varphi) &\le aS - J \\ &\lesssim S + S R^{-1/2} J^{1/2} - J \le S + \tfrac{1}{4} S^2 R^{-1}, \end{aligned}$$

where the last inequality follows from the elementary estimate $\sup_{y \ge 0}(x y^{1/2} - y) = x^2/4$ for any $x \ge 0$. Taking the supremum over φ yields (2.12). □

2.3. Legendre transform of the energy. Here we compute the Legendre transform of the convex functional $\mathrm{J}_\omega = \mathrm{J}_\omega(\cdot, 0)\colon \mathcal{M} \to [0, +\infty]$.

DEFINITION 2.9. *For any $f \in \mathrm{C}^0(X)$ we set*

$$\widetilde{\mathrm{E}}_\omega(f) := \sup_{f \ge \varphi \in \mathcal{D}_\omega} \mathrm{E}_\omega(\varphi).$$

By monotonicity of E_ω on $\mathcal{D}_\omega$ (see (1.26)), the functional $\widetilde{\mathrm{E}}_\omega\colon \mathrm{C}^0(X) \to \mathbb{R}$ so defined restricts to E_ω on $\mathcal{D}_\omega$. Like the latter, $\widetilde{\mathrm{E}}_\omega$ is further concave, monotone increasing, and equivariant with respect to translation, i.e.

$$\widetilde{\mathrm{E}}_\omega(f + c) = \widetilde{\mathrm{E}}_\omega(f) + c \text{ for } c \in \mathbb{R}.$$

PROPOSITION 2.10. *For all $f \in C^0(X)$ and $\mu \in \mathcal{M}$ we have*

$$\widetilde{E}_\omega(f) = \inf_{\nu \in \mathcal{M}} \left\{ J_\omega(\nu) + \int f\,\nu \right\}; \quad J_\omega(\mu) = \sup_{g \in C^0(X)} \left\{ \widetilde{E}_\omega(g) - \int g\,\mu \right\}. \tag{2.13}$$

PROOF. Define the (convex) Legendre transform $\widetilde{E}_\omega^\vee \colon C^0(X)^\vee \to \mathbb{R} \cup \{+\infty\}$ as the right-hand side of (2.13), i.e.

$$\widetilde{E}_\omega^\vee(\mu) := \sup_{g \in C^0(X)} \left\{ \widetilde{E}_\omega(g) - \int g\,\mu \right\}.$$

Since $\widetilde{E}_\omega$ is increasing and equivariant, it is straightforward to see that $\widetilde{E}_\omega^\vee(\mu) < \infty$ implies $\mu \geq 0$ and $\int \mu = 1$, i.e. $\mu \in \mathcal{M}$ (compare [**BoJ22**, Proposition 9.8]). By Legendre duality, the result is thus equivalent to $\widetilde{E}_\omega^\vee(\mu) = J_\omega(\mu)$ for $\mu \in \mathcal{M}$. Since $\widetilde{E}_\omega$ restricts to E_ω on $\mathcal{D}_\omega$, we trivially have $\widetilde{E}_\omega^\vee(\mu) \geq J_\omega(\mu)$. Conversely, pick $f \in C^0(X)$ and $\varphi \in \mathcal{D}_\omega$ with $\varphi \leq f$. Then

$$J_\omega(\mu) \geq E_\omega(\varphi) - \int \varphi\,\mu \geq E_\omega(\varphi) - \int f\,\mu,$$

where the first and second inequality respectively follow from (2.2) and $\varphi \leq f$. Taking the supremum over φ and then over f yields $J_\omega(\mu) \geq \widetilde{E}_\omega(f) - \int f\,\mu$ and $J_\omega(\mu) \geq \widetilde{E}_\omega^\vee(\mu)$. $\square$

2.4. Orthogonality and differentiability.

DEFINITION 2.11. *We say that $\mathcal{D}_\omega$* admits maxima *if, for all $\varphi, \psi \in \mathcal{D}_\omega$ and $f \in \mathcal{D}$ such that* $\max\{\varphi, \psi\} < f$ *pointwise on X, there exists $\tau \in \mathcal{D}_\omega$ with* $\max\{\varphi, \psi\} \leq \tau < f$.

This equivalently means that, for any $f \in \mathcal{D}$, the poset

$$\mathcal{D}_{\omega,<f} := \{\varphi \in \mathcal{D}_\omega \mid \varphi < f\}$$

is inductive. We can then consider limits of nets indexed by $\mathcal{D}_{\omega,<f}$. For instance, note that

$$\widetilde{E}_\omega(f) = \lim_{\varphi \in \mathcal{D}_{\omega,<f}} E_\omega(\varphi). \tag{2.14}$$

EXAMPLE 2.12. *If X is a compact Kähler space, then $\mathcal{D}_\omega$ admits maxima: take $\tau := \widetilde{\max}(\varphi, \psi)$ for an appropriate regularized max function $\widetilde{\max}$.*

EXAMPLE 2.13. *If X is a projective Berkovich space, then $\mathcal{D}_\omega$ also admits maxima, since $\mathbb{Q}$-PL functions in $\mathcal{D}_\omega$ are dense in $\mathcal{D}_\omega$, and stable under max.*

REMARK 2.14. *While we will not pursue this direction here, one can also introduce as in* [**BFJ16a, BE21, BoJ22**] *the space* $\mathrm{PSH}(\omega)$ *of ω-psh functions $\varphi \colon X \to \mathbb{R} \cup \{-\infty\}$, defined as usc functions that can be obtained as pointwise limits of decreasing nets in $\mathcal{D}_\omega$, and such that $\int \varphi\,\mu_\omega > -\infty$. Then $\mathcal{D}_\omega$ admits maxima iff* $\mathrm{PSH}(\omega)$ *(or, equivalently, the subspace* $\mathrm{CPSH}(\omega) := \mathrm{PSH}(\omega) \cap C^0(X)$ *of continuous ω-psh functions) is stable under max.*

DEFINITION 2.15. *We say that ω has the* orthogonality property *if $\mathcal{D}_\omega$ admits maxima and*

$$\lim_{\varphi \in \mathcal{D}_{\omega,<f}} \int (f - \varphi)\, \mathrm{MA}_\omega(\varphi) = 0 \tag{2.15}$$

for all $f \in \mathcal{D}$.

Explicitly, this means for that for any $\varepsilon > 0$ there exists $\varphi_0 \in \mathcal{D}_\omega$ such that $\varphi_0 < f$ and $\int (f - \varphi) \operatorname{MA}_\omega(\varphi) \le \varepsilon$ for all $\varphi \in \mathcal{D}_\omega$ with $\varphi_0 \le \varphi < f$.

REMARK 2.16. *The orthogonality property for ω only depends on $[\omega] \in \operatorname{Pos}(X)$. Indeed, for any $\tau \in \mathcal{D}_\omega$ and $f \in \mathcal{D}$ we have*

$$\varphi \in \mathcal{D}_{\omega_\tau,<f} \Longleftrightarrow \varphi + \tau \in \mathcal{D}_{\omega,<f+\tau} \tag{2.16}$$

This implies that $\mathcal{D}_\omega$ admits finite maxima iff $\mathcal{D}_{\omega_\tau}$ does, and similarly for the orthogonality property, using

$$\operatorname{MA}_{\omega_\tau}(\varphi) = \operatorname{MA}_\omega(\varphi + \tau) \quad \text{and}$$

$$\int (f - \varphi) \operatorname{MA}_{\omega_\tau}(\varphi) = \int ((f + \tau) - (\varphi + \tau)) \operatorname{MA}_\omega(\varphi + \tau).$$

EXAMPLE 2.17. *Assume X is a compact Kähler space. Conjecturally, the orthogonality property always holds. This is known when X is normal, or X is projective and $[\omega] \in \operatorname{Amp}(X)$, see Appendix* B *for a more detailed discussion.*

Recall from Corollary 1.13 that any test function $f \in \mathcal{D}$ can be written as

$$f = f^+ - f^-, \quad f^\pm \in \mathcal{D}_{C\omega} \tag{2.17}$$

for some $C = C(f) > 0$. In line with [**BFJ15**, §7], we show:

PROPOSITION 2.18. *Assume $\mathcal{D}_\omega$ admits maxima. The following properties are then equivalent:*

(i) *ω has the orthogonality property;*
(ii) *for any $f \in \mathcal{D}$ written as (2.17) for a given $C > 0$, we have*

$$\left| \widetilde{\operatorname{E}}_\omega(\varphi + f) - \operatorname{E}_\omega(\varphi) - \int f \operatorname{MA}_\omega(\varphi) \right| \lesssim C \sup |f| \tag{2.18}$$

for all $\varphi \in \mathcal{D}_\omega$;
(iii) *in the setting of (ii), we have*

$$\left| \widetilde{\operatorname{E}}_\omega(\varphi + tf) - \operatorname{E}_\omega(\varphi) - t \int f \operatorname{MA}_\omega(\varphi) \right| \lesssim C t^2 \sup |f| \tag{2.19}$$

for all $\varphi \in \mathcal{D}_\omega$ and $t \in \mathbb{R}$.

Note that the uniform differentiability property (2.19) implies in particular

$$\left. \frac{d}{dt} \right|_{t=0} \widetilde{\operatorname{E}}_\omega(\varphi + tf) = \int f \operatorname{MA}_\omega(\varphi).$$

PROOF. Assume (i). Write $f \in \mathcal{D}$ as in (2.17), and pick $\varphi \in \mathcal{D}_\omega$. For any $\psi \in \mathcal{D}_{\omega,<\varphi+\varepsilon f}$, (1.28) yields

$$\int (\psi - \varphi) \operatorname{MA}_\omega(\psi) \le \operatorname{E}_\omega(\psi) - \operatorname{E}_\omega(\varphi) \le \int (\psi - \varphi) \operatorname{MA}_\omega(\varphi) \le \int f \operatorname{MA}_\omega(\varphi).$$

By Lemma 1.23, we also have $\left| \int f(\operatorname{MA}_\omega(\varphi) - \operatorname{MA}_\omega(\psi)) \right| \le 2nC \sup |\varphi - \psi|$, and we infer

$$\left| \operatorname{E}_\omega(\psi) - \operatorname{E}_\omega(\varphi) - \int f \operatorname{MA}_\omega(\varphi) \right| \le \int ((\varphi + f) - \psi) \operatorname{MA}_\omega(\psi) + 2nC \sup |\varphi - \psi|. \tag{2.20}$$

Now $\lim_{\psi\in\mathcal{D}_{\omega,<\varphi+f}} \mathrm{E}_\omega(\psi) = \widetilde{\mathrm{E}}_\omega(\varphi+f)$ (see (2.14)), while orthogonality yields

$$\lim_{\psi\in\mathcal{D}_{\omega,<\varphi+f}} \int ((\varphi+f)-\psi)\,\mathrm{MA}_\omega(\psi) = 0.$$

Further, any $\psi \in \mathcal{D}_{\omega,<\varphi+f}$ large enough is greater than $\varphi - \sup|f| \in \mathcal{D}_{\omega,<\varphi+f}$, and hence satisfies $\sup|\varphi-\psi| \le \sup|f|$. As a result, (2.20) implies

$$\left|\widetilde{\mathrm{E}}_\omega(\varphi+f) - \mathrm{E}_\omega(\varphi) - \int f\,\mathrm{MA}_\omega(\varphi)\right| \le 2nC\sup|f|,$$

which shows (i)⇒(ii). Next, (ii)⇒(iii), since $f = f^+ - f^-$ with $f \in \mathcal{D}_{C\omega}$ implies $tf = \mathrm{sgn}(t)(|t|f^+ - |t|f^-)$ with $|t|f^\pm \in \mathcal{D}_{C|t|\omega}$.

Finally, assume (iii), and pick $f \in \mathcal{D}$. We need to show that

$$L := \limsup_{\varphi\in\mathcal{D}_{\omega,<f}} \int (f-\varphi)\,\mathrm{MA}_\omega(\varphi) \ge 0$$

vanishes. Write f as in (2.17) for some $C > 0$. Pick also $\varphi \in \mathcal{D}_{\omega,<f}$, and set $g := f - \varphi \in \mathcal{D}$. For any $t \in [0,1]$ we have $\varphi + tg = (1-t)\varphi + tf \le f$, and hence $\widetilde{\mathrm{E}}_\omega(\varphi+tg) \le \widetilde{\mathrm{E}}_\omega(f)$. On the other hand, since $f - \varphi = f^+ - (f^- + \varphi)$ with $f^+, f^- + \varphi \in \mathcal{D}_{(C+1)\omega}$, (iii) yields a constant $A > 0$ only depending on φ such that

$$\begin{aligned} 0 &\le t\int (f-\varphi)\,\mathrm{MA}_\omega(\varphi) \\ &\le \widetilde{\mathrm{E}}_\omega(\varphi+tg) - \mathrm{E}_\omega(\varphi) + t^2 A\sup|g| \le \widetilde{\mathrm{E}}_\omega(f) - \mathrm{E}_\omega(\varphi) + t^2 A\sup|g|. \end{aligned}$$

Since any $\varphi \in \mathcal{D}_{\omega,<f}$ large enough is greater then $f^+ - \sup f^-$, it satisfies

$$0 \le g = f - \varphi \le \sup f^- - f^- \le B$$

with B only depending on f. We infer $0 \le tL \le t^2 AB$. Dividing by $t > 0$ and letting $t \to 0_+$ yields, as desired, $L = 0$. This proves (iii)⇒(i). □

EXAMPLE 2.19. *Assume X is a projective Berkovich space over a non-Archimedean field. Then* [**BE21**, Theorem A] *combined with the uniform differentiability estimate of* [**BGM22**, Lemma 3.2] *shows that* (2.18) *is satisfied (compare* [**BoJ22**, Lemma 8.7]*). By Proposition* 2.18*, it follows that the orthogonality property always holds in this setting.*

2.5. Maximizing sequences.

DEFINITION 2.20. *We say that a sequence (ψ_i) in $\mathcal{D}_\omega$ is* maximizing *for $\mu \in \mathcal{M}^1_\omega$ if it computes the energy of μ* (2.2)*, i.e. $\mathrm{E}_\omega(\psi_i) - \int \psi_i\,\mu \to \mathrm{J}_\omega(\mu)$.*

Equivalently, (ψ_i) is maximizing for μ iff $\mathrm{J}_\omega(\mu,\psi_i) \to 0$, see (2.4).

EXAMPLE 2.21. *For any $\varphi \in \mathcal{D}_\omega$ the constant sequence $\psi_i = \varphi$ is maximizing for $\mu = \mathrm{MA}_\omega(\varphi)$ (see* (2.5)*).*

As a key consequence of Proposition 2.18, we show:

THEOREM 2.22. *Assume ω has the orthogonality property. Pick $\mu \in \mathcal{M}^1_\omega$ and a maximizing sequence $\psi_i \in \mathcal{D}_\omega$. Then the measures $\mu_i := \mathrm{MA}_\omega(\psi_i)$ converge strongly to μ in $\mathcal{M}^1_\omega$, i.e. $\mu_i \to \mu$ weakly and $\mathrm{J}_\omega(\mu_i) \to \mathrm{J}_\omega(\mu)$. In particular, the image of the Monge–Ampère operator*

$$\mathrm{MA}_\omega\colon \mathcal{D}_\omega \to \mathcal{M}^1_\omega$$

is dense in the strong topology.

PROOF. Pick $f \in \mathcal{D}$, and choose $C > 0$ such that $f = f^+ - f^-$ with $f^\pm \in \mathcal{D}_{C\omega}$. Since we assume orthogonality, Proposition 2.18 yields $A > 0$ such that

$$\left|\widetilde{\mathrm{E}}_\omega(\psi_i + tf) - \mathrm{E}_\omega(\psi_i) - t\int f\,\mu_i\right| \le At^2$$

for all i and $t > 0$. By Proposition 2.10 and (2.4), we have, on the other hand,

$$\widetilde{\mathrm{E}}_\omega(\psi_i + tf) \le \mathrm{J}_\omega(\mu) + \int(\psi_i + tf)\mu = \mathrm{J}_\omega(\mu, \psi_i) + \mathrm{E}_\omega(\psi_i) + t\int f\,\mu.$$

Combining these estimates, we get

$$t\int f\,\mu_i \le t\int f\,\mu + \mathrm{J}_\omega(\mu, \psi_i) + At^2.$$

Since $\mathrm{J}_\omega(\mu, \psi_i) \to 0$, we infer

$$t\limsup_i \int f\,\mu_i \le t\int f\,\mu + At^2.$$

Dividing by t and letting $t \to 0_+$ yields $\limsup_i \int f\,\mu_i \le \int f\,\mu$. Replacing f with $-f$, we get $\lim_i \int f\,\mu_i = \int f\,\mu$. By density of $\mathcal{D}$ in $\mathrm{C}^0(X)$, this shows $\mu_i \to \mu$ weakly.

For each i we have $\mathrm{J}_\omega(\mu_i) = \mathrm{E}_\omega(\psi_i) - \int \psi_i\,\mu_i$ (see (2.5)), and $\mathrm{E}_\omega(\psi_i) - \int \psi_i\,\mu \to \mathrm{J}_\omega(\mu)$, since (ψ_i) is maximizing for μ. It only remains to prove $\int \psi_i\,(\mu_i - \mu) \to 0$. Since $\mathrm{J}_\omega(\psi_i)$ is bounded (see (2.7)), (1.37) yields $C > 0$ such that

$$\left|\int \psi_i\,(\mu_i - \mu_j)\right| \lesssim C\mathrm{J}_\omega(\psi_i, \psi_j)^\alpha$$

for all i, j, and hence

$$\left|\int \psi_i\,(\mu_i - \mu_j)\right| \lesssim C\max\{\mathrm{J}_\omega(\mu, \psi_i), \mathrm{J}_\omega(\mu, \psi_j)\}^\alpha,$$

by (2.7). Since $\mu_j \to \mu$ weakly and $\mathrm{J}_\omega(\mu, \psi_j) \to 0$ as $j \to \infty$, we infer

$$\left|\int \psi_i\,(\mu_i - \mu)\right| \lesssim C\mathrm{J}_\omega(\mu, \psi_i)^\alpha,$$

and we conclude, as desired, that the left-hand side tends to 0 as $i \to \infty$. □

2.6. The Dirichlet quasi-metric. From now on, **we assume that the orthogonality property holds** for ω. Recall from Examples 2.17 and 2.19, that this is the case if X is a normal compact Kähler space, or X is any projective Berkovich space.

THEOREM 2.23. *There exists a unique continuous functional*

$$\delta_\omega \colon \mathcal{M}^1_\omega \times \mathcal{M}^1_\omega \to \mathbb{R}_{\ge 0},$$

such that

$$\delta_\omega(\mathrm{MA}_\omega(\varphi), \mathrm{MA}_\omega(\psi)) = \mathrm{J}_\omega(\varphi, \psi) \tag{2.21}$$

for all $\varphi, \psi \in \mathcal{D}_\omega$. Furthermore:

(i) *for all $\mu \in \mathcal{M}^1_\omega$ and $\psi \in \mathcal{D}_\omega$ we have*

$$\delta_\omega(\mu, \mathrm{MA}_\omega(\psi)) = \mathrm{J}_\omega(\mu, \psi), \quad \delta_\omega(\mu, \mu_\omega) = \mathrm{J}_\omega(\mu); \tag{2.22}$$

(ii) *δ_ω is a quasi-metric: for all $\mu, \nu, \rho \in \mathcal{M}^1_\omega$ we have*

$$\delta_\omega(\mu,\nu) = 0 \Leftrightarrow \mu = \nu, \quad \delta_\omega(\mu,\nu) \approx \delta_\omega(\nu,\mu), \quad \delta_\omega(\mu,\nu) \lesssim \delta_\omega(\mu,\rho) + \delta_\omega(\rho,\nu); \tag{2.23}$$

(iii) *the quasi-metric δ_ω satisfies the Hölder continuity property*

$$|\delta_\omega(\mu,\nu) - \delta_\omega(\mu',\nu')| \lesssim \max\{\delta_\omega(\mu,\mu'), \delta_\omega(\nu,\nu')\}^\alpha R^{1-\alpha} \tag{2.24}$$

for all $R > 0$ and $\mu, \mu', \nu, \nu' \in \mathcal{M}^1_{\omega,R}$, with $\alpha := 2^{-n}$;

(iv) *for all $R > 0$ and $\varphi, \psi \in \mathcal{D}_{\omega,R}$, $\mu, \nu \in \mathcal{M}^1_{\omega,R}$, we have the Hölder estimate*

$$\left|\int (\varphi - \psi)(\mu - \nu)\right| \lesssim \mathrm{J}_\omega(\varphi,\psi)^\alpha \delta_\omega(\mu,\nu)^{1/2} R^{1/2-\alpha}. \tag{2.25}$$

We call δ_ω the *Dirichlet quasi-metric* on $\mathcal{M}^1_\omega$.

LEMMA 2.24. *For all $\mu \in \mathcal{M}^1_{\omega,R}$ and $\varphi, \psi, \tau \in \mathcal{D}_{\omega,R}$, we have*

$$|\mathrm{J}_\omega(\mu,\varphi) - \mathrm{J}_\omega(\mathrm{MA}_\omega(\tau),\psi)| \lesssim \max\{\mathrm{J}_\omega(\mu,\tau), \mathrm{J}_\omega(\varphi,\psi)\}^\alpha R^{1-\alpha}; \tag{2.26}$$

$$\left|\int (\varphi - \psi)(\mu - \mathrm{MA}_\omega(\tau))\right| \lesssim \mathrm{J}_\omega(\varphi,\psi)^\alpha \mathrm{J}_\omega(\mu,\tau)^{1/2} R^{1/2-\alpha}. \tag{2.27}$$

PROOF. When μ lies in the image of $\mathrm{MA}_\omega \colon \mathcal{D}_\omega \to \mathcal{M}^1_\omega$, this is equivalent to (1.38) and (1.37), in view of (2.5) and (2.6). By Theorem 2.22 the image of MA_ω is dense in $\mathcal{M}^1_\omega$, and the general case thus follows by continuity in the strong topology of all functions of μ involved. □

PROOF OF THEOREM 2.23. Uniqueness is clear, since (2.21) determines δ_ω on the image of $\mathrm{MA}_\omega \colon \mathcal{D}_\omega \to \mathcal{M}^1_\omega$, which is dense in the strong topology, by Theorem 2.22. To show existence, pick $\mu, \nu \in \mathcal{M}^1_\omega$, and choose a maximizing sequences (ψ_i) for ν. We can then find $R > 0$ such that $\mu \in \mathcal{M}^1_{\omega,R}$ and $\psi_i \in \mathcal{D}_{\omega,R}$ for all i, and (2.26) and (2.7) yield

$$\begin{aligned}|\mathrm{J}_\omega(\mu,\psi_i) - \mathrm{J}_\omega(\mu,\psi_j)| &\lesssim \mathrm{J}_\omega(\psi_i,\psi_j)^\alpha R^{1-\alpha}\\ &\lesssim \max\{\mathrm{J}_\omega(\nu,\psi_i), \mathrm{J}_\omega(\nu,\psi_j)\}^\alpha R^{1-\alpha}.\end{aligned}$$

This estimate implies that $(\mathrm{J}_\omega(\mu,\psi_i))$ is a Cauchy sequence, which thus admits a limit

$$\delta_\omega(\mu,\nu) := \lim_i \mathrm{J}_\omega(\mu,\psi_i). \tag{2.28}$$

The same estimate also shows that the limit is independent of the choice of maximizing sequence (ψ_i), and that the convergence in (2.28) is uniform with respect to $\mu \in \mathcal{M}^1_{\omega,R}$. As a consequence, $\mu \mapsto \delta_\omega(\mu,\nu)$ so defined is continuous on $\mathcal{M}^1_\omega$ for each $\nu \in \mathcal{M}^1_\omega$.

By construction, (2.22) holds, and hence also (2.21), by (2.5). This proves (i).

Next, (2.24) holds when ν, ν' lie in the image of $\mathrm{MA}_\omega \colon \mathcal{D}_\omega \to \mathcal{M}^1_\omega$, by applying (2.26) to a maximizing sequence for μ', and the general case follows by using maximizing sequences for ν, ν'. This shows (iii), which also yields the continuity of δ_ω on $\mathcal{M}^1_\omega \times \mathcal{M}^1_\omega$ (and hence concludes the proof of existence), since $\mu_i \to \mu$ strongly implies $\delta_\omega(\mu_i,\mu) \to \delta_\omega(\mu,\mu) = 0$, by continuity of $\delta_\omega(\cdot,\mu)$.

Similarly, (iv) follows follows by applying (2.27) to a maximizing sequence (τ_i) for ν.

Finally, the first point in (ii) follows from (2.25), since $\mathcal{D}_\omega$ spans the dense subspace $\mathcal{D}$ of $\mathrm{C}^0(X)$ (see Corollary 1.13). By (1.34) and (1.35), the last two properties in (ii) hold when the measures lie in the image of MA_ω, and hence in general, by continuity of δ_ω. □

We next show:

THEOREM 2.25. *The quasi-metric space $(\mathcal{M}^1_\omega, \delta_\omega)$ only depends on the class $[\omega]$. It is complete, and its topology coincides with the strong topology.*

LEMMA 2.26. *For any $\nu \in \mathcal{M}^1$ and $R > 0$, $\delta_\omega(\cdot, \nu)$ is weakly lsc on $\mathcal{M}^1_{\omega,R}$.*

PROOF. When $\nu = \mathrm{MA}_\omega(\psi)$ with $\psi \in \mathcal{D}_\omega$, (2.22) yields $\delta_\omega(\cdot, \nu) = \mathrm{J}_\omega(\cdot, \psi)$, which is weakly lsc on $\mathcal{M}^1_\omega$ (see (2.1)). In the general case, pick a maximizing sequence (ψ_i) for ν, and set $\nu_i := \mathrm{MA}_\omega(\psi_i)$. By (2.24), we have $\delta_\omega(\mu, \nu_i) \to \delta_\omega(\mu, \nu)$ uniformly for $\mu \in \mathcal{M}^1_{\omega,R}$, and the result follows. □

PROOF OF THEOREM 2.25. We already know that $\mathcal{M}^1_\omega$ only depends on $[\omega]$ (see Proposition 2.6). Pick $\tau \in \mathcal{D}_\omega$, $\mu, \nu \in \mathcal{M}^1_\omega = \mathcal{M}^1_{\omega_\tau}$, and choose maximizing sequences (φ_i), (ψ_i) in $\mathcal{D}_{\omega_\tau}$ for μ, ν, so that $\delta_{\omega_\tau}(\mu, \nu) = \lim_i \mathrm{J}_{\omega_\tau}(\varphi_i, \psi_i)$. By (2.8), $(\varphi_i + \tau)$ and $(\psi_i + \tau)$ are maximizing sequences in $\mathcal{D}_\omega$ for μ, ν, and hence $\delta_\omega(\mu, \nu) = \lim_i \mathrm{J}_\omega(\varphi_i, \psi_i)$. Now (1.39) yields $\mathrm{J}_{\omega_\tau}(\varphi_i, \psi_i) = \mathrm{J}_\omega(\varphi_i + \tau, \psi_i + \tau)$, which proves $\delta_{\omega_\tau}(\mu, \nu) = \delta_\omega(\mu, \nu)$. Thus δ_ω only depends on $[\omega]$.

We next show that the topology of $(\mathcal{M}^1_\omega, \delta_\omega)$ is the strong topology, i.e. a net (μ_i) converges strongly to $\mu \in \mathcal{M}^1_\omega$ iff $\delta_\omega(\mu_i, \mu) \to 0$. When the latter holds, (2.25) implies $\mu_i \to \mu$ weakly (since $\mathcal{D}_\omega$ spans the dense subspace $\mathcal{D}$ of $\mathrm{C}^0(X)$), while (2.24) yields $\mathrm{J}_\omega(\mu_i) = \mathrm{J}_\omega(\mu_i, 0) \to \mathrm{J}_\omega(\mu)$. Thus $\mu_i \to \mu$ strongly, and the converse holds by strong continuity of δ_ω.

Finally, consider a Cauchy net (μ_i) in $(\mathcal{M}^1, \delta_\omega)$. Then $\mathrm{J}_\omega(\mu_i) = \delta_\omega(\mu_i, \mu_\omega)$ is eventually bounded. By weak compactness of $\mathcal{M}$, we may assume, after passing to a subnet, that (μ_i) admits a weak limit $\mu \in \mathcal{M}$. Since J_ω is weakly lsc on $\mathcal{M}$, we get $\mathrm{J}_\omega(\mu) \le \liminf_i \mathrm{J}_\omega(\mu_i) < +\infty$, i.e. $\mu \in \mathcal{M}^1$. It remains to show $\delta_\omega(\mu_i, \mu) \to 0$. To see this, pick $\varepsilon > 0$ and i_0 such that $\delta_\omega(\mu_i, \mu_j) \le \varepsilon$ for all $i, j \ge i_0$. Since $\mathrm{J}_\omega(\mu_j)$ is bounded and $\mu_j \to \mu$ weakly, Lemma 2.26 yields $\delta_\omega(\mu_i, \mu) \le \liminf_j \delta_\omega(\mu_i, \mu_j) \le \varepsilon$, and we are done. □

To conclude this section, we show:

PROPOSITION 2.27. *For each $\nu \in \mathcal{M}^1_\omega$, $\delta_\omega(\cdot, \nu) \colon \mathcal{M}^1_\omega \to \mathbb{R}_{\ge 0}$ is strictly convex, and we further have the uniform convexity estimate*

$$(1-t)\delta_\omega(\mu_0, \nu) + t\delta_\omega(\mu_1, \nu) - \delta_\omega((1-t)\mu_0 + t\mu_1, \nu) \gtrsim t(1-t)\delta_\omega(\mu_0, \mu_1).$$

for all $\mu_0, \mu_1 \in \mathcal{M}^1_\omega$ and $t \in [0, 1]$.

PROOF. By density of the image of $\mathrm{MA}_\omega \colon \mathcal{D}_\omega \to \mathcal{M}^1_\omega$ and continuity of δ_ω, we may assume without loss $\nu = \mathrm{MA}_\omega(\psi)$ with $\psi \in \mathcal{D}_\omega$, and hence $\delta_\omega(\cdot, \nu) = \mathrm{J}_\omega(\cdot, \psi)$. Set

$$J_t := (1-t)\mathrm{J}_\omega(\mu_0, \psi) + t\mathrm{J}_\omega(\mu_1, \psi), \quad \mu_t := (1-t)\mu_0 + t\mu_1,$$

and pick $\varphi \in \mathcal{D}_\omega$. Applying (2.4) to μ_0 and μ_1 yields

$$J_t = \mathrm{E}_\omega(\varphi) - \mathrm{E}_\omega(\psi) + \int (\psi - \varphi)\mu_t + (1-t)\mathrm{J}_\omega(\mu_0, \varphi) + t\mathrm{J}_\omega(\mu_1, \varphi),$$

and hence

$$J_t - \mathrm{E}_\omega(\varphi) + \mathrm{E}_\omega(\psi) + \int (\varphi - \psi)\mu_t \geq t(1-t)(\mathrm{J}_\omega(\mu_0, \psi) + \mathrm{J}_\omega(\mu_1, \psi)),$$

using the elementary estimate $(1-t)a + tb \geq t(1-t)(a+b)$ for $a, b \geq 0$ (see for instance [**BoJ22**, Lemma 7.29]). By (2.23), this implies

$$J_t - \mathrm{E}_\omega(\varphi) + \mathrm{E}_\omega(\psi) + \int (\varphi - \psi)\mu_t \gtrsim t(1-t)\delta_\omega(\mu_0, \mu_1),$$

and taking the infimum over φ shows $J_t - \mathrm{J}_\omega(\mu_t, \psi) \gtrsim t(1-t)\,\mathrm{d}_\omega(\mu_0, \mu_1)$, which concludes the proof. □

2.7. An equivalent metric on $\mathcal{M}^1$. One can show that the quasi-metric space $(\mathcal{M}^1_\omega, \delta_\omega)$ is metrizable, by general theory. Here we introduce a concrete metric that defines the strong topology of $\mathcal{M}^1_\omega$. Recall from (1.33) that

$$\mathcal{D}_{\omega,1} = \{\varphi \in \mathcal{D}_\omega \mid \mathrm{J}_\omega(\varphi) \leq 1\}.$$

PROPOSITION 2.28. *Setting*

$$\mathrm{d}_\omega(\mu, \nu) := \sup_{\varphi \in \mathcal{D}_{\omega,1}} \left| \int \varphi(\mu - \nu) \right| \tag{2.29}$$

yields a complete metric on $\mathcal{M}^1_\omega$ that defines the strong topology. Furthermore:

(i) *the metric d_ω and the Dirichlet quasi-metric δ_ω share the same bounded sets;*

(ii) *they are Hölder equivalent on bounded sets; more precisely:*

$$\mathrm{d}_\omega(\mu, \nu) \lesssim \delta_\omega(\mu, \nu)^{1/2} R^{1/2} \quad \textit{and} \quad \delta_\omega(\mu, \nu) \lesssim \mathrm{d}_\omega(\mu, \nu) R^{1/2} \tag{2.30}$$

for all $\mu, \nu \in \mathcal{M}^1_{\omega,R}$ with $R \geq 1$;

(iii) *for all $\varphi \in \mathcal{D}_\omega$ and $\mu, \nu \in \mathcal{M}^1_\omega$ we have*

$$\left| \int \varphi\,(\mu - \nu) \right| \lesssim (\mathrm{J}_\omega(\varphi)^{1/2} + 1)\,\mathrm{d}_\omega(\mu, \mu). \tag{2.31}$$

REMARK 2.29. *In the compact Kähler and non-Archimedean cases, the metric d_ω just constructed is* a priori *unrelated to the usual Darvas-type metric d_1 [**Dar15**, **Reb22**].*

PROOF OF PROPOSITION 2.28. Pick $\varphi \in \mathcal{D}_\omega$. Since $\mathrm{J}_\omega(a^{-1}\varphi) \lesssim a^{-2}\mathrm{J}_\omega(\varphi)$ for $a \geq 1$ (see (1.36)), we can choose $1 \leq a \lesssim \mathrm{J}_\omega(\varphi)^{1/2} + 1$ such that $\mathrm{J}_\omega(a^{-1}\varphi) \leq 1$. Then $\left|\int a^{-1}\varphi\,(\mu - \nu)\right| \leq \mathrm{d}_\omega(\mu, \nu)$, which proves (2.31).

The first part of (2.30) is a direct consequence of (2.25). It shows, in particular, that d_ω is finite valued. It is also clear that d_ω is symmetric, vanishes on the diagonal, and satisfies the triangle inequality. Since $\mathcal{D}_{\omega,1}$ spans the dense subspace $\mathcal{D}$ of $\mathrm{C}^0(X)$ (see Lemma 1.32), d_ω further separates points, and hence defines a metric on $\mathcal{M}^1_\omega$.

The first part of (2.30) also shows that $\mu_i \to \mu$ in $\mathcal{M}^1_\omega$ implies $\mathrm{d}_\omega(\mu_i, \mu) \to 0$, by continuity of δ_ω, and it follows that the metric d_ω is continuous. By density of the image of MA_ω, it is thus enough to show the second half of (2.30) when $\mu = \mathrm{MA}_\omega(\varphi)$ and $\nu = \mathrm{MA}_\omega(\psi)$ with $\varphi, \psi \in \mathcal{D}_\omega$. Then $\mathrm{J}_\omega(\varphi) \approx \mathrm{J}_\omega(\mu) \leq R$ and $\mathrm{J}_\omega(\psi) \approx \mathrm{J}_\omega(\nu) \leq R$ (see (2.6)), while

$$\delta_\omega(\mu, \nu) = \mathrm{J}_\omega(\varphi, \psi) \leq \int (\varphi - \psi)(\mu - \nu),$$

by (2.21) and (1.30). Using (2.31), we get the second half of (2.30). Next pick $\mu \in \mathcal{M}^1_\omega$ and set $R := \max\{1, \delta_\omega(\mu, \mu_\omega)\}$ and $S := \max\{1, \mathrm{d}_\omega(\mu, \mu_\omega)\}$. Applying (2.30) to $\nu = \mu_\omega$ yields $S \lesssim R$, and also $R \lesssim SR^{1/2}$, i.e. $R \lesssim S^2$. This proves (i) and (ii). Since δ_ω defines the strong topology of $\mathcal{M}^1_\omega$ and is complete (see Theorem 2.25), the same therefore holds, as desired, for d_ω. □

By Theorem 2.25, the quasi-metric space $(\mathcal{M}^1_\omega, \delta_\omega)$ only depends on $[\omega] \in \mathrm{Pos}(X)$. Here we show:

LEMMA 2.30. *For each $\tau \in \mathcal{D}_\omega$ we have $\mathrm{d}_{\omega_\tau} \lesssim (\mathrm{J}_\omega(\tau)^{1/2} + 1)\, \mathrm{d}_\omega$. In particular, the Lipschitz equivalence class of the metric space $(\mathcal{M}^1_\omega, \mathrm{d}_\omega)$ only depends on $[\omega]$.*

PROOF. For any $\varphi \in \mathcal{D}_{\omega_\tau}$, (1.39) yields $\mathrm{J}_{\omega_\tau}(\varphi) = \mathrm{J}_{\omega_\tau}(0, \varphi) = \mathrm{J}_\omega(\tau, \varphi + \tau)$. When $\mathrm{J}_{\omega_\tau}(\varphi) \le 1$, the quasi-triangle inequality (1.35) thus yields $\mathrm{J}_\omega(\varphi + \tau) = \mathrm{J}_\omega(0, \varphi + \tau) \lesssim 1 + J$ with $J := \mathrm{J}_\omega(\tau) = \mathrm{J}_\omega(0, \tau)$. By (2.31) we infer

$$\left|\int (\varphi + \tau)(\mu - \nu)\right| \lesssim (1 + J^{1/2})\, \mathrm{d}_\omega(\mu, \nu), \qquad \left|\int \tau(\mu - \nu)\right| \lesssim (1 + J^{1/2})\, \mathrm{d}_\omega(\mu, \nu).$$

Thus $\left|\int \varphi(\mu - \nu)\right| \lesssim (1 + J^{1/2})\, \mathrm{d}_\omega(\mu, \nu)$. Taking the supremum over $\varphi \in \mathcal{D}_{\omega_\tau}$ such that $\mathrm{J}_{\omega_\tau}(\varphi) \le 1$ yields the result. □

3. Lipschitz and Hölder estimates for the energy

In what follows we consider $\omega \in \mathcal{Z}_+$ with $[\omega] \in \mathrm{Pos}(X)$. As above, we assume that the orthogonality property holds. From now on, we further assume the submean value property (see Definition 1.25), and use it to investigate the dependence of $\mathcal{M}^1_\omega$ on ω and establish a Hölder continuity estimate for the energy pairing.

Recall that the standing assumptions hold when X is a normal irreducible compact Kähler space, or any irreducible projective Berkovich space (see Theorem 1.27 and Examples 2.17 and 2.19).

3.1. Lipschitz estimates for the energy. Recall from §2.7 the metric d_ω, which defines the strong topology of $\mathcal{M}^1_\omega$. As a first key consequence of the submean value property, we show:

THEOREM 3.1. *The Lipschitz equivalence class of the metric space $(\mathcal{M}^1_\omega, \mathrm{d}_\omega)$ is independent of ω.*

In particular, the topological space $\mathcal{M}^1_\omega$ is independent of ω (see Proposition 2.28), and will henceforth simply by denoted by $\mathcal{M}^1$.

LEMMA 3.2. *Assume $\omega' \in \mathcal{Z}_+$ is commensurable to ω, with $[\omega'] \in \mathrm{Pos}(X)$. For all $\mu, \nu \in \mathcal{M}$ we then have*

$$\mathrm{d}_{\omega'}(\mu, \nu) \le e^{O(\delta)}(1 + T_\omega)^{1/2}\, \mathrm{d}_\omega(\mu, \nu) \tag{3.1}$$

with $\delta := \mathrm{d}_{\mathrm{T}}(\omega, \omega') \in \mathbb{R}_{\ge 0}$.

PROOF. Pick any $\varphi \in \mathcal{D}_{\omega'}$ such that $\mathrm{J}_{\omega'}(\varphi) = \int \varphi\, \mu_{\omega'} - \mathrm{E}_{\omega'}(\varphi) \le 1$. We need to show

$$\left|\int \varphi(\mu - \nu)\right| \le e^{O(\delta)}(1 + T_\omega)^{1/2}\, \mathrm{d}_\omega(\mu, \nu). \tag{3.2}$$

By translation invariance of J_ω on $\mathcal{D}_\omega$, we may assume without loss $\sup \varphi = 0$. Then

$$-\mathrm{E}_{\omega'}(\varphi) \le 1 + T_{\omega'} \le 1 + e^{O(\delta)} T_\omega,$$

by (1.23). On the other hand,

$$0 \geq e^{(n+1)\delta}(\omega, e^{-\delta}\varphi)^{n+1} = (e^{\delta}\omega, \varphi)^{n+1} \geq e^{O(\delta)}(\omega', \varphi)^{n+1},$$

where the last inequality follows from Lemma 1.20, since $\omega' \leq e^{\delta}\omega \leq e^{2\delta}\omega'$. Dividing by $(n+1)V_\omega = e^{O(\delta)}(n+1)V_{\omega'}$, we get

$$0 \leq \mathrm{J}_\omega(e^{-\delta}\varphi) \leq -\mathrm{E}_\omega(e^{-\delta}\varphi) \leq -e^{O(\delta)}\,\mathrm{E}_{\omega'}(\varphi) \leq e^{O(\delta)}(1+T_\omega).$$

By (2.31) this implies

$$\left|\int e^{-\delta}\varphi(\mu-\nu)\right| \leq e^{O(\delta)}(1+T_\omega)^{1/2}\,\mathrm{d}_\omega(\mu,\nu),$$

and hence (3.2). □

Proof of Theorem 3.1. We argue as in the proof of Proposition 1.24. On the one hand, the Lipschitz equivalence class only depends on the positive class $[\omega]$, by Lemma 2.30. On the other hand, it only depends on the commensurability class of ω, by Lemma 3.2, and we conclude since any two positive classes admit commensurable representatives, see Proposition 1.15. □

3.2. Mixed Monge–Ampère measures. It will be convenient to introduce, for $\varphi \in \mathcal{D}_\omega$ and $\mu \in \mathcal{M}$, the quantities

$$\mathrm{J}^+_\omega(\varphi) := \mathrm{J}_\omega(\varphi) + T_\omega, \quad \text{and} \quad \mathrm{J}^+_\omega(\mu) := \mathrm{J}_\omega(\mu) + T_\omega. \tag{3.3}$$

Lemma 3.3. *For each $\varphi \in \mathcal{D}_\omega$ we have*

$$0 \leq \sup\varphi - \mathrm{E}_\omega(\varphi) \leq \mathrm{J}^+_\omega(\varphi); \tag{3.4}$$

$$\mathrm{J}^+_\omega(\varphi) \approx \mathrm{J}^+_\omega(\mathrm{MA}_\omega(\varphi)). \tag{3.5}$$

Proof. By (1.32), we have

$$\sup\varphi - \mathrm{E}_\omega(\varphi) = \mathrm{J}_\omega(\varphi) + (\sup\varphi - \int \varphi\,\mu_\omega) \leq \mathrm{J}_\omega(\varphi) + T_\omega.$$

This yields (3.4), while (3.5) is a direct consequence of (2.6). □

We next establish a key energy estimate for mixed Monge–Ampère measures.

Theorem 3.4. *For $i = 1, \dots, n$, pick $\omega_i \in \mathcal{Z}_+$ with $[\omega_i] \in \mathrm{Pos}(X)$ and $\varphi_i \in \mathcal{D}_{\omega_i}$, and set*

$$\mu := ([\omega_1]\cdot\ldots\cdot[\omega_n])^{-1}(\omega_1 + \mathrm{dd^c}\varphi_1)\wedge\cdots\wedge(\omega_n + \mathrm{dd^c}\varphi_n) \in \mathcal{M}. \tag{3.6}$$

Then μ lies in $\mathcal{M}^1$, and satisfies:

(i) *if each ω_i is commensurable to ω, then*

$$\mathrm{J}^+_\omega(\mu) \lesssim e^{O(\delta)} \max_i \mathrm{J}^+_{\omega_i}(\varphi_i)$$

with $\delta := \max_i \mathrm{d_T}(\omega_i, \omega)$;

(ii) *in the general case,*

$$\mathrm{J}_\omega(\mu) \lesssim C(\max_i \mathrm{J}_{\omega_i}(\varphi_i) + 1),$$

where $C > 0$ only depends on ω and the ω_i.

PROOF. Assume first that each ω_i is commensurable to ω. Set

$$J := \max_i \mathrm{J}^+_{\omega_i}(\varphi_i), \quad V := V_\omega, \quad V_i := V_{\omega_i},$$

and observe that

$$e^{-n\delta} V \le V_i \le e^{n\delta} V, \quad e^{-n\delta} V \le [\omega_1] \cdot \ldots \cdot [\omega_n] \le e^{n\delta} V. \tag{3.7}$$

Since μ is unchanged when the φ_i's are translated by constants, we may assume without loss that $\sup \varphi_i = 0$. Then

$$0 \ge (\omega_i, \varphi_i)^{n+1} = (n+1) V_i \, \mathrm{E}_{\omega_i}(\varphi_i) \ge -(n+1) V_i \mathrm{J}^+_{\omega_i}(\varphi_i),$$

by (3.4), and hence

$$0 \ge (\omega_i, \varphi_i)^{n+1} \gtrsim -e^{O(\delta)} V J, \tag{3.8}$$

using (3.7). Now pick $\psi \in \mathcal{D}_\omega$ such that $\sup \psi = 0$, and set $R := \mathrm{J}_\omega(\psi)$. On the one hand,

$$0 \ge (\omega, \psi)^{n+1} = (n+1) V \, \mathrm{E}_\omega(\psi) \ge -(n+1) V (R + T_\omega), \tag{3.9}$$

using (3.4) again. On the other hand, (3.7) yields

$$\begin{aligned}
0 &\ge e^{-n\delta} V \int \psi \, \mu \\
&\ge \int \psi \, (\omega_1 + dd^c \varphi_1) \wedge \cdots \wedge (\omega_n + dd^c \varphi_n) \\
&= (\omega, \psi) \cdot (\omega_1, \varphi_1) \cdot \ldots \cdot (\omega_n, \varphi_n) - (\omega, 0) \cdot (\omega_1, \varphi_1) \cdot \ldots \cdot (\omega_n, \varphi_n) \\
&\ge (\omega, \psi) \cdot (\omega_1, \varphi_1) \cdot \ldots \cdot (\omega_n, \varphi_n) \\
&\gtrsim e^{O(\delta)} \min\{(\omega, \psi)^{n+1}, \min_i (\omega_i, \varphi_i)^{n+1}\},
\end{aligned}$$

by (1.16) and Theorem 1.18. Combined with (3.8) and (3.9), this yields

$$\int \psi (\mu_\omega - \mu) \le -\int \psi \, \mu \lesssim e^{O(\delta)} (R + J + T_\omega) \lesssim e^{O(\delta)} (R + J),$$

since $T_\omega \le e^{O(\delta)} T_{\omega_i} \le e^{O(\delta)} J$ by (1.23). By Lemma 2.8, we infer,

$$\mathrm{J}_\omega(\mu) \lesssim e^{O(\delta)} \inf_{R>0} (R + J) \left(1 + R^{-1}(R + J)\right) \le e^{O(\delta)} J,$$

which concludes the proof of (i) (using (1.23)).

We now consider the general case. By Proposition 1.15, we can choose $\tau \in \mathcal{D}_\omega$ and $\tau_i \in \mathcal{D}_{\omega_i}$ such that $\omega' := \omega_\tau$ and $\omega'_i := \omega_{i,\tau_i}$ are commensurable for all i. Then

$$\mu = ([\omega'_1] \cdot \ldots \cdot [\omega'_n])^{-1} (\omega'_1 + dd^c \varphi'_1) \wedge \cdots \wedge (\omega'_n + dd^c \varphi_n)$$

with $\varphi'_i := \varphi_i - \tau_i$, and hence

$$\mathrm{J}_{\omega'}(\mu) \lesssim \max_i (\mathrm{J}_{\omega'_i}(\varphi'_i) + T_{\omega'_i}),$$

by Theorem 3.4. By (2.9), we have $\mathrm{J}_\omega(\mu) \le \mathrm{J}_{\omega'}(\mu) + C$ and $\mathrm{J}_{\omega'_i}(\varphi'_i) \le \mathrm{J}_\omega(\varphi_i) + C$, with $C > 0$ independent of μ, and (ii) follows. □

As a consequence, we get the following estimate for the energy:

COROLLARY 3.5. *Pick $\omega' \in \mathcal{Z}_+$ such that $[\omega'] \in \mathrm{Pos}(X)$, and $\mu \in \mathcal{M}^1$.*

(i) *If ω' is commensurable to ω, then*

$$\mathrm{J}^+_{\omega'}(\mu) \approx e^{O(\delta)} \mathrm{J}^+_\omega(\mu)$$

with $\delta := \mathrm{d_T}(\omega, \omega')$.

(ii) *In the general case, there exists $C > 0$ only depending on ω, ω' such that*

$$\mathrm{J}_{\omega'}(\mu) \le C(\mathrm{J}_\omega(\mu) + 1).$$

We refer to (4.21) below for a more precise estimate when δ is small.

PROOF. Assume first ω' commensurable. Pick a maximizing sequence (φ_j) for μ in $\mathcal{D}_\omega$, and set $\mu_j := \mathrm{MA}_\omega(\varphi_j)$. Then $\mu_j \to \mu$ strongly in $\mathcal{M}^1$ (see Theorem 2.22), and hence $\mathrm{J}_\omega(\mu_j) \to \mathrm{J}_\omega(\mu)$ and $\mathrm{J}_{\omega'}(\mu_j) \to \mathrm{J}_{\omega'}(\mu)$. For each j we have $\mathrm{J}_\omega(\varphi_j) \approx \mathrm{J}_\omega(\mu_j)$ (see (2.6)). Theorem 3.4 thus yields $\mathrm{J}^+_{\omega'}(\mu_j) \lesssim e^{O(\delta)} \mathrm{J}^+_\omega(\mu_j)$, and (i) follows.

In the general case, we can choose $\tau \in \mathcal{D}_\omega$ and $\tau' \in \mathcal{D}_{\omega'}$ such that ω_τ and $\omega_{\tau'}$ are commensurable (see Proposition 1.15). By (2.9), we then have $\mathrm{J}_{\omega'}(\mu) \le \mathrm{J}_{\omega_\tau}(\mu) + C$ and $\mathrm{J}_{\omega_{\tau'}}(\mu) \le \mathrm{J}_{\omega'}(\mu) + C$ with $C > 0$ independent of μ, and (ii) now follows from (i). □

3.3. Hölder continuity of the energy pairing. Recall from §1.3 that

$$\mathcal{Z}_\omega = \{\theta \in \mathcal{Z} \mid \|\theta\|_\omega < \infty\}.$$

Using the above results, we establish a general Hölder continuity property for the energy pairing.

THEOREM 3.6. *For $i = 0, \dots, n$, pick $\theta_i, \theta'_i \in \mathcal{Z}_\omega$ and $\varphi_i, \varphi'_i \in \mathcal{D}_\omega$, normalized by $\int \varphi_i \, \mu_\omega = \int \varphi'_i \, \mu_\omega = 0$. Then*

$$|(\theta_0, \varphi_0) \cdot \ldots \cdot (\theta_n, \varphi_n) - (\theta'_0, \varphi'_0) \cdot \ldots \cdot (\theta'_n, \varphi'_n)| \lesssim A \left(\max_i \|\theta_i - \theta'_i\|_\omega J + \max_i \mathrm{J}_\omega(\varphi_i, \varphi'_i)^\alpha J^{1-\alpha} \right).$$

with $\alpha := 2^{-n}$ and

$$A := V_\omega \prod_i (1 + \|\theta_i\|_\omega + \|\theta'_i\|_\omega), \quad J := \max_i \mathrm{J}^+_\omega(\varphi_i).$$

In particular,

$$|(\theta_1, \varphi_1) \cdot \ldots \cdot (\theta_n, \varphi_n)| \lesssim V_\omega \prod_i (1 + \|\theta_i\|_\omega) \max_i \mathrm{J}^+_\omega(\varphi_i). \tag{3.10}$$

PROOF. Assume first $\theta_i = \theta'_i$ for all i. By symmetry of the energy pairing, we may assume $\varphi_i = \varphi'_i$ for $i \ge 1$. For $i = 1, \dots, n$ set $t_i := 1 + \|\theta_i\|_\omega$. Then $\|t_i^{-1}\theta_i\|_\omega \le 1$, and

$$\mathrm{J}_\omega(t_i^{-1}\varphi_i) \le t_i^{-1} \mathrm{J}_\omega(\varphi_i) \le \mathrm{J}_\omega(\varphi_i),$$

by convexity of J_ω on $\mathcal{D}_\omega$. By homogeneity, we may thus assume $\|\theta_i\|_\omega \le 1$ for all $i = 1, \dots, n$. Thus $-\omega \le \theta_i \le \omega$, and hence $\theta_i = \theta_i^+ - \theta_i^-$ where $\theta_i^+ := \theta_i + 2\omega$ and $\theta_i^- := 2\omega$ both satisfy $\omega \le \theta_i^\pm \le 3\omega$. By multilinearity, we may finally assume $\omega \le \theta_i \le 3\omega$. By Proposition 1.17 (i), it then suffices to show

$$\left| \int (\varphi_0 - \varphi'_0)(\theta_1 + dd^c\varphi_1) \wedge \cdots \wedge (\theta_n + dd^c\varphi_n) \right| \le V_\omega \mathrm{J}_\omega(\varphi_0, \varphi'_0)^\alpha J^{1-\alpha}. \tag{3.11}$$

Since $\omega \le \theta_i \le 3\omega$, Theorem 3.4 shows that

$$\mu := ([\theta_0]\cdot\ldots\cdot[\theta_n])^{-1}(\theta_1 + \mathrm{dd^c}\varphi_1)\wedge\cdots\wedge(\theta_n + \mathrm{dd^c}\varphi_n) \in \mathcal{M}$$

satisfies $\mathrm{J}_\omega(\mu) \lesssim J$. By (2.25), we infer

$$\left|\int(\varphi_0 - \varphi_0)(\mu - \mu_\omega)\right| \lesssim \mathrm{J}_\omega(\varphi_0, \varphi_0)^\alpha J^{1-\alpha},$$

which yields (3.11) since $\int(\varphi_0 - \varphi_0)\mu_\omega = 0$ and $[\theta_0]\cdot\ldots\cdot[\theta_n] \lesssim V_\omega$.

In the general case, we may again assume $(\theta_i, \varphi_i) = (\theta_i', \varphi_i')$ for $i \ge 1$, by symmetry and multilinearity of the energy pairing. Then

$$(\theta_0, \varphi_0)\cdot(\theta_1, \varphi_1)\cdot\ldots\cdot(\theta_n, \varphi_n) - (\theta_0', \varphi_0')\cdot(\theta_1, \varphi_1)\cdot\ldots\cdot(\theta_n, \varphi_n)$$
$$= (\theta_0 - \theta_0', 0)\cdot(\theta_1, \varphi_1)\cdot\ldots\cdot(\theta_n, \varphi_n) + (0, \varphi_0 - \varphi_0')\cdot(\theta_1, \varphi_1)\cdot\ldots\cdot(\theta_n, \varphi_n),$$

where the last term has already been estimated by (3.11). We are thus reduced to showing

$$|(\theta_0, 0)\cdot(\theta_1, \varphi_1)\cdot\ldots\cdot(\theta_n, \varphi_n)| \lesssim AJ\|\theta_0\|_\omega.$$

By homogeneity we may further assume $\|\theta_0\|_\omega = 1$, and the desired estimate now follows from the first step of the proof applied to $\varphi_i' = 0$, using $(\theta_0, 0)\cdot\ldots\cdot(\theta_n, 0) = 0$. $\square$

4. Twisted energy and differentiability

As in §3, we assume that the orthogonality and submean value properties hold, and recall that this is satisfied when X is a normal irreducible compact Kähler space or any irreducible projective Berkovich space. We fix $\omega \in \mathcal{Z}_+$ with $[\omega] \in \mathrm{Pos}(X)$. In this section we introduce and study the twisted energy of a measure, and show that it computes the directional derivatives of the energy.

4.1. The twisted energy of a measure. For any $\theta \in \mathcal{Z}$ and $\varphi \in \mathcal{D}$,

$$\mathrm{E}_{\omega+t\theta}(\varphi) := V_{\omega+t\theta}^{-1}(\omega + t\theta, \varphi)^{n+1}$$

make sense for all $t \in \mathbb{R}$ small enough, and

$$\nabla_\theta \mathrm{E}_\omega(\varphi) := \left.\frac{d}{dt}\right|_{t=0} \mathrm{E}_{\omega+t\theta}(\varphi) \tag{4.1}$$

satisfies

$$\nabla_\theta \mathrm{E}_\omega(\varphi) = \mathrm{E}_\omega^\theta(\varphi) - V_\omega^\theta \mathrm{E}_\omega(\varphi), \tag{4.2}$$

where

$$\mathrm{E}_\omega^\theta(\varphi) := V_\omega^{-1}(\theta, 0)\cdot(\omega, \varphi)^n \tag{4.3}$$

and

$$V_\omega^\theta := nV_\omega^{-1}[\theta]\cdot[\omega]^{n-1}.$$

Note that $\mathrm{E}_\omega^\theta(\varphi)$ is a linear function of $\theta \in \mathcal{Z}$, and

$$\mathrm{E}_\omega^\theta(\varphi + c) = \mathrm{E}_\omega^\theta(\varphi) + cV_\omega^\theta \text{ for } c \in \mathbb{R}, \tag{4.4}$$

while $\nabla_\theta \mathrm{E}_\omega$ is translation invariant. For all $\varphi, \psi \in \mathcal{D}$, we further have

$$\mathrm{E}_\omega^\theta(\varphi) - \mathrm{E}_\omega^\theta(\psi) = \sum_{j=0}^{n-1} V_\omega^{-1}\int(\varphi - \psi)\theta\wedge\omega_\varphi^j\wedge\omega_\psi^{n-1-j}. \tag{4.5}$$

EXAMPLE 4.1. *By (1.24), we have* $\mathrm{E}_{(1+t)\omega}((1+t)\varphi) = (1+t)\,\mathrm{E}_\omega(\varphi)$. *By (1.27), this implies* $\nabla_\omega \mathrm{E}_\omega(\varphi) + \int \varphi\, \mathrm{MA}_\omega(\varphi) = \mathrm{E}_\omega(\varphi)$, *and hence*

$$\nabla_\omega \mathrm{E}_\omega(\varphi) = \mathrm{E}_\omega(\varphi) - \int \varphi\, \mathrm{MA}_\omega(\varphi) = \mathrm{J}_\omega(\varphi, 0) \approx \mathrm{J}_\omega(0,\varphi) = \mathrm{J}_\omega(\varphi). \tag{4.6}$$

EXAMPLE 4.2. *For each* $\psi \in \mathcal{D}$ *and* $t \in \mathbb{R}$, *(1.29) yields* $\mathrm{E}_{\omega + t\mathrm{dd}^c\psi}(\varphi) = \mathrm{E}_\omega(\varphi + t\psi) - \mathrm{E}_\omega(t\psi)$, *and using again (1.27) we get*

$$\nabla_{\mathrm{dd}^c\psi} \mathrm{E}_\omega(\varphi) = \int \psi(\mathrm{MA}_\omega(\varphi) - \mu_\omega). \tag{4.7}$$

EXAMPLE 4.3. *For any* $\tau \in \mathcal{D}_\omega$, *we similarly have* $\mathrm{E}_{\omega_\tau + t\theta}(\varphi) = \mathrm{E}_{\omega+t\theta}(\varphi+\tau) - \mathrm{E}_{\omega+t\theta}(\tau)$ *and*

$$\nabla_\theta \mathrm{E}_{\omega_\tau}(\varphi) = \nabla_\theta \mathrm{E}_\omega(\varphi+\tau) - \nabla_\theta \mathrm{E}_\omega(\tau). \tag{4.8}$$

LEMMA 4.4. *For all* $\theta \in \mathcal{Z}$ *and* $\varphi, \psi \in \mathcal{D}_\omega$, *the following holds:*

(i) *if* $\theta \in \mathcal{Z}_\omega$, *then*

$$|\nabla_\theta \mathrm{E}_\omega(\varphi) - \nabla_\theta \mathrm{E}_\omega(\psi)| \lesssim \mathrm{J}_\omega(\varphi,\psi)^\alpha \max\{\mathrm{J}^+_\omega(\varphi), \mathrm{J}^+_\omega(\psi)\}^{1-\alpha} \|\theta\|_\omega \tag{4.9}$$

with $\alpha := 2^{-n}$;

(ii) *in the general case, there exist* $C > 0$ *only depending on* ω *and* θ *such that*

$$|\nabla_\theta \mathrm{E}_\omega(\varphi) - \nabla_\theta \mathrm{E}_\omega(\psi)| \le C \mathrm{J}_\omega(\varphi,\psi)^\alpha (\max\{\mathrm{J}_\omega(\varphi), \mathrm{J}_\omega(\psi)\} + 1)^{1-\alpha}. \tag{4.10}$$

PROOF. By translation invariance of $\nabla_\theta \mathrm{E}_\omega$, we may assume $\int \varphi\, \mu_\omega = \int \psi\, \mu_\omega = 0$. When θ lies in $\mathcal{Z}_\omega$, Theorem 3.6 applied to (1.24) and (4.3) yields

$$|\mathrm{E}_\omega(\varphi) - \mathrm{E}_\omega(\psi)| \le C \quad \text{and} \quad \left|\mathrm{E}^\theta_\omega(\varphi) - \mathrm{E}^\theta_\omega(\psi)\right| \le C(1 + \|\theta\|_\omega)$$

with $C := \mathrm{J}_\omega(\varphi,\psi)^\alpha \max\{\mathrm{J}^+_\omega(\varphi), \mathrm{J}^+_\omega(\psi)\}^{1-\alpha}$. By homogeneity of E^θ_ω with respect to θ, we may replace $1 + \|\theta\|_\omega$ with $\|\theta\|_\omega$ in the last estimate, and (i) now follows from (4.2) together with $|V^\theta_\omega| \le n\|\theta\|_\omega$.

In the general case, pick $\tau \in \mathcal{D}_\omega$ such that $\theta \in \mathcal{Z}_{\omega_\tau}$ (see Proposition 1.12). Given $\varphi, \psi \in \mathcal{D}_\omega$, we then have $\varphi - \tau, \psi - \tau \in \mathcal{D}_{\omega_\tau}$, and (4.8), (1.39) yield

$$\nabla_\theta \mathrm{E}_{\omega_\tau}(\varphi-\tau) - \nabla_\theta \mathrm{E}_{\omega_\tau}(\psi-\tau) = \nabla_\theta \mathrm{E}_\omega(\varphi) - \nabla_\theta \mathrm{E}_\omega(\psi), \quad \mathrm{J}_{\omega_\tau}(\varphi-\tau, \psi-\tau) = \mathrm{J}_\omega(\varphi,\psi),$$

$$\mathrm{J}_{\omega_\tau}(\varphi-\tau) = \mathrm{J}_{\omega_\tau}(0, \varphi-\tau) = \mathrm{J}_\omega(\tau,\varphi) \lesssim \mathrm{J}_\omega(\varphi) + \mathrm{J}_\omega(\tau), \quad \mathrm{J}_{\omega_\tau}(\psi-\tau) \lesssim \mathrm{J}_\omega(\psi) + \mathrm{J}_\omega(\tau).$$

By (i) we thus get

$$|\nabla_\theta \mathrm{E}_\omega(\varphi) - \nabla_\theta \mathrm{E}_\omega(\psi)| \lesssim \mathrm{J}_\omega(\varphi,\psi)^\alpha \left(\max\{\mathrm{J}^+_\omega(\varphi), \mathrm{J}^+_\omega(\psi)\} + \mathrm{J}^+_\omega(\tau)\right)^{1-\alpha} \|\theta\|_{\omega_\tau},$$

which proves (ii). □

PROPOSITION 4.5. *For any* $\theta \in \mathcal{Z}$, *there exists a unique strongly continuous functional* $\mathrm{J}^\theta_\omega \colon \mathcal{M}^1 \to \mathbb{R}$, *the* θ-twisted energy, *such that*

$$\mathrm{J}^\theta_\omega(\mathrm{MA}_\omega(\varphi)) = \nabla_\theta \mathrm{E}_\omega(\varphi) \tag{4.11}$$

for all $\varphi \in \mathcal{D}_\omega$. *For all* $\mu, \nu \in \mathcal{M}^1$, *we also have:*

(i) $\mathrm{J}^\theta_\omega(\mu)$ *is a linear function of* θ;

(ii) *if* $\theta \in \mathcal{Z}_\omega$, *then*

$$|\mathrm{J}^\theta_\omega(\mu) - \mathrm{J}^\theta_\omega(\nu)| \lesssim \delta_\omega(\mu,\nu)^\alpha \max\{\mathrm{J}^+_\omega(\mu), \mathrm{J}^+_\omega(\nu)\}^{1-\alpha} \|\theta\|_\omega; \tag{4.12}$$

with $\alpha := 2^{-n}$, *and hence*

$$|\mathrm{J}^\theta_\omega(\mu)| \lesssim \mathrm{J}^+_\omega(\mu) \|\theta\|_\omega; \tag{4.13}$$

(iii) *in the general case, there exist $C>0$ only depending on ω and θ such that*

$$|\mathrm{J}^\theta_\omega(\mu)-\mathrm{J}^\theta_\omega(\nu)|\le C\delta_\omega(\mu,\nu)^\alpha(\max\{\mathrm{J}_\omega(\mu),\mathrm{J}_\omega(\nu)\}+1)^{1-\alpha}, \tag{4.14}$$

and hence

$$|\mathrm{J}^\theta_\omega(\mu)|\le C(\mathrm{J}_\omega(\mu)+1). \tag{4.15}$$

PROOF. Assume $\varphi,\psi\in\mathcal{D}_\omega$ satisfy $\mathrm{MA}_\omega(\varphi)=\mathrm{MA}_\omega(\psi)$. Then $\mathrm{J}_\omega(\varphi,\psi)=0$, and hence $\nabla_\theta\,\mathrm{E}_\omega(\varphi)=\nabla_\theta\,\mathrm{E}_\omega(\psi)$, by (4.10). As a result, there exists a unique function J^θ_ω on the image of $\mathrm{MA}_\omega\colon\mathcal{D}_\omega\to\mathcal{M}^1$ such that (4.11) is satisfied. For all $\mu,\nu\in\mathcal{M}^1$ in the image of MA_ω, it further follows from (4.10) that (4.14) holds. This shows that J^θ_ω is uniformly continuous on a dense subspace of the quasi-metric space $(\mathcal{M}^1,\delta_\omega)$, and hence admits a unique continuous extension $\mathrm{J}^\theta_\omega\colon\mathcal{M}^1\to\mathbb{R}$. Finally, (i) holds by linearity of $\nabla_\theta\,\mathrm{E}_\omega$ with respect to θ,(ii) and (iii) follow, by continuity, from Lemma 4.4. □

Using (4.11) (with $\varphi=0$), (4.6), (4.7) and (4.8), we further get, for all $\mu\in\mathcal{M}^1$, $\psi\in\mathcal{D}$ and $\tau\in\mathcal{D}_\omega$,

$$\mathrm{J}^\theta_\omega(\mu_\omega)=0; \tag{4.16}$$

$$\mathrm{J}^\omega_\omega(\mu)=\mathrm{J}_\omega(\mu); \tag{4.17}$$

$$\mathrm{J}^{\mathrm{dd^c}\psi}_\omega(\mu)=\int\psi(\mu-\mu_\omega); \tag{4.18}$$

$$\mathrm{J}^\theta_{\omega_\tau}(\mu)=\mathrm{J}^\theta_\omega(\mu)+c \tag{4.19}$$

with $c\in\mathbb{R}$ uniquely determined by (4.16), i.e. $c=\mathrm{J}^\theta_{\omega_\tau}(\mu_\omega)=-\mathrm{J}^\theta_\omega(\mu_{\omega_\tau})$.

4.2. Hölder continuity of the twisted energy. The following estimates will be the key ingredients for the continuity of coercivity thresholds (see Theorem 5.5 below).

THEOREM 4.6. *Pick $\omega,\omega'\in\mathcal{Z}_+$ with $\delta:=\mathrm{d_T}(\omega,\omega')\le 1$ and $[\omega']\in\mathrm{Pos}(X)$. For all $\theta,\theta'\in\mathcal{Z}_\omega$ and $\mu\in\mathcal{M}^1$, we then have*

$$\left|\mathrm{J}^\theta_\omega(\mu)-\mathrm{J}^{\theta'}_{\omega'}(\mu)\right|\lesssim(\delta^a\|\theta\|_\omega+\|\theta-\theta'\|_\omega)\,\mathrm{J}^+_\omega(\mu) \tag{4.20}$$

and

$$\mathrm{J}^+_\omega(\mu)=(1+O(\delta^\alpha))\mathrm{J}^+_{\omega'}(\mu). \tag{4.21}$$

LEMMA 4.7. *Assume $\omega\le\omega'\le e^\delta\omega$ with $\delta\in[0,2]$. Pick $\varphi\in\mathcal{D}_\omega\subset\mathcal{D}_{\omega'}$, and set $\mu:=\mathrm{MA}_\omega(\varphi)$, $\mu':=\mathrm{MA}_{\omega'}(\varphi)$. Then:*

$$|\nabla_\theta\,\mathrm{E}_\omega(\varphi)-\nabla_\theta\,\mathrm{E}_{\omega'}(\varphi)|\lesssim\delta\mathrm{J}^+_\omega(\mu)\|\theta\|_\omega; \tag{4.22}$$

$$\max\{\mathrm{J}^+_{\omega'}(\mu),\mathrm{J}^+_{\omega'}(\mu')\}\lesssim\mathrm{J}^+_\omega(\mu); \tag{4.23}$$

$$\mathrm{J}_{\omega'}(\mu,\mu')\lesssim\delta\mathrm{J}^+_\omega(\mu). \tag{4.24}$$

PROOF. All three estimates are invariant under translation of φ by a constant, and we shall rely on a different normalization for each of them. We first normalize φ by $\int \varphi\, \mu_\omega = 0$. By homogeneity, we may further assume $\|\theta\|_\omega = 1$. Since

$$\mathrm{E}^\theta_\omega(\varphi) - \mathrm{E}^\theta_{\omega'}(\varphi) = V_\omega^{-1}(\theta, 0) \cdot (\omega, \varphi)^n - V_{\omega'}^{-1}(\theta, 0) \cdot (\omega', \varphi)^n$$

with $V_{\omega'}/V_\omega = 1 + O(\delta)$, Theorem 3.6 yields

$$\left|\mathrm{E}^\theta_\omega(\varphi) - \mathrm{E}^\theta_{\omega'}(\varphi)\right| \lesssim \|\omega' - \omega\|_\omega \mathrm{J}^+_\omega(\mu) \lesssim \delta \mathrm{J}^+_\omega(\mu).$$

We similarly get $|\mathrm{E}_\omega(\varphi) - \mathrm{E}_{\omega'}(\varphi)| \lesssim \delta \mathrm{J}^+_\omega(\mu)$, and (4.22) follows, using the trivial estimate

$$V^\theta_{\omega'} = V^\theta_\omega + O(\delta).$$

Next, we normalize φ by $\sup \varphi = 0$. By Lemma 1.20, we then have

$$\mathrm{J}_{\omega'}(\mu') \approx \mathrm{J}_{\omega'}(\varphi) = \int \varphi\, \mu_{\omega'} - \mathrm{E}_{\omega'}(\varphi) \leq -\mathrm{E}_{\omega'}(\varphi) \lesssim -\mathrm{E}_\omega(\varphi) \leq \mathrm{J}^+_\omega(\mu).$$

On the other hand, Corollary 3.5 yields $\mathrm{J}^+_{\omega'}(\mu) \lesssim \mathrm{J}^+_\omega(\mu)$, and (4.23) follows.

Finally, we normalize φ by $\int \varphi\, \mu_{\omega'} = 0$. Pick a maximizing sequence (ψ_i) in $\mathcal{D}_{\omega'}$ for μ, also normalized by $\int \psi_i\, \mu_{\omega'} = 0$, and set $\mu_i := \mathrm{MA}_{\omega'}(\psi_i)$. By Theorem 2.22, $\mathrm{J}_{\omega'}(\mu, \mu')$ is the limit of

$$\mathrm{J}_{\omega'}(\mu_i, \mu') = \mathrm{J}_{\omega'}(\psi_i, \varphi) \approx \int (\psi_i - \varphi)(\mu' - \mu_i) = \int (\psi_i - \varphi)(\mu' - \mu) + o(1),$$

where the last two points hold by (1.30) and (2.25), respectively. Now

$$\begin{aligned}\int (\psi_i - \varphi)(\mu' - \mu) = &\left[V_{\omega'}^{-1}(0, \psi_i) \cdot (\omega', \varphi)^n - V_\omega^{-1}(0, \psi_i) \cdot (\omega, \varphi)^n\right] \\ &- \left[V_{\omega'}^{-1}(0, \varphi) \cdot (\omega', \varphi)^n - V_\omega^{-1}(0, \varphi) \cdot (\omega, \varphi)^n\right],\end{aligned}$$

where

$$\left|V_{\omega'}^{-1}(0, \psi_i) \cdot (\omega', \varphi)^n - V_\omega^{-1}(0, \psi_i) \cdot (\omega, \varphi)^n\right| \lesssim \|\omega' - \omega\|_{\omega'} \max\{\mathrm{J}^+_{\omega'}(\varphi), \mathrm{J}^+_{\omega'}(\psi_i)\}$$

and

$$\left|V_{\omega'}^{-1}(0, \varphi) \cdot (\omega', \varphi)^n - V_\omega^{-1}(0, \varphi) \cdot (\omega, \varphi)^n\right| \lesssim \|\omega' - \omega\|_{\omega'} \mathrm{J}^+_{\omega'}(\varphi),$$

by Theorem 3.6. Since $\|\omega' - \omega\|_{\omega'} \lesssim \delta$, $\mathrm{J}_{\omega'}(\varphi) \approx \mathrm{J}_{\omega'}(\mu')$ and $\mathrm{J}_{\omega'}(\psi_i) \approx \mathrm{J}_{\omega'}(\mu_i) \to \mathrm{J}_{\omega'}(\mu)$, this proves (4.24), thanks to (4.23). □

PROOF OF THEOREM 4.6. Note first that $\omega'' := e^{-\delta}\omega \leq \omega' \leq e^\delta \omega$, and hence

$$\omega'' \leq \omega' \leq e^{2\delta}\omega'', \quad \omega'' \leq \omega \leq e^\delta \omega''.$$

Arguing successively with ω'', ω', and with ω'', ω, and relying on Corollary 3.5, it is thus enough to prove the result when $\omega \leq \omega' \leq e^\delta \omega$, which we henceforth assume. By density of the image of $\mathrm{MA}_\omega\colon \mathcal{D}_\omega \to \mathcal{M}^1$ and strong continuity of J^θ_ω and $\mathrm{J}^\theta_{\omega'}$ (recall that the strong topology of $\mathcal{M}^1$ is independent of ω), we may assume $\mu = \mathrm{MA}_\omega(\varphi)$ with $\varphi \in \mathcal{D}_\omega \subset \mathcal{D}_{\omega'}$. As in Lemma 4.7, set $\mu' := \mathrm{MA}_{\omega'}(\varphi)$. By (4.11), we have

$$\mathrm{J}^\theta_\omega(\mu) = \nabla_\theta \mathrm{E}_\omega(\varphi), \quad \mathrm{J}^\theta_{\omega'}(\mu') = \nabla_\theta \mathrm{E}_{\omega'}(\varphi),$$

and (4.22) thus yields

$$\left|\mathrm{J}^\theta_\omega(\mu) - \mathrm{J}^\theta_{\omega'}(\mu')\right| \lesssim \delta \mathrm{J}^+_\omega(\mu) \|\theta\|_\omega. \tag{4.25}$$

On the other hand, (4.12), implies

$$\left|\mathrm{J}^{\theta}_{\omega'}(\mu)-\mathrm{J}^{\theta}_{\omega'}(\mu')\right|\lesssim\delta^{\alpha}\mathrm{J}^{+}_{\omega}(\mu)\|\theta\|_{\omega},\tag{4.26}$$

thanks to (4.23) and (4.24). Finally, (4.13) yields

$$\left|\mathrm{J}^{\theta}_{\omega'}(\mu)-\mathrm{J}^{\theta'}_{\omega'}(\mu)\right|=\left|\mathrm{J}^{\theta-\theta'}_{\omega'}(\mu)\right|\lesssim\mathrm{J}^{+}_{\omega'}(\mu)\|\theta-\theta'\|_{\omega'},\tag{4.27}$$

and summing up (4.25), (4.26) and (4.27) yields (4.20). Applying the latter estimate with $\theta:=\omega$, $\theta':=\omega'$, we get

$$|\mathrm{J}_{\omega}(\mu)-\mathrm{J}_{\omega'}(\mu)|\lesssim\delta^{\alpha}\mathrm{J}^{+}_{\omega}(\mu),\tag{4.28}$$

in view of (4.17). Since we also have $T_{\omega'}=(1+O(\delta))T_{\omega}$ (see (1.23)), (4.21) follows. □

4.3. Differentiability of the energy.

THEOREM 4.8. *For any $\theta\in\mathcal{Z}$ and $\mu\in\mathcal{M}^1$, we have*

$$\frac{d}{dt}\Big|_{t=0}\mathrm{J}_{\omega+t\theta}(\mu)=\mathrm{J}^{\theta}_{\omega}(\mu).$$

If we do not assume $\theta\in\mathcal{Z}_{\omega}$, the condition $\omega+t\theta\geq 0$ will fail in general, but one can still make sense of $\mathrm{J}_{\omega+t\theta}(\mu)$, see Remark 2.4.

LEMMA 4.9. *Pick $\theta\in\mathcal{Z}_{\omega}$ and $\varphi\in\mathcal{D}_{\omega}$ normalized by $\int\varphi\,\mu_{\omega}=0$. Then*

$$|\mathrm{E}_{\omega}(\varphi)|\lesssim\mathrm{J}^{+}_{\omega}(\varphi),\quad\left|\int\varphi\,\mathrm{MA}_{\omega}(\varphi)\right|\lesssim\mathrm{J}^{+}_{\omega}(\varphi),\quad\left|\mathrm{E}^{\theta}_{\omega}(\varphi)\right|\lesssim\|\theta\|_{\omega}\mathrm{J}^{+}_{\omega}(\varphi).$$

PROOF. We have

$$\mathrm{E}_{\omega}(\varphi)=\frac{(\omega,\varphi)^{n+1}}{(n+1)V_{\omega}},\quad\int\varphi\,\mathrm{MA}_{\omega}(\varphi)=\frac{(0,\varphi)\cdot(\omega,\varphi)^{n}}{V_{\omega}},\quad\mathrm{E}^{\theta}_{\omega}(\varphi)=\frac{(\theta,0)\cdot(\omega,\varphi)^{n}}{V_{\omega}},$$

and the estimates thus follow from (3.10) (and homogeneity in θ). □

LEMMA 4.10. *Pick $\theta\in\mathcal{Z}_{\omega}$ such that $\|\theta\|_{\omega}<1$ and $[\omega+\theta]\in\mathrm{Pos}(X)$. For all $\mu\in\mathcal{M}^1$ we then have*

$$\mathrm{J}_{\omega+\theta}(\mu)\geq\mathrm{J}_{\omega}(\mu)+\mathrm{J}^{\theta}_{\omega}(\mu)-O(\mathrm{J}^{+}_{\omega}(\mu)\|\theta\|^{2}_{\omega}).$$

Recall that, in this paper, the implicit constant in O only depends on n.

PROOF. Set $J:=\mathrm{J}^{+}_{\omega}(\mu)$ and $\varepsilon:=\|\theta\|_{\omega}$. If $\varepsilon=0$, then $\theta=0$ and the result is clear. We may thus assume $\varepsilon>0$ and write $\theta=\varepsilon\tilde{\theta}$ with $\|\tilde{\theta}\|_{\omega}=1$ and $\varepsilon\in(0,1)$.

By density of the image of $\mathrm{MA}_{\omega}\colon\mathcal{D}_{\omega}\to\mathcal{M}^1$, it is enough to prove the result when $\mu=\mathrm{MA}_{\omega}(\varphi)$ with $\varphi\in\mathcal{D}_{\omega}$, which we normalize by $\int\varphi\,\mu_{\omega}=0$. By (2.5), (4.11) and (4.2), we then have

$$\mathrm{J}_{\omega}(\mu)=\mathrm{E}_{\omega}(\varphi)-\int\varphi\,\mu,\quad\mathrm{J}^{\theta}_{\omega}(\mu)=\mathrm{E}^{\theta}_{\omega}(\varphi)-V^{\theta}_{\omega}\,\mathrm{E}_{\omega}(\varphi).\tag{4.29}$$

Note that

$$(1-\varepsilon)\omega\leq\omega+\theta\leq(1+\varepsilon)\omega\leq(1-\varepsilon)^{-1}\omega,\tag{4.30}$$

and hence

$$(1-\varepsilon)\varphi\in\mathcal{D}_{(1-\varepsilon)\omega}\subset\mathcal{D}_{\omega+\theta}.$$

By (2.2), this yields

$$\mathrm{J}_{\omega+\theta}(\mu) \geq \mathrm{E}_{\omega+\theta}\left((1-\varepsilon)\varphi\right) - (1-\varepsilon)\int \varphi\,\mu. \tag{4.31}$$

Further,

$$\begin{aligned}(n+1)V_{\omega+\theta}\,\mathrm{E}_{\omega+\theta}\left((1-\varepsilon)\varphi\right) &= (\omega+\theta,(1-\varepsilon)\varphi)^{n+1} = \left((\omega,\varphi)+(\theta,-\varepsilon\varphi)\right)^{n+1}\\ &= (\omega,\varphi)^{n+1} + (n+1)\left[(\theta,0)\cdot(\omega,\varphi)^n - \varepsilon(0,\varphi)\cdot(\omega,\varphi)^n\right] + \varepsilon^2 a(\varepsilon)\\ &= (n+1)V_\omega\left[\mathrm{E}_\omega(\varphi)+\mathrm{E}^\theta_\omega(\varphi) - \varepsilon\int\varphi\,\mu\right] + \varepsilon^2 a(\varepsilon)\end{aligned}$$

with

$$a(\varepsilon) := \sum_{j=2}^{n+1}\binom{n+1}{j}\varepsilon^{j-2}(\tilde\theta,-\varphi)^j\cdot(\omega,\varphi)^{n+1-j}.$$

Since $\|\tilde\theta\|_\omega = 1$, (3.10) yields $|a(\varepsilon)| \lesssim V_\omega J$. Combining this with

$$V_{\omega+\theta} = [\omega+\theta]^n = [\omega]^n + n[\theta]\cdot[\omega]^{n-1} + O(\varepsilon^2[\omega]^n) = V_\omega\left(1+V^\theta_\omega + O(\varepsilon^2)\right),$$

and

$$\left|V^\theta_\omega\right| \lesssim \varepsilon, \quad |\mathrm{E}_\omega(\varphi)| \lesssim J, \quad \left|\int\varphi\,\mu\right| \leq J, \quad \left|\mathrm{E}^\theta_\omega(\varphi)\right| \lesssim \varepsilon J,$$

(see Lemma 4.9), we infer

$$\begin{aligned}\mathrm{E}_{\omega+\theta}((1-\varepsilon)\varphi) &\geq \left(1-V^\theta_\omega+O(\varepsilon^2)\right)\left(\mathrm{E}_\omega(\varphi)+\mathrm{E}^\theta(\varphi)-\varepsilon\int\varphi\,\mu - O(\varepsilon^2 J)\right)\\ &= \mathrm{E}_\omega(\varphi)+\mathrm{E}^\theta_\omega(\varphi) - V^\theta_\omega\,\mathrm{E}_\omega(\varphi) - \varepsilon\int\varphi\,\mu - O(\varepsilon^2 J).\end{aligned}$$

Injecting this into (4.31) and using (4.29), we get

$$\mathrm{J}_{\omega+\theta}(\mu) \geq \mathrm{J}_\omega(\mu) + \mathrm{J}^\theta_\omega(\mu) - O(\varepsilon^2 J),$$

which completes the proof. □

Proof of Theorem 4.8. Assume first $\theta \in \mathcal{Z}_\omega$. Set $J := \mathrm{J}^+_\omega(\mu)$. By Lemma 4.10 we then have

$$\mathrm{J}_{\omega+t\theta}(\mu) \geq \mathrm{J}_\omega(\mu) + t\mathrm{J}^\theta_\omega(\mu) - O(t^2\|\theta\|_\omega J).$$

For $|t| \ll 1$, we also have $\delta(\omega+t\theta,\omega) \leq 1$, and hence $\mathrm{J}^+_{\omega+t\theta}(\mu) \lesssim J$, by Corollary 3.5. Reversing the roles of ω and $\omega+t\theta$, Lemma 4.10 thus yields

$$\mathrm{J}_\omega(\mu) \geq \mathrm{J}_{\omega+t\theta}(\mu) - t\mathrm{J}^\theta_{\omega+t\theta}(\mu) - O(t^2\|\theta\|_\omega J).$$

By Theorem 4.6, $\mathrm{J}^\theta_{\omega+t\theta}(\mu) \to \mathrm{J}^\theta_\omega(\mu)$ as $t \to 0$, and we conclude, as desired,

$$\mathrm{J}_{\omega+t\theta}(\mu) = \mathrm{J}_\omega(\mu + t\mathrm{J}^\theta_\omega(\mu) + o(t).$$

as $t \to 0$.

In the general case, pick $\tau \in \mathcal{D}_\omega$ such that $\theta \in \mathcal{Z}_{\omega_\tau}$ (see Proposition 1.12). Then (2.9) yields

$$\mathrm{J}_{\omega+t\theta}(\mu) = \mathrm{J}_{\omega_\tau+t\theta}(\mu) - \int\tau\,\mu + \mathrm{E}_{\omega+t\theta}(\tau),$$

see Remark 2.4. We infer

$$\left.\frac{d}{dt}\right|_{t=0}\mathrm{J}_{\omega+t\theta}(\mu) = \mathrm{J}^\theta_{\omega_\tau}(\mu) + \nabla_\theta\,\mathrm{E}_\omega(\tau) = \mathrm{J}^\theta_{\omega_\tau}(\mu) + \mathrm{J}^\theta_\omega(\mu_{\omega_\tau}) = \mathrm{J}^\theta_\omega(\mu),$$

where the first equality follows from the first step of the proof and (4.2), the second one from (4.11), and the third from (4.19). □

5. Coercivity thresholds and free energy

As in §4, we assume that the orthogonality and submean value properties hold. We will establish a general continuity result for coercivity thresholds, and apply this to the free energy, which induces the Mabuchi K-energy on potentials.

5.1. Continuity of coercivity thresholds. Fix for the moment $\omega \in \mathcal{Z}_+$ with $[\omega] \in \mathrm{Pos}(X)$, and consider an arbitrary functional $F\colon \mathcal{M}^1 \to \mathbb{R} \cup \{+\infty\}$.

DEFINITION 5.1. *We define the* coercivity threshold *of F as*

$$\sigma_\omega(F) := \sup\{\sigma \in \mathbb{R} \mid F \geq \sigma \mathrm{J}_\omega + A \text{ for some } A \in \mathbb{R}\} \in [-\infty, +\infty]. \tag{5.1}$$

We say that F is coercive *if $\sigma_\omega(F) > 0$, i.e. $F \geq \sigma \mathrm{J}_\omega + A$ for some $\sigma > 0$ and $A \in \mathbb{R}$.*

By Corollary 3.5 (ii), the condition that F is coercive (resp. $\sigma_\omega(F) = \pm\infty$) is independent of ω.

Recall from (2.22) that the energy of any $\mu \in \mathcal{M}^1$ coincides with the quasi-distance to the base point μ_ω, i.e. $\mathrm{J}_\omega(\mu) = \delta_\omega(\mu, \mu_\omega)$. As a result, the coercivity threshold measures the linear growth of F with respect to quasi-metric δ_ω. As we next show, any other base point can be used in place of μ_ω (something that could fail for a general quasi-metric).

LEMMA 5.2. *For any $F\colon \mathcal{M}^1 \to \mathbb{R} \cup \{+\infty\}$ and $\nu \in \mathcal{M}^1$ we have*

$$\sigma_\omega(F) = \sup\{\sigma \in \mathbb{R} \mid F \geq \sigma \delta_\omega(\cdot, \nu) + A \text{ for some } A \in \mathbb{R}\}. \tag{5.2}$$

In particular, $\sigma_\omega(F)$ only depends on $[\omega] \in \mathrm{Pos}(X)$.

PROOF. For any $\mu \in \mathcal{M}^1$, (2.24) yields

$$|\delta_\omega(\mu, \nu) - \mathrm{J}_\omega(\mu)| \lesssim \mathrm{J}_\omega(\nu)^\alpha \max\{\mathrm{J}_\omega(\mu), \mathrm{J}_\omega(\nu)\}^{1-\alpha}.$$

For any $\varepsilon > 0$, we can thus find $C_\varepsilon > 0$ (depending on ν) such that

$$(1-\varepsilon)\mathrm{J}_\omega(\mu) - C_\varepsilon \leq \delta_\omega(\mu, \nu) \leq (1+\varepsilon)\mathrm{J}_\omega(\mu) + C_\varepsilon$$

for all $\mu \in \mathcal{M}^1$. This implies (5.2), and the last point follows, since the Dirichlet quasi-metric δ_ω only depends on $[\omega]$ (see Theorem 2.25). □

DEFINITION 5.3. *For each $\theta \in \mathcal{Z}$, we introduce the* twisted coercivity threshold

$$\sigma_\omega^\theta(F) := \sigma_\omega(F + \mathrm{J}_\omega^\theta). \tag{5.3}$$

We refer to §5.2 below for a discussion of the concrete cases we have in mind.

LEMMA 5.4. *The following holds:*

(i) *$\sigma_\omega^0(F) = \sigma_\omega(F)$, and $\sigma_\omega^{\theta+t\omega}(F) = \sigma_\omega^\theta(F) + t$ for all $\theta \in \mathcal{Z}$ and $t \in \mathbb{R}$;*
(ii) *we have $\sigma_\omega(F) \in \mathbb{R}$ (resp. $\sigma_\omega(F) = \pm\infty$) iff $\sigma_\omega^\theta(F) \in \mathbb{R}$ (resp. $\sigma_\omega^\theta(F) = \pm\infty$) for all $\theta \in \mathcal{Z}$.*

PROOF. The first point is a direct consequence of (4.17). For any $\theta \in \mathcal{Z}$, (4.15) yields a constant $C > 0$ such that $|\mathrm{J}_\omega^\theta| \leq C(\mathrm{J}_\omega + 1)$. This implies

$$\sigma_\omega(F) - C \leq \sigma_\omega^\theta(F) \leq \sigma_\omega(F) + C,$$

and (ii) follows. □

We can now state the main result of this section.

THEOREM 5.5. *For any functional* $F\colon \mathcal{M}^1 \to \mathbb{R}\cup\{+\infty\}$, *the twisted coercivity threshold* $\sigma^\theta_\omega(F)$ *is a continuous function of* $([\omega],[\theta]) \in \mathrm{Pos}(X)\times \mathrm{H}_{\mathrm{BC}}(X)$.

We emphasize that F here is a fixed functional, independent of ω and θ. Continuity is understood with respect to the finest vector space topology of $\mathrm{H}_{\mathrm{BC}}(X)$, i.e. for $[\omega]$ and $[\theta]$ constrained to any given finite dimensional subspace.

LEMMA 5.6. *There exists* $\delta_n > 0$ *only depending on* n *such that, for all* $\omega,\omega' \in \mathcal{Z}_+$ *with* $\delta := \mathrm{d}_{\mathrm{T}}(\omega,\omega') \le \delta_n$ *and all* $\theta,\theta' \in \mathcal{Z}_\omega$, *we have*

$$\sigma^{\theta'}_{\omega'}(F) \ge (1+O(\delta^\alpha))\left[\sigma^\theta_\omega(F) + O(\delta^\alpha\|\theta\|_\omega + \|\theta-\theta'\|_\omega)\right]. \tag{5.4}$$

PROOF. Since $\mathrm{J}^+_\omega = \mathrm{J}_\omega + T_\omega$, we can replace J_ω with J^+_ω in (5.1), and hence

$$\sigma^\theta_\omega(F) = \sup\left\{\sigma\in\mathbb{R} \mid F + \mathrm{J}^\theta_\omega \ge \sigma \mathrm{J}^+_\omega + A \text{ for some } A\in\mathbb{R}\right\}. \tag{5.5}$$

Pick $\sigma, A\in\mathbb{R}$ such that $F+\mathrm{J}^\theta_\omega \ge \sigma\mathrm{J}^+_\omega + A$ on $\mathcal{M}^1$. By (4.20) we get

$$\begin{aligned} F + \mathrm{J}^{\theta'}_{\omega'} &\ge F + \mathrm{J}^\theta_\omega + O(\delta^\alpha\|\theta\|_\omega + \|\theta-\theta'\|_\omega)\mathrm{J}^+_\omega \\ &\ge \sigma\mathrm{J}^+_\omega + A + O(\delta^\alpha\|\theta\|_\omega + \|\theta-\theta'\|_\omega)\mathrm{J}^+_\omega \\ &\ge (1+O(\delta^\alpha))\left[\sigma + O(\delta^\alpha\|\theta\|_\omega + \|\theta-\theta'\|_\omega\right]\mathrm{J}^+_{\omega'} + A, \end{aligned}$$

using (4.21), and hence $\sigma^{\theta'}_{\omega'}(F) \ge (1+O(\delta^\alpha))\left[\sigma + O(\delta^\alpha\|\theta\|_\omega + \|\theta-\theta'\|_\omega\right]$. Choosing $\delta_n > 0$ such that $1+O(\delta^\alpha)\ge 0$ for $\delta\le\delta_n$ and taking the supremum over σ yields (5.4). □

PROOF OF THEOREM 5.5. We first show that $\sigma^\theta_\omega(F)$ only depends on the classes $[\omega]\in\mathrm{Pos}(X)$ and $[\theta]\in\mathrm{H}_{\mathrm{BC}}(X)$. For each $\tau\in\mathcal{D}_\omega$, we have

$$\sigma^\theta_{\omega_\tau}(F) = \sigma_{\omega_\tau}(F + \mathrm{J}^\theta_{\omega_\tau}) = \sigma_{\omega_\tau}(F+\mathrm{J}^\theta_\omega) = \sigma_\omega(F+\mathrm{J}^\theta_\omega) = \sigma^\theta_\omega(F),$$

where the second equality follows from (4.19), and the third from Lemma 5.2. This proves that $\sigma^\theta_\omega(F)$ only depends on $[\omega]$. On the other hand, for any $\psi\in\mathcal{D}$, (4.18) yields

$$F + \mathrm{J}^\theta_\omega - C \le F + \mathrm{J}^{\theta+\mathrm{dd^c}\psi}_\omega \le F + \mathrm{J}^\theta_\omega + C$$

with $C := 2\sup|\psi|$. This implies $\sigma^\theta_\omega(F) = \sigma^{\theta+\mathrm{dd^c}\psi}_\omega(F)$, which thus only depends on $[\theta]$.

As noted above, Corollary 3.5 (ii) shows that the condition $\sigma_\omega(F) = \pm\infty$ is independent of ω, and in that case the result trivially holds, since all twisted thresholds are then equal to $\pm\infty$, by Lemma 5.4 (ii).

Assume now that the twisted thresholds are finite valued. Pick $\omega_0\in\mathcal{Z}_+$ with $[\omega_0]\in\mathrm{Pos}(X)$, $\theta_0\in\mathcal{Z}$. Choose a finite dimensional subspace $W\subset\mathrm{H}_{\mathrm{BC}}(X)$ containing $[\omega_0]$, $[\theta_0]$, and a finite dimensional subspace $V\subset\mathcal{Z}$ containing θ_0, and whose image in $\mathrm{H}_{\mathrm{BC}}(X)$ contains W. Since $\sigma^{\theta_0}_{\omega_0}(F)$ only depends on $[\omega_0]$, we can assume without loss $V\subset\mathcal{Z}_{\omega_0}$ (see Proposition 1.12). After enlarging V, we may further assume that it contains ω_0. For each $\omega,\theta\in V\subset\mathcal{Z}_{\omega_0}$ with ω close enough to ω_0, (5.5) yields

$$|\sigma^\theta_\omega(F) - \sigma^{\theta_0}_{\omega_0}(F)| \lesssim \delta^\alpha\left[\sigma^{\theta_0}_{\omega_0}(F) + O(\delta^\alpha\|\theta_0\|_{\omega_0} + \|\theta-\theta_0\|_{\omega_0})\right]$$

with $\delta = \mathrm{d}_{\mathrm{T}}(\omega,\omega_0)$. By Lemma 1.14, this shows that the restriction of $(\omega,\theta)\mapsto\sigma^\theta_\omega(F)$ to V is continuous at (ω_0,θ_0), and the result follows, since the image of V contains W. □

5.2. Free energy vs. Mabuchi K-energy.

5.2.1. *The Kähler case.* Consider first a compact connected Kähler manifold X. Any smooth metric ρ on K_X induces a volume form μ_ρ (normalized to mass one), and hence a *(relative) entropy* functional

$$\mathrm{Ent}_\rho\colon \mathcal{M} \to \mathbb{R} \cup \{+\infty\}, \tag{5.6}$$

such that

$$\mathrm{Ent}_\rho(\mu) := \int \log\left(\frac{\mu}{\mu_\rho}\right)\mu$$

if μ is absolutely continuous with respect to μ_ρ, and $\mathrm{Ent}_\rho(\mu) = +\infty$ otherwise. Note that

$$\mathrm{Ent}_{\rho'}(\mu) = \mathrm{Ent}_\rho(\mu) + \int (\rho - \rho')\,\mu \tag{5.7}$$

for any other metric ρ' on K_X. As is well-known, the relative entropy can also be written as the Legendre transform

$$\mathrm{Ent}_\rho(\mu) := \sup_{f \in \mathcal{D}} \left(\int f\,\mu - \log \int e^f \mu_\rho\right)$$

which shows that (5.6) is convex and lsc. In particular, the restriction $\mathrm{Ent}_\rho\colon \mathcal{M}^1 \to \mathbb{R} \cup \{+\infty\}$ is lsc in the strong topology. While it is not continuous, we nevertheless have (see [**BDL17**, Theorem 4.7]):

LEMMA 5.7. *For each $\mu \in \mathcal{M}^1$, there exists a sequence (φ_i) in $\mathcal{D}_\omega$ such that $\mu_i := \mathrm{MA}_\omega(\varphi_i)$ satisfies $\mu_i \to \mu$ strongly in $\mathcal{M}^1$ and* $\mathrm{Ent}_\rho(\mu_i) \to \mathrm{Ent}_\rho(\mu)$.

Extending the terminology of [**Berm13**], we introduce:

DEFINITION 5.8. *The* free energy $\mathrm{F}_\omega\colon \mathcal{M}^1 \to \mathbb{R} \cup \{+\infty\}$ *is defined by setting*

$$\mathrm{F}_\omega(\mu) := \mathrm{Ent}_\rho(\mu) - \mathrm{Ent}_\rho(\mu_\omega) + \mathrm{J}_\omega^{\theta_\rho}(\mu), \tag{5.8}$$

where $\theta_\rho \in \mathcal{Z}$ denotes the curvature of ρ.

As the notation suggests, F_ω is independent of the choice of ρ; this follows from (5.7) combined with (4.18). Furthermore, (4.16) and (4.19) show that F_ω only depends on the Kähler class $[\omega]$, up to an additive constant uniquely determined by the normalization $\mathrm{F}_\omega(\mu_\omega) = 0$.

The *raison d'être* of the free energy is that its composition with the Monge–Ampère operator coincides with the *Mabuchi K-energy* $\mathrm{M}_\omega\colon \mathcal{D}_\omega \to \mathbb{R}$, i.e. we have

$$\mathrm{F}_\omega(\mathrm{MA}_\omega(\varphi)) = \mathrm{M}_\omega(\varphi) \tag{5.9}$$

for all $\varphi \in \mathcal{D}_\omega$. Indeed, in view of (4.11), (5.9) is equivalent to the well-known *Chen–Tian formula* for the K-energy, which can be written as

$$\mathrm{M}_\omega(\varphi) = \mathrm{Ent}_\rho(\mathrm{MA}_\omega(\varphi)) + \nabla_{\theta_\rho}\mathrm{E}_\omega(\varphi) + c \tag{5.10}$$

in the present formalism, for a constant $c \in \mathbb{R}$ determined by the normalization $\mathrm{M}_\omega^\theta(0) = 0$, i.e. $c = -\mathrm{Ent}_\rho(\mu_\omega)$.

DEFINITION 5.9. *The* coercivity threshold *of (X, ω) is defined as $\sigma(X,\omega) := \sigma_\omega(\mathrm{F}_\omega)$.*

THEOREM 5.10. *The coercivity threshold of (X, ω) is a continuous function of the Kähler class $[\omega]$, and it satisfies*

$$\sigma(X,\omega) = \sup\{\sigma \in \mathbb{R} \mid \mathrm{M}_\omega \geq \sigma \mathrm{J}_\omega + A \text{ on } \mathcal{D}_\omega \text{ for some } A \in \mathbb{R}\}.$$

By [**CC21**], $\sigma(X,\omega)>0$ iff there exists a unique constant scalar curvature Kähler (cscK) metric in $[\omega]$. In particular, we recover the fact, originally proved in [**LS94**], that the set of Kähler classes of X that contain a unique cscK metric is open.

PROOF. Pick a smooth metric ρ on K_X. By (5.8), we have $\sigma(X,\omega)=\sigma_\omega^{\theta_\rho}(\mathrm{Ent}_\rho)$, and the first point thus follows from Theorem 5.5. As to the second point, it is a simple consequence of (5.9) and Lemma 5.7. □

REMARK 5.11. *Given $\theta\in\mathcal{Z}$, one can more generally consider the θ-twisted free energy $\mathrm{F}_\omega^\theta:=\mathrm{F}_\omega+\mathrm{J}_\omega^\theta$, whose composition with the Monge–Ampère operator coincides with the θ-twisted K-energy $\mathrm{M}_\omega^\theta=\mathrm{M}_\omega+\nabla_\theta\mathrm{E}_\omega$ (see* [**BDL17, CC21**]*). Again, Theorem 5.5 shows that the coercivity threshold*

$$\sigma^\theta(X,\omega):=\sigma_\omega(\mathrm{F}_\omega^\theta)=\sigma_\omega^\theta(\mathrm{Ent}_\rho)$$

is a continuous function of $[\omega]$ and $[\theta]$, while Lemma 5.7 shows that

$$\sigma^\theta(X,\omega):=\sup\left\{\sigma\in\mathbb{R}\mid \mathrm{M}_\omega^\theta\geq\sigma\mathrm{J}_\omega+A \text{ on } \mathcal{D}_\omega \text{ for some } A\in\mathbb{R}\right\}$$

as considered for instance in [**SD20**]. *Combining this with* [**CC21**], *and assuming $\theta>0$, this implies that the set of Kähler classes of a compact Kähler manifold that contain a θ-twisted cscK metric is open—something that can also be directly obtained along the lines of* [**LS94**].

5.2.2. *The non-Archimedean case.* Next we consider a smooth, irreducible projective Berkovich space X over a non-Archimedean field k of characteristic 0. Pick a PL metric ρ on K_X, and define the associated *non-Archimedean entropy* $\mathrm{Ent}_\rho\colon\mathcal{M}\to\mathbb{R}\cup\{+\infty\}$ by setting

$$\mathrm{Ent}_\rho(\mu):=\int(\mathrm{A}_X-\rho)\,\mu.$$

Here A_X denotes the Temkin metric (see [**Tem16**] and also [**Ste19**]) on K_X, and $\mathrm{A}_X-\rho\colon X\to\mathbb{R}\cup\{+\infty\}$ is the corresponding lsc function, using additive notation for metrics (see [**BoJ17**, §5.7]). Again, Ent_ρ is convex and lsc, and (5.7) holds for any other choice of PL metric ρ' on K_X. See also [**Ino22**] for a related notion in the trivially valued case.

As above, one can then define the (non-Archimedean) *free energy* $\mathrm{F}_\omega\colon\mathcal{M}^1\to\mathbb{R}\cup\{+\infty\}$ by (5.8), with $\theta_\rho\in\mathcal{Z}$ the curvature of ρ. Its composition with the Monge–Ampère operator coincides with the (non-Archimedean) *Mabuchi K-energy* $\mathrm{M}_\omega\colon\mathcal{D}_\omega\to\mathbb{R}$, defined by (5.10).

However, the major difference in the non-Archimedean case is that the analogue of Lemma 5.7 is only a conjecture (compare [**BoJ18a**, Conjecture 2.5] and [**C.Li22**, Conjecture 4.4]). Explicitly:

CONJECTURE 5.12. *For each $\mu\in\mathcal{M}^1$, there exists a sequence (φ_i) of ω-psh PL functions such that $\mu_i:=\mathrm{MA}_\omega(\varphi_i)$ satisfies $\mu_i\to\mu$ in $\mathcal{M}^1$ and $\mathrm{Ent}_\rho(\mu_i)\to\mathrm{Ent}_\rho(\mu)$.*

As a consequence, the coercivity threshold $\sigma(X,\omega):=\sigma_\omega(\mathrm{F}_\omega)$ only satisfies

$$\sigma(X,\omega)\leq\sup\{\sigma\in\mathbb{R}\mid \mathrm{M}_\omega\geq\sigma\mathrm{J}_\omega+A \text{ on } \mathcal{D}_\omega \text{ for some } A\in\mathbb{R}\}, \tag{5.11}$$

and equality holds if Conjecture 5.12 is valid.

These definitions are especially relevant when k is trivially valued and ω lies in $\mathrm{Amp}(X) \hookrightarrow \mathcal{Z}_+$ (see (1.6)). Indeed, the free energy $\mathrm{F}_\omega(\mu)$ then coincides with the invariant $\beta_\omega(\mu)$ introduced and studied in [**BoJ23**]; see also [**DL23, Liu23**]. By homogeneity with respect to the action of $\mathbb{R}_{>0}$, $\sigma(X,\omega) = \sigma_\omega(\mathrm{F}_\omega)$ is further equal to the *divisorial stability threshold* $\sigma_{\mathrm{div}}(X,\omega)$, which is positive iff (X,ω) is *divisorially stable.*

On the other hand, the right-hand side of (5.11) coincides, by definition, with the *K-stability threshold* $\sigma_K(X,\omega)$, which is positive iff (X,ω) is *uniformly K-stable* in the sense of [**Der16, BHJ17**]). Thus divisorial stability implies uniform K-stability, and the converse holds if Conjecture 5.12 is satisfied.

REMARK 5.13. *Assume that (X,L) is a polarized smooth projective variety over $\mathbb{C}$, and pick a Kähler form $\omega \in c_1(L)$. Using* [**SD18**] *and* [**C.Li22**], *one can then show that the above thresholds satisfy*

$$\sigma_{\mathrm{div}}(X,L) \le \sigma(X,\omega) \le \sigma_K(X,L).$$

Conjecture 5.12 *would further yield $\sigma_K(X,L) = \sigma_{\mathrm{div}}(X,L)$, and hence conclude the proof of the 'uniform' version of the Yau–Tian–Donaldson conjecture, as noted in* [**C.Li22, BoJ23**].

Appendix A. Convexity estimates

We consider the following data:

- a surjective map $\pi\colon V \to \Theta$ of $\mathbb{R}$-vector spaces, with fibers $V_\theta := \pi^{-1}(\theta)$;
- a homogeneous polynomial $F\colon V \to \mathbb{R}$ of degree $n+1$, $n \ge 1$, with associated symmetric multilinear map
$$V^{n+1} \to \mathbb{R} \quad (x_0,\dots,x_n) \mapsto x_0 \cdot \dots \cdot x_n,$$
i.e. $F(x) = x^{n+1}$;
- a convex cone $P \subset V$ such that

$$x^2 \cdot x_2 \cdot \dots \cdot x_n \ge 0 \text{ for all } x \in V_0 \text{ and } x_i \in P. \tag{A.1}$$

EXAMPLE A.1. *The main example that we have in mind is the negative of the energy pairing in* §1.4, *where $V = \mathcal{Z} \times \mathcal{D}$ with its projection to $\Theta = \mathcal{Z}$, and $P = \{(\theta,\varphi) \in V \mid \theta_\varphi \ge 0\}$. Another example is given by the negative of intersection pairing on a flat projective scheme over $\mathbb{Z}$.*

Our goal is to use (A.1) and the resulting *Cauchy-Schwarz* inequality

$$(x \cdot y \cdot x_2 \cdot \dots \cdot x_n)^2 \le (x^2 \cdot x_2 \cdot \dots \cdot x_n)(y^2 \cdot x_2 \cdot \dots \cdot x_n) \tag{A.2}$$

for all $x,y \in V_0$ and $x_i \in P$ to derive various inequalities and convexity statements.

For each $x \in V$ we define the linear form $F'(x) \in V^\vee$ by

$$\langle F'(x), y\rangle := \frac{d}{dt}\bigg|_{t=0} F(x+ty) = (n+1)x^n \cdot y,$$

and we set for all $x,y \in V$

$$\delta(x,y) := F(x) - F(y) - \langle F'(y), x-y\rangle. \tag{A.3}$$

A simple computation yields

$$\delta(x,y) = \sum_{j=0}^{n-1} (j+1)(x-y)^2 \cdot y^j \cdot x^{n-1-j}. \tag{A.4}$$

In what follows, we fix $\theta \in \Theta$ and set $P_\theta := V_\theta \cap P$.

LEMMA A.2. *We have $\delta(x,y) \geq 0$ for $x,y \in P_\theta$. Moreover, F is convex on P_θ, and for every $y \in P_\theta$ we have that $x \mapsto \delta(x,y)$ is convex.*

PROOF. By (A.1) and (A.4), we have $\delta(x,y) \geq 0$ for $x,y \in P_\theta$, and this implies that F is convex on P_θ; it then follows from (A.3) that $x \mapsto \delta(x,y)$ is convex. □

THEOREM A.3. *For all $x,y,z \in P_\theta$ and $t \in [0,1]$, the following holds:*

- quasi-symmetry*:*
$$\delta(x,y) \approx \delta(y,x);$$
- quasi-triangle inequality*:*
$$\delta(x,z) \lesssim \delta(x,y) + \delta(y,z);$$
- quadratic estimate*:*
$$\delta(x,(1-t)x+ty) \lesssim t^2\delta(x,y);$$
- uniform convexity*:*
$$[(1-t)F(x)+tF(y)] - F((1-t)x+ty) \gtrsim t(1-t)\delta(x,y).$$

For any base point $x_* \in P_\theta$ and $x_i,y_i,z_j \in P_\theta$, the following Hölder estimates *further hold:*

$$|(x_0-y_0)\cdot(x_1-y_1)\cdot z_2\cdot\ldots\cdot z_n| \lesssim \delta(x_0,y_0)^\alpha\delta(x_1,y_1)^\alpha M^{1-2\alpha}; \tag{A.5}$$

$$|\langle F'(x_0)-F'(y_0), x_1-y_1\rangle| \lesssim \delta(x_0,y_0)^{1/2}\delta(x_1,y_1)^\alpha M^{1/2-\alpha}; \tag{A.6}$$

$$|\delta(x_0,x_1)-\delta(y_0,y_1)| \lesssim \max\{\delta(x_0,y_0),\delta(x_1,y_1)\}^\alpha M^{1-\alpha}; \tag{A.7}$$

with $\alpha := 2^{-n} \in (0,1/2]$ and $M = \max_\xi \delta(\xi,x_*)$, where in each case ξ ranges over the elements of P_θ appearing in the left-hand side of the inequality.

The strategy to get these types of Hölder estimates goes back to [**BBGZ13, BBEGZ19, BoJ22**], building upon an original idea of [**Blo03**]. In the rest of this section we prove Theorem A.3, largely following [**BoJ22**, §3.3].

Given $x,y \in P_\theta$, we set

$$d(x,y) := \max_{0\leq j\leq n-1} (x-y)^2\cdot y^j\cdot x^{n-1-j}.$$

Using (A.4) it is then clear that

$$\delta(x,y) \approx \delta(y,x) \approx d(x,y) \approx (x-y)^2\cdot(\tfrac{1}{2}(x+y))^{n-1}. \tag{A.8}$$

To prove the inequality $\delta(x,(1-t)x+ty) \lesssim t^2\delta(x,y)$ it suffices to prove the corresponding inequality $d(x,(1-t)x+ty) \lesssim t^2 d(x,y)$. But

$$d(x,(1-t)x+ty) = t^2 \max_{0\leq j\leq n-1}(x-y)^2\cdot((1-t)x+ty)^j\cdot x^{n-1-j} \lesssim t^2 d(x,y)$$

using multilinearity and the binomial theorem.

Next we prove what is essentially a special case of (A.5).

LEMMA A.4. *If $x,y,z \in P_\theta$, then*

$$(x-y)^2\cdot z^{n-1} \lesssim d(x,y)^{2\alpha}\max\{d(x,z),d(y,z)\}^{1-2\alpha}.$$

PROOF. Set $w := \frac{1}{2}(x+y)$, $A := d(x,y)$, $B := \max\{d(x,z), d(y,z)\}$, and

$$b_j := (x-y)^2 \cdot z^j \cdot w^{n-1-j}$$

for $0 \le j \le n-1$. Then $b_0 \approx A$, and our goal is to show $b_{n-1} \lesssim A^{2\alpha} B^{1-2\alpha}$. By the triangle inequality for the seminorm $v \mapsto \sqrt{v^2 \cdot z^{n-1}}$ on V_0, we have

$$b_{n-1} \le (\sqrt{(x-z)^2 \cdot z^{n-1}} + \sqrt{(y-z)^2 \cdot z^{n-1}})^2 \le 4B.$$

If $A \ge B$, then $b_{n-1} \le 4B \le 4A^{2\alpha} B^{1-2\alpha}$ and we are done, so we may assume $A \le B$. In this case, we show by induction that

$$b_j \lesssim A^{2^{-j}} B^{1-2^{-j}}$$

for $0 \le j \le n-1$. The case $j=0$ is clear, so suppose $0 \le j \le n-2$, and note that

$$\begin{aligned} b_{j+1} - b_j &= (x-y)^2 \cdot (z-w) \cdot z^j \cdot w^{n-2-j} \\ &= (x-y) \cdot (z-w) \cdot x \cdot z^j \cdot w^{n-2-j} - (x-y) \cdot (z-w) \cdot y \cdot z^j \cdot w^{n-2-j}. \end{aligned}$$

Here we can use the Cauchy-Schwartz inequality to estimate the last two terms. For example:

$$\begin{aligned} &|(x-y) \cdot (z-w) \cdot x \cdot z^j \cdot w^{n-2-j}|^2 \\ &\qquad \le ((x-y)^2 \cdot x \cdot z^j \cdot w^{n-2-j})((z-w)^2 \cdot x \cdot z^j \cdot w^{n-2-j}). \end{aligned}$$

Using that $2w - x = y \in P$, we can bound the first factor by $2b_j$, and the second factor by $2(z-w)^2 \cdot z^j \cdot w^{n-1-j} \le 2d(z,w)$. Adding the two terms, we get $b_{j+1} - b_j \le 4\sqrt{b_j}\sqrt{d(z,w)}$. Now $d(z,w) \approx \delta(w,z) \le \max\{\delta(x,z), d(y,z)\} \approx B$ using the convexity of $\delta(\cdot, z)$, see Lemma A.2. All in all, this yields

$$b_{j+1} - b_j \lesssim \sqrt{b_j B}.$$

for $0 \le j \le n-2$. Using the induction hypothesis $b_j \lesssim A^{2^{-j}} B^{1-2^{-j}}$, we get

$$b_{j+1} \lesssim b_j + \sqrt{B b_j} \lesssim A^{2^{-j}} B^{1-2^{-j}} + A^{2^{-j-1}} B^{1-2^{-j-1}} \le 2A^{2^{-j-1}} B^{1-2^{-j-1}},$$

where the last inequality follows from our assumption that $A \le B$. We are done. □

Using Lemma A.4, we can now prove the quasi-triangle inequality for δ, or equivalently for d. Fix $x, y, z \in P_\theta$, and set $w := \frac{1}{2}(x+y)$. Then

$$\begin{aligned} d(x,y) &\approx (x-y)^2 \cdot w^{n-1} \lesssim (x-z)^2 \cdot w^{n-1} + (y-z)^2 \cdot w^{n-1} \\ &\lesssim \max\{d(x,z), d(y,z)\})^{2\alpha} \max\{d(x,w), d(y,w), d(z,w)\}^{1-2\alpha}, \end{aligned}$$

by the triangle inequality for the norm $v \mapsto \sqrt{v^2 \cdot w^{n-1}}$ and by Lemma A.4. As noted above, the convexity of $\delta(\cdot, z) \approx d(\cdot, z)$ gives $d(z,w) \lesssim \max\{d(x,z), d(y,z)\}$, as well as $d(x,w), d(y,w) \lesssim d(x,y)$. Thus

$$d(x,y) \lesssim \max\{d(x,z), d(y,z)\})^{2\alpha} \max\{d(x,y), d(x,z), d(y,z)\}^{1-2\alpha},$$

which easily implies $d(x,y) \le \max\{d(x,z), d(y,z)\}$, as desired.

Next we prove (A.5) in general. By the Cauchy–Schwartz inequality we may assume $x_0 = x_1$ and $y_0 = y_1$. We may also assume $n \ge 2$, or else we are done by (A.8). Set $z := \frac{1}{n-1}(z_2 + \cdots + z_n)$. Then

$$\begin{aligned} (x_1 - y_1)^2 \cdot z_2 \cdot \ldots \cdot z_n \lesssim (x_1 - y_1)^2 \cdot z^{n-1} &\lesssim d(x_1, y_1)^{2\alpha} \max\{d(x_1, z), d(y_1, z)\}^{1-2\alpha} \\ &\approx \delta(x_1, y_1)^{2\alpha} \max\{\delta(x_1, z), \delta(y_1, z)\}^{1-2\alpha}. \end{aligned}$$

By the quasi-triangle inequality we have $\delta(x_1, z) \lesssim \max\{\delta(x_1, x_*), \delta(z, x_*)\}$, and by quasi-symmetry and Lemma A.2 we have $\delta(z, x_*) \lesssim \max_{i\geq 2} \delta(z_i, x_*)$. A similar estimate for $\delta(y_1, z)$ completes the proof of (A.5).

Next we prove (A.6), which is equivalent to

$$|(x_0^n - y_0^n) \cdot (x_1 - y_1)| \leq \delta(x_0, y_0)^{1/2} \delta(x_1, y_1)^\alpha M^{1/2-\alpha}.$$

By the Cauchy–Schwartz inequality, we have

$$\begin{aligned} |(x_0^n - y_0^n) \cdot (x_1 - y_1)|^2 &= |(x_0 - y_0)(x_1 - y_1) \sum_{j=0}^{n-1} x_0^j y_0^{n-1-j}|^2 \\ &\leq \left((x_0 - y_0)^2 \sum_{j=0}^{n-1} x_0^j y_0^{n-1-j} \right) \left((x_1 - y_1)^2 \sum_{j=0}^{n-1} x_0^j y_0^{n-1-j} \right). \end{aligned}$$

Here the first factor on the right is $\approx \delta(x_0, y_0)$, whereas the second factor can be bounded above using (A.5).

It only remains to prove (A.7). By the quasi-triangle inequality for δ, it suffices to consider the case when $x_0 = y_0$ or $x_1 = y_1$. Now

$$\begin{aligned} \delta(x_0, x_1) - \delta(x_0, y_1) &= n(x_1^{n+1} - y_1^{n+1}) - (n+1)x_0(x_1^n - y_1^n) \\ &= (x_1 - y_1) \sum_{j,k} (x_1^j y_1^{n-j} - x_0 x_1^k y_1^{n-k-1}). \end{aligned}$$

If $j \leq k$, then

$$x_1^j y_1^{n-j} - x_0 x_1^k y_1^{n-k-1} = x_1^j y_1^{n-k-1}(y_1^{k-j+1} - x_0^{k-j+1}) + x_1^j y_1^{n-k-1} x_0 (x_0^{k-j} - x_1^{k-j}),$$

and by factoring each term of the right-hand side we see from (A.5) that

$$|(x_1 - y_1)(x_1^j y_1^{n-k} - x_0 x_1^j y_1^{n-k-1})| \lesssim \delta(x_1, y_1)^\alpha M^{1-\alpha}.$$

The case when $j > k$ is handled in a similar way, and adding all the terms yields $|\delta(x_0, x_1) - \delta(x_0, y_1)| \lesssim \delta(x_1, y_1)^\alpha M^{1-\alpha}$.

A similar argument shows that $|\delta(x_0, x_1) - \delta(y_0, x_1)| \lesssim \delta(x_0, y_0)^\alpha M^{1-\alpha}$, and completes the proof.

Appendix B. Regularization and orthogonality on Kähler spaces

By relying on a variant of the classical Richberg regularization technique, it was proved in [**BK07**] that any ω-psh function on a compact Kähler manifold (X, ω) can be written as the limit of a decreasing sequence of smooth ω-psh functions. It is natural to hope that this holds in the singular case as well:

CONJECTURE B.1. *Let (X, ω) be a compact Kähler space. Then any ω-psh function φ on X can be written as the pointwise limit of a decreasing sequence (φ_i) of smooth ω-psh functions.*

Note that the conclusion only depends on the Kähler class $[\omega]$. Besides the case when X is nonsingular,[6] mentioned above, we have:

[6]Conjecture B.1 is now known to hold for any normal space X as a consequence of the recent work [**CC24**] combined with [**Ric68**].

EXAMPLE B.2. *Conjecture* B.1 *holds if* X *is projective and* $[\omega]$ *lies in the ample cone, i.e. the open convex cone generated by classes of ample line bundles on* X. *This is a consequence of* [**CGZ22**, Theorem 1.1].

LEMMA B.3. *Let* (X, ω) *be a compact Kähler space for which Conjecture* B.1 *holds. Then* $[\omega]$ *has the orthogonality property (cf. Definition* 2.15*).*

PROOF. Pick a resolution of singularities $\pi\colon Y \to X$, and set $\theta := \pi^\star\omega \geq 0$. Since $\int \theta^n = \int \omega^n > 0$, the $(1,1)$-class $[\theta]$ is semipositive and big. For any $g \in \mathrm{C}^0(Y)$, consider the θ-psh envelope

$$\mathrm{P}_\theta(g) := \sup\{\psi \in \mathrm{PSH}(Y,\theta) \mid \psi \leq g\}.$$

As is well-known, $\mathrm{P}_\theta(g)$ is θ-psh, and $\mathrm{MA}_\theta(\mathrm{P}_\theta(g))$ is supported in $\{\mathrm{P}_\theta(g) = g\}$, i.e.

$$\int_Y (g - \mathrm{P}_\theta(g))\,\mathrm{MA}_\theta(\mathrm{P}_\theta(g)) = 0, \tag{B.1}$$

see for instance [**BB10**, Proposition 2.10]. For any $f \in \mathrm{C}^0(X)$, we next claim that $\mathrm{P}_\theta(\pi^\star f)$ is the limit in $\mathcal{E}^1(Y,\theta)$ (the space of θ-psh functions of finite energy) of the increasing net $\{\pi^\star\varphi\}_{\varphi\in\mathcal{D}_{\omega,<f}}$. Assume this for the moment. By continuity of Monge–Ampère integrals along increasing nets in $\mathcal{E}^1(Y,\theta)$, and using

$$\int_X (f-\varphi)\,\mathrm{MA}_\omega(\varphi) = \int_Y (\pi^\star f - \pi^\star\varphi)\,\mathrm{MA}_\theta(\pi^\star\varphi),$$

for any $\varphi \in \mathcal{D}_\omega$, we infer

$$\lim_{\varphi\in\mathcal{D}_{\omega,<f}} \int_X (f-\varphi)\,\mathrm{MA}_\omega(\varphi) = \int_Y (\pi^\star f - \mathrm{P}_\theta(\pi^\star f))\,\mathrm{MA}_\theta(\mathrm{P}_\theta(\pi^\star f)) = 0, \tag{B.2}$$

by (B.1).

To prove the claim, note first that Conjecture B.1 implies that the increasing net $\{\varphi\}_{\varphi\in\mathcal{D}_{\omega,<f}}$ converges pointwise to

$$\mathrm{P}_\omega(f) := \sup\{\psi \in \mathrm{PSH}(X,\omega) \mid \psi \leq f\}.$$

Indeed, given $\delta > 0$ and a function $\psi \in \mathrm{PSH}(X,\omega)$ with $\psi \leq f$, Conjecture B.1 and a Dini-type argument guarantees the existence of a function $\psi' \in \mathcal{D}_{\omega,<f}$ with $\psi \leq \psi' < f + \delta$. The claim is thus equivalent to the statement that $\pi^\star\,\mathrm{P}_\omega(f)$ coincides a.e. with $\varphi := \mathrm{P}_\theta(\pi^\star f)$. To prove this, pick $\tau \in \mathrm{PSH}(X,\omega)$ with $\tau \leq f$ and $\{\tau = -\infty\} = X_{\mathrm{sing}}$ (compare the proof of Lemma 1.28). Since $\psi_\varepsilon := (1-\varepsilon)\varphi + \varepsilon\pi^\star\tau$ is $\pi^\star\omega$-psh outside $\pi^{-1}(X_{\mathrm{sing}})$, and $\psi_\varepsilon \equiv -\infty$ on the latter, we have $\psi_\varepsilon = \pi^\star\varphi_\varepsilon$ for a unique $\varphi_\varepsilon \in \mathrm{PSH}(X,\omega)$ (see [**Dem85**, Théorème 1.10]). Since $\pi^\star\varphi_\varepsilon \leq \pi^\star f$, and hence $\varphi_\varepsilon \leq f$, we have $\varphi_\varepsilon \leq \mathrm{P}_\omega(f)$. Thus

$$(1-\varepsilon)\varphi + \varepsilon\pi^\star\tau = \pi^\star\varphi_\varepsilon \leq \pi^\star\,\mathrm{P}_\omega(f) \leq \varphi,$$

and hence $\pi^\star\,\mathrm{P}_\omega(f) = \mathrm{P}_\theta(\pi^\star f)$ on $Y \setminus \pi^{-1}(X_{\mathrm{sing}})$. □

References

[Berk90] Vladimir G. Berkovich, *Spectral theory and analytic geometry over non-Archimedean fields*, Mathematical Surveys and Monographs, vol. 33, American Mathematical Society, Providence, RI, 1990, DOI 10.1090/surv/033. MR1070709

[Berm13] Robert J. Berman, *A thermodynamical formalism for Monge-Ampère equations, Moser-Trudinger inequalities and Kähler-Einstein metrics*, Adv. Math. **248** (2013), 1254–1297, DOI 10.1016/j.aim.2013.08.024. MR3107540

[BB10] Robert Berman and Sébastien Boucksom, *Growth of balls of holomorphic sections and energy at equilibrium*, Invent. Math. **181** (2010), no. 2, 337–394, DOI 10.1007/s00222-010-0248-9. MR2657428

[BBGZ13] Robert J. Berman, Sébastien Boucksom, Vincent Guedj, and Ahmed Zeriahi, *A variational approach to complex Monge-Ampère equations*, Publ. Math. Inst. Hautes Études Sci. **117** (2013), 179–245, DOI 10.1007/s10240-012-0046-6. MR3090260

[BBEGZ19] Robert J. Berman, Sebastien Boucksom, Philippe Eyssidieux, Vincent Guedj, and Ahmed Zeriahi, *Kähler-Einstein metrics and the Kähler-Ricci flow on log Fano varieties*, J. Reine Angew. Math. **751** (2019), 27–89, DOI 10.1515/crelle-2016-0033. MR3956691

[BBJ21] Robert J. Berman, Sébastien Boucksom, and Mattias Jonsson, *A variational approach to the Yau-Tian-Donaldson conjecture*, J. Amer. Math. Soc. **34** (2021), no. 3, 605–652, DOI 10.1090/jams/964. MR4334189

[BDL20] Robert J. Berman, Tamás Darvas, and Chinh H. Lu, *Regularity of weak minimizers of the K-energy and applications to properness and K-stability* (English, with English and French summaries), Ann. Sci. Éc. Norm. Supér. (4) **53** (2020), no. 2, 267–289, DOI 10.24033/asens.2422. MR4094559

[BDL17] Robert J. Berman, Tamás Darvas, and Chinh H. Lu, *Convexity of the extended K-energy and the large time behavior of the weak Calabi flow*, Geom. Topol. **21** (2017), no. 5, 2945–2988, DOI 10.2140/gt.2017.21.2945. MR3687111

[BGS95] S. Bloch, H. Gillet, and C. Soulé, *Non-Archimedean Arakelov theory*, J. Algebraic Geom. **4** (1995), no. 3, 427–485. MR1325788

[Blo03] Zbigniew Błocki, *Uniqueness and stability for the complex Monge-Ampère equation on compact Kähler manifolds*, Indiana Univ. Math. J. **52** (2003), no. 6, 1697–1701, DOI 10.1512/iumj.2003.52.2346. MR2021054

[BK07] Zbigniew Błocki and Sławomir Kołodziej, *On regularization of plurisubharmonic functions on manifolds*, Proc. Amer. Math. Soc. **135** (2007), no. 7, 2089–2093, DOI 10.1090/S0002-9939-07-08858-2. MR2299485

[BE21] Sébastien Boucksom and Dennis Eriksson, *Spaces of norms, determinant of cohomology and Fekete points in non-Archimedean geometry*, Adv. Math. **378** (2021), Paper No. 107501, 124, DOI 10.1016/j.aim.2020.107501. MR4192993

[BFJ15] Sébastien Boucksom, Charles Favre, and Mattias Jonsson, *Solution to a non-Archimedean Monge-Ampère equation*, J. Amer. Math. Soc. **28** (2015), no. 3, 617–667, DOI 10.1090/S0894-0347-2014-00806-7. MR3327532

[BFJ16a] Sébastien Boucksom, Charles Favre, and Mattias Jonsson, *Singular semipositive metrics in non-Archimedean geometry*, J. Algebraic Geom. **25** (2016), no. 1, 77–139, DOI 10.1090/jag/656. MR3419957

[BG13] Sébastien Boucksom and Vincent Guedj, *Regularizing properties of the Kähler-Ricci flow*, An introduction to the Kähler-Ricci flow, Lecture Notes in Math., vol. 2086, Springer, Cham, 2013, pp. 189–237, DOI 10.1007/978-3-319-00819-6_4. MR3185334

[BGM22] Sébastien Boucksom, Walter Gubler, and Florent Martin, *Differentiability of relative volumes over an arbitrary non-Archimedean field*, Int. Math. Res. Not. IMRN **8** (2022), 6214–6242, DOI 10.1093/imrn/rnaa314. MR4406129

[BHJ17] Sébastien Boucksom, Tomoyuki Hisamoto, and Mattias Jonsson, *Uniform K-stability, Duistermaat-Heckman measures and singularities of pairs* (English, with English and French summaries), Ann. Inst. Fourier (Grenoble) **67** (2017), no. 2, 743–841. MR3669511

[BoJ17] Sébastien Boucksom and Mattias Jonsson, *Tropical and non-Archimedean limits of degenerating families of volume forms* (English, with English and French summaries), J. Éc. polytech. Math. **4** (2017), 87–139, DOI 10.5802/jep.39. MR3611100

[BoJ18a] S. Boucksom and M. Jonsson, *Singular semipositive metrics on line bundles on varieties over trivially valued fields*, `arXiv:1801.08229`, 2018.

[BoJ22] Sébastien Boucksom and Mattias Jonsson, *Global pluripotential theory over a trivially valued field* (English, with English and French summaries), Ann. Fac. Sci. Toulouse Math. (6) **31** (2022), no. 3, 647–836, DOI 10.5802/afst.170. MR4452253

[BoJ23] Sébastien Boucksom and Mattias Jonsson, *A non-Archimedean approach to K-stability, II: Divisorial stability and openness*, J. Reine Angew. Math. **805** (2023), 1–53, DOI 10.1515/crelle-2023-0062. MR4669034

[BGJK21] J. I. Burgos Gil, W. Gubler, P. Jell, K. Künnemann, *Pluripotential theory for tropical toric varieties and non-archimedean Monge-Ampère equations*, `arXiv:2102.07392`, 2021.

[Cha06] Antoine Chambert-Loir, *Mesures et équidistribution sur les espaces de Berkovich* (French), J. Reine Angew. Math. **595** (2006), 215–235, DOI 10.1515/CRELLE.2006.049. MR2244803

[CD12] A. Chambert-Loir and A. Ducros, *Formes différentielles réelles et courants sur les espaces de Berkovich*, `arXiv:1204.6277`, 2012.

[CC21] Xiuxiong Chen and Jingrui Cheng, *On the constant scalar curvature Kähler metrics (II)—Existence results*, J. Amer. Math. Soc. **34** (2021), no. 4, 937–1009, DOI 10.1090/jams/966. MR4301558

[CC24] Y. W. L. Cho and Y. J. Choi, *Continuity of solutions to complex Monge–Ampére equations on compact Kähler spaces*, `arXiv:2401.03935`, 2024.

[Dar15] Tamás Darvas, *The Mabuchi geometry of finite energy classes*, Adv. Math. **285** (2015), 182–219, DOI 10.1016/j.aim.2015.08.005. MR3406499

[CGZ22] D. Coman, V. Guedj, and A. Zeriahi, *On the extension of quasiplurisubharmonic functions*, Anal. Math. **48** (2022), no. 2, 411–426, DOI 10.1007/s10476-022-0153-7. MR4440751

[DaR17] Tamás Darvas and Yanir A. Rubinstein, *Tian's properness conjectures and Finsler geometry of the space of Kähler metrics*, J. Amer. Math. Soc. **30** (2017), no. 2, 347–387, DOI 10.1090/jams/873. MR3600039

[DXZ23] T. Darvas, Mi. Xia, K. Zhang, *A transcendental approach to non-Archimedean metrics of pseudoeffective classes*, `arXiv:2302.02541`, 2023.

[Dem85] Jean-Pierre Demailly, *Mesures de Monge-Ampère et caractérisation géométrique des variétés algébriques affines* (French, with English summary), Mém. Soc. Math. France (N.S.) **19** (1985), 124. MR813252

[Dem92] Jean-Pierre Demailly, *Regularization of closed positive currents and intersection theory*, J. Algebraic Geom. **1** (1992), no. 3, 361–409. MR1158622

[DPS94] Jean-Pierre Demailly, Thomas Peternell, and Michael Schneider, *Compact complex manifolds with numerically effective tangent bundles*, J. Algebraic Geom. **3** (1994), no. 2, 295–345. MR1257325

[Der16] Ruadhaí Dervan, *Uniform stability of twisted constant scalar curvature Kähler metrics*, Int. Math. Res. Not. IMRN **15** (2016), 4728–4783, DOI 10.1093/imrn/rnv291. MR3564626

[DeR17] Ruadhaí Dervan and Julius Ross, *K-stability for Kähler manifolds*, Math. Res. Lett. **24** (2017), no. 3, 689–739, DOI 10.4310/MRL.2017.v24.n3.a5. MR3696600

[DL23] Ruadhaí Dervan and Eveline Legendre, *Valuative stability of polarised varieties*, Math. Ann. **385** (2023), no. 1-2, 357–391, DOI 10.1007/s00208-021-02313-4. MR4542718

[DN15] Eleonora Di Nezza, *Finite pluricomplex energy measures*, Potential Anal. **44** (2016), no. 1, 155–167, DOI 10.1007/s11118-015-9503-4. MR3455214

[DGL21] Eleonora Di Nezza, Vincent Guedj, and Chinh H. Lu, *Finite entropy vs finite energy*, Comment. Math. Helv. **96** (2021), no. 2, 389–419, DOI 10.4171/cmh/515. MR4277276

[Fuj19a] Kento Fujita, *A valuative criterion for uniform K-stability of $\mathbb{Q}$-Fano varieties*, J. Reine Angew. Math. **751** (2019), 309–338, DOI 10.1515/crelle-2016-0055. MR3956698

[GJR21] W Gubler, P. Jell, J. Rabinoff, *Forms on Berkovich spaces based on harmonic tropicalizations*, `arXiv:2111.05741`, 2021.

[GK15] Walter Gubler and Klaus Künnemann, *Positivity properties of metrics and delta-forms*, J. Reine Angew. Math. **752** (2019), 141–177, DOI 10.1515/crelle-2016-0060. MR3975640

[GK17] Walter Gubler and Klaus Künnemann, *A tropical approach to nonarchimedean Arakelov geometry*, Algebra Number Theory **11** (2017), no. 1, 77–180, DOI 10.2140/ant.2017.11.77. MR3602767

[GM16] Walter Gubler and Florent Martin, *On Zhang's semipositive metrics*, Doc. Math. **24** (2019), 331–372. MR3960125

[GZ05] Vincent Guedj and Ahmed Zeriahi, *Intrinsic capacities on compact Kähler manifolds*, J. Geom. Anal. **15** (2005), no. 4, 607–639, DOI 10.1007/BF02922247. MR2203165

[GZ07] Vincent Guedj and Ahmed Zeriahi, *The weighted Monge-Ampère energy of quasi-plurisubharmonic functions*, J. Funct. Anal. **250** (2007), no. 2, 442–482, DOI 10.1016/j.jfa.2007.04.018. MR2352488

[Ino22] E. Inoue, *Entropies in μ-framework of canonical metrics and K-stability, II – Non-archimedean aspect: non-archimedean μ-entropy and μK-semistability*, `arXiv:2202.12168`, 2022.

[Jell16] P. Jell, *Differential forms on Berkovich analytic spaces and their cohomology*, PhD thesis, Universität Regensburg, 2016. `urn:nbn:de:bvb:355-epub-347884`.

[LS94] C. LeBrun and S. R. Simanca, *Extremal Kähler metrics and complex deformation theory*, Geom. Funct. Anal. **4** (1994), no. 3, 298–336, DOI 10.1007/BF01896244. MR1274118

[Li17] Chi Li, *K-semistability is equivariant volume minimization*, Duke Math. J. **166** (2017), no. 16, 3147–3218, DOI 10.1215/00127094-2017-0026. MR3715806

[C.Li22] Chi Li, *Geodesic rays and stability in the cscK problem* (English, with English and French summaries), Ann. Sci. Éc. Norm. Supér. (4) **55** (2022), no. 6, 1529–1574. MR4517682

[Y.Li20] Yang Li, *Metric SYZ conjecture and non-Archimedean geometry*, Duke Math. J. **172** (2023), no. 17, 3227–3255, DOI 10.1215/00127094-2022-0099. MR4688155

[Liu23] Yaxiong Liu, *Openness of uniformly valuative stability on the Kähler cone of projective manifolds*, Math. Z. **303** (2023), no. 2, Paper No. 52, 26, DOI 10.1007/s00209-023-03209-6. MR4543809

[MP24] P. Mesquita-Piccione, *A non-Archimedean theory of complex spaces and the cscK problem*, `arXiv:2409.06221`, 2024.

[Reb21] Rémi Reboulet, *The asymptotic Fubini-Study operator over general non-Archimedean fields*, Math. Z. **299** (2021), no. 3-4, 2341–2378, DOI 10.1007/s00209-021-02770-2. MR4329290

[Reb22] Rémi Reboulet, *Plurisubharmonic geodesics in spaces of non-Archimedean metrics of finite energy*, J. Reine Angew. Math. **793** (2022), 59–103, DOI 10.1515/crelle-2022-0059. MR4513163

[Ric68] R. Richberg, *Stetige streng pseudokonvexe Funktionen*, Math. Ann. **175** (1968), 257–286.

[SD18] Zakarias Sjöström Dyrefelt, *K-semistability of cscK manifolds with transcendental cohomology class*, J. Geom. Anal. **28** (2018), no. 4, 2927–2960, DOI 10.1007/s12220-017-9942-9. MR3881961

[SD20] Zakarias Sjöström Dyrefelt, *Optimal lower bounds for Donaldson's J-functional*, Adv. Math. **374** (2020), 107271, 37, DOI 10.1016/j.aim.2020.107271. MR4157574

[Ste19] Matthew Stevenson, *Applications of Canonical Metrics on Berkovich Spaces*, ProQuest LLC, Ann Arbor, MI, 2019. Thesis (Ph.D.)–University of Michigan. MR4071620

[Tem16] Michael Temkin, *Metrization of differential pluriforms on Berkovich analytic spaces*, Nonarchimedean and tropical geometry, Simons Symp., Springer, [Cham], 2016, pp. 195–285. MR3702313

[Tho63] A. C. Thompson, *On certain contraction mappings in a partially ordered vector space*, Proc. Amer. Math. Soc. **14** (1963), 438–443, DOI 10.2307/2033816. MR149237

[YZ17] Xinyi Yuan and Shou-Wu Zhang, *The arithmetic Hodge index theorem for adelic line bundles*, Math. Ann. **367** (2017), no. 3-4, 1123–1171, DOI 10.1007/s00208-016-1414-1. MR3623221

[Zha95] Shou-Wu Zhang, *Positive line bundles on arithmetic varieties*, J. Amer. Math. Soc. **8** (1995), no. 1, 187–221, DOI 10.2307/2152886. MR1254133

Sorbonne Université and Université Paris Cité, CNRS, IMJ-PRG, F-75005 Paris, France

Email address: `sebastien.boucksom@imj-prg.fr`

Dept of Mathematics, University of Michigan, Ann Arbor, Michigan 48109-1043

Email address: `mattiasj@umich.edu`

Contemporary Mathematics
Volume **810**, 2025
https://doi.org/10.1090/conm/810/16209

Duality of Hessian manifolds and optimal transport

Jakob Hultgren

Dedicated to Professor Bo Berndtsson on the occasion of his 70th birthday

Abstract. This is an expository paper describing how duality theory for Hessian manifolds provides a natural setting for optimal transport. We explain how this can be used to solve Monge-Ampère equations and survey recent results along these lines with applications to the SYZ-conjecture in mirror symmetry.

Contents

1. Introduction

An affine manifold is a smooth manifold with a distinguished atlas whose transition functions are affine. These arise naturally in symplectic geometry as the base space of any Lagrangian bundle with compact connected fibers. In particular, this puts them at the center of the SYZ-conjecture in mirror symmetry, which predicts that Calabi-Yau manifolds close to large complex structure limits admits (singular) special Lagrangian torus fibrations. The base spaces of these special Lagrangian torus fibrations are expected to be equipped with distinguished atlases as above (usually referred to as affine structures) extending to the complement of a co-dimension 2 subset.

A Hessian metric is a Riemannian metric on an affine manifold which locally (in coordinates of the distinguished atlas) can be expressed as the Hessian of a convex function. An affine manifold equipped with a Hessian metric is a Hessian manifold. When the transition functions in the distinguished atlas are volume preserving, the affine structure is said to be special. In this case, the manifold admits a well-defined real Monge-Ampère operator associating to each Hessian metric a Borel measure. Solutions to equations involving this operator are conjectured to form

2020 *Mathematics Subject Classification.* Primary 35J96, 32Q25, 14J33, 53A15.

asymptotic models of the Kähler potentials of Ricci flat metrics on Calabi-Yau manifolds close to large complex structure limits. Moreover, constructing solutions to real Monge-Ampère equations on integral affine manifolds with singularities has been central in recent progress on the SYZ-conjecture initiated by Y. Li (see [**Li22a, Li20, HJMM22, PS22, Li23, AH23**]) and arguably constitute one of the main obstacle to further progress along similar lines.

Optimal transport is a collection of ideas surrounding an optimization problem introduced independently by Monge in the 18th century and Kantorovich in the 1940's. This problem is concerned with transporting sand/resources/fluid/probability distributions from one configuration (modeled by a *source* probability measure) to another configuration (modeled by a *target* probability measure). It is well known that in 'model cases' of optimal transport (when the source and target measures are defined on $\mathbb{R}^d$ and has convex support) solving the optimal transport problem with respect to the squared distance cost function is equivalent to solving a real Monge-Ampère equation (see Section 2.4 for details).

Similarly, part of the recent progress on the SYZ-conjecture in [**HJMM22, Li23, AH23**] relies on a variational approach to Monge-Ampère equations on the boundary of reflexive polytopes introduced in [**HJMM22**] which is related to a particular optimal transport problem. Moreover, [**HO19**] uses a variational approach to solve Monge-Ampère equations on compact affine manifolds related to a different optimal transport problem. The aim of this expository paper is to explain how both these optimal transport problem, at least in a heuristic manner, arise from classical duality theory for Hessian manifolds.

We will begin by recalling some basic facts about convex functions, Legendre transform and Monge-Ampère equations, in particular their weak interpretation in terms of Aleksandrov. Then we will give a very brief background to optimal transport, focusing on the central fact that optimal transport plans are (in a generalized sense) supported on graphs of gradients of convex functions.

In Section 3 we will provide some background to affine and Hessian manifolds, including examples and a structural theorem. In Section 4 we will explain how each Hessian manifold has a dual Hessian manifold. We will also compute this dual for two classes of Hessian manifolds:

- General compact Hessian manifolds
- Affine structures induced on dense subsets of the boundaries of reflexive polytopes

This is possibly the only part of this paper which is not readily available in the literature. A crucial point is that given an affine manifold in one of the classes above, for reasonable choices of Hessian metrics the affine structure in the dual Hessian manifold is almost fixed. This allows for a variational approach where the Hessian metric varies but the affine structure in the dual Hessian manifold is fixed.

In Section 5, we will explain how completing this picture with a 'pairing like' object between (X, ∇) and $(X^\vee, \nabla^\vee)$ defines a optimal transport problem which coincides with the functionals used in [**HO19, HJMM22, Li23, AH23**]. Finally, in Section 6 we survey the results in [**HO19, HJMM22, Li23, AH23**] with applications to mirror symmetry.

An important part of this topic which is not directly in the scope of this paper is that affine manifolds conjecturally arise as the essential skeletons of Berkovich spaces associated to maximally degenerate families of Calabi-Yau manifolds. This setting allows for an abstractly defined analog of the Monge-Ampère operator: the Chambert-Loir measure [**CL06**]. See [**BoJ17, BFJ15, Li20, OO21, CLD12, BGJKM21, NX16, NXY19, MP21**] for a very incomplete list of the great progress recent decades has seen along these lines. In particular, the latter two addresses the question of how to assign an integral affine structure to the essential skeleton. Moreover, see [**Li23**] for an account of how optimal transport arise directly out of this framework and [**Lag12**] for an influential idea on how to set up a tropical pluripotential theory.

Finally, the idea to use Legendre transform type constructions to set up a variational framework for Monge-Ampère equations goes back at least to Aleksandrov's solution of the Minkowski problem. More recently, the variational approach in [**HO19, HJMM22, Li23, AH23**] can be compared to the variational approach used to produce Kähler-Einstein metrics on K-stable toric Fano manifolds in [**BB13, San16**], which also plays a major role in [**Ber13, Hul19a**], and the variational approach studied very recently in [**TY23**].

2. Background

2.1. Convex functions. Given an open convex subset U of $\mathbb{R}^d$, a function $\Phi : U \to \mathbb{R}$ is convex if its epigraph $\{(x, h) \in U \times \mathbb{R} : h \geq \Phi(x)\}$ is convex. If Φ is smooth, then Φ is convex if and only if its Hessian is positive semi-definite. In the general case, convexity implies some strong regularity properties. In particular, any convex $\Phi : U \to \mathbb{R}$ is automatically locally Lipschitz continuous on U.

The (multivalued) gradient map $\partial\Phi : U \to \mathbb{R}^d$ is defined by writing $y \in \partial\Phi(x)$ if

$$\Phi(x') \geq \Phi(x) + \langle x' - x, y\rangle$$

for all $x' \in U$. Often it is useful to identify Φ with an $(\mathbb{R}\cup\{\infty\})$-valued extension to $\mathbb{R}^d$, defined in the following way: Extend Φ by lower semi-continuity to the closure of U and set $\Phi \equiv +\infty$ outside the closure of U. It can be checked that this extension does not affect $\partial\Phi(U)$. Conversely, if $\Phi : \mathbb{R}^d \to \{+\infty\}$ is a lower semi-continuous convex function, then $\{\Phi < +\infty\}$ is a convex subset whose interior correspond to U above. A lower semi-continuous convex function Φ is differentiable at x if and only if $\partial\Phi(x)$ consists of a single element. Any convex function Φ is differentiable almost everywhere on the interior of $\{\Phi < +\infty\}$.

2.2. Legendre transform. Given a lower semi-continuous function $\Phi : \mathbb{R}^d \to \mathbb{R}\cup\{\infty\}$, the Legendre transform $\Phi^* : (\mathbb{R}^d)^* \to \mathbb{R}\cup\{+\infty\}$ of Φ is the convex lower semi-continuous function

$$\Phi^*(y) = \sup_{x\in\mathbb{R}^d} \langle x, y\rangle - \Phi(x).$$

Similarly, given a lower semi-continuous function $\Psi : (\mathbb{R}^d)^* \to \mathbb{R} \cup \{+\infty\}$, the Legendre transform $\Psi^* : \mathbb{R}^d \to \mathbb{R}\cup\{+\infty\}$ of Ψ is the convex lower semi-continuous function $\mathbb{R}^d \to \mathbb{R} \cup \{\infty\}$

$$\Psi^*(y) = \sup_{y\in(\mathbb{R}^d)^*} \langle x, y\rangle - \Psi(y).$$

An application of the Hahn-Banach supporting hyperplane theorem gives that

$$(\Phi^*)^* = \Phi \text{ and } (\Psi^*)^* = \Psi$$

if $\Phi : \mathbb{R}^d \to \mathbb{R} \cup \{+\infty\}$ and $\Psi : (\mathbb{R}^d)^* \to \mathbb{R} \cup \{+\infty\}$ are lower semi-continuous and convex.

This duality is also manifested in the following ways.

- If Φ is strictly convex and differentiable on the interior of $\{\Phi < +\infty\}$, then Φ^* is strictly convex and differentiable on the interior of $\{\Phi < +\infty\}$ and
$$\partial\Phi^* \circ \partial\Phi(x) = x$$
$$\partial\Phi \circ \partial\Phi^*(y) = y$$
for all x in the interior of $\{\Phi < +\infty\}$ and y in the interior of $\{\Phi^* < +\infty\}$.
- More generally, if $\Phi : R^d \to \mathbb{R} \cup \{+\infty\}$ is convex and $(x, y) \in \mathbb{R}^d \times (\mathbb{R}^d)^*$, then
$$y \in \partial\Phi(x) \text{ if and only if } x \in \partial\Phi^*(y).$$
Moreover, this holds if and only if
$$\Phi(x) + \Phi^*(y) = \langle x, y\rangle.$$
- If Φ is strictly convex and C^2 on the interior of $\{\Phi < +\infty\}$, then
$$(\Phi^*)_{ij} = \Phi^{ij} \circ \partial\Phi^* \tag{2.1}$$
on the interior of $\{\Phi^* < +\infty\}$.

2.3. The Monge-Ampère operator. Let $\Phi : \mathbb{R}^d \to \mathbb{R} \cup \{+\infty\}$ be lower semi-continuous and assume Φ is smooth and strictly convex on the interior of $\{\Phi < +\infty\}$. The Monge-Ampère measure of Φ on the interior of $\{\Phi < +\infty\}$ is
$$\mathrm{MA}(\Phi) = \det\left(\frac{\partial^2\Phi}{\partial x_i \partial x_j}\right) dVol$$
where $dVol$ is the Lebesgue measure on $(\mathbb{R}^d)^*$. Since $\det\left(\frac{\partial^2\Phi}{\partial x_i\partial x_j}\right)$ is the Jacobian determinant of $\partial\Phi$, we get the following equality for all measurable subsets A of the interior of $\{\Phi < +\infty\}$:
$$\int_{\nabla\Phi(A)} dVol = \int_A \mathrm{MA}(\Phi).$$

This means that $\mathrm{MA}(\Phi)$ is the pullback of the Lebesgue measure on $\partial\Phi(\{\Phi < +\infty\}^\circ)$ by the diffeomorphism $\partial\Phi$. Equivalently, $\mathrm{MA}(\Phi)$ is the push-forward of the Lebesgue measure on $\partial\Phi(\{\Phi < +\infty\}^\circ)$ by $\partial\Phi^*$.

The latter description of $\mathrm{MA}(\Phi)$ generalises to lower semi-continuous convex functions Φ, since Φ^* is convex and hence differentiable almost everywhere on the interior of $\{\Phi^* < +\infty\}$.

DEFINITION 2.1 (The Monge-Ampère operator in the sense of Aleksandrov). Let $\Phi : \mathbb{R}^d \to \mathbb{R}\cup\{+\infty\}$ be lower semi-continuous and convex. The Monge-Ampère measure of Φ is
$$\mathrm{MA}(\Phi) = (\partial\Phi^*)_{\#} d\,\mathrm{Vol} \tag{2.2}$$
where $dVol$ denotes the Lebesgue measure on the interior of $\{\Phi^* < +\infty\} \subset (\mathbb{R}^d)^*$. Equivalently, for every measurable set $A \subset \mathbb{R}^d$,
$$\mathrm{MA}(\Phi)(A) = \int_{\partial\Phi(A)} dVol. \tag{2.3}$$

REMARK 2.2. Even though $\partial\Phi^*$ is a multivalued map, Φ^* is differentiable and $\partial\Phi^*$ single valued almost everywhere. Consequently, the push forward in (2.2) is well-defined. To see that (2.2) implies (2.3), note that by (2.2)

$$\begin{aligned}\mathrm{MA}(\Phi)(A) &= |(\partial\Phi^*)^{-1}(A)| \\ &= |\{y : \Phi^* \text{ is differentiable at } y \text{ and } \partial\Phi^*(y) \in A\}| \\ &= |\{y : \Phi^* \text{ is differentiable at } y \text{ and } y \in \partial\Phi(A)\}|\end{aligned}$$

and that the difference between $\partial\Phi(A)$ and the set in the last row is contained in the non-differentiable locus of Φ^* which is of zero volume.

Note that $\mathrm{MA}(\Phi) = 1$ on an open set $V \subset \{\Phi < +\infty\}$ if and only if the multivalued map $\partial\Phi$ is volume preserving. With a bit of work, this provides another nice duality principle:

$$\mathrm{MA}(\Phi) = 1$$

is satisfied on a set V if and only if

$$\mathrm{MA}(\Phi^*) = 1.$$

on $\partial\Phi(V) \subset (\mathbb{R}^d)^*$.

The Monge-Ampère operator is invariant under volume preserving affine transformations. The formula (2.3) above provides a convenient way to see this: If $B \in \mathrm{SL}_d(\mathbb{R}), b \in \mathbb{R}^d$ and λ is the affine transformation given by $x \to Bx + b$, then

$$|\partial(\Phi \circ \lambda)(A)| = |B^{-1}\partial\Phi(A)| = |\partial\Phi(A)|$$

See for example [**Fig99**] for a more detailed account of the Monge-Ampère equation.

2.4. Optimal transport. We will now give a brief account of basic optimal transport theory. For details, see for example [**AG13, Vil09**]. Given two probability spaces (Y, μ) and (Z, ν) and a continuous *cost function* $c : X \times Y \to \mathbb{R}$, optimal transport seeks to minimize the functional

$$I(\gamma) = \int_{X\times Y} c\gamma \tag{2.4}$$

over all couplings γ of μ and ν, in other words all probability measures γ on $X \times Y$ whose marginals are given by μ and ν, respectively. The *source measure* μ can be interpreted as a distributions of goods, and the *target measure* as a distribution of consumers. A *transport plan* γ then describes a way to transport the goods to the consumers in the sense that $\gamma(A \times B)$ for two measurable subsets $A \subset X$ and $B \subset Y$ describes how much resources should be transported from A to B. If we interpret $c(x, y)$ as the cost of transporting one unit of goods from $x \in X$ to $y \in Y$, then (2.4) is exactly the total cost of transporting the goods to the consumers according to the plan specified by γ.

Under very mild conditions an *optimal transport plan*, i.e a minimizer of (2.4), exists. In particular, if (X, μ) and (Y, ν) are Polish probability spaces and c continuous, then existence of an optimal transport plan follows from simple compactness arguments. However, there may be more than one optimal transport plan.

2.4.1. *Dual formulation.* As a linear optimization problem, optimal transport admits a dual formulation. This is given by maximizing

$$J(\Phi) = -\int \Phi\mu - \int \Phi^c\nu \tag{2.5}$$

over all $L^1(\mu) \ni \Phi : X \to \mathbb{R}$, where $\Phi^c : Y \to \mathbb{R}$ is the c-transform of Φ defined by

$$\Phi^c(y) = \sup_{x\in X} -c(x,y) - \Phi(x).$$

REMARK 2.3. A symmetric version of (2.5) is normally used in the literature. This is given by maximizing

$$-\int \Phi\mu - \int \Psi\nu \tag{2.6}$$

over all $L^1(\mu) \ni \Phi : X \to \mathbb{R}$ and $L^1(\nu) \ni \Psi : Y \to \mathbb{R}$ such that $\Phi(x) + \Psi(y) \geq -c(x,y)$ for all $x \in X$ and $y \in Y$. To get (2.5) from (2.6), simply note that if Φ is fixed, the best possible Ψ is given by Φ^c. Note also that the signs in (2.6) and in the definition of c-transform differs from the conventional ones in the optimal transport literature. The signs in the definitions here are chosen so that the c-transform agrees with the classical Legendre transform when $X = \mathbb{R}^d, Y = (\mathbb{R}^d)^*$ and $c(x,y) = -\langle x,y\rangle$.

We will give a heuristic explanation of how to arrive at (2.5) below. Before that we note that as usual the following inequality is satisfied by construction:

$$\sup J(\Phi) \leq \min I(\gamma).$$

Moreover, with some extra conditions, strong duality holds and the dual admits a maximizer:

THEOREM 2.4 (see Theorem 5.10 in [**Vil09**]). *Let $(X,\mu),(Y,\nu)$ be polish spaces. Assume $c : X \times Y \to \mathbb{R}$ is continuous and bounded from below and that there is some transport plan of finite cost. Then J admits a maximizer and*

$$\max J = \min I. \tag{2.7}$$

A striking feature of this is the following corollary

COROLLARY 2.5. *A transport plan γ is optimal if and only if it is supported on the set*

$$\{(x,y) \in X \times Y : \Phi(x) + \Phi^c(y) = -c(x,y)\}. \tag{2.8}$$

Moreover, if this holds, then Φ is a maximizer of J.

PROOF. By (2.7), γ is optimal if and only if $I(\gamma) = \max J = J(\Phi)$ for some Φ. Moreover, for any $\Phi \in L^1(\mu)$

$$\begin{aligned} 0 &\leq I(\gamma) - J(\Phi) \\ &= \int c\gamma + \int \Phi\mu + \int \Phi^c\nu \\ &= \int (c(x,y) + \Phi(x) + \Phi^c(y))\gamma. \end{aligned} \tag{2.9}$$

Since $c(x,y) + \Phi(x) + \Phi^c(y) \geq 0$ and $\gamma \geq 0$, equality holds in (2.9) if and only if γ is supported on (2.8). □

COROLLARY 2.6. *Let* $X = \mathbb{R}^n$, Y *a bounded convex subset of* $(\mathbb{R}^d)^*$ *and* $c(x,y) = -\langle x, y\rangle$. *Assume* μ *has finite second moment and* ν *is proportional to the Lebesgue measure on* Y. *Then* $J(\Phi) = \max J$ *if and only if there is a constant* $\rho > 0$ *such that*

$$\mathrm{MA}(\Phi) = \rho\mu \tag{2.10}$$

in the sense of Aleksandrov and $\overline{\partial\Phi(\mathbb{R}^d)} = \overline{Y}$.

REMARK 2.7. The constant ρ is picked so that the mass of $\rho\mu$ equals the volume of Y.

REMARK 2.8. When μ is absolutely continuous with positive density, Corollary 2.6 is essentially a special case of results due to Knott-Smith [**KS84**] and Brenier [**Bre91**]. Identifying convexity as the crucial assumption in order for Φ to define an Aleksandrov solution is due to Caffarelli [**Caf97**].

PROOF SKETCH. Assume Φ is a minimizer. Since $\Phi = \Psi^c$ for some $\Psi : Y \to \mathbb{R}$ we have $\partial\Phi(\mathbb{R}^d) \subset Y$. Since $\partial\Phi^*$ is single valued almost everywhere with respect to ν and γ is supported on the graph of $\partial\Phi^*$ it follows that $\gamma = (\partial\Phi^*, id)_{\#}\nu$ where $(\partial\Phi^*, id) : Y \to X \times Y$ is the map given by $y \mapsto (\partial\Phi^*(y), y)$. Combining this with $(p_X)_{\#}\gamma = \mu$ we get $(\partial\Phi^*)_{\#}\nu = \mu$, hence $\rho(\partial\Phi^*)_{\#}dVol = \mu$ for some $\rho > 0$, which proves the corollary. □

Before proceeding, we will give a brief indication on how J arise as the dual of I. A general heuristic form formulating a dual optimization problem is to introduce Lagrange multipliers for each condition and consider the resulting unconstrained, but penalized, problem. We have three conditions: $\gamma > 0$, $(p_X)_{\#}\gamma - \mu = 0$ and $(p_Y)_{\#}\gamma - \nu = 0$. Pick Lagrange multipliers $h > 0, \Phi$ and Ψ. These are functions on $X \times Y$, X and Y, respectively.

The unconstrained problem is then

$$\begin{aligned}\inf_\gamma L_{h,\Phi,\Psi}(\gamma) &:= \inf_\gamma \int_{X\times Y} c\gamma - \int_{X\times Y} h\gamma + \int_Y \Phi((p_X)_{\#}\gamma - \mu) + \int_Y \Psi((p_Y)_{\#}\gamma - \nu) \\ &= \inf_\gamma \int_{X\times Y} (c(x,y) + \Phi(x) + \Psi(y) - h(x,y))\gamma - \int_X \Phi\mu - \int_Y \Psi\nu\end{aligned}$$

where the infimum is taken over all signed finite measures γ on $X \times Y$. The dual optimization problem seeks to find h, Φ, Ψ which maximize the infimum of the unconstrained problem. It is clear that $\inf_\gamma L_{h,\Phi,\Psi}(\gamma) = -\infty$ unless $c(x,y) + \Phi(x) + \Psi(y) - h(x,y) = 0$ for all $(x,y) \in X \times Y$. Fixing $h(x,y) = c(x,y) + \Phi(x) + \Psi(y)$, $L_{h,\Phi,\Psi}(\gamma)$ reduces to $\int \Phi\mu + \int \Psi\nu$ and since $h \geq 0$ we arrive at the dual problem

$$\sup\left(-\int \Phi\mu - \int \Psi\nu\right)$$

under the condition $c(x,y) \geq -\Phi(x) - \Psi(y)$ for all $(x,y) \in X \times Y$.

Finally, $\Phi : X \to \mathbb{R} \cup \{+\infty\}$ and $\Psi : Y \to \mathbb{R} \cup \{+\infty\}$ are c-convex if $(\Phi^c)^c = \Phi$ and $(\Psi^c)^c = \Psi$. Analogously to the classical multivalued gradient for convex functions, we define the c-gradient by $y \in \partial^c\Phi(x)$ and $x \in \partial^c\Phi^c$ if and only if

$$\Phi(x) + \Phi^c(y) = -c(x,y)$$

and similarly for $\Psi \in L^1(\nu)$.

3. Affine and Hessian structures

As explained in the introduction, an affine manifold is a smooth manifold equipped with an affine structure. We will give two equivalent definition of affine structures below:

DEFINITION 3.1. Let X be a smooth manifold of dimension $d \geq 1$. An affine structure on X is a distinguished atlas $\{(U_i, \alpha_i), \alpha_i : U_i \to \mathbb{R}^d\}$ whose transition functions are affine, i.e. for all i, j, $\alpha_i \circ \alpha_j^{-1} \in \mathrm{GL}_d(\mathbb{R}) \rtimes \mathbb{R}^d$.

DEFINITION 3.2. Let X be a smooth manifold of dimension $d \geq 1$. An affine structure on X is a flat torsion free connection ∇ on the tangent bundle of X.

Given a flat torsion free connection, the exponential map can be used to define coordinates for a distinguished atlas. Conversely, given a distinguished atlas, a flat torsion free connection can be constructed by pulling back the standard connection on $\mathbb{R}^d$ to U_i by α_i.

The affine structure defines a notion of geodesics and parallel transport of tensor fields on X. However, as ∂ is not necessarily the Levi-Civita connection of a Riemnannian metric there is no notion of distance or (globally) unit speed geodesics (see Example 3.5 below). Parallel transport on (X, ∇) is locally constant, i.e. the parallel transport along a curve only depend on the homotopy type of the curve. The global holonomy of the connection is related to the linear part of the transition functions in the distinguished atlas. Essentially, if H is a subgroup of $\mathrm{GL}_d(\mathbb{R})$ then there is a distinguished atlas such that $\alpha_i \circ \alpha_j \in H \rtimes \mathbb{R}^d$ if and only if the holonomy of ∇ lies in H. If the holonomy of ∇ lies in $\mathrm{SL}_d(\mathbb{R})$, then (X, ∇) is said to be *special affine.* If the holonomy of ∇ lies in $\mathrm{SL}_d(\mathbb{Z})$, then (X, ∇) is said to be *integral affine.* The holonomy of ∇ lies in $\mathrm{O}_d(\mathbb{R})$ if and only if ∇ is the Levi-Civita connection of a Riemannian metric.

Any special affine manifold admits a parallel volume form, unique up to scaling. We will use μ_X to denote the unique parallel volume form of total volume 1.

EXAMPLE 3.3. Let G be a subgroup of $\mathrm{GL}_d(\mathbb{R}) \rtimes \mathbb{R}^d$ and $\Omega \subset \mathbb{R}^d$ be a G-invariant open convex set with non-empty interior such that $X = \Omega/G$ is a smooth manifold. Then (X, ∇) where ∇ is the connection inherited from $\mathbb{R}^d$ is an affine manifold. If in addition, $G \subset \mathrm{SL}_d(\mathbb{R}) \otimes \mathbb{R}^d$ or $G \subset \mathrm{SL}_d(\mathbb{Z}) \otimes \mathbb{R}^d$, then (X, ∇) is special affine or integral affine, respectively. Whenever G is not commutative (see for example Example 3.6 below), X is not diffeomorphic to a torus.

EXAMPLE 3.4. As a special case of Example 3.3, let $G \subset O_d(\mathbb{R})$. Then X inherits the flat Euclidean metric on $\mathbb{R}^d$ and ∇ is the Levi-Civita connection of this metric. When G is cocompact, this case is categorized by Bieberbach's Theorem, saying that (X, ∇) is a finite quotient of a flat torus.

EXAMPLE 3.5. Also a special case of Example 3.3, but contrasting Example 3.4, let $\Omega = \mathbb{R}_+$ be the positive real numbers and G be the group generated by $2 \in \mathrm{GL}_1(\mathbb{R})$, i.e. the map that sends $x \in \mathbb{R}_+$ to $2x$. The geodesics on (X, ∇) are exactly the images of geodesics on $\mathbb{R}_+$. In particular, starting at $1 \in R_+$, moving in the direction of $-\partial/\partial x$, we get a geodesic γ on (X, ∇) which is only defined for $t \in [0, 1)$. This means that, although X is compact, (X, ∇) is not geodesically complete. The holonomy of the generator of $\pi_1(X)$ defined by $\gamma|_{[0,1/2]}$ is $2 \in \mathrm{GL}_1(\mathbb{R})$, and an intuitive picture of what is happening is that each time the geodesics traverses X

its speed doubles (measured as the length of its tangent vector in a fixed frame near a point in X). Consequently, the speed diverges as $t \to 1$. Note that (X, ∇) is not special. A conjecture by Markus (see for example [**Gol22**, Chapter 11]) states that a compact affine manifold is geodesically complete if and only if it is special.

EXAMPLE 3.6. Let Ω be the cone in $\mathbb{R}^2 \times \mathbb{R}$ defined by

$$\Omega = \{(x_1, x_2, t) \in \mathbb{R}^2 \times \mathbb{R} : Q(x_1, x_2, t) := x_1^2 + x_2^2 - t^2 < 0\}$$

and consider the action on Ω generated by

$$\begin{pmatrix} 2 & 0 & 0 \\ 0 & 2 & 0 \\ 0 & 0 & 2 \end{pmatrix}, \begin{pmatrix} \cosh(1) & 0 & \sinh(1) \\ 0 & 1 & 0 \\ \sinh(1) & 0 & \cosh(1) \end{pmatrix} \text{ and } \begin{pmatrix} 0 & -1 & 0 \\ 1 & 0 & 0 \\ 0 & 0 & 1 \end{pmatrix}.$$

The latter two matrices preserve the level set of Q and $X = \Omega/G$ is a compact affine manifold.

The topic of compact affine manifolds has been a very active research area at least since the 1970's (see for example [**Gol22**]). However, in the context of the special Lagrangian torus fibrations in the SYZ-conjecture, the base spaces are expected to have singularities. More precisely, (in the generic case) the base is expected to be a sphere equipped with an integral affine structure which extends to the complement of a co-dimension 2 set. We will now look closer at a family of such objects.

3.1. Singular integral affine structures on reflexive polytopes. Let $M_{\mathbb{R}}$ be a real vector spaces of dimension $d+1$, $N_{\mathbb{R}} = (M_{\mathbb{R}})^*$, $M \subset M_{\mathbb{R}}$ a lattice and $N \subset N_{\mathbb{R}}$ the dual lattice. A reflexive polytope $\Delta \subset M_{\mathbb{R}}$ is a lattice polytope (i.e. the convex hull of a set of lattice points) such that its dual

$$\Delta^\vee = \{n \in N_{\mathbb{R}} : \langle m, n\rangle \leq 1 \text{ for all } x \in \Delta\} \subset N_{\mathbb{R}}$$

is also a lattice polytope. For each facet σ of Δ we will let n_σ denote the corresponding vertex of $\Delta^\vee$ and for each facet τ of $\Delta^\vee$ we will let m_τ denote the corresponding vertex of Δ. We will use σ° and τ° to denote the relative interior of σ and τ, respectively and $\operatorname{Star}(m)$, for $m \in \partial\Delta$ to denote the union of all closed faces in $\partial\Delta$ containing m.

EXAMPLE 3.7. Note that Δ is homeomorphic to the unit sphere. We will now explain how to a define a natural atlas on a large subsets of the boundary $\partial\Delta$ of Δ. First of all, for each facet τ of $\Delta^\vee$, pick an open set U_τ in $\operatorname{Star}(m_\tau)$ such that the sets in the collection $\{U_\tau\}_\tau$ are pairwise disjoint and $\cup U_\tau$ covers a large part (say everything except for a (d-2)-dimensional subset) of the $d-1$ skeleton in $\partial\Delta$. The distinguished atlas will consist of one coordinate chart on U_τ for each facet τ of $\Delta^\vee$ and one coordinate chart on σ° for each pair of facets σ and τ of Δ and $\Delta^\vee$ such that $m_\tau \in \sigma$ or, equivalently, $n_\sigma \in \tau$.

For each facet τ, let $n_1, \ldots, n_d$ be a set of generators of the lattice $N \cap m_\tau^\perp$ in the orthogonal complement of m_τ and let $\alpha_\tau : \operatorname{Star}(m_\tau) \to \mathbb{R}^d$ be defined by

$$\alpha_\tau(m) = (\langle m, n_1\rangle, \ldots, \langle m, n_d\rangle).$$

We define the distinguished atlas as the set of coordinate charts $\{(U_\tau, \alpha_\tau|_{U_\tau})\}_\tau$ together with the coordinate charts $\{(\sigma^\circ, \alpha_\tau|_{\sigma^\circ})\}_{\sigma,\tau}$ where the latter family is indexed by all pairs of facets σ and τ of Δ and $\Delta^\vee$, respectively, such that $m_\tau \in \sigma$.

LEMMA 3.8. *The coordinate charts in Example* 3.7 *define an integral affine structure.*

PROOF. (See also [**AH23**, Lemma 3]) It suffice to show that if $n_\sigma \in \tau\cap\tau'$, then $\alpha_\tau|_{\sigma^\circ} \circ (\alpha_{\tau'}|_{\sigma^\circ})^{-1} \in \mathrm{SL}_d(\mathbb{Z}) \rtimes \mathbb{R}^d$. First of all, let $n_1, \ldots, n_d$ be the generators for $N \cap m_\tau^\perp$ used in the definition of α_τ. Since $\langle m_\tau, n_\sigma\rangle = 1 \neq 0$ we get that $n_\sigma \notin m_\tau^\perp$, hence if $\langle m, n_\sigma\rangle = 0$ and $\langle m, n_i\rangle = 0$ for all i, then $m = 0$. It follows that α_τ define a non-degenerate affine map from the affine subspace spanned by σ to $\mathbb{R}^d$. The same argument can be made for $\alpha_{\tau'}$ Consequently, $\alpha_\sigma \circ \alpha_{\sigma'}$ is affine.

To prove the lemma it thus suffices to verify that its linear part is in $\mathrm{SL}_d(\mathbb{Z})$. To do this, we will show that both α_σ and $\alpha_{\sigma'}$ maps $M \cap \mathrm{affspan}(\sigma)$ to $\mathbb{Z}^d \in \mathbb{R}^d$. First of all, since $n_1, \ldots, n_d \in N$, we get that $\alpha_\tau(m) \in \mathbb{Z}^d$ if $m \in M$. For the converse, assume $m \in M \cap \mathrm{affspan}(\sigma)$ and $\alpha_\tau(m) \in \mathbb{Z}^d$. Let $e_0, \ldots, e_d$ be generators of M. We have

$$n_\sigma \wedge n_1 \wedge \ldots \wedge n_d = \langle m_\tau, n_\sigma\rangle e_0 \wedge \ldots \wedge e_d = e_0 \wedge \ldots \wedge e_d,$$

hence $n_\sigma, n_1, \ldots, n_d$ generate N. By assumption $\langle m, n_\sigma\rangle = 1 \in \mathbb{Z}$ and $\langle m, n_i\rangle \in \mathbb{Z}$ for each i. Consequently, $m \in M$. The corresponding statement for $\alpha_{\tau'}$ is proved in the same way. □

REMARK 3.9. The definition of the integral affine structure in Example 3.7 involves a choice of sets $\{U_\tau\}_\tau$. However, note that this choice only influence the integral affine structure on the $(d-1)$-skeleton of $\partial\Delta$. More precisely, the choice of sets $\{U_\tau\}_\tau$ is equivalent to picking a "boundary" in each $(d-1)$-face of Δ which divides the face into a number of open regions, each one of these regions corresponding to one of the vertices of the face. In particular, the parallel volume form of the integral affine structures defines a probability measure μ_M on $\partial\Delta$ independent of this choice. Similarly, we get a probability measure ν_N on $\partial\Delta^\vee$.

REMARK 3.10. Assume Φ is a convex function on $M_\mathbb{R}$ such that

$$|\sup_{n\in\Delta^\vee} \langle m, n\rangle - \Phi|$$

is bounded. Then, as argued in [**AH23**], Φ dictates a natural choice for $\{U_\tau\}_\tau$, namely

$$U_\tau = \Delta \setminus \partial^c\Phi^c(\Delta^\vee \setminus \tau^\circ)$$

where Φ^c is the c-transform of Φ and $\partial^c\Phi^c$ its c-gradient with respect to the cost function $c(m,n) = -\langle m, n\rangle_{M,N}$ and $\langle\cdot,\cdot\rangle_{M,N}$ is the pairing of $M_\mathbb{R}$ and $N_\mathbb{R}$. In sufficiently regular cases, this is equivalent to putting $U_\tau = \partial^c\Phi^c(\tau^\circ)$. However, the complement of $\cup U_\tau$ does not in general have co-dimension 2. An interesting open problem is whether it does have co-dimension 2 when $\Psi = \Phi^c$ is the solution in Theorem 6.8 and Theorem 6.9 below (see also Section 6.1).

REMARK 3.11. The integral affine structure in Example 3.7 relates directly to a class of families of Calabi-Yau manifolds relevant to SYZ mirror symmetry. When Δ is Delzant, it defines a toric Fano manifold Z_Δ of complex dimension $d+1$. Let s_0 be the $(\mathbb{C}^*)^{d+1}$ invariant section of its anticanonical bundle and s be a generic section of its anticanonical bundle and Y be the family of Calabi-Yau manifolds given by

$$Y = \{ts + s_0 : t \in \mathbb{C}^*\} \subset Y. \tag{3.1}$$

Then $\Delta^\vee$ arise as the bounded component of a limiting amoeba of Y and the coordinate charts on $\Delta^\vee$ described by the procedure above correspond to a set of holomorphic coordinate charts on Y which cover a large part of the Calabi-Yau manifolds near the central fiber (see [**Li22a**], and also [**HZ01, GS06**] for more details). Moreover, solvability of a certain Monge-Ampère equation on these has direct implications on the SYZ conjecture (see Corollary 6.10 below).

3.2. Hessian metrics.

DEFINITION 3.12. Let (X, ∇) be an affine manifold. A Hessian metric on (X, ∇) is a Riemannian metric g on X which is locally the Hessian of a convex function, i.e. there is a collection of open sets $\{U_i\}$ covering X and convex functions $\{\phi_i : U_i \to \mathbb{R}\}$ such that

$$g|_{U_i} = \nabla d\phi_i = \sum_{jk} \frac{\partial^2 \phi_i}{\partial x_j \partial x_k} dx_j \otimes dx_k$$

for each i, where $x_1, \ldots, x_d$ are coordinates on U_i from the distinguished atlas.

We will say that a collection of convex functions $\{\phi_i : U_i \to \mathbb{R}\}$ (not necessarily smooth) such that $\phi_i - \phi_j$ is affine, i.e. $\nabla d(\phi_i - \phi_j) = 0$, for all i is a *weak* Hessian metric. If (X, ∇, g) is a special Hessian manifold and u is a function on X such that $\phi_i + u$ is convex for each i, then we define the Monge-Ampère measure $\mathrm{MA}(g + \nabla du)$ as the Monge-Ampère measure of $\phi_i + u$ in the sense of Aleksandrov.

EXAMPLE 3.13. Let Ω and G be as in Example 3.3 and Φ a smooth strictly convex function on Ω such that $\Phi \circ g - \Phi$ is affine for all $g \in G$. Then the Riemannian metric on Ω defined by $\Phi_{ij} dx_i \otimes dx_j$ is G invariant and its pushforward under the quotient map defines Hessian metric on X. In particular, if $\Omega = \mathbb{R}$, and $G = \mathbb{Z}$ acting on $\mathbb{R}$ by translations, we may pick $\Phi = x^2/2$, since $\Phi(x - m) - \Phi(x) = -mx + m^2/2$ is affine in x for each $m \in \mathbb{Z}$. In Example 3.5 we can pick $\Phi(x) = -\log(x)$, since $\log(2^m x) - \log(x) = m \log(2)$ is affine in x. In Example 3.6 we can pick $\Phi = -\log(-Q)$.

EXAMPLE 3.14. In Example 3.7, any Φ defines a weak Hessian metric. On the charts given by $(\sigma^\circ, \alpha_\tau)$ we use $\Phi|_{\sigma^\circ}$ and on the charts given by (U_τ, α_τ) we use $(\Phi - n)|_{\sigma^\circ}$ for any vertex n of τ.

THEOREM 3.15. *Let (X, ∇, g) be a compact Hessian manifold. Then X is a quotient Ω/G, for some convex $\Omega \subset \mathbb{R}^d$ and G as in Example* 3.3. *Moreover, $g = \nabla d\Phi$ for some proper function $\Phi : \Omega \to \mathbb{R}$.*

The proof of Theorem 3.15 is mostly elementary, but it helps to introduce some terminology: Let $p : \Omega \to X$ be the universal covering space of X. We may pull back the affine structure and the Hessian metric on X to get a Hessian structure on Ω. The fact that Ω is simply connected implies the following:

- The pullback of g can be written as $\nabla d\Phi$ for some globally defined $\Phi : X \to \mathbb{R}$.
- Ω admits d linearly independent parallel 1-forms.

Fix $x_0 \in \Omega$. The first of these points can be showed by defining $\Phi(x) = \int_{\gamma_{x_0,x}} \int_{\gamma_{x_0,x}} g$, where $\gamma_{x_0,x}$ is some curve in Ω from x_0 to x and $\int_{\gamma_{x_0,x}} g$ is the 1-form given by integrating g along $\gamma_{x_0,x}$. The second point can be showed by fixing a basis of the co-tangent space of Ω at x_0 and extending this by paralell transport. Both of these procedures only depend on the homotopy type of the $\gamma_{x_0,x}$. Since Ω is simply connected they are thus independent of $\gamma_{x_0,x}$.

Let $m_1,\ldots,m_d$ be linearly independent parallel 1-forms on Ω and, as above, $x_0 \in \Omega$. The *developing map* at x_0 defined by $m_1,\ldots,m_d$ is the map $dev : \Omega \to \mathbb{R}$ given by

$$dev(x) = \left(\int_{\gamma_{x_0,x}} m_1, \ldots, \int_{\gamma_{x_0,x}} m_d \right)$$

where as above $\gamma_{x_0,x}$ is any curve in Ω from x_0 to x.

We will say that a geodesic segment $\gamma : [0,1] \to X$ (with respect to ∇) is maximal if it can't be continued beyond time 1, i.e. there is no other geodesic $\gamma' : [0, 1+\epsilon] \to X$, $\epsilon > 0$ such that $\gamma = \gamma|_{[0,1]}$. The following is an important property of compact Hessian manifolds:

LEMMA 3.16. *Assume $\gamma : [0,1] \to \Omega$ is a maximal geodesic segment. Then $\frac{d}{dt}\Phi(\gamma(t)) \to +\infty$ as $t \to 1$.*

PROOF. Let $f(t) = \Phi(\gamma(t))$. By Cauchy-Schwarz, we have

$$\begin{aligned}
\frac{d}{dt}\Phi(\gamma(t)) &= f'(t) \\
&= f'(0) + \int_0^t f''(s)ds \\
&\geq f'(0) + \frac{1}{t}\left(\int_0^t (f''(s))^{1/2} ds \right)^2 \\
&\to |\gamma|_g^2 \\
&= +\infty
\end{aligned}$$

where $|\gamma|_g$ denotes the length of γ with respect to g. To see that this is infinite, note that since (X, g) is a compact Riemannian manifold, $|\gamma|_g$ is either infinite, or $\gamma(t)$ has a limit x as $t \to 1$. But in the latter case, we can pick affine coordinates around x and continue γ as a straight line in these coordinates. This contradict the maximality of γ. $\square$

PROOF OF THEOREM 3.15, FOLLOWING [**SY97**]. The theorem is proved by showing that the developing map is injective and that its image is convex. To do this, it suffices to prove that if $x_1, x_2, x_3 \in \Omega$, and there are geodesics γ_{x_1,x_2} and γ_{x_2,x_3} connecting x_1 with x_2 and x_2 with x_3, respectively, then there is a geodesics connecting x_1 with x_3. Since every pair of points can be connected by a piecewise geodesic segment the theorem then follows by repeated applications of this statement. More preciesly, if $\gamma : [0,1] \to \Omega$ is a piece geodesic segment which is geodesic on $[t_1, t_2]$ and $[t_2, t_3]$ then the statement above can be used to modify γ on $[t_1, t_3]$ to a piecewise geodesic which is geodesic on $[t_1, t_3]$.

Assume for a contradiction that there is no geodesic from x_1 to x_3. Without loss of generality we may assume that there is a geodesic γ_t from $\gamma_{x_1,x_2}(1-t)$ to $\gamma_{x_2,x_3}(t)$ for all $t < 1$. Moreover, by considering the geodesic emanating from x_1 in the direction given by $\dot\gamma_{x_2,x_3} - \dot\gamma_{x_1,x_2}$ and the geodesic emanating from x_3 in the direction $\dot\gamma_{x_1,x_2} - \dot\gamma_{x_2,x_3}$ we get two maximal geodesic segments $\gamma_1 : [0,a) \to \mathbb{R}$, $\gamma_2 : [0,b) \to \mathbb{R}$ where $a+b \leq 1$. By Lemma 3.16,

$$\langle d\Phi, \dot\gamma_1\rangle = \langle d\Phi, \dot\gamma_{x_2,x_3} - \dot\gamma_{x_1,x_2}\rangle >> 0$$

close to $\gamma_1(a-\epsilon)$ in Ω and

$$\langle d\Phi, \dot\gamma_2\rangle = \langle d\Phi, \dot\gamma_{x_1,x_2} - \dot\gamma_{x_2,x_3}\rangle >> 0$$

close to $\gamma_2(b-\epsilon)$ in Ω. This contradicts convexity of Φ. □

REMARK 3.17. An alternative way of thinking about Theorem 3.15 is that if (X,∇,g) is a compact Hessian manifold, then any curve on X can be continuously deformed to a geodesic (with respect to ∇) while keeping its endpoints fixed.

4. Duality of Hessian manifolds

Let U be a convex subset of $\mathbb{R}^d$, $\Phi : U \to \mathbb{R}$ be smooth and strictly convex and put $g := \sum_{ij} \frac{\partial^2\Phi}{\partial x_i \partial x_j} dx_i \otimes dx_j$. Let $\Psi : \partial\Phi(U) \to \mathbb{R}$ be the Legendre transform of Φ and $y_1,\dots,y_d$ be the (non-affine) coordinates on U defined by the diffeomorphism $\partial\Phi$, i.e. $(y_1,\dots,y_d) = \partial\Phi(x_1,\dots,x_d)$. Let $\Phi_{ij} = \frac{\partial^2\Phi}{\partial x_i\partial x_j}$, $\Psi_{ij} = \frac{\partial^2\Psi}{\partial y_i \partial y_j}$ and (Φ^{ij}) denote the inverse of the matrix (Φ_{ij}). Using (2.1) we get

$$\begin{aligned} g &= \sum_{ij} \Phi_{ij} dx_i \otimes dx_j \\ &= \sum_{ij} \Phi^{ij} \circ \partial\Phi^* dy_i \otimes dy_j \\ &= \sum_{ij} \Psi_{ij} \circ \partial\Phi dy_i \otimes dy_j. \end{aligned} \tag{4.1}$$

This suggest there is another affine structure on U, different from the one inherited from $\mathbb{R}^d$, and for which g is still a Hessian metric. More precisely, if ∇' is the affine structure on U defined by pulling back the affine structure on $(\mathbb{R}^d)^*$ by $\partial\Phi$, then (U,∇',g) is a Hessian metric and a potential for g is given by $\Psi \circ \partial\Phi$.

This operation can be performed locally on a Hessian manifold (X,∇,g) to define a *dual Hessian structure* (X,∇',g).

LEMMA 4.1. *Let (X,∇,g) be a Hessian manifold and assume $\{U_i,\alpha_i\}$ is a set of affine coordinate charts covering X such that for each i $\alpha_i(U_i) \subset \mathbb{R}^d$ is convex and there is $\phi_i : U_i \to \mathbb{R}$ such that $g|_{U_i} = \nabla d\phi_i$. Then*

(1) *The affine structure ∇_i on U_i defined by pulling back the affine structure on $(\mathbb{R}^d)^*$ by $\partial(\phi_i \circ \alpha_i^{-1})$ has the following form: $\nabla'_i = 2\nabla_g - \nabla$, where ∇_g is the Levi-Civita connection of g. Consequently, the locally defined connections $\{\nabla'_i\}$ glue together to form a global connection ∇' on X.*
(2) *The object (X,∇',g) is a Hessian manifold and a set of potentials of g is given by $\{\phi_i^* \circ \partial\phi_i : U_i \to \mathbb{R}\}$.*

We will refer to [**SY97**] or [**Shi07**] for a proof of the first bullet point above. Assuming the first bullet point, the second bullet point then follows from the discussion preceding Lemma 4.1.

The Hessian manifold (X,∇',g) is called the dual of (X,∇,g).

4.1. Duals of compact Hessian manifolds. Let (X, ∇, g) be a compact Hessian manifold. By Theorem 3.15 we can assume (X, ∇, g) is of the form Ω/G for some convex Ω and G as in Example 3.3, equipped with a smooth, strictly convex function $\Phi : \Omega \to \mathbb{R}$ whose Hessian is the pullback of g. Note that $\partial\Phi$ defines a diffeomorphism from Ω to $\partial\Phi(\Omega)$. We can define an action of G on $\partial\Phi(\Omega) \subset (\mathbb{R}^d)^*$ by pushing forward the action of G on Ω, in other words: $g(\partial\Phi(m)) = \partial\Phi(m')$ if $g(m) = m'$. By construction, $\partial\Phi$ defines an equivivariant diffeomorphism from Ω to $\partial\Phi(\Omega)$, hence $\Phi(\Omega)/G$ is a compact smooth manifold. We can equip this with the affine structure inherited from $(\mathbb{R}^d)^*$.

LEMMA 4.2. *$\partial\Phi(\Omega)/G$ is affinely isomorphic to (X, ∇').*

PROOF. Let $\{U_i\}$ be a set of open sets in Ω such that the quotient map $p : \Omega \to X$ restricts to a homeomorphism on U_i for each i. For each i, we may pick $\phi_i := \Phi|_{U_i} \circ (p|_{U_i})^{-1}$ as a potential for g on $p(U_i)$. As noted above, $\partial\Phi$ defines a diffeomorphism from X to $\partial\Phi(\Omega)/G$. Restricting to U_i, this is exactly the map defined by $\partial\phi_i$. Consequently, the pullback of the connection on $\partial\Phi(\Omega)/G$ is exactly ∇'. □

We will say that two Hessian metrics g and g' on (X, ∇) lies in the same *class* if $g - g' = \nabla du$ for some smooth function $u : X \to \mathbb{R}$. It turns out that both the set $\partial\Phi(\Omega)$ and the action above only depend on the class of g.

THEOREM 4.3. *Assume (X, ∇, g) is a compact Hessian manifold and g' is a Hessian metric on (X, ∇) in the same class as g. Let $\Phi : \Omega \to \mathbb{R}$ and $\Phi' : \Omega \to \mathbb{R}$ be potentials of g and g', respectively. Then $\partial\Phi(\Omega) = \partial\Phi'(\Omega)$ and the action of G induced on $\partial\Phi(\Omega)$ by Φ is the same as the action induced by Φ'.*

PROOF. By properness of Φ and convexity of Ω we get that Φ extends to a lsc convex function on $\mathbb{R}^d$ by putting $\Phi(x) = +\infty$ if $x \notin \Omega$. It follows that $\overline{\partial\Phi(\Omega)} = \overline{\{\Phi^* < +\infty\}}$. Similarly, $\overline{\partial\Phi'(\Omega)} = \overline{\{(\Phi')^* < +\infty\}}$. However, since $\Phi - \Phi'$ is bounded (it descends to a smooth function on X), $\overline{\{\Phi^* < +\infty\}} = \overline{\{(\Phi')^* < +\infty\}}$. Since two convex open sets are the same if their closures are the same, we get that $\partial\Phi(\Omega) = \partial\Phi'(\Omega)$.

Assume $m \in \Omega$, $g \in G$, $n_0 = \partial\Phi(m)$ and $n_1 = \partial\Phi(gm)$. In other words, with respect to the action induced by Φ, g maps n_0 to n_1. Let $m_0' \in \Omega$ satisfy $n_0 = \partial\Phi'(m_0')$. We need to show that $n_1 = \partial\Phi'(gm_0')$, and hence that with respect to the the action induced by Φ', g maps n_0 to n_1.

To do this, let γ be a curve from m_0 to gm_0, γ_0 be a curve from m_0' to m_0 and note that $g \circ \gamma_0$ is a curve from gm_0' to gm_0. Let γ' be the curve from m_0' to gm_0' given by first traversing γ_0, then traversing γ and then traversing $g \circ \gamma_0$ backwards. The fact that $\sum_{ij} \Phi_{ij} dx_i \otimes dx_j$ is invariant under G means

$$\int_{\gamma_0} \sum_{ij} \Phi_{ij} dx_i \otimes dx_j = \int_{g\circ\gamma_0} \sum_{ij} \Phi_{ij} dx_i \otimes dx_j$$

and the fact that $\Phi-\Phi'$ is invariant under G means that $\int_\gamma \sum_{ij}(\Phi_{ij}-\Phi'_{ij})dx_i\otimes dx_j = 0$. We have

$$\begin{aligned}
\partial\Phi'(gm'_0) &= \partial\Phi'(m'_0) + \int_{\gamma'} \Phi'_{ij}dx_i\otimes dx_j \\
&= n_0 + \int_{\gamma_0} \Phi'_{ij}dx_i\otimes dx_j + \int_\gamma \Phi'_{ij}dx_i dx_j - \int_{g\circ\gamma_0} \Phi'_{ij}dx_i\otimes dx_j \\
&= \partial\Phi(m_0) + \int_\gamma \Phi_{ij}dx_i\otimes dx_j \\
&= \partial\Phi(gm_0) \\
&= n_1.
\end{aligned}$$

□

COROLLARY 4.4. *Let (X,∇,g) be a compact Hessian manifold and (X,∇',g) its dual. Up to affine isomorphism, ∇' only depends on the class of g. In other words, if g' is another Hessian metric on (X,∇) such that $g-g'=\nabla du$ for a smooth function u on X, then the dual of (X,∇,g') is (X,∇',g').*

PROOF. Follows directly from Lemma 4.2 and Theorem 4.3. □

REMARK 4.5. In analogy with the complex geometric case, if (X,∇,g) is a compact Hessian manifold, then it is possible to construct an 'affine line bundle' over X such that the collection of local potentials of any Hessian metric in the same class as g can be interpreted as a metric on the affine line bundle. See [**HO19**] for details.

4.2. Duals of reflexive polytopes. In Section 3.1 we described natural integral affine structures on large subsets of the boundary of reflexive polytopes. We will now describe their duals. The result we will present relies on some subtle regularity properties of the Hessian metric, hence might not be directly applicable, but arguably it gives a good intuitive picture.

First of all, we fix a reflexive polytope $\Delta\subset M_\mathbb{R}$ and a convex function Φ on $M_\mathbb{R}$ such that $|\sup_{n\in\Delta^\vee}\langle m,n\rangle-\Phi(m)|$ is bounded. As in Remark 3.10, we will let $U_\tau=\Delta\setminus\partial^c\Phi^{-1}(\Delta^\vee\setminus\tau^\circ)$.

Define $\Phi^c: N_\mathbb{R}\to\mathbb{R}$ by

$$\Psi(n)=\Phi^c(n)=\sup_{m\in\partial\Delta}\langle m,n\rangle-\Phi(m). \tag{4.2}$$

Note that $\Delta^\vee$ is also a reflexive polytope and $(\Delta^\vee)^\vee=\Delta$, hence

$$|\sup_{m\in(\Delta^\vee)^\vee}\langle m,n\rangle-\Psi(n)|$$

is bounded. Following Remark 3.10 again, we set $U_\sigma=\Delta^\vee\setminus\partial^c\Psi^{-1}(\Delta\setminus\sigma^\circ)$. Together with the recipe in Example 3.7, this determines integral affine structures on large subsets of $\partial\Delta$ and $\partial\Delta^\vee$.

We will assume that the singular Hessian metrics induced by Φ and Ψ are smooth on $\cup U_\tau$ and $\cup U_\sigma$. In addition, we will assume that $\partial^c\Phi:\Delta\to\Delta^\vee$ is a homeomorphism. The latter property has some very useful consequences:

- $\partial^c\Psi=(\partial^c\Phi)^{-1}$ is also a homeomorphism.
- The definition of U_τ and U_σ in Remark 3.10 reduces to $U_\tau=\partial^c\Psi(\tau)$ and $U_\sigma=\partial^c\Phi(\sigma)$.

- $\partial^c\Phi(\sigma) \subset \text{Star}(n_\sigma)$ and $\partial^c\Psi(\tau) \subset \text{Star}(m_\tau)$ for all facets σ of Δ and τ of $\Delta^\vee$ (see [**AH23**, Lemma 8]).
- Let $m \in \sigma^\circ$. Then the generators of $N \cap m_\tau^\perp$ and $M \cap n_\sigma^\perp$ in the definitons of α_τ and α_σ can be picked so that

$$\alpha_\sigma \circ \partial^c\Phi = \partial(\Phi \circ \alpha_\tau^{-1}) \circ \alpha_\tau$$

and if $m \in U_\tau$ then the generators can be picked so that

$$\alpha_\sigma \circ \partial^c\Phi = \partial((\Phi - n_\sigma) \circ \alpha_\tau^{-1}) \circ \alpha_\tau$$

(see [**AH23**, Corollary 3 and Lemma 14]).

THEOREM 4.6. *Let (X, ∇, g) be the integral affine structure on $\partial\Delta$ induced by Φ and (X, ∇', g) its dual. Then, under the assumptions above, (X, ∇) is affinely isomorphic to the integral affine structure on $\Delta^\vee$ induced by Ψ and g coincides with the Hessian metric defined by Ψ.*

PROOF. We may use the homeomorphism $\partial^c\Phi$ to identify the domain of the affine structure on $\partial\Delta$ with the domain of the affine structure of $\partial\Delta^\vee$. Moreover, by the last bullet point preceding this theorem, locally $\partial^c\Phi$ is the classical gradient of the potential of the Hessian metric, hence the theorem follows in the same way as Theorem 4.3. □

REMARK 4.7. The assumption that $\partial^c\Phi$ defines a homeomorphism is rather subtle. There does not seem to be any obvious ways to write down explicit examples Φ satisfying this. On the other hand, the second and third bullet point, and hence also the fourth, can be showed directly for the solution to important Monge-Ampère equations when Δ is the standard unit simplex or the unit cube ([**AH23**, Lemma 16 and Lemma 17]).

5. Pairings and optimal transport problems

The purpose of the detailed discussion about dual Hessian structures in Section 4.1 and Section 4.2 are to provide heuristic optimal transport theoretic motivations for the functionals in [**HO19, HJMM22, Li23, AH23**]. In order to do this, the picture need to be completed with an extra structure in the form of a pairing between (X, ∇, g) and its dual. We will do this in the two cases discussed in Section 4.1 and Section 4.2.

5.1. Compact Hessian manifolds. Assume (X, ∇, g) is a compact Hessian manifold. Pick Ω and G as in Example 3.3 so that $X = \Omega/G$ (c.f. Theorem 3.15) and let $\Phi : \Omega \to \mathbb{R}$ be a potential of the pullback of g. Identify (X, ∇') with $X^\vee := \partial\Phi(\Omega)/G$ (c.f. Lemma 4.2) and let Ψ be a potential of g (Ψ is the Legendre transform of Φ). We define the pairing $\langle \cdot, \cdot \rangle_\Phi : X \times X^\vee \to \mathbb{R}$ by

$$-\langle x, y \rangle_\Phi = \sup_{g \in G} \Phi(gx) + \Psi(y) - \langle gx, y \rangle = \sup_{g \in G} \Phi(x) + \Psi(gy) - \langle x, gy \rangle \tag{5.1}$$

where $\langle \cdot, \cdot \rangle$ is the standard pairing of $\mathbb{R}^d$ and $(\mathbb{R}^d)^*$ (note that the formula (5.1) describes a $G \times G$-invariant function on $\Omega \times \partial\Phi(x)$). Let μ be a probability measure on X, ν_X be the unique parallel probability measure on $X^\vee$ and consider the optimal transport problem defined by $(X, \mu), (X^\vee, \nu)$ and $c(x, y) = -\langle x, y \rangle_{\Phi,\Psi}$ we arrive at the dual functional

$$\int_X u\mu + \int_{X^\vee} u^c\nu \tag{5.2}$$

where

$$u^c(y) = \sup\langle x, y\rangle_\Phi - u(x)$$

which is the functional used in [**HO19**] to produce solutions to the equation

$$\mathrm{MA}(g + \nabla du) = \mu \text{ on } X.$$

5.2. Reflexive polytopes. Assume $\Delta \subset M_{\mathbb{R}}$ is a reflexive polytope. Let μ_M be the parallel volume form on $\partial\Delta$ and ν a measure on $\partial\Delta^\vee$. Let $\langle\cdot,\cdot\rangle_{M,N}$ be the pairing of $M_{\mathbb{R}}$ and $N_{\mathbb{R}}$. Considering the optimal transport problem given by $(\partial\Delta_{reg}, \mu_M)$, $(\partial\Delta_{reg}, \nu)$ and $c(m, n) = -\langle m, n\rangle_{M,N}$, we arrive at the dual functional

$$\int_{\partial\Delta} \Phi\mu_M + \int_{\partial\Delta^\vee} \Phi^c\nu$$

or, equivalently,

$$\int_{\partial\Delta} \Psi^c\mu_M + \int_{\partial\Delta^\vee} \Psi\nu$$

which is the functional used in [**HJMM22, AH23, Li23**].

6. Monge-Ampère equations

The first general existence result for Monge-Ampère equations on special affine manifolds is due to Cheng and Yau:

THEOREM 6.1 ([**CY82**], see also [**Del89**]). *Let (X, ∇, g) be a compact special Hessian manifold and μ a volume form on X, then there is a unique smooth function u on X such that $g + \nabla du$ is positive and*

$$\det(g + \nabla du) = c\mu \tag{6.1}$$

for a positive constant c.

REMARK 6.2. Theorem 6.1 has some interesting consequences. Let $\mu = \mu_X$ and let u be a solution to (6.1). Pull back $g + \nabla du$ to the covering space $\Omega \subset \mathbb{R}^d$ and let Φ be a global potential. By Theorem 3.15, Φ is a proper function which solves the equation

$$\det(D^2\Phi) = c$$

on Ω. It then follows from a theorem by Jurgen, Pogorelov and Calabi that $\Omega = \mathbb{R}^d$ and Φ is a quadratic polynomial. It follows that $g + \nabla du$ is a flat metric, ∇ is the Levi-Civita connection of $g + \nabla du$ and G acts on $\Omega = \mathbb{R}^d$ by rigid motions, hence (X, ∇) is of the type described by Example 3.4.

Using the variational approach described in Section 5 Theorem 6.1 was generalized by Önnheim and the author to:

THEOREM 6.3 ([**HO19**]). *Let (X, ∇, g) be a compact special Hessian manifold and μ a probability measure on X, then there is a unique function u on X such that* (6.1) *holds.*

REMARK 6.4. See also [**GT21**] for an approach to degenerate Monge-Ampère equation on compact Hessian manifolds based on the Perron method.

As before, let Δ be a reflexive polytope and $\Psi : N_{\mathbb{R}}$ satisfy $|\sup_{m\in\Delta}\langle m, n\rangle - \Psi|$ is bounded. We will use $\mathrm{MA}(\Psi)$ to denote the Monge-Ampère of the weak Hessian metric induced by Ψ (c.f. Example 3.14)

THEOREM 6.5 ([**HJMM22**]). *Let Δ be the $(d+1)$-dimensional unit simplex*

$$\Delta = \mathrm{Conv}\{-\sum e_i, (d+1)e_0, \dots, (d+1)e_d\}$$

and assume ν is a probability measure on $\partial\Delta^\vee$ which is invariant under the symmetry group of $\Delta^\vee$, then there is a unique $\Psi : N_\mathbb{R} \to \mathbb{R}$ satisfying $|\sup_{m\in\Delta}\langle m,n\rangle - \Psi|$ such

$$\mathrm{MA}(\Psi) = \nu. \tag{6.2}$$

REMARK 6.6. For $\nu = \nu_N$, the existence part of Theorem 6.5 was proved in [**Li22a**], where Ψ is constructed as a limit of the potentials of Calabi-Yau metrics on the associated family (3.1).

DEFINITION 6.7. Let Δ be a reflexive polytope and assume ν is a probability measure on $\partial\Delta^\vee$ supported on the open faces. We will say that ν is stable if the optimal transport problem given by $(\partial\Delta, \mu_M), (\partial\Delta^\vee, \nu)$ and cost function $c(m,n) = -\langle m,n\rangle_{M,N}$ admits an optimal transport plan supported on

$$\cup_\sigma \sigma^\circ \times \mathrm{Star}(n_\sigma)^\circ = \cup_\tau \mathrm{Star}(m_\tau)^\circ \times \tau^\circ.$$

THEOREM 6.8 ([**AH23**]). *Let Δ be a reflexive polytope and assume ν is a probability measure on $\partial\Delta^\vee$. Then there is a unique $\Psi : N_\mathbb{R} \to \mathbb{R}$ satisfying $|\sup_{m\in\Delta}\langle m,n\rangle - \Psi|$ and* (6.2) *if and only if ν is stable.*

When $\nu = \nu_N$, the stability condition above can be verified in cases where Δ have a lot of symmetries, for example the standard unit simplex (recovering Theorem 6.5) and the unit cube). However, given a general reflexive polytope Δ, it does not seem easy to check the stability of ν_N. On the other hand, a necessary condition for stability of ν_N is given by the volume estimates

$$\nu_N(\tau) \le \mu(\mathrm{Star}(\mu_M)) \text{ and } \mu_M(\sigma) \le \nu(\mathrm{Star}(n_\sigma)) \tag{6.3}$$

(see [**AH23**, Lemma 19]) which are easily computed given Δ. Computer aided computations show that out of the 4319 existing reflexive 3-dimensional polytopes, 1542 violate (6.3), hence are not stable (see [**AH23**, Table 2]).

Conversely, a sufficient condition for existence of solutions (and hence stability) was given in [**Li23**]:

THEOREM 6.9 ([**Li23**]). *Let Δ be a reflexive polytope such that $\langle m,n\rangle \neq 0$ for all vertices m and n of Δ and $\Delta^\vee$, respectively. Then there is a unique $\Psi : N_\mathbb{R} \to \mathbb{R}$ satisfying $|\sup_{m\in\Delta}\langle m,n\rangle - \Psi|$ and*

$$\mathrm{MA}(\Psi) = \nu_N.$$

Theorem 6.9 is striking since it applies to some polytopes with small symmetry groups. In particular, it applies when Δ is the polytope associated to $\mathbb{P}^3$ blown up in a torus invariant point. However, out of the 4319 existing reflexive 3-dimensional polytopes, only 238 satisfy this condition (see [**AH23**, Table 2]). Together with (6.3), this means that for the majority of reflexive polytopes Δ stability/existence of solutions is not known.

By the breakthrough by Yang Li in [**Li22a, Li20**], the existence results in Theorem 6.8 and Theorem 6.9 have the following consequence:

COROLLARY 6.10. *Assume Δ is a reflexive Delzant Polytope and either ν_N is stable or the condition in Theorem 6.9 holds, then* (3.1) *satisfies the weak metric*

SYZ conjecture, i.e. for every $\delta > 0$, the Calabi-Yau manifolds close to the central fiber admit a special Lagrangian torus fibration on a subset of normalized Calabi-Yau volume at least $1 - \delta$ (see [**Li22a**] *for details).*

REMARK 6.11. When Δ is the standard unit simplex and $\{s_0 + ts\}$ is the Fermat family, the weak metric SYZ conjecture was established (modulo passing to subsequence) by Yang Li in the seminal work [**Li22a**]. The requirement to pass to subsequence was eliminated and the scope extended to more general families of hypersurfaces of $\mathbb{P}^{d+1}$ in [**HJMM22, PS22**].

REMARK 6.12. In dimension 2, singular integral affine structures on the 2-sphere and solutions to Monge-Ampère equations where constructed by Loftin in [**Lof05**].

6.1. Open problems.

- Find sufficient conditions for stability (and thus existence of solutions) which are more general than the assumption in Theorem 6.9. Find necessary conditions for stability/existence of solutions which are less general than (6.3).
- Under what conditions are $\partial^c\Phi : \partial\Delta \to \partial\Delta^\vee$ a homeomorphism, and under what conditions does the image of the $(d-1)$-skeleton of Δ under $\partial^c\Phi$ intersect the $(d-1)$-skeleton of $\Delta^\vee$ in a $(d-2)$-dimensional set? The latter, but not the former, is known for the standard unit simplex and the unit cube by [**AH23**].
- Given a Hessian manifold (X, ∇, g) with dual (X, ∇', g). Find a natural pairing of X with itself, which generalize the pairings discussed in Section 5. Incidentally, since a pairing induce a notion of c-convex functions when used as a cost function, this also addresses a question concerning good positivity notions for Hessian metrics asked in [**Li22a, Li22b**].

In this exposition, we've intentionally avoided the question of regularity for the solutions to the Monge-Ampère equations. On compact Hessian manifolds, Pogorelov type counterexamples where ruled out by Caffarelli and Viaclovski in [**CV01**]. For the standard unit simplex and the unit cube, the solution is known to be smooth by [**AH23**]. However, for reflexive polytopes in general, the question of regularity is open. See for example [**Moo21, MR22**] for some pathological behaviour in related settings.

Acknowledgments

The author would like to thank Rolf Andreasson, Mattias Jonsson, Enrica Mazzon, Nick McCleerey, Yang Li and Magnus Önnheim for many discussions on the subject.

References

[AG13] L. Ambrosio and N. Gigli, *A user's guide to optimal transport*, Modelling and optimisation of flows on networks, Lecture Notes in Math., vol. 2062, Springer, Heidelberg, 2013, pp. 1–155, DOI 10.1007/978-3-642-32160-3_1. MR3050280

[AH23] R. Andreasson, J. Hultgren, *Solvability of Monge-Ampère equations and integral affine structures on reflexive polytopes*, arXiv:2303.05276, 2023.

[Ber13] R. Berman, *Statistical mechanics of permanents, real-Monge-Ampere equations and optimal transport*, arXiv:1302.4045, 2013.

[BB13] R. J. Berman and B. Berndtsson, *Real Monge-Ampère equations and Kähler-Ricci solitons on toric log Fano varieties* (English, with English and French summaries), Ann. Fac. Sci. Toulouse Math. (6) **22** (2013), no. 4, 649–711, DOI 10.5802/afst.1386. MR3137248

[BFJ15] S. Boucksom, C. Favre, and M. Jonsson, *Solution to a non-Archimedean Monge-Ampère equation*, J. Amer. Math. Soc. **28** (2015), no. 3, 617–667, DOI 10.1090/S0894-0347-2014-00806-7. MR3327532

[BoJ17] S. Boucksom and M. Jonsson, *Tropical and non-Archimedean limits of degenerating families of volume forms* (English, with English and French summaries), J. Éc. polytech. Math. **4** (2017), 87–139, DOI 10.5802/jep.39. MR3611100

[Bre91] Y. Brenier, *Polar factorization and monotone rearrangement of vector-valued functions*, Comm. Pure Appl. Math. **44** (1991), no. 4, 375–417, DOI 10.1002/cpa.3160440402. MR1100809

[BGJKM21] J. I. Burgos Gil, W. Gubler, P. Jell, K. Künnemann, and F. Martin, *Pluripotential theory for tropical toric varieties and non-Archimedean Monge–Ampére equations*, `arXiv:2102.07392`, 2021.

[Caf97] L. A. Caffarelli, *The regularity of mappings with a convex potential*, J. Amer. Math. Soc. **5** (1992), no. 1, 99–104, DOI 10.2307/2152752. MR1124980

[CV01] L. A. Caffarelli and J. A. Viaclovsky, *On the regularity of solutions to Monge-Ampère equations on Hessian manifolds*, Comm. Partial Differential Equations **26** (2001), no. 11-12, 2339–2351, DOI 10.1081/PDE-100107825. MR1876421

[CL06] A. Chambert-Loir, *Mesures et équidistribution sur les espaces de Berkovich* (French), J. Reine Angew. Math. **595** (2006), 215–235, DOI 10.1515/CRELLE.2006.049. MR2244803

[CLD12] A. Chambert-Loir and A. Ducros, *Formes différentielles réelles et courants sur les espaces de Berkovich*, `arXiv:1204.6277`, 2012.

[CY82] S. Y. Cheng and S.-T. Yau, *The real Monge-Ampère equation and affine flat structures*, Proceedings of the 1980 Beijing Symposium on Differential Geometry and Differential Equations, Vol. 1, 2, 3 (Beijing, 1980), Sci. Press Beijing, Beijing, 1982, pp. 339–370. MR714338

[Del89] P. Delanoë, *Remarques sur les variétés localement hessiennes* (French), Osaka J. Math. **26** (1989), no. 1, 65–69. MR991282

[Fig99] A. Figalli, *The Monge-Ampère equation and its applications*, Zurich Lectures in Advanced Mathematics, European Mathematical Society (EMS), Zürich, 2017, DOI 10.4171/170. MR3617963

[Gol22] W. M. Goldman, *Geometric structures on manifolds*, Graduate Studies in Mathematics, vol. 227, American Mathematical Society, Providence, RI, [2022] ©2022, DOI 10.1090/gsm/227. MR4500072

[GS06] M. Gross and B. Siebert, *Mirror symmetry via logarithmic degeneration data. I*, J. Differential Geom. **72** (2006), no. 2, 169–338. MR2213573

[GT21] V. Guedj and T. D. Tô, *Monge–Ampère equations on compact Hessian manifolds*, `arXiv:2106.14740`, 2021.

[HZ01] C. Haase and I. Zharkov. *Integral affine structures on spheres and torus fibrations of Calabi-Yau toric hypersurfaces I*, `arXiv:math/0205321`, 2002.

[Hul19a] J. Hultgren, *Permanental point processes on real tori, theta functions and Monge-Ampère equations* (English, with English and French summaries), Ann. Fac. Sci. Toulouse Math. (6) **28** (2019), no. 1, 11–65, DOI 10.5802/afst.1592. MR3940791

[HJMM22] J. Hultgren, M. Jonsson, E. Mazzon, and N. McCleerey, *Tropical and non-Archimedean Monge-Ampère equations for a class of Calabi-Yau hypersurfaces*, Adv. Math. **439** (2024), Paper No. 109494, 42, DOI 10.1016/j.aim.2024.109494. MR4692438

[HO19] J. Hultgren and M. Önnheim, *An optimal transport approach to Monge-Ampère equations on compact Hessian manifolds*, J. Geom. Anal. **29** (2019), no. 3, 1953–1990, DOI 10.1007/s12220-018-0068-5. MR3969419

[KS84] M. Knott and C. S. Smith, *On the optimal mapping of distributions*, J. Optim. Theory Appl. **43** (1984), no. 1, 39–49, DOI 10.1007/BF00934745. MR745785

[KS06] M. Kontsevich and Y. Soibelman, *Affine structures and non-Archimedean analytic spaces*, The unity of mathematics, Progr. Math., vol. 244, Birkhäuser Boston, Boston, MA, 2006, pp. 321–385, DOI 10.1007/0-8176-4467-9_9. MR2181810

[Lag12] A. Lagerberg, *Super currents and tropical geometry*, Math. Z. **270** (2012), no. 3-4, 1011–1050, DOI 10.1007/s00209-010-0837-8. MR2892935

[Li20] Y. Li, *Metric SYZ conjecture and non-Archimedean geometry*, Duke Math. J. **172** (2023), no. 17, 3227–3255, DOI 10.1215/00127094-2022-0099. MR4688155

[Li22a] Y. Li, *Strominger-Yau-Zaslow conjecture for Calabi-Yau hypersurfaces in the Fermat family*, Acta Math. **229** (2022), no. 1, 1–53. MR4460593

[Li22b] Y. Li, *Survey on the metric SYZ conjecture and non-Archimedean geometry*, Internat. J. Modern Phys. A **37** (2022), no. 17, Paper No. 2230009, 44, DOI 10.1142/S0217751X22300095. MR4468137

[Li23] Y. Li, *Metric SYZ conjecture for certain toric Fano hypersurfaces*, Camb. J. Math. **12** (2024), no. 1, 223–252. MR4701493

[Li23] Y. Li, *Intermediate complex structure limit for Calabi-Yau metrics* `arXiv:2305.02258`, 2023.

[Lof05] J. C. Loftin, *Singular semi-flat Calabi-Yau metrics on S^2*, Comm. Anal. Geom. **13** (2005), no. 2, 333–361. MR2154822

[MP21] E. Mazzon and L. Pille-Schneider, *Toric geometry and integral affine structures in non-archimedean mirror symmetry*, `arXiv:2110.04223`, 2021.

[Moo21] C. Mooney, *Solutions to the Monge-Ampère equation with polyhedral and Y-shaped singularities*, J. Geom. Anal. **31** (2021), no. 10, 9509–9526, DOI 10.1007/s12220-021-00615-2. MR4303931

[MR22] C. Mooney and A. Rakshit, *Singular structures in solutions to the Monge–Ampère equations with point masses*, `arXiv:2204.11365`, 2022.

[NX16] J. Nicaise and C. Xu, *The essential skeleton of a degeneration of algebraic varieties*, Amer. J. Math. **138** (2016), no. 6, 1645–1667, DOI 10.1353/ajm.2016.0049. MR3595497

[NXY19] J. Nicaise, C. Xu, and T. Y. Yu, *The non-archimedean SYZ fibration*, Compos. Math. **155** (2019), no. 5, 953–972, DOI 10.1112/s0010437x19007152. MR3946280

[OO21] Y. Odaka and Y. Oshima, *Collapsing K3 surfaces, tropical geometry and moduli compactifications of Satake, Morgan-Shalen type*, MSJ Memoirs, vol. 40, Mathematical Society of Japan, Tokyo, 2021, DOI 10.1142/e071. MR4290084

[PS22] L. Pille-Schneider, *Hybrid toric varieties and the non-archimedean SYZ fibration on Calabi-Yau hypersurfaces*. `arXiv:2210.05578`, 2022.

[San16] F. Santambrogio, *Dealing with moment measures via entropy and optimal transport*, J. Funct. Anal. **271** (2016), no. 2, 418–436, DOI 10.1016/j.jfa.2016.04.009. MR3501852

[Shi07] H. Shima, *The geometry of Hessian structures*, World Scientific Publishing Co. Pte. Ltd., Hackensack, NJ, 2007, DOI 10.1142/9789812707536. MR2293045

[SY97] H. Shima and K. Yagi, *Geometry of Hessian manifolds*, Differential Geom. Appl. **7** (1997), no. 3, 277–290, DOI 10.1016/S0926-2245(96)00057-5. MR1480540

[TY23] F. Tong, S.-T. Yau, *Generalized Monge-Ampère functionals and related variational problems*. `arXiv:2306.01636`, 2023.

[Vil09] C. Villani, *Optimal transport: Old and new*, Grundlehren der mathematischen Wissenschaften [Fundamental Principles of Mathematical Sciences], vol. 338, Springer-Verlag, Berlin, 2009, DOI 10.1007/978-3-540-71050-9. MR2459454

DEPT OF MATHEMATICS AND MATHEMATICAL STATISTICS, UMEØA UNIVERSITY, 901 87 UMEØA, SWEDEN

Email address: `jakob.hultgren@umu.se`

Contemporary Mathematics
Volume **810**, 2025
https://doi.org/10.1090/conm/810/16212

To the geometry of spaces of plurisubharmonic functions on a Kähler manifold

László Lempert

To Bo Berndtsson, a friend from long ago. His mathematics has inspired and continues to inspire many; the reader will find his imprint in this paper, too.

ABSTRACT. Consider a compact Kähler manifold (X,ω) and the space $\mathcal{E}(X,\omega)=\mathcal{E}$ of ω–plurisubharmonic functions of full Monge–Ampère mass on it. We introduce a quantity $\rho[u,v]$ to measure the distance between $u,v\in\mathcal{E}$; $\rho[u,v]$ is not a number but rather a decreasing function on a certain interval $(0,V)\subset\mathbb{R}$. We explore properties of $\rho[u,v]$, and using them we study Lagrangians and associated energy spaces of ω–plurisubharmonic functions. Many results here generalize Darvas's findings about his metrics d_χ.

1. Introduction

Consider an n dimensional connected compact Kähler manifold (X,ω) and the space $\mathrm{PSH}(X,\omega)$ of ω–plurisubharmonic functions $X\to[-\infty,\infty)$. (Its definition, as well as other background material will be reviewed in section 2.) Over thirty years now this space and its subspaces have been studied by endowing them with various metrics, and introducing special classes of paths, (weak) geodesics, in them. In this paper we focus on the space $\mathcal{E}(X,\omega)=\mathcal{E}\subset\mathrm{PSH}(X,\omega)$ of functions of full Monge–Ampère mass, and study a quantity $\rho[u,v]$, that we call the rise between $u,v\in\mathcal{E}$, a notion that is related to the metrics, but is more fundamental. To define it we have to review the idea of geodesics in $\mathcal{E}$ and some of their properties.

Given real numbers $a<b$, consider the strip

$$S_{ab}=\{s\in\mathbb{C}: a<\mathrm{Re}\, s<b\}$$

and let $\pi: S_{ab}\times X\to X$ be the projection. Following Berndtsson and Darvas [**Be1**], [**Da3**, section 3.3] we make this definition:

DEFINITION 1.1. (a) A path (i.e., a map) $\varphi:(a,b)\to\mathrm{PSH}(X,\omega)$ is a subgeodesic if $\Phi: S_{ab}\times X\to[-\infty,\infty)$ defined by $\Phi(s,x)=\varphi(\mathrm{Re}\, s)(x)$ is $\pi^*\omega$–plurisubharmonic.

2020 *Mathematics Subject Classification.* Primary 32Q15, 32U15, 53C35, 58B20, 58E30, 70H99.

Research partially supported by NSF grant DMS 1764167.

(b) Given $u, v \in \mathrm{PSH}(X, \omega)$, the geodesic $\psi : (a, b) \to \mathrm{PSH}(X, \omega)$ joining them is

$$\psi = \sup\{\varphi \mid \varphi : (a, b) \to \mathrm{PSH}(X, \omega) \text{ is subgeodesic, } \lim_a \varphi \le u, \lim_b \varphi \le v\}, \tag{1.1}$$

provided the supremum is not identically $-\infty$.

In (1.1) the limits are understood pointwise on X; they exist because $\pi^*\omega$-plurisub-harmonicity of the associated function Φ implies that $\varphi(\cdot)(x)$ is convex (or $\equiv -\infty$) for all $x \in X$. We will only be interested in geodesics when $u, v \in \mathcal{E}$. In this case the connecting geodesic ψ is also a subgeodesic that maps into $\mathcal{E}$; furthermore $\lim_a \psi = u$, $\lim_b \psi = v$ in $L^1(X)$ and in capacity (and if u, v are bounded, the limits are uniform) [**Da2**, Corollaries 5.3, 5.4, and (23)], [**Be2**, pp.156–157]. Accordingly, it is natural to define $\psi(a) = u$, $\psi(b) = v$, and call the function $\psi : [a, b] \to \mathcal{E}$ thus extended a geodesic.

Traditionally, what we call geodesics here are termed weak (or plurisubharmonic or maximal) geodesics, and the name geodesic is reserved for ψ in (1.1) that define a smooth path into the space $\mathcal{H}$ of Kähler potentials,

$$\mathcal{H} = \{u \in C^\infty(X) : \omega + i\partial\overline{\partial}u > 0\} \subset \mathrm{PSH}(X, \omega). \tag{1.2}$$

Such ψ have velocity $\dot{\psi} : [a, b] \to T\mathcal{H}$ that is parallel for a certain connection ∇ on the tangent bundle $T\mathcal{H}$ (Mabuchi's connection [**M**]). However, since what tradition calls weak geodesics turned out to be rather more important than smooth solutions of the geodesic equation $\nabla_{\dot{\psi}}\dot{\psi} = 0$, they earned the right to a short name, geodesic, and this is the name we will be using.

Over the years geodesics in various subspaces of $\mathcal{E}$ have been found to be subject to conservation laws [**S**, Corollary 3.19], [**Be2**, Proposition 2.2], [**Da1**], [**DNL**, Theorem 4.4]. The conserved quantities can be expressed in terms of the velocity of the geodesic and its decreasing rearrangement. As said, if $\psi : [a, b] \to \mathcal{E}$ is a geodesic and $x \in X$, the function $\psi(\cdot)(x)$ is convex on (a, b), and is either everywhere finite or $\equiv -\infty$. In the former case it has left and right derivatives, that we denote $\partial_t^-\psi(t)(x), \partial_t^+\psi(t)(x)$ or $\partial^\pm\psi$. We define the derivatives $\partial_t^+\psi(a)(x), \partial_t^-\psi(b)(x) \in [-\infty, \infty]$ as limits of $\partial_t^\pm\psi(t)(x)$ at $t = a, b$. If $\psi(\cdot)(x) \equiv -\infty$, we define $\partial_t^\pm\psi(t)(x) = 0$, but this convention will be of little importance. Both $\partial_t^\pm\psi(t) : X \to [-\infty, \infty]$ are Borel functions; these are the left and right velocities of ψ. For $u \in \mathcal{E}$ we denote by μ_u the Borel measure on X induced by the non–pluripolar product $(\omega + i\partial\overline{\partial}u)^n$; the full mass condition means that $\mu_u(X) = \mu_0(X) = \int_X \omega^n$. We will abbreviate

$$\mu_0(X) = V.$$

If $a \le t < b$, resp. $a < t \le b$, we view $\partial_t^+\psi(t), \partial_t^-\psi(t)$ as functions on the measure space $(X, \mu_{\psi(t)})$, and form their decreasing rearrangements

$$(\partial_t^+\psi(t))^\star,\ (\partial_t^-\psi(t))^\star : (0, V) \to \mathbb{R}.$$

It follows from [**Da1**, Lemma 4.10] that for geodesics ψ that map into

$$\mathcal{H}^{1\bar{1}} = \{u \in \mathrm{PSH}(X, \omega) : \text{the current } \partial\overline{\partial}u \text{ is represented by a bounded form}\}, \tag{1.3}$$

these rearrangements are conserved: $(\partial_t^-\psi(t))^\star$ and $(\partial_s^+\psi(s))^\star$ agree if $a < t \le b$, $a \le s < b$. (In Darvas's lemma $\dot{u}_t$ should be interpreted as right or left derivative.) At the same time, Darvas [**Da2**, section 7] argues that for geodesics in the somewhat larger space of Lipschitz continuous functions, right and left derivatives at the endpoints a, b may have different rearrangements; and [**L2**, Example 5.4] features a geodesic ψ in the same space for which $(\partial_t^- \psi(t))^\star \neq (\partial_t^+ \psi(t))^\star$ for all t (see [**L2**, (5.5) and the following discussion]).

Nevertheless, it is possible to generalize the conservation law for geodesics in $\mathcal{H}^{1\bar{1}}$ to geodesics in $\mathcal{E}$. This is based on the following:

THEOREM 1.2 (See Theorem 5.7). *Consider geodesics $\psi : [a, b] \to \mathcal{E}$ and $\psi_j : [a, b] \to \mathcal{H}^{1\bar{1}}$. If $\psi_j(t)$ decreases to $\psi(t)$ for all $t \in [a, b]$, then there is a decreasing upper semicontinuous (usc) function $\rho : (0, V) \to \mathbb{R}$ to which the decreasing rearrangements $(\partial_t^- \psi_j(t))^\star$ and $(\partial_t^+ \psi_j(t))^\star$ converge Lebesgue almost everywhere as $j \to \infty$, for all $t \in (a, b)$. This ρ depends only on ψ and not on the choice of ψ_j.*

Note that for monotone functions a.e. convergence is the same as convergence at each point of continuity of the limit function, and the same as convergence on a dense set. Furthermore, any geodesic $\psi : [a, b] \to \mathcal{E}$ can be obtained as the decreasing limit of geodesics $\psi_j : [a, b] \to \mathcal{H}^{1\bar{1}}$. This is so because one can choose $u_j, v_j \in \mathcal{H}$ that decrease to $\psi(a)$, resp. $\psi(b)$ (the customary references are [**De**, **DP**] and especially [**BK**]); the geodesics $\psi_j : [a, b] \to \mathcal{E}$ joining u_j, v_j map in fact into $\mathcal{H}^{1\bar{1}}$ by the work of Chen, with complements by Błocki [**Bl**, **C**]; and ψ_j will decrease to ψ by [**Da3**, Proposition 3.15].

DEFINITION 1.3. We call $\rho = \rho_\psi : (0, V) \to \mathbb{R}$ of Theorem 1.2 the rise of the geodesic ψ.

At first sight, ρ_ψ has nothing to do with conserved quantities; but it is conserved in the sense that for $[c, d] \subset [a, b]$, the rise of the geodesic $\psi|[c, d]$ is independent of the choice of c, d. This is obvious from the way ρ is obtained in Theorem 1.2.

Once we have the notion of the rise of a geodesic, we can also talk about the rise $\rho[u, v]$ between $u, v \in \mathcal{E}$: it is the rise of the geodesic $\psi : [0, 1] \to \mathcal{E}$ joining u, v,

$$\rho[u, v] = \rho_\psi. \tag{1.4}$$

This is a distance–like quantity, except it measures distance between u, v not by a number, but by a decreasing function $(0, V) \to \mathbb{R}$.

The notion of the rise is useful because it contains metric information. The distance between u, v measured in any of the Orlicz–type metrics d_χ that Darvas studies in [Da1–3], and the length of a geodesic measured in d_χ, are easily recovered from the rise, see Definition 10.4 and (10.4). For example, L^p distance $d_p(u, v)$ is just the L^p norm of $\rho[u, v]$, and similarly for d_χ and the more general notion of action $\mathcal{L}_T$ of [**L2**], associated with a Lagrangian. (The simplest instance of action is the increment of Monge–Ampère energy between u and v: it is $\int_0^V \rho[u, v]$, if u, v belong to the energy space $\mathcal{E}^1 \subset \mathcal{E}$, that we will introduce later.) At the same time, properties of the rise can be formulated independently of any choice of weight function χ or Lagrangian L. It is in this sense that the rise is more fundamental than the metrics.

The aim of this paper is to explore properties of the rise, of geodesics and between points in $\mathcal{E}$; and to use them to study Lagrangians on spaces of ω–plurisubharmonic functions. The reader will notice that many of our results correspond to Darvas's results on d_χ in [Da 1–3]. Even if the results here appear to be more general, often our proofs can simply be extracted from Darvas's proofs.

Contents. After reviewing background material in section 2, we dicuss rise in $\mathcal{H}^{1\bar{1}}$ and its relation with envelopes of functions in sections 3, 4. Section 5 defines the rise in $\mathcal{E}$ and formulates its main properties. In section 6 we show how, using decreasing rearrangements, one can compare velocities of geodesics with velocities of rectilinear paths. Section 7 is an admission that even if we can work with the notion of rise, its true meaning escapes us. The last three sections bring back Lagrangians to study metric properties of certain subspaces of $\mathcal{E}$, show how metric notions can be directly reduced to the notion of rise, and formulate a Principle of Least Action, generalizing [**L2**, Theorem 8.1].

2. Background

If Y is a complex manifold and Ω a real $(1,1)$ form on it, $d\Omega = 0$, a function $u : Y \to [-\infty, \infty)$ is said to be ω–plurisubharmonic if for every open set $U \subset Y$ on which Ω can be written as $i\partial\overline{\partial} f$, the function $u + f$ is plurisubharmonic on U. We write $\mathrm{PSH}(Y, \Omega)$ for the set of Ω–plurisubharmonic functions. Most of the time we will be interested in ω–plurisubharmonic functions on a connected compact Kähler manifold (X, ω) of dimension n. With any $u \in \mathrm{PSH}(X, \omega)$ one can associate a Borel measure μ_u on X, induced by $(\omega + i\partial\overline{\partial} u)^n$, see [GZ1–2]. We let $V = \int_X \omega^n = \mu_0(X)$, and define the space $\mathcal{E} \subset \mathrm{PSH}(X, \omega)$ consisting of those $u \in \mathrm{PSH}(X, \omega)$ for which

$$\mu_u(X) = V. \tag{2.1}$$

By $\mathcal{E}^\infty$ we denote the space of bounded functions in $\mathrm{PSH}(X, \omega)$; they are all contained in $\mathcal{E}$. In (1.2), (1.3) we have already introduced the subspaces $\mathcal{H} \subset \mathcal{H}^{1\bar{1}} \subset \mathcal{E}^\infty$. He, Berndtsson, and Darvas proved that geodescics between points in $\mathcal{H}^{1\bar{1}}, \mathcal{E}^\infty$, resp. $\mathcal{E}$ stay entirely in those spaces [**H**], [**Be1**, Section 2.2], [**Da2**, Corollary 5.4]. Not much is known about the regularity of these geodesics, but by work of Chen and Błocki, geodesics with endpoints in $\mathcal{H}$ are C^1 as maps into $C(X)$, [**Bl**, **C**].

Next we turn to rearrangements. If (X, μ) and (Y, ν) are measure spaces, almost everywhere defined measurable functions $\xi : X \to [-\infty, \infty]$ and $\eta : Y \to [-\infty, \infty]$ are said to be equidistributed, or strict rearrangements of one another, if $\mu(\xi^{-1}B) = \nu(\eta^{-1}B)$ for every Borel set $B \subset [-\infty, \infty]$. If $\mu(X) = \nu(Y) < \infty$, this is equivalent to $\mu(\xi > t) = \nu(\eta > t)$ for all $t \in \mathbb{R}$. Our notation for equidistribution will be

$$(\xi, \mu) \sim (\eta, \nu), \quad \text{or} \quad \xi \sim \eta, \tag{2.2}$$

if the measures are understood. The decreasing rearrangement of ξ is the decreasing usc function $\xi^\star : (0, \mu(X)) \to [-\infty, \infty]$ that is equidistributed with ξ, when $(0, \mu(X))$ is endowed with Lebesgue measure. Thus $\mu(s \le \xi \le t)$ is the length of the longest interval on which $s \le \xi^\star \le t$, $s, t \in [-\infty, \infty]$. If ξ is μ–almost everywhere finite, then $\xi^\star$ is finite everywhere. We have (see for example [**L2**, (3.2), (3.3)])

$$\mu(\xi > \xi^\star(s)) \le s \le \mu(\xi \ge \xi^\star(s)) \qquad \text{and} \tag{2.3}$$

$$\mu(\xi \ge \tau) < s \text{ implies } \xi^\star(s) < \tau, \qquad \mu(\xi > \tau) > s \text{ implies } \xi^\star(s) > \tau, \tag{2.4}$$

because, e.g., the Lebesgue measure of $\big(\xi^\star > \xi^\star(s)\big) \subset (0, s)$ equals $\mu\big(\xi > \xi^\star(s)\big)$.

LEMMA 2.1. *Suppose $\mu(X) < \infty$ and a.e. defined measurable functions $\xi_j : X \to [-\infty, \infty]$ converge a.e. to $\xi : X \to \mathbb{R}$. If $\xi^\star$ is continuous at some $s \in \big(0, \mu(X)\big)$, then $\lim_{j\to\infty} \xi_j^\star(s) = \xi^\star(s)$.*

PROOF. We can assume $\xi_j \to \xi$ everywhere. If $G \subset [-\infty, \infty]$ is open, then $(\xi \in G) \subset \bigcup_{i\geq 1} \bigcap_{j\geq i}(\xi_j \in G)$. Therefore, given $\delta > 0$, for sufficiently large j

$$\mu(\xi_j \in G) > \mu(\xi \in G) - \delta \quad \text{and} \quad \mu(\xi_j \notin G) < \mu(\xi \notin G) + \delta.$$

Let $0 < s < \mu(X)$ and apply these estimates with $G = (\xi^\star(s+\delta) - \delta, \infty]$, resp. $G = [-\infty, \xi^\star(s-\delta) + \delta)$, in conjunction with (2.3) to obtain

$$\mu\big(\xi_j > \xi^\star(s+\delta) - \delta\big) > \mu\big(\xi \geq \xi^\star(s+\delta)\big) - \delta \geq s,$$
$$\mu\big(\xi_j \geq \xi^\star(s-\delta) + \delta\big) < \mu\big(\xi > \xi^\star(s-\delta)\big) + \delta \leq s.$$

(2.4) now implies $\xi^\star(s+\delta) - \delta < \xi_j^\star(s) < \xi^\star(s-\delta) + \delta$ for sufficiently large j. If $\xi^\star$ is continuous at s, then $\xi_j^\star(s) \to \xi^\star(s)$ follows by letting $\delta \to 0$. □

LEMMA 2.2. *Suppose $\mu(X) < \infty$ and $\xi, \eta : X \to \mathbb{R}$ are a.e. defined measurable functions. If $F : \mathbb{R}^2 \to \mathbb{R}$ is convex, and increasing in both variables, then*

$$F(\xi,\eta)^\star(s) \leq F\big(\xi^\star(\sigma), \eta^\star(s-\sigma)\big), \qquad 0 < \sigma < s < \mu(X). \tag{2.5}$$

PROOF. It suffices to prove (2.5) for a dense set of (σ, s), because then to arbitrary (σ, s) we can converge by (σ_j, s_j) in this set in such a way that $\sigma_j < \sigma$, $s_j < s$, $s_j - \sigma_j < s - \sigma$; and (2.5) is obtained in the limit.

Consider first $F(x,y) = x + y$. If $\xi, \eta \geq 0$ are bounded, then $(\xi+\eta)^\star(s) \leq \xi^\star(\sigma) + \eta^\star(s-\sigma)$ holds by [**BS**, p. 41, (1.16)]. (Bennet and Sharpley use a different notion of rearrangement, but for nonnegative functions the two agree Lebesgue a.e.) The same also holds for arbitrary bounded ξ, η, because adding a constant will make them ≥ 0. Finally, (2.5) is obtained for general ξ, η by approximating them with bounded ξ_j, η_j, letting $j \to \infty$, and applying Lemma 2.1.

From this (2.5) follows when $F(x,y) = ax + by + c$, with $a, b \geq 0$. Indeed,

$$F(\xi,\eta)^\star(s) = (a\xi + b\eta)^\star(s) + c \leq (a\xi)^\star(\sigma) + (b\eta)^\star(s-\sigma) + c = F\big(\xi^\star(\sigma), \eta^\star(s-\sigma)\big).$$

Since a general $F(x,y)$ can be represented as $\sup_{a,b,c} ax + by + c$ with a certain family of triples $a, b \geq 0$, $c \in \mathbb{R}$, (2.5) holds in complete generality. □

We will talk about decreasing rearrangements of Borel functions ξ defined on our Kähler manifold (X, ω), and the measure will be μ_u for some $u \in \mathcal{E}$. If ξ is naturally associated with a $u \in \mathcal{E}$, for example because it arises as left or right velocity $\partial_t^\mp \varphi(t)$ of a path $\varphi : [a,b] \to \mathcal{E}$ and $u = \varphi(t)$, the measure will be μ_u, and notation, as before

$$\xi^\star = (\partial_t^\mp \varphi(t))^\star : (0, V) \to [-\infty, \infty], \quad \text{cf. (2.1).} \tag{2.6}$$

Yet if ξ is not clearly associated with some $u \in \mathcal{E}$, we will have to indicate with respect to which measure μ_u we rearrange, and the notation will be

$$\xi^{\star u} : (0, V) \to [-\infty, \infty], \qquad u \in \mathcal{E}. \tag{2.7}$$

As said, [**Da1**, Lemma 4.10] implies:

LEMMA 2.3. *If $\varphi : [a,b] \to \mathcal{H}^{1\bar{1}}$ is a geodesic, then $(\partial_t^- \varphi(t))^\star = (\partial_s^+ \varphi(s))^\star$ for $a < t \leq b$, $a \leq s < b$.*

Among the subspaces of $\mathcal{E}$ of interest, $\mathcal{H}$ is a manifold. Indeed, as an open subset of the Fréchet space $C^\infty(X)$, it inherits the structure of a Fréchet manifold; its tangent bundle $T\mathcal{H}$ has a canonical trivialization $T\mathcal{H} \approx \mathcal{H} \times C^\infty(X) \to \mathcal{H}$. As in [**L2**], we will consider certain functions $L : T\mathcal{H} \to \mathbb{R}$, Lagrangians, that give rise to geometric notions on $\mathcal{H}$ such as metrics, and more generally, actions. The Lagrangians of interest extend to Banach bundles larger than $T\mathcal{H}$. We write $B(X)$ for the Banach space of bounded Borel functions $X \to \mathbb{R}$, endowed with sup norm, and $T^\infty\mathcal{E} = \mathcal{E} \times B(X)$. This is a set theoretical Banach bundle over $\mathcal{E}$, into which $T\mathcal{H}$ embeds via the trivialization $T\mathcal{H} = \mathcal{H} \times C^\infty(X)$.

DEFINITION 2.4. By an invariant convex Lagrangian we mean a function $L : T^\infty\mathcal{E} \to \mathbb{R}$ that is convex and continuous on the fibers $T_u^\infty\mathcal{E} \approx B(X)$, $u \in \mathcal{E}$, and is strict rearrangement invariant in the sense that $L(\xi) = L(\eta)$ if $\xi \in T_u^\infty\mathcal{E}$, $\eta \in T_v^\infty\mathcal{E}$ are equidistributed as functions on $(X, \mu_u), (X, \mu_v)$. We say L is strongly continuous if the following holds: whenever $\xi_j \in T_u^\infty\mathcal{E}$ are uniformly bounded and converge μ_u–almost everywhere, then $L(\xi_j)$ is also convergent.

For example, $L(\xi) = \int_X |\xi|^p \, d\mu_u$, $\xi \in T_u^\infty\mathcal{E}$, defines an invariant, convex, strongly continuous Lagrangian, and so does its p'th root, $1 \le p < \infty$. But structurally the simplest Lagrangians, that generate all invariant, convex, strongly continuous Lagrangians in a precise sense, are obtained as follows.

LEMMA 2.5. *If $f : (0, V) \to \mathbb{R}$ is a decreasing integrable function, then*

$$L(\xi) = \int_0^V \xi^{\star u} f, \quad \xi \in T_u^\infty\mathcal{E} \approx B(X), \tag{2.8}$$

defines an invariant, convex, strongly continuous Lagrangian on $T^\infty\mathcal{E}$.

In (2.8) the integral is against Lebesgue measure.

PROOF. If $\xi \in T_u^\infty\mathcal{E}$ and $\eta \in T_v^\infty\mathcal{E}$ are equidistributed, then $\xi^{\star u} = \eta^{\star v}$ and $L(\xi) = L(\eta)$. For the rest of the proof we fix $u \in \mathcal{E}$, consider functions $\xi \in T_u^\infty\mathcal{E} \approx B(X, \mu_u)$, and will omit the u from the notation $\xi^{\star u}$ of decreasing rearrangement.

To prove that L is convex, we can assume f in (2.8) is usc. We claim

$$L(\xi) = \sup\Big\{ \int_X \gamma\xi \, d\mu_u \mid \gamma : X \to \mathbb{R} \text{ is Borel}, \gamma^\star = f \Big\}. \tag{2.9}$$

Indeed, let $\theta : X \to (0, V)$ preserve measure, when X is endowed with μ_u and $(0, V)$ with Lebesgue measure, restricted to the Borel sets. The existence of such θ follows e.g. from [**BS**, Proposition 7.4, p. 81]. Then $\gamma^\star = f$ is equivalent to $\gamma \sim f \circ \theta$. This shows that it suffices to prove (2.9) when f is bounded, the general case will follow by approximation. Since adding a constant c to f changes both sides of (2.9) by $c \int_X \xi \, d\mu_u$, we can even assume $f \ge 0$. Similarly, we can assume $\xi \ge 0$. But then our claim

$$L(\xi) = \int_0^V (f \circ \theta)^\star \xi^\star = \sup\Big\{ \int_X \gamma\xi \, d\mu_u : \gamma \sim f \circ \theta \Big\}$$

is an instance of [**BS**, Theorem 2.6, p. 49]. By (2.9) L is the supremum of linear forms, and must be convex.

Finally, strong continuity follows from Lemma 2.1 and dominated convergence. □

For the rest of the section we fix an invariant, convex, strongly continuous Lagrangian $L : T^\infty\mathcal{E} \to \mathbb{R}$. The Principle of Least Action [**L2**, Theorem 8.1] then says:

THEOREM 2.6. *If $\varphi : [a, b] \to \mathcal{E}^\infty$ is piecewise C^1, as a map into the Banach space $B(X)$, and $\psi : [a, b] \to \mathcal{E}^\infty$ is a geodesic of class C^1, connecting $\varphi(a)$ and $\varphi(b)$, then*

$$\int_a^b L(\partial_t \varphi(t))\, dt \geq \int_a^b L(\partial_t \psi(t))\, dt.$$

Given $T \in (0, \infty)$, the (least) action between $u, v \in \mathcal{E}^\infty$ is

$$\mathcal{L}_T(u, v) = \inf \int_0^T L(\partial_t \varphi(t))\, dt, \tag{2.10}$$

the infimum taken over all piecewise C^1 paths $\varphi : [0, T] \to \mathcal{E}^\infty$ joining u, v. If the geodesic $\psi : [0, T] \to \mathcal{E}^\infty$ joining u, v is C^1, Theorem 2.6 implies.

$$\mathcal{L}_T(u, v) = \int_0^T L(\partial_t \psi(t))\, dt. \tag{2.11}$$

We will need the following continuity properties of $\mathcal{L}_T, L$:

LEMMA 2.7. *If $u_j, v_j \in C(X) \cap \mathrm{PSH}(X, \omega)$ decrease, or converge uniformly, to $u \in \mathcal{E}^\infty$, resp. $v \in \mathcal{E}^\infty$, then*

$$\lim_{j\to\infty} \mathcal{L}_T(u_j, v_j) = \mathcal{L}_T(u, v) \in \mathbb{R}.$$

This is [**L2**, Lemma 9.4]. By (the proof of) Lemma 10.5 below and the discussion around it, it also holds if $u_j, v_j \in \mathcal{E}^\infty$, and even more generally.

Let $B(0, V)$ stand for the Banach space of bounded Borel functions on $(0, V)$, with the sup norm. Any strict rearrangement invariant function $L : T^\infty\mathcal{E} \to \mathbb{R}$ determines (and is determined by) a function $L_\star : B(0, V) \to \mathbb{R}$,

$$L_\star(\zeta) = L(\xi) \qquad \text{if } \zeta \in B(0, V) \text{ and } \xi \in T^\infty_u\mathcal{E} \text{ are equidistributed.} \tag{2.12}$$

In particular, $L(\xi) = L_\star(\xi^\star)$. Indeed, if $u \in \mathcal{E}$ and $\theta : (X, \mu_u) \to \big((0, V), \text{Lebesgue}\big)^1$ preserves measure, we can define $L_\star(\zeta) = L(\zeta \circ \theta)$.

LEMMA 2.8. *If uniformly bounded $\zeta_j \in B(0, V)$ converge a.e. to $\zeta \in B(0, V)$, then $L_\star(\zeta_j) \to L_\star(\zeta)$.*

PROOF. With a measure preserving $\theta : (X, \mu_u) \to (0, V)$ as before,

$$\lim_j L_\star(\zeta_j) = \lim_j L(\zeta_j \circ \theta) = L(\zeta \circ \theta) = L_\star(\zeta).$$

□

3. Rise in $\mathcal{H}^{1\bar{1}}$

In this section we introduce the notion of rise in $\mathcal{H}^{1\bar{1}}$.

DEFINITION 3.1. The rise $\rho_\varphi : (0, V) \to \mathbb{R}$ of a geodesic $\varphi : [a, b] \to \mathcal{H}^{1\bar{1}}$ is $(\partial_t^- \varphi(t))^\star = (\partial_s^+ \varphi(s))^\star$ for any $a < t \leq b$, $a \leq s < b$, cf. Lemma 2.3. The rise between $u, v \in \mathcal{H}^{1\bar{1}}$, denoted $\rho[u, v]$, is the rise ρ_φ of the geodesic $\varphi : [0, 1] \to \mathcal{H}^{1\bar{1}}$ joining u and v.

[1] Here and in what follows, we always restrict Lebesgue measure on $(0, V)$ to the Borel σ–algebra.

An affine reparametrization $\varphi(pt+q)$ of a geodesic φ, where $p, q \in \mathbb{R}$, is also a geodesic, because the same holds for subgeodesics. This implies that with an arbitrary geodesic $\varphi : [a,b] \to \mathcal{H}^{1\bar{1}}$

$$\rho[\varphi(a), \varphi(b)] = (b-a)\rho_\varphi. \tag{3.1}$$

A geodesic $\varphi : [a,b] \to \mathcal{H}^{1\bar{1}}$ can be reversed to produce a geodesic $\psi : [-b,-a] \to \mathcal{H}^{1\bar{1}}$, $\psi(t) = \varphi(-t)$. It follows that the function $-\rho_\varphi(V - \cdot)$ is a decreasing function, equidistributed with ρ_ψ. In general, it will not be usc, so all we can say is that

$$\rho_\psi(\lambda) = -\rho_\varphi(V-\lambda) \quad \text{for a.e. } \lambda \in (0,V). \tag{3.2}$$

LEMMA 3.2. *Suppose $u, v, w \in \mathcal{H}^{1\bar{1}}$.*

(a) *If $v \le w$ then $\rho[u,v] \le \rho[u,w]$ and $\rho[v,u] \ge \rho[w,u]$.*
(b) *If $c \in \mathbb{R}$ then $\rho[u, v+c] = \rho[u,v] + c = \rho[u-c, v]$.*

PROOF. Let $\varphi_c : [0,1] \to \mathcal{H}^{1\bar{1}}$ be the geodesic that joins u with $v+c$, and $\psi : [0,1] \to \mathcal{H}^{1\bar{1}}$ the geodesic that joins u with w. Definition 1.1 implies that $\varphi_0(t) \le \psi(t)$, and $\varphi_c(t) = \varphi_0(t) + ct$. Since $\varphi_0(0) = \psi(0)$,

$$\partial^+\varphi_0(0) \le \partial^+\psi(0) \quad \text{and} \quad \partial_t^\pm \varphi_c(t) = \partial_t^\pm \varphi_0(t) + c.$$

Hence the corresponding decreasing rearrangements are related in the same way, $\rho[u,v] \le \rho[u,w]$ and $\rho[u,v+c] = \rho[u,v] + c$. The rest of the lemma can be proved similarly. ☐

LEMMA 3.3. *Consider geodesics $\varphi, \varphi_j : [a,b] \to \mathcal{H}^{1\bar{1}}$, $j \in \mathbb{N}$, and assume that the $(1,1)$ forms $\partial\bar\partial\varphi_j(a), \partial\bar\partial\varphi_j(b)$ on X are uniformly bounded. If $\varphi_j(a), \varphi_j(b)$ decrease, or converge uniformly, to $\varphi(a)$, resp. $\varphi(b)$, then $\rho_{\varphi_j} \to \rho_\varphi$ a.e. More precisely, the latter convergence holds at all points of continuity of ρ_φ.*

Equivalently, if $u_j, v_j \in \mathcal{H}^{1\bar{1}}$ decrease or converge uniformly to $u, v \in \mathcal{H}^{1\bar{1}}$, and $\partial\bar\partial u_j, \partial\bar\partial v_j$ are uniformly bounded, then $\rho[u_j, v_j] \to \rho[u,v]$ at all points of continuity of $\rho[u,v]$.

PROOF. Consider the case when $\varphi_j(a), \varphi_j(b)$ decrease. As $j \to \infty$, by [**Da3**, Proposition 3.15] $\varphi_j(t)$ decrease to $\varphi(t)$ for every $t \in [a,b]$. Further, as explained in the proof of [**L2**, Lemma 11.2], it follows from He's work [**H**] that the family $\mu_{\varphi_j(t)}$, $\mu_{\varphi(t)}$, $j \in \mathbb{N}$, $t \in [a,b]$, of measures is hereditarily tight, in the sense that given an open $U \subset X$ and $\varepsilon > 0$, there exists a compact $K \subset U$ such that $\mu_{\varphi_j(t)}(U \setminus K)$, $\mu_{\varphi(t)}(U \setminus K) < \varepsilon$.

Fix $t \in (a,b)$. By convexity $\partial^-\varphi(t) \le \partial^+\varphi(t)$ on X. Since the two functions are equidistributed, it follows that $\partial^-\varphi(t) = \partial^+\varphi(t)$ $\mu_{\varphi(t)}$–almost everywhere. In other words, for $\mu_{\varphi(t)}$–almost every $x \in X$ the function $\varphi(\cdot)(x)$ is differentiable at t. For brevity, we denote the derivative $\dot\varphi(t)(x)$. It follows that $\partial^+\varphi_j(t)(x) \to \dot\varphi(t)(x)$ at such x. Indeed, given $\varepsilon > 0$, choose $s > 0$ so that

$$\Big|\frac{\varphi(t \pm s)(x) - \varphi(t)(x)}{\pm s} - \dot\varphi(t)(x)\Big| < \varepsilon. \tag{3.3}$$

By convexity

$$\frac{\varphi_j(t-s)(x) - \varphi_j(t)(x)}{-s} \le \partial^+\varphi_j(t)(x) \le \frac{\varphi_j(t+s)(x) - \varphi_j(t)(x)}{s}.$$

As $j \to \infty$, the upper and lower bounds of $\partial^+\varphi_j(t)(x)$ tend to $\big(\varphi(t\pm s)(x) - \varphi(t)(x)\big)/\pm s$. Hence $\partial^+\varphi_j(t)(x) \to \dot\varphi(t)(x)$ by (3.3), i.e., $\partial^+\varphi_j(t) \to \dot\varphi(t)$ $\mu_{\varphi(t)}$–a.e. on X.

This and hereditary tightness imply by [**L2**, Lemma 3.6]

$$\rho_{\varphi_j} = (\partial^+\varphi_j(t))^{\star\varphi_j(t)} \to \dot\varphi(t)^{\star\varphi(t)} = \rho_\varphi, \qquad j\to\infty,$$

away from a countable subset of $(0,V)$. As the functions involved are monotone, the exceptional set will be included in the set of discontinuity of ρ_φ, as claimed.

The case of uniformly convergent $\varphi_j(a), \varphi_j(b)$ can be reduced in a standard way to the decreasing case, see for example the last paragraph of the proof of [**L2**, Lemma 3.4]. □

LEMMA 3.4. *Suppose $L: T^\infty\mathcal{E} \to \mathbb{R}$ is an invariant, convex, strongly continuous Lagrangian. If $T\in(0,\infty)$ and $u,v\in\mathcal{H}^{1\bar1}$, then (cf. (2.10), (2.12))*

$$\mathcal{L}_T(u,v) = TL_\star(\rho[u,v]/T).$$

PROOF. First assume $u,v\in\mathcal{H}$, so that the geodesic $\psi:[0,T]\to\mathcal{H}^{1\bar1}$ between them is C^1, viewed as a map into $B(X)$. By (2.11), (2.12), and (3.1)

$$\mathcal{L}_T(u,v) = \int_0^T L(\dot\psi(t))\,dt = \int_0^T L_\star(\rho_\psi)\,dt = TL_\star(\rho[u,v]/T).$$

For general $u,v\in\mathcal{H}^{1\bar1}$ it is easy to find sequences $u_j,v_j\in\mathcal{H}$ with $\partial\bar\partial u_j, \partial\bar\partial v_j$ uniformly bounded, that converge uniformly to u,v. (One approximates, say, $(1-1/j)u$ by convolutions on coordinate neighborhoods, then patches together the local approximants by a smooth partition of unity.) By Lemmas 2.7, 3.2, 3.3, and 2.8, and by dominated convergence

$$\mathcal{L}_T(u,v) = \lim_j \mathcal{L}_T(u_j,v_j) = \lim_j TL_\star(\rho[u_j,v_j]/T) = TL_\star(\rho[u,v]/T).$$

□

4. Envelopes and rise in $\mathcal{H}^{1\bar1}$

Various spaces of ω–plurisubharmonic functions on (X,ω) form lattices. We will use $\vee,\wedge$ to denote the lattice operations, envelopes. In general, if $u,v\in\mathrm{PSH}(X,\omega)$, we define $u\vee v, u\wedge v: X\to[-\infty,\infty)$ by

$$\begin{aligned}(u\vee v)(x) &= \max(u(x),v(x)) \qquad \text{and}\\ (u\wedge v)(x) &= \sup\{w(x) : w\in\mathrm{PSH}(X,\omega),\ w\le u,v\}.\end{aligned} \tag{4.1}$$

Thus $u\vee v\in\mathrm{PSH}(X,\omega)$ and, according to [**Da3**, Section 2.4], $u\wedge v\in\mathrm{PSH}(X,\omega)$ or $\equiv -\infty$. Darvas proves that if $u,v\in\mathcal{E}$, then $u\vee v, u\wedge v\in\mathcal{E}$ [**Da2**, Corollaries 2.7, 3.5]. Also, if $u,v\in\mathcal{H}^{1\bar1}$, then $u\wedge v\in\mathcal{H}^{1\bar1}$ by [**DR**, Theorem 2.5]. Darvas and Rubinstein write $P(u,v)$ for what is denoted $u\wedge v$ here. That $\wedge$ is used both for envelope of functions and for exterior product of forms might confuse the careless reader; but since these two occur in different contexts, and $u\wedge v$ is rather simpler than $P(u,v)$, adopting it is worth the risk.

LEMMA 4.1. *Let $u,v\in\mathcal{H}^{1\bar1}$, and $\varphi,\psi,\theta:[a,b]\to\mathcal{H}^{1\bar1}$ the geodesics joining u with $u\wedge v$, $u\wedge v$ with v, and u with v. Then*

$$\partial^+\varphi(a) = \min\big(0,\partial^+\theta(a)\big) \quad \textit{and} \quad \partial^-\psi(b) = \max\big(0,\partial^-\theta(b)\big). \tag{4.2}$$

Hence $\rho_\varphi = \min(0,\rho_\theta)$, $\rho_\psi = \max(0,\rho_\theta)$, *and*

$$\rho[u, u\wedge v] + \rho[u\wedge v, v] = \rho[u,v].$$

The lemma corresponds to Darvas's Pythagorean Theorem, [**Da1**, Proposition 4.13], [**Da2**, Proposition 8.1].

PROOF. We show that two functions are equal by checking that their level sets coincide. By [**Da3**, Lemma 3.17], for any $\tau\in\mathbb{R}$

$$(\partial^+\theta(a)\geq\tau) = \big(u\wedge(v-\tau)=u\big), \tag{4.3}$$

and replacing v by $u\wedge v$

$$\big(\partial^+\varphi(a)\geq\tau\big) = \big(u\wedge(u\wedge v-\tau)=u\big) = \big(u\wedge(u-\tau)\wedge(v-\tau)=u\big).$$

This latter set is $\big(u\wedge(v-\tau)=u\big)$ if $\tau\leq 0$, and empty otherwise. Comparison with (4.3) gives the first formula in (4.2). The second follows by applying what we proved but with the roles of u, v interchanged. The rest of the claim is immediate from (4.2). □

Next we look at how enveloping interacts with rise. The result corresponds to [**Da2**, Proposition 8.2].

LEMMA 4.2. *Let* $u,v,w\in\mathcal{H}^{1\bar 1}$. *If* $u\leq v$ *then* $\rho[u\wedge w, v\wedge w]\leq\rho[u,v]$; *if* $u\geq v$ *then* $\rho[u\wedge w, v\wedge w]\geq\rho[u,v]$.

PROOF. Assume $u\leq v$, and let $\varphi,\psi:[0,1]\to\mathcal{H}^{1\bar 1}$ be the geodesics between $u=u\wedge v$ and v, respectively $u\wedge w$ and $v\wedge w$. Lemma 4.1 implies $\rho_\varphi\geq 0$, and similarly $\rho_\psi\geq 0$. Using again [**Da3**, Lemma 3.17], for $\tau\in\mathbb{R}$

$$\mu_u\big(\partial^+\varphi(0)\geq\tau\big) = \mu_u\big(u\wedge(v-\tau)=u\big), \tag{4.4}$$

$$\mu_{u\wedge w}\big(\partial^+\psi(0)\geq\tau\big) = \mu_{u\wedge w}\big(u\wedge w\wedge(v\wedge w-\tau)=u\wedge w\big). \tag{4.5}$$

If $\tau\leq 0$, both measures equal V. If $\tau>0$ then $w\wedge(v\wedge w-\tau)=v\wedge w-\tau$, and so the set on the right of (4.5) is $\big(u\wedge(v\wedge w-\tau)=u\wedge w\big)$. According to Darvas [**Da2**, Proposition 2.2], for Borel sets $Y\subset X$

$$\mu_{u\wedge w}(Y) = \mu_u\big(Y\cap(u=u\wedge w)\big) + \mu_w\big(Y\cap(u>u\wedge w=w)\big). \tag{4.6}$$

In light of (4.5), then

$$\begin{aligned}\mu_{u\wedge w}\big(\partial^+\psi(0)\geq\tau\big) \leq \mu_u\big(u\wedge(v\wedge w-\tau)&=u\wedge w=u\big)\\ &+\mu_w\big(u\wedge(v\wedge w-\tau)=u\wedge w=w\big).\end{aligned} \tag{4.7}$$

Since $v\wedge w-\tau<w$, the second term on the right is 0. Furthermore

$$u\wedge(v\wedge w-\tau)\leq u\wedge(v-\tau)\leq u,$$

whence the first term on the right of (4.7) is $\leq\mu_u(u\wedge(v-\tau)=u)$. Comparing (4.4) and (4.7) we therefore obtain $\mu_{u\wedge w}(\partial^+\psi(0)\geq\tau)\leq\mu_u(\partial^+\varphi(0)\geq\tau)$, or

$$\rho[u\wedge w, v\wedge w] = (\partial^+\psi(0))^\star \leq (\partial^+\varphi(0))^\star = \rho[u,v].$$

A similar argument gives the lemma when $u\geq v$. □

Both lemmas support the view of $\rho[u,v]$ as measuring some distance between u,v. The question arises if ρ satisfies the triangle inequality. It is easy to see that $\rho[u,v]+\rho[v,w]\ge\rho[u,w]$ will not hold for a general triple $u,v,w\in\mathcal{H}$. Indeed, if it did, it would also hold for the triple w,v,u. But in light of (3.2), the two triangle inequalities would simply imply $\rho[u,v]+\rho[v,w]=\rho[u,w]$, something that fails if $u=w$ but $\rho[u,v]$ is not a constant.

However, the triangle inequality does hold in an integrated form:

LEMMA 4.3. *If $u,v,w\in\mathcal{H}^{1\bar{1}}$ and $\lambda\in[0,V]$, then*

$$\int_0^\lambda \rho[u,v]+\int_0^\lambda \rho[v,w]\ge\int_0^\lambda \rho[u,w]. \tag{4.8}$$

PROOF. The proof will exhibit (4.8) as an instance of the Principle of Least Action, Theorem 2.6. In view of Lemma 3.3 it suffices to prove when $u,v,w\in\mathcal{H}$. Let $\varphi:[0,2]\to\mathcal{H}^{1\bar{1}}$ be such that $\varphi|[0,1]$ is the geodesic connecting u,v and $\varphi|[1,2]$ is the geodesic connecting v,w. Let $\psi:[0,2]\to\mathcal{H}^{1\bar{1}}$ be the geodesic connecting u,w. For Lagrangians L as in Theorem 2.6 we have $\int_0^2 L(\partial_t\varphi(t))\,dt\ge\int_0^2 L(\partial_t\psi(t))\,dt$, or

$$L_\star(\rho[u,v])+L_\star(\rho[v,w])\ge 2L_\star(\rho[u,w]/2) \tag{4.9}$$

by (2.11) and Lemma 3.4. We apply this with the Lagrangian

$$L(\xi)=\int_0^\lambda \xi^\star=\int_0^V f\xi^\star,\quad \xi\in T^\infty\mathcal{E},$$

where $f=1_{(0,\lambda)}$ is the characteristic function of $(0,\lambda)$, cf. Lemma 2.5. Since $L_\star(\xi^\star)=\int_0^\lambda\xi^\star$, (4.9) reduces to (4.8). □

If we introduce a partial order $\succeq$ on decreasing integrable functions on $(0,V)$ whereby $f\succeq g$ means $\int_0^\lambda f\ge\int_0^\lambda g$ for $0<\lambda<V$ (known in harmonic analysis as the Hardy–Littlewood–Pólya relation), then (4.8) becomes $\rho[u,v]+\rho[v,w]\succeq\rho[u,w]$.

5. The rise in $\mathcal{E}$

We are ready to extend the notion of rise to geodesics φ and functions u,v in $\mathcal{E}$.

THEOREM 5.1. *If sequences u_j, $v_j\in\mathcal{H}^{1\bar{1}}$ decrease to $u,v\in\mathcal{E}$, then the functions $\rho[u_j,v_j]:(0,V)\to\mathbb{R}$ converge a.e. to a decreasing usc function $r:(0,V)\to\mathbb{R}$. The function r depends on u,v but not on the approximating sequences u_j,v_j.*

DEFINITION 5.2. We call the function r above the rise between u,v, and denote it by $\rho[u,v]$. If $\psi:[a,b]\to\mathcal{E}$ is a geodesic, we define its rise by

$$\rho_\psi=\frac{1}{b-a}\rho[\psi(a),\psi(b)].$$

Thus both types of rise are decreasing usc function $(0,V)\to\mathbb{R}$. Since if $u,v\in\mathcal{H}^{1\bar{1}}$, we can choose $u_j=u,v_j=v$, within $\mathcal{H}^{1\bar{1}}$ the new notion of the rise agrees with the notion in Definition 3.1.

For the proof of Theorem 5.1 we have to recall some of Darvas's results in [**Da4**]. With a concave smooth diffeomorphism $\chi:[0,\infty)\to[0,\infty)$ Guedj and Zeriahi associate the space

$$\mathcal{E}_\chi=\Big\{u\in\mathcal{E}:\int_X\chi(|u|)\,d\mu_u<\infty\Big\}$$

and Darvas defines a metric d_χ on $\mathcal{H}^{1\bar{1}}$ (which he subsequently extends to all of $\mathcal{E}_\chi$). When $u,v\in\mathcal{H}^{1\bar{1}}$ and $\varphi:[0,1]\to\mathcal{H}^{1\bar{1}}$ is the geodesic to join them,

$$d_\chi(u,v)=\int_X \chi\big(|\partial_t^\pm\varphi(t)|\big)\,d\mu_{\varphi(t)},\qquad 0<t<1,$$

see [**Da4**, (24)]. He proves [**Da4**, Lemma 5.2.]:

LEMMA 5.3. *If $u_j,v_j\in\mathcal{H}^{1\bar{1}}$ decrease to $u,v\in\mathcal{E}_\chi$, then $d_\chi(u_j,v_j)$ is convergent and its limit depends only on u,v, not on the choice of u_j,v_j. This limit is denoted $d_\chi(u,v)$.*

In $\mathcal{H}^{1\bar{1}}$, d_χ can be expressed through $\rho[u,v]$,

$$d_\chi(u,v)=\int_0^V \chi\big(|(\partial^+\varphi(0))^\star|\big)=\int_0^V\chi(|\rho[u,v]|),\quad u,v\in\mathcal{H}^{1\bar{1}}. \tag{5.1}$$

LEMMA 5.4. *Suppose $u_j,v_j\in\mathcal{H}^{1\bar{1}}$ decrease to $u,v\in\mathcal{E}_\chi$, $u_j\le v_j$, and $\rho[u_j,v_j]$ converge a.e. to a function $r:(0,V)\to[0,\infty)$. Then*

$$d_\chi(u,v)=\lim_{j\to\infty}d_\chi(u_j,v_j)=\int_0^V\chi\circ r.$$

PROOF. Since $\rho[u_j,v_j]\ge 0$, in light of (5.1) we need to show

$$\lim_{j\to\infty}\int_0^V\chi(\rho[u_j,v_j])=\int_0^V\chi\circ r. \tag{5.2}$$

As in the proof of [**GZ2**, Proposition 10.16], one can construct a smooth increasing function $f:[0,\infty)\to[1,\infty)$ such that $\lim_\infty f=\infty$, $\overline{\chi}=f\chi:[0,\infty)\to[0,\infty)$ is still a concave diffeomorphism, and $u,v\in\mathcal{E}_{\overline{\chi}}$. Thus the sequence $d_{\overline{\chi}}(u_j,v_j)=\int_0^V\overline{\chi}(\rho[u_j,v_j])$ is still convergent by Lemma 5.3; let $C=\sup_j d_{\overline{\chi}}(u_j,v_j)$.

With a positive number M write

$$\rho_M[u_j,v_j]=\min(M,\rho[u_j,v_j])\quad\text{and}\quad r_M=\min(M,r).$$

By dominated resp. monotone convergence

$$\lim_{j\to\infty}\int_0^V\chi(\rho_M[u_j,v_j])=\int_0^V\chi\circ r_M,\qquad \lim_{M\to\infty}\int_0^V\chi\circ r_M=\int_0^V\chi\circ r. \tag{5.3}$$

Furthermore, using also (5.1)

$$\begin{aligned}0\le\int_0^V\chi(\rho[u_j,v_j])-\int_0^V\chi(\rho_M[u_j,v_j])&\le\int_{\rho[u_j,v_j]>M}\chi(\rho[u_j,v_j])\\&\le\frac{1}{f(M)}\int_{\rho[u_j,v_j]>M}\overline{\chi}(\rho[u_j,v_j])\le\frac{1}{f(M)}\int_0^V\overline{\chi}(\rho[u_j,v_j])\le\frac{C}{f(M)}.\end{aligned}$$

Since $\lim_{M\to\infty}f(M)=\infty$, this estimate together with (5.3) gives (5.2). $\square$

PROOF OF THEOREM 5.1. By [**GZ2**, Proposition 10.16] there is a concave diffeomorphism $\chi:[0,\infty)\to[0,\infty)$ such that $u\in\mathcal{E}_\chi$; and inspecting the proof we see that we can arrange $\chi''<0$ and $v\in\mathcal{E}_\chi$ as well. By Lemma 5.3 the sequence $d_\chi(u_j,v_j)$ is convergent, hence bounded. Assume first that $u_j\le v_j$ for all j. If $0<\lambda<V$, by (5.1)

$$0\le\chi\big(\rho[u_j,v_j](\lambda)\big)\le\frac{1}{\lambda}\int_0^\lambda\chi(\rho[u_j,v_j])\le\frac{1}{\lambda}d_\chi(u_j,v_j).$$

In particular, the sequences $\chi(\rho[u_j, v_j])$ and $\rho[u_j, v_j]$ are pointwise bounded. By Helly's theorem a subsequence $\rho[u_{j_k}, v_{j_k}]$ will converge everywhere to a decreasing function. By modifying the limit function at its points of discontinuity we obtain a decreasing usc function $r : (0, V) \to [0, \infty)$ such that $\lim_{k\to\infty} \rho[u_{j_k}, v_{j_k}] = r$ at points where r is continuous. Fix a $\tau \in (0, V)$ where r is continuous. We claim that if $\overline{u}_j, \overline{v}_j \in \mathcal{H}^{1\bar{1}}$ decrease to u, v, and $\overline{u}_j \le \overline{v}_j$, then

$$\rho[\overline{u}_j, \overline{v}_j](\tau) \to r(\tau); \tag{5.4}$$

this will then prove the theorem whenever $u_j \le v_j$.

Indeed, suppose (5.4) fails, and $|\rho[\overline{u}_j, \overline{v}_j](\tau) - r(\tau)| \ge \delta$ with some $\delta > 0$ and infinitely many j. Using Helly's theorem as before, we would then have a sequence $i_1 < i_2 < \dots$ and a decreasing usc function $\overline{r} : (0, V) \to [0, \infty)$ such that

$$\lim_{k\to\infty} \rho[\overline{u}_{i_k}, \overline{v}_{i_k}] = \overline{r} \quad \text{wherever } \overline{r} \text{ is continuous; but } \overline{r}(\tau) \ne r(\tau). \tag{5.5}$$

By Lemma 5.4 $\int_0^V \chi \circ r = d_\chi(u, v) = \int_0^V \chi \circ \overline{r}$. But, if $f \in C^\infty[0, \infty)$ has compact support contained in $(0, \infty)$ and $\varepsilon > 0$ is sufficiently small, then $\chi + \varepsilon f : [0, \infty) \to [0, \infty)$ is also a concave diffeomorphism, and $u, v \in \mathcal{E}_{\chi+\varepsilon f}$. Therefore

$$\int_0^V (\chi + \varepsilon f) \circ r = \int_0^V (\chi + \varepsilon f) \circ \overline{r}, \qquad i.e., \qquad \int_0^V f \circ r = \int_0^V f \circ \overline{r}.$$

The latter even holds for characteristic functions $f = 1_{[p,q]}$, where $0 < p < q$, since these characteristic functions are decreasing limits of smooth functions f with compact support in $(0, \infty)$. Therefore r and $\overline{r}$ are equidistributed. As both decrease and are usc, $r = \overline{r}$ follows, in contradiction with (5.5). This proves the theorem when $u_j \le v_j$. Clearly, the theorem also holds if $u_j \ge v_j$, cf. (3.2).

Now consider general u_j, v_j. By [**Da2**, Corollary 3.5] and [**DR**, Theorem 2.5] $u \wedge v \in \mathcal{E}$ and $u_j \wedge v_j \in \mathcal{H}^{1\bar{1}}$. A moment's thought gives that the ω–plurisubharmonic function $w = \lim_j u_j \wedge v_j$ is equal to $u \wedge v$. Indeed, $u_j \wedge v_j \ge u \wedge v$ implies $w \ge u \wedge v$. Also $w \le u_j, v_j$, whence $w \le u, v$ and $w \le u \wedge v$. From Lemma 4.1

$$\rho[u_j, u_j \wedge v_j] = \min(0, \rho[u_j, v_j]), \quad \rho[u_j \wedge v_j, v_j] = \max(0, \rho[u_j, v_j]), \tag{5.6}$$

and we conclude that $\rho[u_j, v_j] = \rho[u_j, u_j \wedge v_j] + \rho[u_j \wedge v_j, v_j]$ indeed converges a.e.; the limit depends only on $u, u \wedge v$ and v, i.e., on u, v. □

THEOREM 5.5. *If $u, v \in \mathcal{E}$, then*

$$\rho[u, u \wedge v] = \min(0, \rho[u, v]), \quad \rho[u \wedge v, v] = \max(0, \rho[u, v]).$$

In particular, $\rho[u, v] = \rho[u, u \wedge v] + \rho[u \wedge v, v]$.

This is immediate from (5.6) and Definition 5.2.—Theorem 5.1 has the following generalization:

THEOREM 5.6. *If $u_j, v_j \in \mathcal{E}$ decrease to $u, v \in \mathcal{E}$, then $\rho[u_j, v_j] \to \rho[u, v]$ at all points in $(0, V)$ where $\rho[u, v]$ is continuous.*

PROOF. It will suffice to prove convergence along a subsequence. For each $j \in \mathbb{N}$ let $u_{ij}, v_{ij} \in \mathcal{H}, i = 1, 2, \dots$, strictly decrease to u_j, v_j (i.e., $u_{i+1,j} < u_{ij}$ and $v_{i+1,j} < v_{ij}$ everywhere). Similarly, choose $U^i, V^i \in \mathcal{H}$ strictly decreasing to u, v. Let $\tau_1, \tau_2, \dots \in (0, V)$ be a dense sequence of points where each $\rho[u, v], \rho[u_j, v_j]$ is continuous. By Theorem 5.1/Definition 5.2

$$\rho[u_{ij}, v_{ij}] \to \rho[u_j, v_j] \qquad \text{at each } \tau_k \text{ as } i \to \infty. \tag{5.7}$$

We define natural numbers $1 = i(1) < i(2) < \dots$ and $1 = j(1) < j(2) < \dots$ recursively, and set $U_k = u_{i(k)j(k)}$, $V_k = v_{i(k)j(k)}$, as follows. Suppose we already have $i(k-1), j(k-1)$. If j is sufficiently large, then $u_j < U_{k-1}, U^k$ and $v_j < V_{k-1}, V^k$ by Dini's theorem. Fix such $j = j(k) > j(k-1)$. Since $\lim_{i\to\infty} u_{ij} = u_j < U_{k-1}, U^k$ and $\lim_{i\to\infty} v_{ij} = v_j < V_{k-1}, V^k$, for sufficiently large $i = i(k) > i(k-1)$ again by Dini's theorem and by (5.7)

$$u_{ij} < U_{k-1}, U^k, \quad v_{ij} < V_{k-1}, V^k, \quad \text{and} \quad |\rho[u_{ij}, v_{ij}] - \rho[u_j, v_j]| < 1/k \tag{5.8}$$

at $\tau_1, \dots, \tau_k$. Thus $U_k = u_{i(k)j(k)}$ decrease to u and $V_k = v_{i(k)j(k)}$ decrease to v. By Theorem 5.1/Definition 5.2 therefore $\rho[U_k, V_k] \to \rho[u, v]$ a.e.; and in fact by monotonicity, at each τ_l. Hence (5.8) gives $\rho[u_{j(k)}, v_{j(k)}] \to \rho[u, v]$ at each τ_l. But this implies convergence at each point of continuity of $\rho[u, v]$. □

Theorem 5.6 has an obvious consequence concerning geodesics.

THEOREM 5.7. *If $\varphi_j : [a, b] \to \mathcal{E}$ are geodesics that decrease to a geodesic $\varphi : [a, b] \to \mathcal{E}$, then $\rho_{\varphi_j} \to \rho_\varphi$ a.e.*

Indeed, $\rho_{\varphi_j} = \rho[\varphi_j(a), \varphi_j(b)]/(b - a)$ and $\rho_\varphi = \rho[\varphi(a), \varphi(b)]/(b - a)$. This then proves Theorem 1.2.

The results of section 4 easily generalize from $\mathcal{H}^{1\bar{1}}$ to $\mathcal{E}$.

THEOREM 5.8. *Let $u, v, w \in \mathcal{E}$. If $u \le v$ then $\rho[u, v] \ge \rho[u \wedge w, v \wedge w]$; if $u \ge v$ then $\rho[u, v] \le \rho[u \wedge w, v \wedge w]$.*

The statement follows from Lemma 4.2 and Theorem 5.6 upon representing u, v, w as limits of decreasing sequences $u_j, v_j, w_j \in \mathcal{H}^{1\bar{1}}$.

THEOREM 5.9. *If $u, v, w \in \mathcal{E}$ and $0 \le \lambda < V$, then*

$$\int_0^\lambda \rho[u, v] + \int_0^\lambda \rho[v, w] \ge \int_0^\lambda \rho[u, w]. \tag{5.9}$$

One should keep in mind that the integrals in (5.9) may be infinite. In this case a stronger version of the triangle inequality

$$\int_0^\lambda \big(\rho[u, v] + \rho[v, w] - \rho[u, w]\big) \ge 0$$

would be more meaningful; but we were not able to prove it. The difficulty is how to prove that the negative part of the integrand is integrable near 0.

In light of Theorem 5.6 the following is an immediate consequence of Lemma 3.2:

LEMMA 5.10. *Suppose $u, v, w \in \mathcal{E}$.*

(a) *If $v \le w$ then $\rho[u, v] \le \rho[u, w]$ and $\rho[v, u] \ge \rho[w, u]$.*
(b) *If $c \in \mathbb{R}$, then $\rho[u, v + c] = \rho[u, v] + c = \rho[u - c, v]$.*

PROOF OF THEOREM 5.9. Observe that if $u_i, w_i \in \mathcal{E}$ decrease to $u, w \in \mathcal{E}$, then on $(0, \lambda)$ the functions $\rho[u_i, w_i]$ have a common lower bound. Indeed, if $\rho[u, w]$ is continuous at some $\tau \in (\lambda, V)$, then by Theorem 5.6 $\rho[u_i, w_i](\tau)$ is convergent, hence bounded below by some $c \in \mathbb{R}$. Since $\rho[u_i, w_i]$ is a decreasing function, all $\rho[u_i, w_i] \ge c$ on $(0, \tau) \supset (0, \lambda)$. This has the consequence that Fatou's lemma applies and gives

$$\liminf_{i\to\infty} \int_0^\lambda \rho[u_i, w_i] \ge \int_0^\lambda \rho[u, w]. \tag{5.10}$$

Represent the given $u, v, w \in \mathcal{E}$ as decreasing limits of $u_j, v_j, w_j \in \mathcal{E}$ for which (5.9) holds, for example because they are in $\mathcal{H}$, see Lemma 4.3. Using Lemma 5.10

$$\begin{aligned}\int_0^\lambda \rho[u_j, w] \le \int_0^\lambda \rho[u_j, w_j] &\le \int_0^\lambda \rho[u_j, v_j] + \int_0^\lambda \rho[v_j, w_j] \\ &\le \int_0^\lambda \rho[u, v_j] + \int_0^\lambda \rho[v, w_j].\end{aligned} \tag{5.11}$$

The last two integrands decrease to $\rho[u, v]$, $\rho[v, w]$, and are uniformly bounded above if u, v happen to be bounded. If we let $j \to \infty$, (5.9) then follows by (5.10) and by monotone convergence. This takes care of the theorem if u, v are bounded; w can be arbitrary.

But this means that in our choice of u_j, v_j, w_j we can take $w_j = w$. In the limit (5.11) then gives (5.9) under the sole assumption that u is bounded. Hence in (5.11) we can take $v_j = v$ as well; letting $j \to \infty$ we then obtain (5.9) in complete generality. □

6. Comparison with the flat geometry

The geometry of $\mathcal{E}$ that we have studied so far, through its geodesics, originated from Mabuchi's connection on $\mathcal{H}$. But $\mathcal{H}$ has a simpler, flat geometry, too, inherited as an open subset of the Fréchet space $C^\infty(X)$. The geodesics in this geometry are straight line segments in $\mathcal{H} \subset C^\infty(X)$, and just like Mabuchi's geodesics, the notion extends to $\mathrm{PSH}(X, \omega)$ and $\mathcal{E}$: for any u, v in $\mathrm{PSH}(X, \omega)$ or in $\mathcal{E}$, the connecting segment is contained in the space, [**GZ1**, Proposition 1.6]. While the geodesics of Definition 1.1 are rather different from straight lines, it turns out that various quantities associated with the two are quite comparable. This is the theme we develop in this section. The reader again will notice connections with Darvas's work, in this case with his estimates of various distances in $\mathcal{E}$ by simple integrals, [**Da1**, Theorem 5.5], [**Da4**, Theorem 6.1].

THEOREM 6.1. *If $u, v \in \mathcal{E}$, then*

$$(v-u)^{\star v} \le \rho[u, v] \le (v-u)^{\star u}. \tag{6.1}$$

If also $u \le v$, then

$$\frac{(v-u)^{\star u}(s)}{n+1} \le \rho[u, v]\Big(\frac{s}{e}\Big), \quad 0 < s < V. \tag{6.2}$$

Here $v - u$ is defined and finite away from a pluripolar set, which is of μ_u, μ_v measure 0. It is the velocity vector of the rectilinear path $u + t(v-u)$, $0 \le t \le 1$, joining u and v. Thus the theorem compares the decreasing rearrangements of this velocity vector, when its footpoint is placed at one or the other endpoint, with the decreasing rearrangement of the velocity of the connecting geodesic.

LEMMA 6.2. (a) *If $u_j, v_j \in \mathcal{E}^\infty$ decrease to $u, v \in \mathcal{E}^\infty$, then a.e.*

$$\lim_{j\to\infty} (v_j - u_j)^{\star u_j} = (v-u)^{\star u}, \qquad \lim_{j\to\infty} (v_j - u_j)^{\star v_j} = (v-u)^{\star v}.$$

(b) *If $u, v \in \mathcal{E}$ then a.e. (cf. (4.1))*

$$\limsup_{k\to-\infty} \lim_{j\to-\infty} (v \vee j - u \vee k)^{\star u \vee k} \le (v-u)^{\star u},$$

$$\liminf_{k\to-\infty} \lim_{j\to-\infty} (v \vee k - u \vee j)^{\star v \vee k} \ge (v-u)^{\star v}.$$

PROOF. (a) [**GZ2**, Proposition 9.11] implies that $v_j - u_j \to v - u$ in capacity, whence the claim follows by [**L2**, Lemma 3.4].

(b) If $Y \subset X$ is Borel, by [**GZ2**, Proposition 10.5] $\mu_{u\vee k}(Y) \to \mu_u(Y)$ as $k \to -\infty$. This implies for every Borel function $\xi : X \to \mathbb{R}$, defined a.e. with respect to $\mu_u, \mu_{u\vee k}$,

$$\lim_{k\to-\infty} \xi^{\star u\vee k} = \xi^{\star u} \quad \text{a.e.}. \tag{6.3}$$

Indeed, in $\lim_k \mu_{u\vee k}(\xi \geq \tau) = \mu_u(\xi \geq \tau)$ set $\tau = \xi^{\star u}(s-\varepsilon)+\varepsilon$ with $\varepsilon \in (0,s)$. Since $\mu_u(\xi \geq \tau) \leq \mu_u\big(\xi > \xi^{\star u}(s-\varepsilon)\big) \leq s - \varepsilon < s$ by (2.3), for sufficiently negative k

$$\mu_{u\vee k}(\xi \geq \tau) < s, \quad \text{and} \quad \xi^{\star u\vee k}(s) < \xi^{\star u}(s-\varepsilon)+\varepsilon$$

by (2.4). Working with $\tau = \xi^{\star u}(s+\varepsilon)-\varepsilon$ and the set $(\xi > \tau)$, we obtain similarly that $\xi^{\star u\vee k}(s) > \xi^{\star u}(s+\varepsilon) - \varepsilon$ when k is sufficiently negative. This proves the limit in (6.3) at points of continuity of $\xi^{\star u}$. Of course, (6.3) holds with u replaced by v, too.

Now

$$(v\vee j - u \vee k)^{\star u\vee k} \leq (v\vee j - u)^{\star u\vee k}, \qquad (v\vee k - u\vee j)^{\star v\vee k} \geq (v - u\vee j)^{\star v\vee k}.$$

Letting $j \to -\infty$, then $k\to -\infty$, the estimates follow by Lemma 2.1 and (6.3). □

PROOF OF THEOREM 6.1. Since all functions involved are left continuous, it suffices to prove (6.1), (6.2) on a dense subset of $(0,V)$. (6.1) will be proved by reduction to the special cases $u, v \in \mathcal{H}^{1\bar{1}}$, respectively $u, v \in \mathcal{E}^\infty$, and (6.2) will then be obtained, perhaps paradoxically, from (6.1).

We start by assuming $u, v \in \mathcal{H}^{1\bar{1}}$. Let $\varphi : [0,1] \to \mathcal{H}^{1\bar{1}}$ be the rectilinear path joining u, v, so $\varphi(t) = u + t(v-u)$, and $\psi : [0,1] \to \mathcal{H}^{1\bar{1}}$ the geodesic between u and v. Since $\varphi(\cdot)(x)$ is linear, $\psi(\cdot)(x)$ is convex for every $x \in X$, and the two agree at 0 and 1,

$$\partial^+\psi(0) \leq \partial^+\varphi(0) = v - u = \partial^-\varphi(1) \leq \partial^-\psi(1).$$

Passing to decreasing rearrangements we obtain (6.1).

Next assume u, v are bounded. Choose $u_j, v_j \in \mathcal{H}^{1\bar{1}}$ that decrease to u, v; then

$$(v_j - u_j)^{\star v_j} \leq \rho[u_j, v_j] \leq (v_j - u_j)^{\star u_j}.$$

Hence Lemma 6.2a and Theorem 5.6 imply (6.1).

Finally, with general $u, v \in \mathcal{E}$, by what has been proved already, if $j \leq k$

$$(v\vee k - u\vee j)^{\star v\vee k} \leq \rho[u\vee j, v\vee k], \qquad \rho[u\vee k, v\vee j] \leq (v\vee j - u\vee k)^{\star u\vee k}.$$

Letting $j \to -\infty$ then $k \to -\infty$, (6.1) follows from Lemmas 2.1, 6.2b and Theorem 5.6.

To prove (6.2), when $u \leq v$ we introduce $w = (nu+v)/(n+1)$, the greedy version of a trick that in this context goes back to proofs of [**GZ2**, Proposition 10.7] and [**Da1**, Theorem 5.5]. Thus $u \leq w \leq v$. We claim (independently of the assumption $u \leq v$)

$$\mu_w \geq \Big(\frac{n}{n+1}\Big)^n \mu_u. \tag{6.4}$$

If u, v, w are smooth, this follows from

$$\omega + i\partial\bar{\partial} w = \frac{n(\omega + i\partial\bar{\partial}u)}{n+1} + \frac{\omega + i\partial\bar{\partial}v}{n+1} \geq \frac{n(\omega+i\partial\bar{\partial}u)}{n+1}.$$

In general we choose $u_j, v_j \in \mathcal{H}$ that decrease to u, v as $j \to \infty$. Then μ_u, μ_w, as weak limits of the corresponding μ_{u_j}, μ_{w_j}, also satisfy (6.4).

(6.4) and (2.3) imply for Borel functions $\xi : X \to \mathbb{R}$, $0 < s < V$, and $\varepsilon > 0$

$$\mu_w\big(\xi > \xi^{\star u}(s) - \varepsilon\big) \geq \mu_w\big(\xi \geq \xi^{\star u}(s)\big) \geq \Big(\frac{n}{n+1}\Big)^n \mu_u\big(\xi \geq \xi^{\star u}(s)\big) > \frac{s}{e}.$$

Hence by (2.4) $\xi^{\star w}(s/e) > \xi^{\star u}(s) - \varepsilon$, and so $\xi^{\star w}(s/e) \geq \xi^{\star u}(s)$. Applying this with $\xi = w - u$, (6.1) and Lemma 5.10 yield

$$\frac{(v-u)^{\star u}(s)}{n+1} = (w-u)^{\star u}(s) \leq (w-u)^{\star w}\Big(\frac{s}{e}\Big) \leq \rho[u,w]\Big(\frac{s}{e}\Big) \leq \rho[u,v]\Big(\frac{s}{e}\Big),$$

as claimed. □

COROLLARY 6.3. *If $\varphi : [a,b] \to \mathcal{E}$ is a geodesic, then for $a \leq t < b$, $a < \tau \leq b$*

$$(\partial_\tau^- \varphi(\tau))^\star \leq \rho_\varphi \leq (\partial_t^+ \varphi(t))^\star. \tag{6.5}$$

If φ increases ($\varphi(t) \leq \varphi(\tau)$ when $t \leq \tau$), then

$$(\partial_\tau^+ \varphi(\tau))^\star(s) \leq (n+1)\rho_\varphi(s/e), \quad 0 < s < V.$$

PROOF. By Theorem 6.1, if $a \leq \sigma < \tau$

$$\Big(\frac{\varphi(\tau) - \varphi(\sigma)}{\tau - \sigma}\Big)^{\star\varphi(\tau)} = \frac{(\varphi(\tau) - \varphi(\sigma))^{\star\varphi(\tau)}}{\tau - \sigma} \leq \frac{\rho[\varphi(\sigma), \varphi(\tau)]}{\tau - \sigma} = \rho_\varphi.$$

Letting $\sigma \to \tau$ we obtain the first inequality in (6.5) in light of Lemma 2.1. The other estimates are obtained similarly. □

COROLLARY 6.4. *If $u, v \in \mathcal{E}$ then $u \leq v$ is equivalent to $\rho[u,v] \geq 0$.*

PROOF. By Theorem 5.5 $\rho[u,v] \geq 0$ is equivalent to $\rho[u, u \wedge v] = 0$. This latter certainly holds if $u \leq v$. Conversely, suppose $\rho[u, u \wedge v] = 0$. By Theorem 6.1

$$0 = (n+1)\rho[u \wedge v, u](s/e) \geq (u - u \wedge v)^{\star u \wedge v}(s).$$

In other words, $u \leq u \wedge v$ holds $\mu_{u \wedge v}$–a.e. According to S. Dinew, this implies $u \leq u \wedge v$ everywhere, i.e. $u \leq v$. Dinew does not seem to have published his proof, but communicated it to Levenberg, and the standard reference to the result is [**BL**, Proposition 5.9]. □

The following estimate we will need in section 10.

LEMMA 6.5. *If $u, v, w \in \mathcal{E}$ and $0 < \sigma < s < V$, then*

$$\rho[u \wedge v, w](s) \leq \max\big((w-u)^{\star u}(\sigma), (w-v)^{\star v}(s - \sigma)\big). \tag{6.6}$$

PROOF. As in Lemma 2.2, it suffices to prove for a dense set of (σ, s). First assume $u, v \in \mathcal{H}^{1\bar{1}}$ and $w \in \mathcal{E}^\infty$. Since adding a constant to w does not affect the validity of (6.6), we will also assume $u, v \leq w$. Define

$$\xi = (w - u \wedge v)1_{u \leq v}, \qquad \eta = (w - u \wedge v)1_{u \geq v}.$$

Thus $w - u \wedge v = \max(\xi, \eta)$. By the partition formula (4.6), if $\tau \geq 0$,

$$\begin{aligned}\mu_{u \wedge v}(\xi > \tau) &= \mu_u(\xi > \tau, u \wedge v = u) + \mu_v(\xi > \tau, u > u \wedge v = v)\\ &= \mu_u(w - u \wedge v > \tau, u \wedge v = u) + 0 \leq \mu_u(w - u > \tau).\end{aligned}$$

Let $s \in (0, V)$ and $\tau = (w-u)^{\star u}(s)$. With $\delta > 0$ (2.3) implies

$$\mu_{u \wedge v}(\xi \geq \tau + \delta) \leq \mu_u\big(w - u > (w-u)^{\star u}(s)\big) \leq s < s + \delta.$$

Hence $\xi^{\star u\wedge v}(s+\delta) < (w-u)^{\star u}(s)+\delta$ by (2.4). Letting $\delta\to 0$, $\xi^{\star u\wedge v}(s)\le (w-u)^{\star u}(s)$ follows if $\xi^{\star u\wedge v}$ is continuous at s, so on a dense subset of $(0,V)$. In fact

$$\xi^{\star u\wedge v}\le (w-u)^{\star u}\qquad\text{and similarly}\qquad \eta^{\star u\wedge v}\le (w-v)^{\star v}$$

everywhere, since the functions involved are left continuous. Lemma 2.2 implies

$$\begin{aligned}(w-u\wedge v)^{\star u\wedge v}(s) &= \max(\xi,\eta)^{\star u\wedge v}(s)\\ &\le \max\big(\xi^{\star u\wedge v}(\sigma),\,\eta^{\star u\wedge v}(s-\sigma)\big)\le \max\big((w-u)^{\star u}(\sigma),\,(w-v)^{\star v}(s-\sigma)\big).\end{aligned}\tag{6.7}$$

In this form it is hard to carry over the estimate to $u,v\notin\mathcal H^{1\bar 1}$, but (6.7) implies (6.6) in view of Theorem 6.1. Thus we proved (6.6) when $u,v\in\mathcal H^{1\bar 1}$, $w\in\mathcal E^\infty$.

From this we obtain (6.6) for $u,v,w\in\mathcal E^\infty$ by decreasing to u,v with $u_j,v_j\in\mathcal H^{1\bar 1}$, and using Theorem 5.6 and Lemma 6.2a. Thus for arbitrary $u,v,w\in\mathcal E$ and $j,k\in\mathbb Z$

$$\rho[(u\vee k)\wedge(v\vee k),w\vee j](s)\le \max\big((w\vee j-u\vee k)^{\star u\vee k}(\sigma),\,(w\vee j-v\vee k)^{\star v\vee k}(s-\sigma)\big).$$

If we let $j\to-\infty$, then $k\to-\infty$, in the limit we obtain (6.6) by virtue of Theorem 5.6 and Lemma 6.2b. □

7. The meaning of the rise

This section will be short. The meaning of the rise of a geodesic $\varphi:[a,b]\to\mathcal H^{1\bar 1}$ is clear, it is $(\partial_t^\pm\varphi(t))^\star$ for any t, by definition. However, for a geodesic in $\mathcal E$ we do not understand the meaning of its rise directly. One could hope that the rise of a general geodesic can still be expressed in terms of velocities $\partial^\pm\varphi$, but we do not know how. Corollary 6.3 estimates ρ_φ in terms of these velocities; this is the most we can say in general.

Still, if a geodesic $\varphi:[a,b]\to\mathcal E$ passes through a single point in $\mathcal H^{1\bar 1}$, its rise can be expressed by the velocities there.

THEOREM 7.1. *Let $\varphi:[a,b]\to\mathcal E$ be a geodesic. If $\varphi(\alpha)\in\mathcal H^{1\bar 1}$ with some $\alpha\in[a,b]$, then $\rho_\varphi=(\partial^\pm\varphi(\alpha))^\star$.*

PROOF. We will prove when $\alpha=a$. The general result can be obtained from this special case by cutting the geodesic in two (and perhaps reversing it).

Construct geodesics $\varphi_j:[a,b]\to\mathcal H^{1\bar 1}$ such that $\varphi_j(a)=\varphi(a)$ and $\varphi_j(b)$ decreases to $\varphi(b)$. By [**Da3**, Proposition 3.15] $\varphi_j(t)$ decreases to $\varphi(t)$ for all $t\in[a,b]$. If $a<s<t\le b$, by convexity

$$\frac{\varphi(s)-\varphi(a)}{s-a}\le\frac{\varphi_j(s)-\varphi_j(a)}{s-a}\le\frac{\varphi_j(t)-\varphi_j(a)}{t-a};\qquad\text{letting } s\to a,$$

$$\partial^+\varphi(a)\le\partial^+\varphi_j(a)\le\frac{\varphi_j(t)-\varphi_j(a)}{t-a}.$$

Letting next $j\to\infty$, then $t\to a$ we obtain

$$\partial^+\varphi(a)\le\liminf_{j\to\infty}\partial^+\varphi_j(a)\le\limsup_{j\to\infty}\partial^+\varphi_j(a)\le\partial^+\varphi(a),$$

or $\partial^+\varphi(a)=\lim_j\partial^+\varphi_j(a)$. Thus, by Theorem 5.7 and Lemma 2.1

$$\rho_\varphi=\lim_j\rho_{\varphi_j}=\lim_j(\partial^+\varphi_j(a))^\star=(\partial^+\varphi(a))^\star,$$

first almost everywhere, but as a consequence, in fact everywhere on $(0,V)$. □

The following is immediate:

COROLLARY 7.2. *If $\varphi : [a,b] \to \mathcal{E}$ is a geodesic and $\varphi(\alpha) \in \mathcal{H}^{1\bar{1}}$ for some $\alpha \in (a,b)$, then $\partial^-\varphi(\alpha) = \partial^+\varphi(\alpha)$ holds $\mu_{\varphi(\alpha)}$–almost everywhere.*

A similar result follows from a recent paper by Di Nezza and Lu. A special case of [**DNL**, Corollary 3.3] together with Corollary 6.3 above imply that if $\varphi : [a,b] \to \mathcal{E}^\infty$ is a geodesic and $\mu_{\varphi(\alpha)}$ has finite entropy for some $\alpha \in [a,b]$, then $\rho_\varphi = \big(\partial^\pm\varphi(\alpha)\big)^\star$; if $\alpha \in (a,b)$, then $\partial^-\varphi(\alpha) = \partial^+\varphi(\alpha)$ holds $\mu_{\varphi(\alpha)}$–a.e.

8. Lagrangians

The rest of the paper is concerned with Lagrangians on bundles larger than $T^\infty\mathcal{E}$ and the relationship between rise and the induced action. This section will be about the structure and general properties of Lagrangians.

Our starting point is an invariant convex Lagrangian $L : T^\infty\mathcal{E} \to \mathbb{R}$ in the sense of Definition 2.4. [**L2**, Theorem 2.4] describes the structure of such Lagrangians:

THEOREM 8.1. *Given an invariant convex Lagrangian $L : T^\infty\mathcal{E} \to \mathbb{R}$, there is a family of $\mathcal{A}_u \subset \mathbb{R} \times B(X)$, $u \in \mathcal{H}$, such that*

$$L(\xi) = \sup_{(a,f)\in\mathcal{A}_u} a + \int_X \xi f\, d\mu_u, \quad \xi \in T_u\mathcal{H} \approx C^\infty(X). \tag{8.1}$$

The $\mathcal{A}_u$ can be chosen invariant in the sense of the following definition.

DEFINITION 8.2. Let $\mathcal{U} \subset \mathcal{E}$. A family of $\emptyset \neq \mathcal{A}_u \subset \mathbb{R} \times B(X)$, $u \in \mathcal{U}$, is (strict rearrangement) invariant if whenever $u, v \in \mathcal{U}$, $f \in B(X,\mu_u)$ and $g \in B(X,\mu_v)$ are equidistributed, and $(a,f) \in \mathcal{A}_u$, then $(a,g) \in \mathcal{A}_v$.

Since there are measure preserving bijections $(X,\mu_u) \to (X,\mu_v)$ for any $u, v \in \mathcal{E}$ [**R**, p.409, Theorem 16], an invariant family $\mathcal{A}_u, u \in \mathcal{U}$, is uniquely determined by any of its members $\mathcal{A}_u$.

We will investigate functionals given by formulas like (8.1). Most of the material here will be needed in sections 9 and 10.

THEOREM 8.3. *Let $\emptyset \neq \mathcal{A}_u \subset \mathbb{R} \times B(X)$, $u \in \mathcal{E}$, be an invariant family and*

$$L(\xi) = \sup_{(a,f)\in\mathcal{A}_u} a + \int_X \xi f\, d\mu_u \in (-\infty,\infty], \qquad \xi \in T_u^\infty\mathcal{E}. \tag{8.2}$$

(a) *L is finite on $T\mathcal{H}$ if and only if*

$$\sup_{(a,f)\in\mathcal{A}_u} a + \lambda\int_X |f|\, d\mu_u < \infty \tag{8.3}$$

for all $\lambda \in (0,\infty)$ and for all (equivalently: for some) $u \in \mathcal{E}$. In this case L is finite, fiberwise convex and continuous in the sup norm topology on $T_u^\infty\mathcal{E} \approx B(X)$.

(b) *In addition to being finite on $T\mathcal{H}$, L is strongly continuous (Definition 2.4) on the fibers $T_u\mathcal{H}$, $u \in \mathcal{H}$, if and only if in addition to (8.3) $\mathcal{A}_u$ is uniformly integrable in the sense that for every $\varepsilon, \lambda > 0$ there is a $\delta > 0$ such that whenever $u \in \mathcal{H}$ and $E \subset X$ has μ_u measure $\le \delta$, then*

$$\sup_{(a,f)\in\mathcal{A}_u} a + \lambda\int_E |f|\, d\mu_u < L(0) + \varepsilon. \tag{8.4}$$

In this case L is strongly continuous on all fibers $T_u^\infty\mathcal{E}$.

Clearly, for any invariant family $\mathcal{A}_u$, the Lagragian L defined by (8.2) itself will be invariant under strict rearrangements.—We will need a result from [**L1**]. Recall that given equidistributed Borel functions f on (X,μ_u) and g on (X,μ_v), we write $(f,\mu_u)\sim(g,\mu_v)$, or just $f\sim g$ if the measures are understood. Another way to express equidistribution is $f^{\star u}=g^{\star v}$. If $\mu_u(Y)>0$ we write $⨍_Y \xi\,d\mu_u$ for the average $\int_Y \xi\,d\mu_u/\mu_u(Y)$. If $\mu_u(Y)=0$, we just set $⨍_Y \xi\,d\mu_u=0$. Further, we put $\xi_+=\max(0,\xi)$, $\xi_-=\max(0,-\xi)$. The measure μ below is μ_0; but it can be any μ_u because, as said, all measures μ_u are isomorphic [**R**, p.409, Theorem 16].

LEMMA 8.4 ([**L1**], Lemma 4.4, simplified). *Let $f\in L^1(X,\mu)$, $\xi\in B(X)$, and $S,T\subset X$ of equal measure. If $\xi\ge 0$ on T and $\xi\le 0$ on $X\setminus T$, then*

$$\sup_{g\sim f}\int_X \xi g\ge ⨍_S f\,d\mu\int_X \xi_+\,d\mu-⨍_{X\setminus S} f\,d\mu\int_X \xi_-\,d\mu.$$

A further result we will use concerns convergence properties of convex functions on general vector spaces.

LEMMA 8.5. *Let Ξ be a real vector space and $l:\Xi\to(-\infty,\infty]$ convex. If $\xi,\xi_j\in\Xi$, $j\in\mathbb{N}$, satisfy*

$$\limsup_{\alpha\to\pm\infty}\limsup_{j\to\infty} l\big((1-\alpha)\xi+\alpha\xi_j\big)<\infty,$$

then $l(\xi)<\infty$ and $\lim_{j\to\infty} l(\xi_j)=l(\xi)$.

PROOF. Write $\xi_j(\alpha)=(1-\alpha)\xi+\alpha\xi_j$, $\alpha\in\mathbb{R}$, and note that $\xi_j(0)=\xi$, $\xi_j(1)=\xi_j$. Fix a positive number $C>\limsup_{\alpha\to\pm\infty}\limsup_{j\to\infty} l\big(\xi_j(\alpha)\big)$. Choose $\alpha_0>1$ and $j_\beta\in\mathbb{N}$ for $\beta>\alpha_0$ such that $l\big(\xi_j(\alpha)\big)<C$ when $|\alpha|>\alpha_0$, $j>j_{|\alpha|}$. Since $l\big(\xi_j(\alpha)\big)$ is a convex function of α and it is finite if $j>j_{|\alpha|}$, with such j it follows that $l(\xi)=l\big(\xi_j(0)\big)<\infty$. Convexity also implies with $j>j_{|\alpha|}$

$$l(\xi_j)-l(\xi)\begin{cases}\le \dfrac{l(\xi_j(\alpha))-l(\xi)}{\alpha}<\dfrac{C-l(\xi)}{\alpha} & \text{if } \alpha>\alpha_0\\[2ex] \ge \dfrac{l(\xi_j(\alpha))-l(\xi)}{\alpha}>\dfrac{C-l(\xi)}{\alpha} & \text{if } \alpha<-\alpha_0.\end{cases}$$

Therefore $\limsup_{j\to\infty}|l(\xi_j)-l(\xi)|\le|C-l(\xi)|/|\alpha|$, which completes the proof. □

PROOF OF THEOREM 8.3. It suffices to work with $u=0$, since all measure spaces (X,μ_u) are isomorphic. We continue writing μ for μ_0. For simplicity we assume $\mu(X)=1$. We can also assume $L(0)=0$, since this can be arranged if we subtract $L(0)$ from all a that occur in $\mathcal{A}_u$.

(a) If (8.3) holds, then with $\xi\in T_0^\infty\mathcal{H}$ and $\lambda>\sup|\xi|$,

$$L(\xi)\le\sup_{(a,f)\in\mathcal{A}_0} a+\lambda\int_X|f|\,d\mu<\infty \tag{8.5}$$

shows L is locally bounded above on $T_0^\infty\mathcal{H}\approx B(X)$. As the supremum of affine functions, L is convex; convexity and local upper boundedness imply continuity, see e.g. [**L1**, Lemma 4.2].

Conversely, suppose L is finite on $T\mathcal{H}$. Given $\lambda\in(0,\infty)$, choose a nonnegative $\xi\in T_0\mathcal{H}\approx C^\infty(X)$ such that $\int_X\xi\,d\mu=\lambda$, but such that $T'=(\xi>0)$ has measure

$\le 1/2$. Let $(a,f)\in\mathcal{A}_0$, and assume first that $\mathcal{S}=(f\ge 0)$ has measure $\ge 1/2$. Choose $T\supset T'$ with $\mu(S)=\mu(T)$. By (8.2) and by Lemma 8.4

$$L(3\xi)\ge a+\sup_{g\sim f}3\int_X \xi g\,d\mu\ge a+3⨍_S f\,d\mu\int_X\xi\,d\mu\ge a+3\lambda\int_X f_+\,d\mu. \tag{8.6}$$

Viewing the constant functions $\pm 3\lambda$ as elements of $T_0^\infty\xi$,

$$\begin{aligned} L(3\lambda)&\ge a+3\lambda\int_X f\,d\mu=a+3\lambda\int_X(f_+-f_-)\,d\mu,\\ L(-3\lambda)&\ge a-3\lambda\int_X f\,d\mu=a+3\lambda\int_X(f_--f_+)\,d\mu.\end{aligned}$$

Therefore

$$a+\lambda\int_X|f|\,d\mu=a+\lambda\int_X(f_++f_-)\,d\mu\le\frac{2L(3\xi)+L(-3\lambda)}{3}. \tag{8.7}$$

If, instead of $(f\ge 0)$, the set $(f\le 0)$ has measure $\ge 1/2$, we apply (8.6) with ξ replaced by $-\xi$, to obtain

$$L(-3\xi)\ge a+\sup_{g\sim -f}3\int_X\xi g\,d\mu\ge a+3\lambda\int_X(-f)_+\,d\mu=a+3\lambda\int_X f_-\,d\mu,\quad\text{and}$$

$$a+\lambda\int_X|f|\,d\mu=a+\lambda\int_X(f_++f_-)\,d\mu\le\frac{2L(-3\xi)+L(3\lambda)}{3}. \tag{8.8}$$

To sum up, for every $(a,f)\in\mathcal{A}_0$ either (8.7) or (8.8) holds, and so

$$\sup_{(a,f)\in\mathcal{A}_0}a+\lambda\int_X|f|\,d\mu\le\max\Big(\frac{2L(3\xi)+L(-3\lambda)}{3},\frac{2L(-3\xi)+L(3\lambda)}{3}\Big)<\infty. \tag{8.9}$$

(b) Suppose $\mathcal{A}_0$ is uniformly integrable in the sense of (8.4). Strong continuity of $L|T_0^\infty\mathcal{E}$ will follow from Lemma 8.5. First we derive a better bound on L than (8.5).

If $\lambda,\delta\in(0,\infty)$, let

$$C(\lambda,\delta)=\sup\Big\{a+\lambda\int_E|f|\,d\mu:(a,f)\in\mathcal{A}_0,\mu(E)\le\delta\Big\}. \tag{8.10}$$

Our assumption is $\lim_{\delta\to 0}C(\lambda,\delta)=0$. Let $\eta\in T_0^\infty\mathcal{E}$, $\lambda>0$. If $(a,f)\in\mathcal{A}_0$

$$\begin{aligned}2\Big(a+\int_X\eta f\,d\mu\Big)&\le a+2\sup|\eta|\int_{|\eta|>\lambda}|f|\,d\mu+a+2\lambda\int_{|\eta|\le\lambda}|f|\,d\mu\\ &\le C\big(2\sup|\eta|,\mu(|\eta|>\lambda)\big)+C(2\lambda,1),\qquad\text{whence}\end{aligned}$$

$$2L(\eta)\le C\big(2\sup|\eta|,\mu(|\eta|>\lambda)\big)+C(2\lambda,1). \tag{8.11}$$

This can be an improvement on (8.5), which says $L(\eta)\le C(\sup|\eta|,1)$.

Now consider uniformly bounded $\xi_j\in T_0^\infty\mathcal{E}$, $j\in\mathbb{N}$, that μ–a.e. tend to $\xi\in T_0^\infty\mathcal{E}$. To show $L(\xi_j)\to L(\xi)$, with $\alpha\in\mathbb{R}$ let

$$\xi_j(\alpha)=(1-\alpha)\xi+\alpha\xi_j;\qquad\lim_{j\to\infty}\xi_j(\alpha)=\xi\quad\mu\text{–a.e.}$$

Let $\lambda=1+\sup_j\sup_X|\xi_j|$ and note that $\sup_X|\xi_j(\alpha)|<(2|\alpha|+1)\lambda$. Given α, choose $\delta>0$ so that $C\big((4|\alpha|+2)\lambda,\delta\big)<1$. By Egorov's theorem $\lim_{j\to\infty}\xi_j(\alpha)=\xi$ uniformly outside a set of measure $\le\delta$. Choose j_0 such that

$$\mu(|\xi_j(\alpha)|>\lambda)\le\delta\quad\text{for}\quad j>j_0.$$

For such j, by (8.11)

$$2L(\xi_j(\alpha)) \le C\big((4|\alpha|+2)\lambda,\delta\big)+C(2\lambda,1)<1+C(2\lambda,1).$$

Lemma 8.5 applies, and gives $L(\xi_j)\to L(\xi)$. We conclude L is strongly continuous on $T_0^\infty\mathcal{E}$.

To prove the converse implication we assume L is strongly continuous on $T_0\mathcal{H}$. It will be convenient to pass to decreasing rearrangements of $f,\xi\in B(X,\mu)$; then (8.2) can be rewritten

$$L(\xi)=\sup_{(a,f)\in\mathcal{A}_0} a+\int_0^1 \xi^\star f^\star,\quad \xi\in T_0^\infty\mathcal{E},$$

the integral with respect to Lebesgue measure. Indeed, if $\theta:(X,\mu)\to\big((0,1),\text{Lebesgue}\big)$ preserves measure, then e.g. [**L1**, Lemma 7.2] implies

$$\sup_{g\sim f}\int_X \xi f\,d\mu=\int_X(\xi^\star\circ\theta)(f^\star\circ\theta)\,d\mu=\int_0^1\xi^\star f^\star.$$

Accordingly consider $M: B(0,1)\to\mathbb{R}$ given by

$$M(\zeta)=\sup_{(a,f)\in\mathcal{A}_0} a+\int_0^1\zeta f^\star.$$

Thus $M(\zeta)=L(\zeta\circ\theta)$, and part (a) implies that M is continuous if $B(0,1)$ is endowed with the supremum norm. We show that it is even strongly continuous on $C[0,1]$: if uniformly bounded $\zeta_j\in C[0,1]$ converge almost everywhere, then $M(\zeta_j)$ also converges.

For this purpose we need a continuous measure preserving $\theta:(X,\mu)\to[0,1]$. E.g. by [**L1**, Lemma 5.5] such θ exists. Thus $\zeta_j\circ\theta\in C(X)$. Choose $\xi_j\in C^\infty(X)\approx T_0\mathcal{H}$ such that

$$\sup_X|\xi_j-\zeta_j\circ\theta|<1/j,\qquad |L(\xi_j)-L(\zeta_j\circ\theta)|<1/j.$$

Then $\zeta_j\circ\theta$ and ξ_j converge μ–a.e. Hence $L(\xi_j)$ converge and so do $L(\zeta_j\circ\theta)=M(\zeta_j)$.

We can also rewrite $C(\lambda,\delta)$ of (8.10) as

$$C(\lambda,\delta)=\sup_{(a,f)\in\mathcal{A}_0} a+\lambda\int_0^\delta|f^\star|, \tag{8.12}$$

and we need to show $\lim_{\delta\to0}C(\lambda,\delta)=0$ for all $\lambda>0$. Part (a) at least implies

$$0=L(0)=\sup_{(a,f)\in\mathcal{A}_0} a\le C(\lambda,\delta)\le C(\lambda,1)<\infty.$$

For any $(a,f)\in\mathcal{A}_0$, the negative part $f_-^\star$ of $f^\star$ is increasing. Therefore by (8.12)

$$C(\lambda,1)\ge a+\lambda\int_{1/2}^1 f_-^\star\ge a+\frac{\lambda}{2}f_-^\star\Big(\frac12\Big),\qquad\text{whence}$$

$$\lambda f_-^\star(\tau)\le 2\big(C(\lambda,1)-a\big)\quad\text{if}\quad 0<\tau<1/2. \tag{8.13}$$

For $j=2,3,\dots$ consider the functions $\zeta_j\in C[0,1]$,

$$\zeta_j=\begin{cases}\lambda & \text{on}\quad [0,1/j]\\ 0 & \text{on}\quad [2/j,1]\\ \text{linear in between,}\end{cases}$$

uniformly bounded and tending to 0 a.e. as $j \to \infty$; whence $M(\zeta_j) \to M(0) = 0$. If $(a, f) \in \mathcal{A}_0$ and $j \geq 4$, using (8.13) and that $L(0) = 0$ implies $a \leq 0$,

$$M(\zeta_j) \geq a + \int_0^1 \zeta_j f^\star = a + \int_0^{2/j} \zeta_j\big(|f^\star| - 2f^\star_-\big) \geq a + \int_0^{2/j} \zeta_j |f^\star| - 2\lambda \int_0^{2/j} f^\star_-$$
$$\geq a + \lambda \int_0^{1/j} |f^\star| - \frac{8}{j}\big(C(\lambda, 1) - a\big) \geq 3\Big(a + \frac{\lambda}{3}\int_0^{1/j} |f^\star|\Big) - \frac{8C(\lambda, 1)}{j}.$$

In view of (8.12), therefore

$$0 \leq 3C(\lambda/3, 1/j) \leq M(\zeta_j) + 8C(\lambda, 1)/j \to 0, \quad j \to \infty,$$

which completes the proof of Theorem 8.3. □

Now consider an invariant family of nonempty $\mathcal{A}_u \subset \mathbb{R} \times B(X)$, $u \in \mathcal{E}$. Such a family defines a function $L : T^\infty\mathcal{E} \to (-\infty, \infty]$ by formula (8.2), and in fact on a larger (set theoretical) Banach bundle $T^1\mathcal{E} \to \mathcal{E}$. The fibers of this latter are

$$T^1_u\mathcal{E} = L^1(X, \mu_u), \quad \text{and} \quad T^1\mathcal{E} = \coprod_{u\in\mathcal{E}} T^1_u\mathcal{E}.$$

Occasionally it will be convenient to view $T^\infty\mathcal{E}$ as a subbundle of $T^1\mathcal{E}$. Although in reality it is only a quotient of $T^\infty_u\mathcal{E}$, modulo equality μ_u–a.e., that is a subspace of $T^1_u\mathcal{E}$, this should not cause confusion, since all the Lagrangians we work with here take the same value on functions that agree a.e.—The invariant family $\mathcal{A}_u$ thus determines $L : T^1\mathcal{E} \to (-\infty, \infty]$,

$$L(\xi) = \sup_{(a,f)\in\mathcal{A}_u} a + \int_X \xi f \, d\mu_u, \qquad \xi \in T^1_u\mathcal{E}, \tag{8.14}$$

a fiberwise lower semicontinuous convex function that is strict rearrangement invariant. (The topology on $T^1_u\mathcal{E}$ is the L^1 topology.) With $\mathcal{A}_u$ we can associate three more invariant families

$$\begin{gathered}\mathcal{A}^+_u = \{(a, f_+) : (a, f) \in \mathcal{A}_u\}, \quad \mathcal{A}^-_u = \{(a, f_-) : (a, f) \in \mathcal{A}_u\},\\ \mathcal{A}^{||}_u = \{(a, g) \in \mathbb{R} \times B(X) : \text{ there is } (a, f) \in \mathcal{A}_u \text{ such that } |g| = |f|\},\end{gathered} \tag{8.15}$$

and Lagrangians $L^+, L^-, L^{||} : T^1\mathcal{E} \to (-\infty, \infty]$,

$$\begin{aligned} L^\pm(\xi) &= \sup_{(a,g)\in\mathcal{A}^\pm_u} a + \int_X \xi g \, d\mu_u = \sup_{(a,f)\in\mathcal{A}_u} a + \int_X \xi f_\pm \, d\mu_u,\\ L^{||}(\xi) &= \sup_{(a,g)\in\mathcal{A}^{||}_u} a + \int_X \xi g \, d\mu_u = \sup_{(a,f)\in\mathcal{A}_u} a + \int_X |\xi f| \, d\mu_u,\end{aligned} \tag{8.16}$$

$\xi \in T^1_u\mathcal{E}$. These Lagrangians are comparable:

THEOREM 8.6. *If $\xi \in T^1_u\mathcal{E}$ then*

(8.17)

$$2L^+(\xi) \leq L(2\xi) + L^-\Big(2\fint_X \xi \, d\mu_u\Big), \quad 2L^-(\xi) \leq L(-2\xi) + L^+\Big(-2\fint_X \xi \, d\mu_u\Big),$$
$$2L^{||}(\xi) \leq \max\big(L(8\xi), L(-8\xi)\big) + L^{||}\Big(6\fint_X |\xi| \, d\mu_u\Big),$$
$$|L(\xi) - L(0)| \leq L^{||}(\xi) - L^{||}(0).$$

Furthermore, if $L|T^\infty\mathcal{E}$ is finite and strongly continuous on the fibers $T_u^\infty\mathcal{E}$, then the same holds for $L^\pm, L^{||}$.

In (8.17) averages such as $⨍_X \xi\, d\mu_u = \int_X \xi\, d\mu_u/V$ are constant functions on X, and are viewed as $\in T_u^1\mathcal{E}$. The interest of the theorem is that in various situations it allows one to replace L by $\Lambda = L^{||}$, which is absolutely monotone in the following sense:

DEFINITION 8.7. A function $\Lambda : T^1\mathcal{E} \to (-\infty, \infty]$ is absolutely monotone if $\xi, \eta \in T_u^1\mathcal{E}$, $|\xi| \le |\eta|$ imply $\Lambda(\xi) \le \Lambda(\eta)$.

PROOF. The proof depends on certain estimates proved in [**L1**].[2] Again, it suffices to work with $u = 0$; we write $\mu_0 = \mu$. Let $(a, f_0) \in \mathcal{A}_0$. By [**L1**, (7.4)]

$$\begin{aligned} 2a + 2\sup_{f\sim f_0}\int_X \xi f_+ &\le a + 2\sup_{f\sim f_0}\int_X \xi f\, d\mu + a + 2\int_X f_{0-}\, d\mu ⨍_X \xi\, d\mu \\ &\le L(2\xi) + L^-\Big(2⨍_X \xi\, d\mu\Big). \end{aligned}$$

Passing to the supremum over all $(a, f_0) \in \mathcal{A}$ gives the first estimate in (8.17). The second estimate follows by the same computation, with f replaced by $g = -f$.

To prove the third estimate we use [**L1**, Lemma 7.5]:

$$\sup_{f\sim f_0}\int_X |\xi f|\, d\mu \le 4\sup_{f\sim f_0}\Big|\int_X \xi f\, d\mu\Big| + 3⨍_X |\xi|\, d\mu \int_X |f_0|\, d\mu.$$

Therefore

$$2\sup_{f\sim f_0} a + \int_X |\xi f|\, d\mu \le \sup_{f\sim f_0}\Big(a + \Big|\int_X 8\xi f\, d\mu\Big|\Big) + a + 6⨍_X |\xi|\, d\mu \int |f_0|\, d\mu. \tag{8.18}$$

Now

$$a + \Big|\int_X 8\xi f\, d\mu\Big| = \max\Big(a + \int_X 8\xi f\, d\mu, a - \int_X 8\xi f\, d\mu\Big) \le \max\big(L(8\xi), L(-8\xi)\big);$$

taking the supremum over all $(a, f_0) \in \mathcal{A}_0$ in (8.18) thus gives the third estimate in (8.17). As to the last estimate, from the definition $L(\xi) \le L^{||}(\xi)$ and $L(\xi) - L(0) \le L^{||}(\xi) - L^{||}(0)$. By convexity $2L(0) \le L(\xi) + L(-\xi)$, whence

$$L(0) - L(\xi) \le L(-\xi) - L(0) \le L^{||}(-\xi) - L(0) = L^{||}(\xi) - L^{||}(0),$$

and so indeed $|L(\xi) - L(0)| \le L^{||}(\xi) - L^{||}(0)$.

The last claim of Theorem 8.6 follows from Theorem 8.3b, as $L^\pm(0) = L^{||}(0) = L(0)$. □

There is another way to introduce the class of Lagrangians of (8.14).

LEMMA 8.8. *A function $L : T^1\mathcal{E} \to (-\infty, \infty]$ can be represented in the form (8.14) with an invariant family $\mathcal{A}_u \subset \mathbb{R} \times B(X)$, $u \in \mathcal{E}$, if and only if it is strict rearrangement invariant, and on the fibers $T_u^1\mathcal{E}$ it is convex and lower semicontinuous.*

[2][**L1**] typically works with bounded functions, and for this reason the results we cite in this proof and later directly apply only to bounded ξ. Nonetheless, the estimates follow for general $\xi \in L^1(X, \mu_u)$ if we replace ξ by $\max\big(\min(\xi, c), -c\big)$, then let the constant $c \to \infty$.

PROOF. If L is representable in the form (8.14), then as the supremum of continuous affine functions on $T^1_u\mathcal{E}$, it is convex and lsc. If the family $\mathcal{A}_u$, $u \in \mathcal{E}$, is invariant and $\xi \in T^1_u\mathcal{E}(\omega)$, then for every $(a, f_0) \in \mathcal{A}_u$

$$a + \sup_{(f,\mu_u)\sim(f_0,\mu_u)} \int_X \xi f \, d\mu_u = a + \int_0^V \xi^{\star u} f^{\star u},$$

since the sup above occurs when f and ξ are 'similarly ordered', see e.g. [**L1**, Lemma 7.2], whence ξf and $\xi^{\star u} f^{\star u}$ are equidistributed. Hence

$$L(\xi) = \sup_{(a,f)\in\mathcal{A}_u} a + \int_0^V \xi^{\star u} f^{\star u},$$

obviously invariant under strict rearrangements.

Conversely, suppose L is strict rearrangement invariant, and convex and lsc on the fibers $T^1_u\mathcal{E}$. Let L^* denote its fiberwise conjugate convex function (Fenchel–Legendre transform)

$$L^*(f) = \sup\Big\{ \int_X \xi f \, d\mu_u - L(\xi) : \xi \in T^1_u\mathcal{E} \Big\}, \qquad f \in B(X, \mu_u).$$

By Fenchel's theorem [**IT**, p.175]

$$L(\xi) = \sup\Big\{ \int_X \xi f \, d\mu_u - L^*(f) : f \in B(X, \mu_u) \Big\}, \qquad \xi \in T^1_u\mathcal{E},$$

and so

$$\mathcal{A}_u = \big\{ \big(- L^*(f), f\big) : f \in B(X, \mu_u) \big\} \tag{8.19}$$

gives the required representation (8.14). □

9. The bundle $T^L\mathcal{E} \to \mathcal{E}$

With the Lagrangians $L : T^1\mathcal{E} \to (-\infty, \infty]$ studied in the previous section here we will associate a subbundle $T^L\mathcal{E} \subset T^1\mathcal{E}$, and through it, in the next section, an energy space $\mathcal{E}^L \subset \mathcal{E}$. In the simplest cases, such as $L(\xi) = \int_X |\xi|^p \, d\mu_u$, the subbundle consists of ξ for which $L(\xi)$ is finite. In general the definition of $T^L\mathcal{E}$ has to be more involved for two reasons. First, $L(\xi) < \infty$ in general does not imply $L(\alpha\xi) < \infty$ for values of $\alpha \in \mathbb{R}$ other than $\alpha \in [0, 1]$. Second, to work efficiently with L we need to restrict it to a subbundle on which it has a certain fine continuity property.

DEFINITION 9.1. Consider a function $L : T^1\mathcal{E} \to (-\infty, \infty]$ with $L(0)$ finite. It is absolutely continuous on $\xi \in T^1_u\mathcal{E}$ if for every $\varepsilon > 0$ there is a $\delta > 0$ such that

$$|L(\xi 1_E) - L(0)| < \varepsilon \quad \text{whenever} \quad \mu_u(E) \le \delta. \tag{9.1}$$

If $\Xi \subset T^1\mathcal{E}$, we say L is absolutely equicontinuous on Ξ if for every $\varepsilon > 0$ there is a $\delta > 0$ such that (9.1) holds whenever $u \in \mathcal{E}$, $\xi \in \Xi \cap T^1_u\mathcal{E}$, and $\mu_u(E) \le \delta$.

The term 'absolutely continuous' is borrowed from the theory of rearrangement invariant Banach spaces, see e.g. [**BS**, p.13]. As an example, $L(\xi) = \int_X |\xi|^p \, d\mu_u$, $1 \le p < \infty$, is absolutely continuous on ξ if and only if $L(\xi) < \infty$; but $L(\xi) = \operatorname{ess\,sup}_X |\xi|$ is absolutely continuous only on $\xi = 0$. In general, if $\xi, \eta \in T^1\mathcal{E}$ are equidistributed, and an invariant L is absolutely continuous on ξ, it is absolutely continuous on η, too.

Henceforward we will work with strict rearrangement invariant Lagrangians $L : T^1\mathcal{E} \to (-\infty, \infty]$ that are convex and lsc on the fibers, and we assume L is finite on $T^\infty\mathcal{E}$. Starting with such L, we can represent it through the family $\mathcal{A}_u$ constructed in (8.19); then we can form $\mathcal{A}_u^{||}$ and $L^{||}$ of (8.15), (8.16). Because of Theorem 8.3a, $L^{||}$ is also finite on $T^\infty\mathcal{E}$. It is of course strict rearrangement invariant, and fiberwise convex and lsc.

Now to the definition of $T^L\mathcal{E}$. Whether to include $\xi \in T^1\mathcal{E}$ into $T^L\mathcal{E}$ will depend not only on how L interacts with ξ, but also on how it interacts with η that are allied to ξ, in the following sense.

DEFINITION 9.2. We say that $\eta \in T_v^1\mathcal{E}$ is allied to $\xi \in T_u^1\mathcal{E}$ if there are $\alpha \in \mathbb{R}$ and $\beta \in (0, 1]$ such that $\eta^{\star v}(s) = (\alpha\xi)^{\star u}(\beta s)$ for $0 < s < V$.

If this relation holds, we will say $\eta^{\star v}$ is the (α, β)-rescaling of $\xi^{\star u}$.—In particular, $\alpha = \beta = 1$ corresponds to equidistribution.

DEFINITION 9.3. We let $T_u^L\mathcal{E}$ consist of $\xi \in T_u^1\mathcal{E}$ such that L is finite and absolutely continuous on any $\eta \in T^1\mathcal{E}$ (or only on any $\eta \in T_u^1\mathcal{E}$) allied to it. Further, we let $T^L\mathcal{E} = \coprod_{u\in\mathcal{E}} T_u^L\mathcal{E}$, a subbundle of $T^1\mathcal{E} \to \mathcal{E}$.

If $L(\xi) = \int_X \chi(\xi)\, d\mu_u$ with some convex function $\chi : \mathbb{R} \to \mathbb{R}$, then the definition of $T^L\mathcal{E}$ simplifies to requiring $L(\alpha\xi) < \infty$ for all real α; and if χ is quasihomogeneous, $\chi(t)/c \le \chi(\pm 2t) \le c\chi(t)$ with some $c > 0$, then simply $T^L\mathcal{E} = \{\xi \in T^1\mathcal{E} : L(\xi) < \infty\}$.

One could also associate with L another, in general larger subbundle $T^{(L)}\mathcal{E} \to \mathcal{E}$, consisting of $\xi \in T^1\mathcal{E}$ for which there is an $\alpha_0 > 0$ such that L is finite and absolutely continuous on η whenever $\eta^\star(s) = (\alpha\xi)^\star(\beta s)$ with $\alpha \in (-\alpha_0, \alpha_0)$, $\beta \in (0, 1]$. The theory would be similar, but we will not pursue it here.

LEMMA 9.4. *(a) $T_u^L\mathcal{E} \subset T_u^1\mathcal{E}$ is a vector subspace.*
(b) If $\xi \in T_u^1\mathcal{E}$, $\xi', \xi'' \in T_u^L\mathcal{E}$, and $\xi' \le \xi \le \xi''$ μ_u-a.e., then $\xi \in T_u^L\mathcal{E}$ and $|\xi| \in T_u^L\mathcal{E}$.
(c) If $\Lambda = L^{||}$, then $T^\Lambda\mathcal{E} = T^L\mathcal{E}$.

PROOF. If we add a constant to L, the bundles $T^L\mathcal{E}, T^\Lambda\mathcal{E}$ are not going to change. For this reason we can assume $L(0) = 0 = \Lambda(0)$. (c) now follows from the estimates in Theorem 8.6. Therefore in the rest of the proof, at the price of replacing L by $L^{||}$, we can assume that L is absolutely monotone (Definition 8.7). In particular, $L(\xi) = L(|\xi|) \ge 0$.

To prove (b) it now suffices to show that if $\xi, \xi', \xi'' \in T_u^1\mathcal{E}$, L is finite and absolutely continuous on $\alpha\xi', \alpha\xi''$ for all $\alpha \in \mathbb{R}$, and $\xi' \le \xi \le \xi''$, then L is finite and absolutely continuous on ξ. But this follows since $|\xi| \le |\xi'| + |\xi''|$, and so with any Borel set $E \subset X$

$$L(\xi 1_E) \le L(|\xi'|1_E + |\xi''|1_E) \le \big(L(2\xi' 1_E) + L(2\xi'' 1_E)\big)/2.$$

Finally, to prove (a) we need to show that if $\xi, \eta \in T_u^L\mathcal{E}$, and $\zeta \in T_u^1\mathcal{E}$ is allied to $\xi + \eta$, then L is finite and absolutely continuous on ζ or, equivalently, on $|\zeta|$. Suppose $\zeta^\star(s) = \big(\alpha(\xi+\eta)\big)^\star(\beta s)$. Upon multiplying ξ, η with a suitable unimodular function, we can arrange that $\zeta^\star \ge 0$. Then

$$\zeta^\star(s) \le |\alpha|(|\xi| + |\eta|)^\star(\beta s) \le |\alpha\xi|^\star(\beta s/2) + |\alpha\eta|^\star(\beta s/2),$$

the last inequality by Lemma 2.2, say. Choose $\xi_1, \eta_1 \in T_u^1\mathcal{E}$ such that $\xi_1^\star(s) = |\alpha\xi|^\star(\beta s/2)$, $\eta_1^\star(s) = |\alpha\eta|^\star(\beta s/2)$. These ξ_1, η_1 can be obtained for example by pulling back $|\alpha\xi|^\star(\beta s/2)$ by a measure preserving $\theta : (X, \mu_u) \to (0, V)$. Let ζ_1 be the pull back of $\zeta^\star$ by the same θ. Then $0 \le \zeta_1 \le \xi_1 + \eta_1$. Since with any Borel set $E \subset X$

$$L(\zeta_1 1_E) \le L(\xi_1 1_E + \eta_1 1_E) \le \big(L(2\xi_1 1_E) + L(2\eta_1 1_E)\big)/2,$$

L is finite and absolutely continuous on ζ_1, hence on the equidistributed ζ as well. □

In $T^L\mathcal{E}$ a version of the Dominated Convergence Theorem holds:

LEMMA 9.5. *If $\xi_j, \eta_1, \eta_2 \in T^L_u\mathcal{E}$ for $j = 1, 2, \dots$, $\eta_1 \le \xi_j \le \eta_2$, and $\xi_j \to \xi$ holds μ_u–a.e., then $L(\xi_j) \to L(\xi)$.*

PROOF. Let $\xi_j(\alpha) = (1-\alpha)\xi + \alpha\xi_j$, $\alpha \in \mathbb{R}$. The claim will follow from Lemma 8.5 once we show that

$$\sup_{\alpha\in\mathbb{R}} \limsup_{j\to\infty} |L(\xi_j(\alpha))| < \infty. \tag{9.2}$$

At the price of replacing L by $L^{||} - L(0)$, we can assume L is absolutely monotone and $L(0) = 0$, cf. Theorem 8.6, Lemma 9.4. By this lemma $\eta = |\eta_1| + |\eta_2| \in T^L_u\mathcal{E}$. Given α, choose $\delta > 0$ so that $L((4\alpha + 2)\eta 1_E) < 1$ when $\mu_u(E) \le \delta$, and choose such an E outside which ξ_j converges uniformly. By convexity

$$0 \le 4L\big(\xi_j(\alpha)\big) \le 2L\big(2\xi_j(\alpha)1_E\big) + L\big(4(\xi_j(\alpha) - \xi)1_{X\setminus E}\big) + L(4\xi 1_{X\setminus E}). \tag{9.3}$$

Since $|\xi_j(\alpha)| \le (2|\alpha| + 1)\eta$, the first term on the right is $\le 2L\big((4|\alpha| + 2)\eta 1_E\big) < 2$. The second term tends to 0 as $j \to \infty$ for the following reason. The function $\mathbb{R} \ni \lambda \mapsto L(\lambda) \in \mathbb{R}$ is convex, hence continuous. Therefore letting $\lambda_j = \sup_{X\setminus E} 4|\xi_j(\alpha) - \xi|$,

$$L\big(4(\xi_j(\alpha) - \xi)1_{X\setminus E}\big) \le L(\lambda_j) \to 0 \quad \text{as } j \to \infty,$$

since $\xi_j(\alpha) \to \xi$ uniformly on $X \setminus E$. Finally, the last term in (9.3) is bounded by $L(4\xi) \le L(4\eta) < \infty$. Hence $\limsup_j 4L(\xi_j(\alpha)) \le 2 + L(4\xi)$, and (9.2) indeed holds. □

Absolute continuity of Definition 9.1 and strong continuity of Definition 2.4 are related.

LEMMA 9.6. *The following are equivalent:*
(i) $L|T^\infty\mathcal{E}$ is strongly continuous;
(ii) $T^\infty\mathcal{E} \subset T^L\mathcal{E}$;
(iii) $1 \in T^L\mathcal{E}$.

PROOF. (i) $\Rightarrow$ (ii). We need to show that if $\xi \in T^\infty_u\mathcal{E}$ and $E_j \subset X$ satisfy $\mu_u(E_j) \to 0$, then $L(\xi 1_{E_j}) \to L(0)$. If this failed, we could find $\varepsilon > 0$ and E_j such that $\sum_{j=1}^\infty \mu_u(E_j) < \infty$ and $|L(\xi 1_{E_j}) - L(0)| > \varepsilon$. But $\xi_j = \xi 1_{E_j}$ are uniformly bounded and tend μ_u–a.e. to 0, hence $L(\xi_j) \to L(0)$, a contradiction.

(ii) $\Rightarrow$ (i) is a special case of Lemma 9.5, and (ii) $\Leftrightarrow$ (iii) is partly obvious, partly follows from Lemma 9.4. □

Henceforward we assume L satisfies the equivalent conditions of Lemma 9.6.

Our Lagrangian L induces a functional $L_\star$ on $L^1(0, V)$. If $u \in \mathcal{E}$ and $\theta : (X, \mu_u) \to \big((0, V), \text{Lebesgue}\big)$ preserves measure, then with $\zeta \in L^1(0, V)$ (cf. (2.12))

$$L_\star(\zeta) = L(\zeta \circ \theta) \in (-\infty, \infty], \quad \zeta \circ \theta \text{ viewed as } \in T^1_u\mathcal{E}. \tag{9.4}$$

We denote by $\mathcal{E}^L(0,V) \subset L^1(0,V)$ the subspace of those ζ for which $\zeta\circ\theta \in T^L_u\mathcal{E}$. The invariance of L implies that $L_\star$ and $\mathcal{E}^L(0,V)$ are independent of the choice of u and θ. We have for $\xi \in T^1\mathcal{E}$

$$L(\xi) = L_\star(\xi^\star), \quad \text{and} \quad \xi \in T^L\mathcal{E} \quad \text{if and only if} \quad \xi^\star \in \mathcal{E}^L(0,V).$$

Part (b) of the next result suggests that $T^L\mathcal{E}$ is complete in some sense:

LEMMA 9.7. *(a) Let $\xi_j \in T^1_u\mathcal{E}$, $j \in \mathbb{N}$. If L is absolutely continuous on $\alpha\xi_j$ and $\lim_{i,j\to\infty} L(\alpha(\xi_i - \xi_j)) = L(0)$ for all $\alpha \in \mathbb{R}$, then L is absolutely equicontinuous on the family $\{\xi_j : j \in \mathbb{N}\}$.*

(b) Let $\xi_j \in T^L_u\mathcal{E}$, $j \in \mathbb{N}$. Suppose for every $\alpha \in \mathbb{R}$, $\beta \in (0,1]$ and $\zeta_j(s) = \alpha\xi_j^\star(\beta s)$, $s \in (0,V)$, we have $\lim_{i,j\to\infty} L_\star(\zeta_i - \zeta_j) = L(0)$. If μ_u-a.e. $\xi_j \to \xi$, then $\xi \in T^L_u\mathcal{E}$, $\lim_{j\to\infty} L(\xi_j - \xi) = L(0)$ and $\lim_{j\to\infty} L(\xi_j) = L(\xi)$.

The condition in (b) that ξ_j converge almost everywhere often can be replaced by requiring $\lim_{i,j\to\infty} L(\xi_i - \xi_j) = L(0)$, and the existence of ξ as in the conclusion still follows. For example, if L is a norm on $T^L_u\mathcal{E}$, then one can show that

$$L(\eta) \geq c\int_X |\eta|\, d\mu_u, \quad \eta \in T^L_u\mathcal{E}, \tag{9.5}$$

with some $c > 0$. [**L1**, Lemma 6.1] gives this for $\eta \in T^\infty_u\mathcal{E}$, but the same proof works for $\eta \in T^L_u\mathcal{E}$. Alternatively, one can approximate $\eta \in T^L_u\mathcal{E}$ by bounded $\eta_j \in T^\infty_u\mathcal{E}$ and obtain (9.5) by dominated convergence, Lemma 9.4. Once (9.5) is known, $L(\xi_i - \xi_j) \to L(0) = 0$ implies ξ_j is a Cauchy sequence in $T^1_u\mathcal{E}$, hence converges to some $\xi \in T^1_u\mathcal{E}$. Applying Lemma 9.7b to subsequences ξ_{j_k} then gives $\xi \in T^L_u\mathcal{E}$ and the rest.—In this case the norms $\|\xi\|_\beta = L_\star(\xi^\star(\beta\,\cdot))$, $0 < \beta \leq 1$, endow $T^L_u\mathcal{E}$ with the structure of a metrizable topological vector space, which is complete (i.e., Fréchet) according to the above discussion.

PROOF. We can assume $L(0) = 0$, and except for the very last limit in (b), $\lim_j L(\xi_j) = L(\xi)$, in light of Theorem 8.6 also that L is absolutely monotone.

(a) Given $\varepsilon > 0$, choose j so that $L\big(2(\xi_i - \xi_j)\big) < \varepsilon$ when $i > j$. Next choose $\delta > 0$ so that $L(2\xi_i 1_E) < \varepsilon$ if $i = 1, 2, \dots, j$ and $\mu_u(E) \leq \delta$. With such E and $i > j$

$$2L(\xi_i 1_E) \leq L\big(2(\xi_i - \xi_j)1_E\big) + L(2\xi_j 1_E) < 2\varepsilon.$$

As $L(\xi_i 1_E) \leq L(2\xi_i 1_E) < \varepsilon$ if $i \leq j$, L is indeed absolutely equicontinuous on $\{\xi_j : j \in \mathbb{N}\}$.

(b) First observe that $\sup_j L(\xi_j) = M < \infty$: as before, we choose j so that $L_\star\big(2(\xi_i^\star - \xi_j^\star)\big) < 1$ when $i > j$, then

$$2L(\xi_i) = 2L_\star(\xi_i^\star) \leq L_\star(2\xi_j^\star) + L_\star\big(2(\xi_i^\star - \xi_j^\star)\big) < L(2\xi_j) + 1, \quad i > j.$$

This also implies $L(\xi) \leq M$ as follows. If $k \in \mathbb{N}$, let $\xi_{jk} = \min(k, |\xi_j|) \in T^\infty_u\mathcal{E}$. Then μ_u-a.e. $\lim_j \xi_{jk} = \min(k, |\xi|)$, and by strong continuity

$$L(\min(k, |\xi|)) = \lim_{j\to\infty} L(\xi_{jk}) \leq M.$$

We represent L with a suitable $\mathcal{A}_u \subset \mathbb{R} \times B(X)$ as

$$L(\eta) = \sup_{(a,f)\in\mathcal{A}_u} a + \int_X |\eta f|\, d\mu_u, \qquad \eta \in T^1_u\mathcal{E}.$$

If $(a, f) \in \mathcal{A}_u$, by monotone convergence

$$a + \int_X |\xi f|\, d\mu_u = \lim_{k\to\infty} a + \int_X \min(k, |\xi|)|f|\, d\mu_u \le \limsup_{k\to\infty} L\big(\min(k, |\xi|)\big) \le M,$$

and so $L(\xi) \le M$.

Next, let $\theta : (X, \mu_u) \to \big((0, V), \text{Lebesgue}\big)$ preserve measure. By part (a), L is absolutely equicontinuous on $\{\xi_j^\star \circ \theta, j \in \mathbb{N}\}$, whence also on $\{\xi_j : j \in \mathbb{N}\}$. Therefore the computation above, with ξ replaced by $\xi 1_E$ gives that L is absolutely continuous on ξ. But we can also replace ξ by any of its allies η, to conclude that $\xi \in T_u^L\mathcal{E}$, indeed.

To prove the rest of (b), we no longer assume L is absolutely monotone. Let

$$\xi_j(\alpha) = (1-\alpha)\xi + \alpha\xi_j \in T_u^L\mathcal{E}, \qquad \alpha \in \mathbb{R},$$

and estimate $L\big(\xi_j(\alpha)\big)$. For fixed α, by part (a), L and $L^{||}$ are absolutely equicontinuous on the family $\{\xi_j(\alpha)^\star \circ \theta : j \in \mathbb{N}\}$, hence on $\{\xi_j(\alpha) : j \in \mathbb{N}\}$. Choose $\delta > 0$ so that $L^{||}\big(2\xi_j(\alpha)1_E\big) < 1$ when $\mu_u(E) \le \delta$; and choose such an E outside which $\xi_j \to \xi$ uniformly. By (8.17) and by convexity,

$$\begin{aligned} 4|L\big(\xi_j(\alpha)\big)| &\le 4L^{||}\big(\xi_j(\alpha)\big) \\ &\le 2L^{||}\big(2\xi_j(\alpha)1_E\big) + L^{||}\big(4(\xi_j(\alpha) - \xi)1_{X\setminus E}\big) + L^{||}(4\xi 1_{X\setminus E}). \end{aligned}$$

The first term on the right is < 2, the last is $\le L^{||}(4\xi) < \infty$ and, as in the proof of Lemma 9.5, since $\xi_j(\alpha) \to \xi$ uniformly on $X \setminus E$, the middle term goes to 0 as $j \to \infty$. Therefore $\limsup_j |L(\xi_j(\alpha))| \le 2 + L^{||}(4\xi)$, and Lemma 8.5 implies $\lim_j L(\xi_j) = L(\xi)$. Finally we note that since $\xi \in T_u^L\mathcal{E}$, L is absolutely equicontinuous on $\{\alpha(\xi_j - \xi) : j \in \mathbb{N}\}$ for fixed $\alpha \in \mathbb{R}$. Therefore in the above estimate we can replace ξ_j by $\xi_j - \xi \in T_u^L\mathcal{E}$, to obtain $\lim_j L(\xi_j - \xi) = 0$. □

10. Rise, energy, action

In this section we fix a strict rearrangement invariant Lagrangian $L : T^1\mathcal{E} \to (-\infty, \infty]$ that is convex and lower semicontinuous on the fibers $T_u^1\mathcal{E}$. We assume that L is finite and absolutely continuous on elements of $T^\infty\mathcal{E}$, i.e., $T^\infty\mathcal{E} \subset T^L\mathcal{E}$, cf. Definitions 9.1–3 and Lemma 9.6. Generalizing Guedj–Zeriahi's high energy classes and Mabuchi's and Darvas's metrics [Da1–2, M, GZ1–2], we will introduce energy classes $\mathcal{E}^L \subset \mathcal{E}$ of ω–plurisubharmonic functions and define action between elements of $\mathcal{E}^L$. We will show that this action can be computed from the rise, and as an application, we prove a Principle of Least Action in $\mathcal{E}^L$.

Recall that $u \in \mathcal{E}$ belongs to the energy space $\mathcal{E}^1$ if $u \in L^1(X, \mu_u)$, or $u \in T_u^1\mathcal{E}$.

DEFINITION 10.1. The energy space[3] $\mathcal{E}^L$ consists of $u \in \mathcal{E}^1$ that, viewed as elements of $T_u^1\mathcal{E}$, are in $T_u^L\mathcal{E}$.

LEMMA 10.2. *A function $u \in \mathcal{E}^1$ is in $\mathcal{E}^L$ if and only if $\rho[u, v] \in \mathcal{E}^L(0, V)$ for some (equivalently: for every) $v \in \mathcal{E}^\infty$.*

---

[3]The reader will notice an inconsistency in our notation, and should keep it in mind to avoid confusion. $\mathcal{E}^1, \mathcal{E}^\infty$ are not $\mathcal{E}^L$ when $L \equiv 1$, resp. ∞. Instead, $\mathcal{E}^1$ corresponds to $L(\xi) = \int_X |\xi|\, d\mu_u$, while the space $\mathcal{E}^\infty$ is not of the type introduced here, because the Lagrangian $L(\xi) = \operatorname{ess\,sup}_X |\xi|$ is not allowed in this section. The same inconsistency already occurs in the notation $T^L\mathcal{E}$.

Recall the definition (9.4) of the function $L_\star : L^1(0,V) \to (-\infty,\infty]$ and of the associated space $\mathcal{E}^L(0,V) \subset L^1(0,V)$ in the paragraph after. Lemma 9.4 implies $\mathcal{E}^L(0,V)$ is a vector space of functions, contains $L^\infty(0,V)$, and if $\zeta \in L^1(0,V)$, $\zeta', \zeta'' \in \mathcal{E}^L(0,V)$, and $\zeta' \le \zeta \le \zeta''$ a.e., then $\zeta, |\zeta| \in \mathcal{E}^L(0,V)$. If $\zeta, \zeta' \in L^1(0,V)$ are equidistributed, or $\zeta'^\star$ is the (α,β) rescaling of $\zeta^\star$, and $\zeta \in \mathcal{E}^L(0,V)$, then $\zeta' \in \mathcal{E}^L(0,V)$.

PROOF. Suppose first $u \in \mathcal{E}^L$. Thus u and so $-u \in T_u^L\mathcal{E}$, whence $(-u)^{\star u} \in \mathcal{E}^L(0,V)$. If $v \in \mathcal{E}^\infty$, then $(v-u)^{\star u} = (-u)^{\star u} + O(1) \in \mathcal{E}^L(0,V)$. With $c = \inf(v-u) > -\infty$

$$c \le \rho[u,v] \le (v-u)^{\star u}$$

by Theorem 6.1, and so $\rho[u,v] \in \mathcal{E}^L(0,V)$.

Conversely, suppose $\rho[u,v] \in \mathcal{E}^L(0,V)$, where $v \in \mathcal{E}^\infty$. With c above, $u+c \le v$. Taking into account that $\mu_{u+c} = \mu_u$, Theorem 6.1 implies for $0 < s < V$

$$\rho[u+c,v](s) \le (v-u-c)^{\star u}(s) \le (n+1)\rho[u+c,v](s/e). \tag{10.1}$$

Here $\rho[u+c,v] = \rho[u,v] - c \in \mathcal{E}^L(0,V)$, see Lemma 5.10. If $\xi \in \mathcal{E}^L$ is chosen so that $\xi^{\star u} = \rho[u+c,v]$, then the function on the right of (10.1) is the decreasing rearrangement of some $\eta \in \mathcal{E}^L$ allied with ξ, Definition 9.2. Thus both lower and upper estimates in (10.1) are in $\mathcal{E}^L(0,V)$, hence so must be $(v-u-c)^{\star u}$. In other words $v-u-c \in T_u^L\mathcal{E}$. Since $v-c$ is bounded, $\pm u \in T_u^L\mathcal{E}$ and $u \in \mathcal{E}^L$ follow. □

LEMMA 10.3. *If $u \in \mathcal{E}^L$, $v \in \mathcal{E}$ and $u \le v$, then $v \in \mathcal{E}^L$.*

PROOF. This follows from Lemmas 5.10, 10.2. If $c = \max v$, then $-c = \rho[c,0] \le \rho[v,0] \le \rho[u,0]$. Since $-c, \rho[u,0] \in \mathcal{E}^L(0,V)$, also $\rho[v,0] \in \mathcal{E}^L(0,V)$. □

Consider now $u, v \in \mathcal{E}^L$ and pick a constant $c \ge u, v$. By Lemma 5.10

$$\rho[c,v] \le \rho[u,v] \le \rho[u,c]. \tag{10.2}$$

Since $\rho[c,v]$ and $-\rho[v,c] \in \mathcal{E}^L(0,V)$ are equidistributed, the two extremes in (10.2) are in $\mathcal{E}^L(0,V)$, hence so is $\rho[u,v]$. In particular, $L_\star(\lambda\rho[u,v]) < \infty$ with any $\lambda \in \mathbb{R}$.

DEFINITION 10.4. If $T > 0$, the action between $u, v \in \mathcal{E}^L$ is

$$\mathcal{L}_T(u,v) = TL_\star(\rho[u,v]/T) < \infty.$$

[**L2**] already defined action between $u, v \in \mathcal{E}^\infty$, see (2.10). That the current definition is consistent with (2.10) will follow from the next result, in which $\mathcal{L}_T$ is used in the sense of Definition 10.4.

LEMMA 10.5. *If $u_j, v_j \in \mathcal{E}^L$ decrease to $u, v \in \mathcal{E}^L$, then*

$$\mathcal{L}_T(u,v) = \lim_{j\to\infty} \mathcal{L}_T(u_j,v_j) \tag{10.3}$$

PROOF. By Theorem 5.6 $\rho[u_j,v_j] \to \rho[u,v]$ almost everywhere. With a constant $c \ge u_1, v_1$ we have $\rho[c,v] \le \rho[u_j,v_j] \le \rho[u,c]$, and dominated convergence (Lemma 9.5), together with (9.4) yields $L_\star(\rho[u_j,v_j]/T) \to L_\star(\rho[u,v]/T)$, i.e., (10.3). □

It follows that Definition 10.4 and (2.10) give the same notion of action on $\mathcal{E}^\infty$. This is so over $\mathcal{H}^{1\bar{1}}$ by Lemma 3.4; in general we take $u_j, v_j \in \mathcal{H}^{1\bar{1}}$ that decrease to $u, v \in \mathcal{E}^\infty$, and note that (10.3) holds with the definition (2.3) too, by [**L2**, Lemma 9.4].

In the special case when L is Orlicz norm associated with a convex, 'normalized', even $\chi : \mathbb{R} \to \mathbb{R}$, as in [**Da1**, section 1.1],

$$L(\xi) = \inf \Big\{ r > 0 : \int_X \chi(\xi/r)\, d\mu_u \le \chi(1) \Big\}, \quad \xi \in T^1_u \mathcal{E},$$

$\mathcal{L}_T$ is independent of T and coincides with Darvas's distance function d_χ if χ is in one of the classes $\mathcal{W}_p^+$, $1 \le p < \infty$. That

$$d_\chi(u, v) = \mathcal{L}_T(u, v) \tag{10.4}$$

first follows from [**Da1**, Theorem 1] when $u, v \in \mathcal{H}$; and for general $u, v \in \mathcal{E}^L$ from Darvas's definition [**Da1**, (5)] of d_χ as a limit, and from (10.3).

A version of the triangle inequality holds for general L:

$$\mathcal{L}_S(u, v) + \mathcal{L}_T(v, w) \ge \mathcal{L}_{S+T}(u, w), \qquad u, v, w \in \mathcal{E}^L. \tag{10.5}$$

When $u, v, w \in \mathcal{E}^\infty$, [**L2**, (5.3)] implies with any $a \in \mathbb{R}$, e.g.,

$$\mathcal{L}_S(u, v) = \inf \int_a^{a+S} L(\partial_t \psi(t)) dt,$$

the infimum taken over all piecewise C^1 paths $\psi : [a, a+S] \to \mathcal{E}^\infty \subset B(X)$; then (10.5) follows by concatenating paths. To general $u, v, w \in \mathcal{E}^L$ we can decrease by $u_j, v_j, w_j \in \mathcal{E}^\infty$. Then $\mathcal{L}_S(u_j, v_j) + \mathcal{L}_T(v_j, u_j) \ge \mathcal{L}_{S+T}(u_j, w_j)$, and (10.5) follows by passing to the limit, cf. (10.3).

The notion of action of a path and the principle of least action of [**L2**] can be extended from $\mathcal{E}^\infty$ to $\mathcal{E}^L$. Let $\varphi : [a, b] \to \mathcal{E}^L$ be an arbitrary map. Its action is

$$\mathcal{L}(\varphi) = \sup \sum_{i=1}^m \mathcal{L}_{t_i - t_{i-1}}\big(\varphi(t_{i-1}), \varphi(t_i)\big) \le \infty,$$

the supremum taken over all partitions $a = t_0 < t_1 < \ldots < t_m = b$. [**L2**, Theorem 10.1] identifies this quantity with $\int_a^b L(\partial_t \varphi(t)) dt$ when φ is a C^1-map into $\mathcal{E}^\infty \subset B(X)$.

THEOREM 10.6 (Principle of Least Action). *If $\varphi : [a, b] \to \mathcal{E}$ is a geodesic and $\varphi(a), \varphi(b) \in \mathcal{E}^L$, then $\varphi(t) \in \mathcal{E}^L$ for all $t \in (a, b)$. If, furthermore, $\psi : [a, b] \to \mathcal{E}^L$ is any map with $\psi(a) = \varphi(a)$, $\psi(b) = \varphi(b)$, then $\mathcal{L}(\psi) \ge \mathcal{L}(\varphi)$.*

This generalizes parts of [**Da2**, Theorem 6], [**Da1**, Theorem 4.11], and [**L2**, Theorem 1.1].

PROOF. As in [Da1-2], we start by showing that $u, v \in \mathcal{E}^L$ implies $u \wedge v \in \mathcal{E}^L$. By [**Da2**, Corollary 3.5] $u \wedge v \in \mathcal{E}^1$. Choose a constant $c \ge u, v$. Lemma 6.5 implies

$$0 \le \rho[u \wedge v, c](s) \le (c - u)^{\star u}(s/2) + (c - v)^{\star v}(s/2), \qquad 0 < s < V.$$

Since the function on the right is in $\mathcal{E}^L(0, V)$, as is 0, so must be the function in the middle; whence by Lemma 10.2 $u \wedge v \in \mathcal{E}^L$.

Now with φ of the theorem the constant path $\widetilde{\varphi}(t) \equiv \varphi(a) \wedge \varphi(b)$ is a subgeodesic, and $\widetilde{\varphi}(a) \le \varphi(a)$, $\widetilde{\varphi}(b) \le \varphi(b)$. Therefore

$$\varphi(a) \wedge \varphi(b) \le \varphi(t), \qquad t \in [a, b],$$

and $\varphi(a) \wedge \varphi(b) \in \mathcal{E}^L$, whence by Lemma 10.3 $\varphi(t) \in \mathcal{E}^L$.

If $a \le t < t' \le b$ then

$$\frac{\rho[\varphi(t),\varphi(t')]}{t'-t} = \rho_{\varphi|[t,t']} = \rho_\varphi = \frac{\rho[\varphi(a),\varphi(b)]}{b-a}.$$

Hence, with any partition $a = t_0 < t_1 < \cdots < t_m = b$

$$\sum_{i=1}^m \mathcal{L}_{t_i - t_{i-1}}\big(\varphi(t_{i-1}),\varphi(t_i)\big) = \sum_{i=1}^m (t_i - t_{i-1}) L_\star\Big(\frac{\rho[\varphi(t_{i-1}),\varphi(t_i)]}{t_i - t_{i-1}}\Big)$$
$$= \sum_{i=1}^m (t_i - t_{i-1}) L_\star(\rho_\varphi) = (b-a)L_\star(\rho_\varphi) = \mathcal{L}_{b-a}\big(\varphi(a),\varphi(b)\big).$$

Therefore $\mathcal{L}(\varphi) = \mathcal{L}_{b-a}\big(\varphi(a),\varphi(b)\big)$. As for the action of ψ, by definition $\mathcal{L}(\psi) \ge \mathcal{L}_{b-a}\big(\psi(a),\psi(b)\big) = \mathcal{L}_{b-a}\big(\varphi(a),\varphi(b)\big) = \mathcal{L}(\varphi)$, q.e.d. $\square$

References

[BS] Colin Bennett and Robert Sharpley, *Interpolation of operators*, Pure and Applied Mathematics, vol. 129, Academic Press, Inc., Boston, MA, 1988. MR928802

[Be1] Bo Berndtsson, *A Brunn-Minkowski type inequality for Fano manifolds and some uniqueness theorems in Kähler geometry*, Invent. Math. **200** (2015), no. 1, 149–200, DOI 10.1007/s00222-014-0532-1. MR3323577

[Be2] Bo Berndtsson, *Probability measures associated to geodesics in the space of Kähler metrics*, Algebraic and analytic microlocal analysis, Springer Proc. Math. Stat., vol. 269, Springer, Cham, 2018, pp. 395–419, DOI 10.1007/978-3-030-01588-6_6. MR3903321

[BL] Thomas Bloom and Norman Levenberg, *Pluripotential energy*, Potential Anal. **36** (2012), no. 1, 155–176, DOI 10.1007/s11118-011-9224-2. MR2886457

[Bl] Zbigniew Błocki, *On geodesics in the space of Kähler metrics*, Advances in geometric analysis, Adv. Lect. Math. (ALM), vol. 21, Int. Press, Somerville, MA, 2012, pp. 3–19. MR3077245

[BK] Zbigniew Błocki and Sławomir Kołodziej, *On regularization of plurisubharmonic functions on manifolds*, Proc. Amer. Math. Soc. **135** (2007), no. 7, 2089–2093, DOI 10.1090/S0002-9939-07-08858-2. MR2299485

[C] Xiuxiong Chen, *The space of Kähler metrics*, J. Differential Geom. **56** (2000), no. 2, 189–234. MR1863016

[Da1] Tamás Darvas, *The Mabuchi geometry of finite energy classes*, Adv. Math. **285** (2015), 182–219, DOI 10.1016/j.aim.2015.08.005. MR3406499

[Da2] Tamás Darvas, *The Mabuchi completion of the space of Kähler potentials*, Amer. J. Math. **139** (2017), no. 5, 1275–1313, DOI 10.1353/ajm.2017.0032. MR3702499

[Da3] Tamás Darvas, *Geometric pluripotential theory on Kähler manifolds*, Advances in complex geometry, Contemp. Math., vol. 735, Amer. Math. Soc., [Providence], RI, [2019] ©2019, pp. 1–104, DOI 10.1090/conm/735/14822. MR3996485

[Da4] Tamás Darvas, *The Mabuchi geometry of low energy classes*, Math. Ann. **389** (2024), no. 1, 427–450, DOI 10.1007/s00208-023-02648-0. MR4735951

[DR] Tamás Darvas and Yanir A. Rubinstein, *Kiselman's principle, the Dirichlet problem for the Monge-Ampère equation, and rooftop obstacle problems*, J. Math. Soc. Japan **68** (2016), no. 2, 773–796, DOI 10.2969/jmsj/06820773. MR3488145

[De] Jean-Pierre Demailly, *Regularization of closed positive currents and intersection theory*, J. Algebraic Geom. **1** (1992), no. 3, 361–409. MR1158622

[DP] Jean-Pierre Demailly and Mihai Păun, *Numerical characterization of the Kähler cone of a compact Kähler manifold*, Ann. of Math. (2) **159** (2004), no. 3, 1247–1274, DOI 10.4007/annals.2004.159.1247. MR2113021

[DNL] Eleonora Di Nezza and Hoang-Chinh Lu, *Geodesic distance and Monge-Ampère measures on contact sets*, Anal. Math. **48** (2022), no. 2, 451–488, DOI 10.1007/s10476-022-0159-1. MR4440753

[GZ1] Vincent Guedj and Ahmed Zeriahi, *The weighted Monge-Ampère energy of quasi-plurisubharmonic functions*, J. Funct. Anal. **250** (2007), no. 2, 442–482, DOI 10.1016/j.jfa.2007.04.018. MR2352488

[GZ2] Vincent Guedj and Ahmed Zeriahi, *Degenerate complex Monge-Ampère equations*, EMS Tracts in Mathematics, vol. 26, European Mathematical Society (EMS), Zürich, 2017, DOI 10.4171/167. MR3617346

[H] Weiyong He, *On the space of Kähler potentials*, Comm. Pure Appl. Math. **68** (2015), no. 2, 332–343, DOI 10.1002/cpa.21515. MR3298665

[IT] A. D. Ioffe and V. M. Tihomirov, *Theory of extremal problems*, Studies in Mathematics and its Applications, vol. 6, North-Holland Publishing Co., Amsterdam-New York, 1979. Translated from the Russian by Karol Makowski. MR528295

[L1] László Lempert, *On the adjoint action of the group of symplectic diffeomorphisms*, Pure Appl. Math. Q. **18** (2022), no. 2, 657–682. MR4429225

[L2] László Lempert, *The principle of least action in the space of Kähler potentials*, Math. Res. Lett. **29** (2022), no. 3, 785–833. MR4516039

[M] Toshiki Mabuchi, *Some symplectic geometry on compact Kähler manifolds. I*, Osaka J. Math. **24** (1987), no. 2, 227–252. MR909015

[R] Halsey Royden, *Real analysis, third edition*, Prentice Hall, Englewood Cliffs, NJ, 1988.

[S] Stephen Semmes, *Complex Monge-Ampère and symplectic manifolds*, Amer. J. Math. **114** (1992), no. 3, 495–550, DOI 10.2307/2374768. MR1165352

DEPARTMENT OF MATHEMATICS, PURDUE UNIVERSITY, WEST LAFAYETTE, INDIANA 47907-2067

Contemporary Mathematics
Volume **810**, 2025
https://doi.org/10.1090/conm/810/16214

Harmonic interpolation and a Brunn-Minkowski theorem for random determinants

Julius Ross and David Witt Nyström

Abstract. We describe the harmonic interpolation of convex bodies, and prove a strong form of the Brunn-Minkowski inequality and characterize its equality case. As an application we improve a theorem of Berndtsson on the volume of slices of a pseudoconvex domain. We furthermore apply this to prove subharmonicity of the expected absolute value of the determinant of a matrix of random vectors through the connection with zonoids.

1. Introduction

Let A and B be convex subsets of $\mathbb{R}^n$. The Minkowski sum of A and B is defined as

$$A + B := \{a + b : a \in A, b \in B\},$$

and the famous Brunn-Minkowski inequality says that

$$|A + B|^{1/n} \geq |A|^{1/n} + |B|^{1/n},$$

where $|\cdot|$ denotes the Euclidean volume.

We wish to consider the interpolation of convex sets. Given convex A and B there is a natural interpolating family $A_t := (1-t)A + tB$, $t \in [0,1]$, and it follows from the Brunn-Minkowski inequality that the map

$$t \mapsto |A_t|^{1/n}$$

is concave in $t \in [0,1]$.

For an infinite family of convex sets, there are many possible interpolations. To consider this in more detail, suppose Ω is a smoothly bounded domain in $\mathbb{R}^m$ and that we have a continuous family of convex bodies (i.e. compact convex set with non-empty interior) $A_\tau \subset \mathbb{R}^n$ parametrized by $\tau \in \partial\Omega$. If Ω is itself convex, a natural interpolation can be obtained by considering

$$A = \text{Convexhull}\left(\bigcup_{\tau \in \partial\Omega} A_\tau \times \{\tau\}\right) \subseteq \mathbb{R}^{n+m}$$

and letting A_x be the fiber of A over $x \in \Omega$. We call this the *convex interpolation* of $\{A_\tau\}$. Then directly from the Brunn-Minkowski inequality it follows that the map $x \mapsto |A_x|^{1/n}$ is concave in $x \in \Omega$.

2020 *Mathematics Subject Classification.* Primary 32J27, 52A40, 52A21; Secondary 32U05, 14C17, 52A40.

If Ω is not convex, the convex interpolation is not suitable since it will not necessarily agree with the given boundary data $\{A_\tau\}$ on $\partial\Omega$. For general Ω a natural interpolation was proposed in our recent paper [**3**] that we now describe.

First note that if A_y is a continuous family of convex bodies in $\mathbb{R}^n$ over some parameter set $D \subseteq \mathbb{R}^m$ and μ is a Radon measure on D, then there is a set-integral

$$\int_D A_y d\mu(y)$$

which is itself a subset of $\mathbb{R}^n$. To define this precisely recall that the support function of a convex set A is given by

$$h_A(\xi) := \sup_{\zeta \in A}(\zeta \cdot \xi)$$

and has the property that

$$h_A \text{ is convex and } h_A(t\xi) = t h_A(\xi) \text{ for } t \in \mathbb{R}_+. \tag{1}$$

On the other hand, if h is a function with those two properties then h is the support function of a unique closed convex set, which we denote by $A(h)$.

It is an elementary exercise to see that $h_{A+B} = h_A + h_B$ and more generally

$$h_{t_1 A_1 + \cdots + t_k A_k} = t_1 h_{A_1} + \cdots + t_k h_{A_k}.$$

Furthermore if A_t are convex sets such that $A_t \to A$ in the Hausdorff topology, then for each ξ, $h_{A_t}(\xi) \to h_A(\xi)$.

Definition 1.1. Let $d\mu$ be a measure on a measurable set D in $\mathbb{R}^m$, and A_y be a convex set for each $y \in D$. We define the *Minkowski integral* $\int_D A_y d\mu(y)$ as

$$\int_D A_y d\mu(y) := A\left(\int_D h_{A_y} d\mu(y)\right).$$

Such set-valued integrals have been considered in various places, for example [**1**, **5**]. As one would expect, some conditions are needed to ensure that the Minkowski integral is well-defined. For our purpose the following is sufficient: assume D is compact, $d\mu$ is a Radon measure and $y \mapsto A_y$ is a continuous family of convex bodies. Then for each ζ the map $y \mapsto h_{A_y}(\zeta)$ is continuous, so $\int_D h_{A_y} d\mu(y)$ exists and has properties (1), and thus $\int_D A_y d\mu(y)$ exists.

Definition 1.2. Let $\Omega \subset \mathbb{R}^m$ be a smoothly bounded domain. The *harmonic interpolation* of a continuous family $\{A_\tau\}_{\tau \in \partial\Omega}$ of convex bodies is defined as

$$A_x := \int_{\partial\Omega} A_\tau d\mu_x(\tau),$$

where $d\mu_x$ is the harmonic measure on $\partial\Omega$ with respect to $x \in \Omega$.

The harmonic interpolation and convex interpolation may differ, even when Ω is convex. We argue that the former is better suited in some contexts, one of which is the theory of zonoids.

A *zonotope* is a convex set that can be written as the Minkowski sum of line segments. Clearly any zonotope is a convex polytope, but it is easy to see that not all convex polytopes are zonotopes. A *zonoid* is a convex set which can be approximated arbitrarily well (in the Hausdorff topology) by zonotopes, or equivalently a convex set that can be written as the Minkowski integral of line segments (see for example [**4**] for a introduction to zonoids). The harmonic interpolation has the property that it preserves zonoids; i.e. if each boundary set A_τ is a zonoid then each member of the interpolating family A_x will also be a zonoid (and this is not true for the convex interpolation, even when Ω is convex). In section 4 we will use this to apply our results to interpolation of families of random (n, n) matrices whose columns are independent identically distributed.

2. A Brunn-Minkowski Inequality for Harmonic Interpolation

We continue to assume $\Omega \subset \mathbb{R}^m$ is a smoothly bounded domain (which by convention is also bounded), and A_τ for $\tau \in \partial\Omega$ is a continuous family of convex bodies in $\mathbb{R}^n$. In [**3**] we proved the following weak version of a Brunn-Minkowski inequality for the harmonic interpolation.

THEOREM 2.1. *If A_x is the harmonic interpolation of $\{A_\tau\}$ then $x \mapsto \log|A_x|$ is superharmonic in x.*

Our main result in this short note is a direct proof of the following stronger version:

THEOREM 2.2. *If A_x is the harmonic interpolation of $\{A_\tau\}$ then $x \mapsto |A_x|^{1/n}$ is superharmonic in x.*

PROOF. Let $B_\epsilon(x)$ denote the Euclidean ball of radius ϵ centered at x. We need to show that if $B_\epsilon(x) \subseteq \Omega$ then

$$|A_x|^{1/n} \geq \int_{\partial B_\epsilon(x)} |A_y|^{1/n} dS(y),$$

where dS denotes the normalized Euclidean surface measure on $\partial B_\epsilon(x)$.

A standard property of harmonic measures is that

$$\mu_x = \int_{\partial B_\epsilon(x)} \mu_y dS(y),$$

and this clearly implies that

$$A_x = \int_{\partial B_\epsilon(x)} A_y dS(y).$$

Now we approximate the surface measure dS with a sequence of atomic measure $\nu_k = \sum_{i=1}^{N_k} \lambda_{i,k}\delta_{y_{i,k}}$ chosen so $\nu_k \to dS$ weakly as $k \to \infty$. Then for k sufficiently large $\sum_{i=1}^{N_k} \lambda_{i,k} A_{y_i}$ is arbitrarily close (in the Hausdorff distance) to A_x. Thus for any $\epsilon > 0$ and k sufficiently large we have

$$|A_x|^{1/n} + \epsilon \geq |\sum_{i=1}^{N_k} \lambda_{i,k} A_{y_{i,k}}|^{1/n} \geq \sum_{i=1}^{N_k} \lambda_{i,k} |A_{y_{i,k}}|^{1/n} \geq \int_{\partial B_\epsilon(x)} |A_y|^{1/n} dS(y) - \epsilon,$$

where the second follows from the classical Brunn-Minkowski inequality. Letting $\epsilon \to 0$ we have

$$|A_x|^{1/n} \geq \int_{\partial B_\epsilon(x)} |A_y|^{1/n} dS(y),$$

which completes the proof. □

DEFINITION 2.3. We say that a continuous family of convex sets $A_x \subseteq \mathbb{R}^n$ over some domain $\Omega \subseteq \mathbb{R}^m$ is *subharmonic* over Ω if whenever $B_\epsilon(x) \subseteq \Omega$ we have that

$$A_x \supseteq \int_{\partial B_\epsilon(x)} A_y dS(y).$$

The then get the following Corollary of Theorem 2.2.

COROLLARY 2.4. *If A_x is subharmonic over Ω then $x \mapsto |A_x|^{1/n}$ is superharmonic.*

As an application we can give a strengthening of the following theorem of Berndtsson [**2**]

THEOREM 2.5. *Let $U \subseteq \mathbb{C}^{n+m}$ be a pseudoconvex domain with the property that if $(x_1 + iy_1, \ldots, x_n + iy_n, w) \in U$ then $(x_1 + iy_1', \ldots, x_n + iy_n', w) \in U$ for all $y_1', \ldots, y_n'$, and let $U_w := \{x \in \mathbb{R}^n : (x, w) \in U\}$.*

Then the map $w \mapsto -\log |U_w|$ is plurisubharmonic in w.

THEOREM 2.6. *In the same setting as Theorem 2.5, the map $w \mapsto -|U_w|^{1/n}$ is plurisubharmonic.*

PROOF. Without loss of generality we can assume that $m = 1$. Note that the pseudoconvexity and symmetry of U implies that U_w is convex for all w. By approximation we can also without loss of generality assume that the family U_w is bounded and continuous. We claim that the family U_w is subharmonic. Note that for two closed convex sets A and B we have that $A \supseteq B$ if and only if $h_A \geq h_B$, so U_w is subharmonic if and only if for any $\xi \in \mathbb{R}^n$, $h_{U_w}(\xi) = \sup_{x \in U_w}(x \cdot \xi)$ is superharmonic in w.

Let ϕ be a plurisubharmonic exhaustion function for U which we can assume to be independent of $\mathrm{Im}(\mathbb{C}^n)$, just as U itself. Note that $\phi_R := \max(\phi - R, 0)$ is also plurisubharmonic and independent of $\mathrm{Im}(\mathbb{C}^n)$, and that the same is true for $\psi_R(x + iy, w) := \phi_R(x + iy, w) - x \cdot \xi$. Thus by Kiselman's minimum principle $\inf_{x \in U_w} \psi_R(x, w)$ is subharmonic in w. We now note that

$$\sup_{x \in U_w}(x \cdot \xi) = -\lim_{R \to \infty} \inf_{x \in U_w} \psi_R(x, w),$$

and hence it follows that $h_{U_w}(\xi)$ is superharmonic and thus U_w is subharmonic. That $-|U_w|^{1/n}$ is subharmonic now follows from Corollary 2.4. □

3. Characterization of the extremal case

By Corollary 2.4 we know that if $\{A_x\}_{x \in \Omega}$ is a subharmonic family of convex bodies in $\mathbb{R}^n$ over a domain Ω then $x \mapsto |A_x|^{1/n}$ is superharmonic. Our next result characterizes when this map is in fact harmonic.

THEOREM 3.1. *The map $x \mapsto |A_x|^{1/n}$ is harmonic if and only if we can write $A_x = c_x B + d_x$ where $B \subset \mathbb{R}^n$ is a fixed convex body, and c_x and d_x are harmonic functions on Ω taking values in $\mathbb{R}_+$ and $\mathbb{R}^n$ respectively.*

PROOF. Let Ω' be a relatively compact subdomain of Ω with smooth boundary. Since A_x is subharmonic it must dominate the harmonic interpolation of A_y restricted to $\partial\Omega'$, but since $|A_x|^{1/n}$ is assumed to be harmonic we must have that A_x is equal to the harmonic interpolation.

Write $\partial\Omega'$ as the disjoint union of a finite number of measurable subsets D_i and let

$$B_i := \int_{D_i} A_y d\mu_x(y),$$

where μ_x is the harmonic measure on $\partial\Omega'$ with respect to x. Then $A_x = \sum_i B_i$, and by the Brunn-Minkowski inequality we have

$$|A_x|^{1/n} \geq \sum_i |B_i|^{1/n}.$$

On the other hand, as in the proof of Theorem 2.2 one sees that

$$|B_i|^{1/n} \geq \int_{D_i} |A_y|^{1/n} d\mu_x(y).$$

But $|A_x|^{1/n}$ being harmonic then implies the equality

$$|A_x|^{1/n} = \sum_i |B_i|^{1/n}.$$

The well-known characterization of the equality case of the Brunn-Minkowski inequality then says that we can write

$$B_i = c_i C_x + d_i$$

where C_x is a convex set (independent of i), for some $c_i \in \mathbb{R}_+$ and $d_i \in \mathbb{R}^n$. We may normalize C_x to have volume one and center of gravity at the origin, at which point c_i and d_i are uniquely determined. Note in particular we then have that

$$A_x = c_x C_x + d_x$$

where $c_x := \sum_i c_i$ and $d_x := \sum_i d_i$.

We now show that C_x is independent of x. First, for fixed x we may decompose a fixed D_i further into disjoint pieces E_j, and then the same argument yields that for each j there are $c'_j \in \mathbb{R}_+$ and $d'_j \in \mathbb{R}^n$ such that $\int_{E_j} A_y d\mu_x(y) = c'_j C_x + d'_j$.

As we can make the decomposition arbitrarily fine the continuity of A_y implies that there are continuous functions c_y and d_y on $\partial\Omega'$ such that $A_y = c_y C_x + d_y$ for all $y \in \partial\Omega'$. Since this is true now for all x, y, we get that C_x is independent of x, and so we can write

$$A_x = c_x C + d_x$$

where C is independent of x. Now since A_x is the harmonic interpolation, this implies c_x and d_x are the harmonic extensions of c_y and d_y as claimed.

As this can be done for any for relatively compact subdomain of Ω with smooth boundary, the result follows. □

4. A Brunn-Minkowski theorem for expected absolute random determinants

In this section we relate harmonic interpolation to random matrices with independently identically distributed columns, which is possible through work of Vitale [**5**, Thm 3.1] that connects such matrices to special classes of convex sets called zonoids.

To discuss this further, consider a random (n, n) matrix M_Y whose columns are iid copies of a random vector Y, corresponding to a Borel probability measure ν_v on $\mathbb{R}^n$. One is then interested in the expected absolute value of the determinant (ead) $E|\det M_Y|$ of M_Y. The connection with convex sets is given in [**5**, Thm 3.1] which states that to any random vector Y with finite expectation we may associate a zonoid

$$Z(Y) := \int_{\mathbb{R}^n} [0, y] d\nu_v(y),$$

and the main result [**5**, Thm. 3.2] gives

$$E|\det M_Y| = n!|Z(Y)|. \tag{2}$$

Suppose now that Y_τ is a family of random vectors parametrized by the boundary of a smoothly bounded domain $\Omega \subseteq \mathbb{R}^m$. We assume that each Y_τ has finite expectation. Then a natural interpolating family Y_x over Ω is given by letting

$$\nu_{Y_x} := \int_{\partial\Omega} \nu_{Y_\tau} d\mu_x(\tau),$$

where as before $d\mu_x$ denotes the harmonic measure with respect to x.

THEOREM 4.1. *The map $x \mapsto (E|\det M_{Y_x}|)^{1/n}$ is superharmonic in x.*

PROOF OF THEOREM 4.1. Note that

$$\begin{aligned} Z(Y_x) &= \int_{\mathbb{R}^n} [0, y] d\nu_{Y_x}(y) = \int_{\mathbb{R}^n} \int_{\partial\Omega} [0, y] d\mu_x(\tau) d\nu_{Y_\tau}(y) \\ &= \int_{\partial\Omega} \int_{\mathbb{R}^n} [0, y] d\nu_{Y_\tau}(y) d\mu_x(\tau) = \int_{\partial\Omega} Z(Y_\tau) d\mu_x(\tau), \end{aligned}$$

i.e. $Z(Y_x)$ is the harmonic interpolation of $Z(Y_\tau)$. Thanks to the volume equality (2) the result follows immediately from Theorem 2.2. □

Acknowledgments

This material is based upon work supported by the National Science Foundation under Grant No. DMS-1749447. The second named author is supported by the Swedish Research Council and the Göran Gustafsson Foundation for Research in Natural Sciences and Medicine. The authors thank Bo Berndtsson and Dario Cordero-Erausquin for conversations on this topic.

References

[1] Robert J. Aumann, *Integrals of set-valued functions*, J. Math. Anal. Appl. **12** (1965), 1–12, DOI 10.1016/0022-247X(65)90049-1. MR185073

[2] Bo Berndtsson, *Prekopa's theorem and Kiselman's minimum principle for plurisubharmonic functions*, Math. Ann. **312** (1998), no. 4, 785–792, DOI 10.1007/s002080050246. MR1660227

[3] Julius Ross and David Witt Nyström, *Interpolation, Prekopa and Brunn-Minkowski for F-subharmonicity*, Adv. Math. **436** (2024), Paper No. 109405, 41, DOI 10.1016/j.aim.2023.109405. MR4669327

[4] Rolf Schneider and Wolfgang Weil, *Zonoids and related topics*, Convexity and its applications, Birkhäuser, Basel, 1983, pp. 296–317. MR731116

[5] Richard A. Vitale, *Expected absolute random determinants and zonoids*, Ann. Appl. Probab. **1** (1991), no. 2, 293–300. MR1102321

MATHEMATICS STATISTICS AND COMPUTER SCIENCE, UNIVERSITY OF ILLINOIS AT CHICAGO, CHICAGO ILLINOIS

Email address: `juliusro@uic.edu`

DEPARTMENT OF MATHEMATICAL SCIENCES, CHALMERS UNIVERSITY OF TECHNOLOGY AND THE UNIVERSITY OF GOTHENBURG, SWEDEN

Email address: `wittnyst@chalmers.se, danspolitik@gmail.com`

Contemporary Mathematics
Volume **810**, 2025
https://doi.org/10.1090/conm/810/16216

Operations on transcendental non-Archimedean metrics

Mingchen Xia

Dedicated to Bo Berndtsson.

ABSTRACT. We define the basic pluripotential-theoretic operations in terms of the transcendental theory of non-Archimedean metrics introduced in [**DXZ23**]. In particular, we establish that the analogue of Boucksom–Jonsson's envelope conjecture holds in our theory.

CONTENTS

1. Introduction

Let X be a reduced irreducible unibranch compact Kähler space of dimension n and θ be a closed real smooth $(1,1)$-form on X. Assume that $\mathrm{PSH}(X,\theta) \neq \emptyset$. The purpose of this paper is to study the pluripotential theory of non-Archimedean metrics in the cohomology class of θ following the approach in [**DXZ23**].

1.1. Background. The non-Archimedean pluripotential theory has been developed by many authors for different purposes in the last three decades. The study of semipositive metrics on line bundles on non-Archimedean spaces dates back to the breakthrough of Zhang [**Zha95**]. In [**CL06**], Chambert-Loir initiated the study of the non-Archimedean Monge–Ampère measures. The pluripotential theory on non-Archimedean curves was developed in detail in the thesis of Thuillier [**Thu05**]. All these developments together lead to the natural question: to what extent is there a pluripotential theory on Berkovich spaces similar to its complex counterpart?

2020 *Mathematics Subject Classification.* Primary 32Q99.

The author benefited from discussions with Sébastien Boucksom. In particular, the possibility of obtaining Theorem 4.21 was suggested by him. The author would like to thank Tamás Darvas and Rémi Reboulet for their comments on the draft and the referee for numerous valuable suggestions. The author is supported by Knut och Alice Wallenbergs Stiftelse grant KAW 2021.0231.

This question has drawn the attention of many authors. In particular, so far we know that there are two general approaches in the literature.

The first approach is the local approach, initiated by Chambert-Loir–Ducros [**CLD12**] and Gubler [**Gub16**]. In this approach, one defines the notions of differential forms and currents on a Berkovich space based on local tropicalizations. Then one can define the Monge–Ampère operator using the usual differential formula. The problem is that this approach only works for regular plurisubharmonic functions and completely fails when a singularity is presented. This is an essential obstacle for applications to complex geometry.

The second approach was initiated by Boucksom–Favre–Jonsson in [**BFJ15, BFJ16**], following the original ideas of Zhang. In the trivially valued case, this theory was further developed by Boucksom–Jonsson [**BJ22b**]. In this approach, one first considers metrics obtained from semipositive models and then defines general global plurisubharmonic metrics using approximations. This approach allows singularities but does not work in the local situation. It does not yield a sheaf of plurisubharmonic metrics. Both approaches have their own drawbacks. We do not have a completely satisfactory non-Archimedean pluripotential theory yet.

1.2. The transcendental theory. The non-Archimedean pluripotential theory is closely related to complex geometry. We refer to [**BBJ21**] and the references therein. The basic idea is that the degeneration in the space of Kähler potentials can be characterized using global plurisubharmonic metrics on the non-Archimedean space. This idea has been intensively studied in [**BBJ21, Li20, DX21, Xia23b**].

We observe that there is a strong obstacle in applying the non-Archimedean geometry to complex geometry: the envelope conjecture [**BJ22b**, Conjecture 5.14] of Boucksom–Jonsson remains open. This conjecture together with its various consequences is indispensable for various applications.

In our previous work [**DXZ23**], we introduced a different framework for the non-Archimedean global pluripotential theory based on the series of works starting from [**BBJ21, RWN14**]. Roughly speaking, we represent a non-Archimedean metric in $\mathrm{PSH}^{\mathrm{NA}}(X,\theta)$ by a *test curve* $\Gamma = (\Gamma_\tau)_\tau$, a concave curve of $\mathcal{I}$-model potentials. Due to the pathology caused by zero-mass, the rigorous definition requires a projective limit construction as well, we refer to Definition 3.5 for the rigorous definition. In this short introduction, we will only talk about test curves to simplify our presentation.

In the algebraic setting, when Boucksom–Jonsson's theory is applicable, there is an obvious comparison map sending Γ to a function on the non-Archimedean space. See Definition 3.11 for the details. The main result in [**DXZ23**] asserts that when X is smooth, this comparison map is in fact a bijection to Boucksom–Jonsson's theory, hence giving a complex representation of the non-Archimedean pluripotential theory. When X is only unibranch, Boucksom–Jonsson's theory is still applicable. In this case, we prove the following result:

THEOREM 1.1 (=Corollary 5.3). *Assume that X is a reduced irreducible unibranch projective variety over $\mathbb{C}$ and L is a pseudoeffective line bundle on X. Then the following are equivalent:*

(1) *Boucksom–Jonsson's envelope conjecture holds for (X, L);*
(2) *The comparison map extends to a bijection from the space of non-Archimedean potentials in the sense of* [**DXZ23**] *to the corresponding space in the sense of* [**BJ22b**].

TABLE 1. Correspondence non-Archimedean/convex

NA potentials	**Test curves**
Addition by a constant	Translation
Maximum	Concave envelope of maximum
Decreasing limit	Decreasing limit
Increasing limit	Increasing limit
Regularized supremum	Increasing limit of concave envelopes of maxima
Rescaling	Rescaling and reparametrization
Addition	Infimal involution
Restriction to subvarieties	Trace operator

Note that Boucksom–Jonsson's envelope conjecture is known to hold when X is smooth, as confirmed in [**BJ22a**]. A different proof was given in [**DXZ23**].

Using the comparison map, we can effectively translate the pluripotential-theoretic operations in Boucksom–Jonsson's theory to our transcendental theory. It turns out that in most cases, these translations can be extended to more general situations where the comparison fails or when Boucksom–Jonsson's theory is not applicable. This is exactly what we do in this paper. The final result is summarized in Table 1. It can be regarded as a duality between non-Archimedean geometry and convex geometry.

Most results from this table are quite straightforward to experts. The only exception is perhaps the last line (=Theorem 4.21), where we establish a quantitative version of the fact that a non-Archimedean plurisubharmonic metric is determined by its restriction to divisorial points. We hope that this result, after suitable extension, could serve to give a proof of Boucksom–Jonsson's envelope conjecture in general.

As a particular case from this table, we have

THEOREM 1.2 (=Theorem 4.13). *The transcendental analogue of the envelope conjecture holds in the theory of* [**DXZ23**].

This result slightly generalizes [**DXZ23**, Theorem 1.2] in that the latter only concerns the regularized supremum of increasing nets while the former works for a general bounded from above family.

Given the fact that Boucksom–Jonsson's envelope conjecture is widely open in general, we expect that our theory is more useful on normal varieties. In particular, we hope that the non-Archimedean theory of K-stability could be effectively translated to our theory. The very first step in this direction is carried out in [**Xia23b**]. On the other hand, our theory works more generally for transcendental classes on compact Kähler manifolds. It should be useful to the transcendental applications as well, for example in the study of cscK metrics as in [**SD18**] or probably in the study of the transcendental Morse inequality.

Most results from the current paper have been announced in several conferences. The author hopes that the details presented in the current paper are of interest to general readers as well. We assume that the readers are familiar with the complex pluripotential theory and Berkovich geometry.

1.3. Conventions. In the whole paper, the Monge–Ampère type products are always taken in the non-pluripolar sense. We adopt the convention that $\mathrm{dd^c} = \frac{\mathrm{i}}{2\pi}\partial\bar{\partial}$. A variety refers to a separated scheme of finite type over $\mathbb{C}$.

2. Pluripotential theory on unibranch spaces

Let X be a reduced irreducible unibranch compact Kähler space of dimension n and θ be a closed real $(1,1)$-form on X. Here and in the sequel, a $(1,1)$-form is always assumed to be smooth. Assume that $\mathrm{PSH}(X,\theta) \neq \emptyset$. There is a well-developed pluripotential theory in this setting. For the Bedford–Taylor theory, we refer to [**Dem85**] and for the non-pluripolar theory, we refer to [**Xia21a**].

Note that we have assumed X to be irreducible. This is not a very severe restriction: since X is unibranch, its irreducible components are just its connected components. Every result in this paper extends in the obvious way to general reduced unibranch compact Kähler spaces.

2.1. Divisorial valuations.

DEFINITION 2.1. A *prime divisor* over a reduced irreducible compact complex space Z is a connected smooth hypersurface $E \subseteq X'$, where $X' \to Z$ is a projective resolution.

Two prime divisors $E_1 \subseteq X_1'$ and $E_2 \subseteq X_2'$ over Z are *equivalent* if there is a common resolution $X'' \to X$ dominating both X_1' and X_2' such that the strict transforms of E_1 and E_2 coincide.

The set Z^{div} is the set of pairs (c,E), where $c \in \mathbb{Q}_{>0}$ and E is an equivalence class of a prime divisor over Z. For simplicity, we will denote the pair (c,E) by $c\,\mathrm{ord}_E$, although one should not really think of this object as a valuation unless Z is projective and irreducible.

Note that a prime divisor on Z does not always define a prime divisor over Z if Z is singular.

DEFINITION 2.2. Given $\varphi \in \mathrm{PSH}(X,\theta)$, we define its *associated non-Archimedean potential* as the map $\varphi^{\mathrm{an}} \colon X^{\mathrm{div}} \to \mathbb{R}$ given by

$$\varphi^{\mathrm{an}}(c\,\mathrm{ord}_E) := -c\nu(\varphi, E).$$

Here associated with E, there is a projective resolution $\pi \colon Y \to X$ such that E is a connected smooth hypersurface in Y. The notation $\nu(\varphi,E)$ means the generic Lelong number of $\pi^*\varphi$ along E. By Siu's semicontinuity theorem, this quantity is independent of the choice of π.

2.2. The pluripotential theory.

DEFINITION 2.3. Assume that X is smooth. Given $\varphi \in \mathrm{PSH}(X,\theta)$, we define

$$P_\theta[\varphi] = \sup\left\{\psi \in \mathrm{PSH}(X,\theta) : \psi \leq 0, \int_X \theta_\varphi^n = \int_X \theta_\psi^n, \varphi \leq \psi + C \text{ for some } C \in \mathbb{R}\right\},$$

$$P_\theta[\varphi]_{\mathcal{I}} = \sup\left\{\psi \in \mathrm{PSH}(X,\theta) : \psi \leq 0, \mathcal{I}(k\varphi) = \mathcal{I}(k\psi) \text{ for all } k \in \mathbb{Z}_{>0}\right\}.$$

A potential $\varphi \in \mathrm{PSH}(X,\theta)$ is *model* (resp. *$\mathcal{I}$-model*) if $\varphi = P_\theta[\varphi]$ (resp. $\varphi = P_\theta[\varphi]_{\mathcal{I}}$).

A potential $\varphi \in \mathrm{PSH}(X,\theta)$ is $\mathcal{I}$-good if $\int_X \theta_\varphi^n > 0$ and $P_\theta[\varphi] = P_\theta[\varphi]_{\mathcal{I}}$.

Here $\mathcal{I}(k\varphi)$ denotes the multiplier ideal sheaf of $k\varphi$. For a detailed study of these notions, we refer to [**DX21, DX22, Xia22**].

DEFINITION 2.4. We say a potential $\varphi \in \mathrm{PSH}(X, \theta)$ is *model* (resp. *$\mathcal{I}$-model, $\mathcal{I}$-good*) if for a projective resolution $\pi\colon Y \to X$, $\pi^*\varphi \in \mathrm{PSH}(Y, \pi^*\theta)$ is model (resp. $\mathcal{I}$-model, $\mathcal{I}$-good).

All three notions are independent of the choice of π.

DEFINITION 2.5. Given $\varphi \in \mathrm{PSH}(X, \theta)$, we define $P_\theta[\varphi], P_\theta[\varphi]_\mathcal{I} \in \mathrm{PSH}(X, \theta)$ as follows: let $\pi\colon Y \to X$ be a projective resolution, then we require the following holds:

$$\pi^* P_\theta[\varphi] = P_{\pi^*\theta}[\pi^*\varphi], \quad \pi^* P_\theta[\varphi]_\mathcal{I} = P_{\pi^*\theta}[\pi^*\varphi]_\mathcal{I}.$$

By Zariski's main theorem, both $P_\theta[\varphi]$ and $P_\theta[\varphi]_\mathcal{I}$ are uniquely determined.

Both operations are clearly independent of the choice of π.

DEFINITION 2.6. Two quasi-psh functions φ and ψ on X are *P-equivalent* if for some Kähler form ω on X such that $\varphi, \psi \in \mathrm{PSH}(X, \omega)$ and both ω_φ and ω_ψ have positive masses, we have $P_\omega[\varphi] = P_\omega[\psi]$. We write $\varphi \sim_P \psi$ in this case.

Here and in the sequel, when we say a potential has positive mass, we always refer to the non-pluripolar mass.

DEFINITION 2.7. Two quasi-psh functions φ and ψ on X are *$\mathcal{I}$-equivalent* if for some Kähler form ω on X such that both $\varphi, \psi \in \mathrm{PSH}(X, \omega)$, we have $P_\omega[\varphi]_\mathcal{I} = P_\omega[\psi]_\mathcal{I}$. We write $\varphi \sim_\mathcal{I} \psi$ in this case.

Both conditions are independent of the choice of ω: for the $\mathcal{I}$-equivalence relation, this follows from [**DX22**, Theorem 2.13]. The statement for the P-equivalence relation is trickier. As we do not need it in the paper, we omit the proof.

PROPOSITION 2.8. *Suppose that $\varphi^i \in \mathrm{PSH}(X, \theta)$ $(i \in I)$ is a decreasing net of $\mathcal{I}$-model potentials and $\inf_i \varphi^i$ has positive mass. Then for all semipositive closed real $(1,1)$-form ω on X, we have*

$$P_{\theta+\omega}\left[\inf_{i\in I} \varphi^i\right]_\mathcal{I} = \inf_{i\in I} P_{\theta+\omega}[\varphi^i]_\mathcal{I}. \tag{2.1}$$

The proof relies on the d_S-pseudometric introduced in [**DDNL21**].

PROOF. We may assume that X is smooth. In fact, we can always take a projection resolution of singularities $\pi : Y \to X$, then as explained in [**Xia21a**, Section 3.5], we have a canonical bijection $\mathrm{PSH}(X, \theta) \cong \mathrm{PSH}(Y, \pi^*\theta)$. Moreover, under this bijection, the pluripotential-theoretic operations on both spaces correspond to each other. So it suffices to prove (2.1) with Y, $\pi^*\theta$, $\pi^*\omega$ and $\pi^*\varphi^i$ in place of X, θ, ω and φ^i. This reduces the problem to the case where X is smooth. In the sequel, we will omit this kind of arguments.

Observe that both sides of (2.1) are $\mathcal{I}$-model in $\mathrm{PSH}(X, \theta + \omega)$, so it suffices to show that they have the same generic Lelong numbers. Fix a prime divisor E over X. As shown in [**DDNL21**, Proposition 4.8][1], $P_{\theta+\omega}[\varphi^i]_\mathcal{I} \xrightarrow{d_S} \inf_{i\in I} P_{\theta+\omega}[\varphi^i]_\mathcal{I}$. Hence, by [**Xia22**, Theorem 0.10],

$$\lim_{i\in I} \nu(\varphi^i, E) = \nu\left(\inf_{i\in I} P_{\theta+\omega}[\varphi^i]_\mathcal{I}, E\right).$$

[1] The result is stated for a sequence, but the proof works for a net as well.

By [**DDNL21**, Proposition 4.8] again, $\varphi^i \xrightarrow{d_S} \inf_{j\in I}\varphi^j$. It follows from [**Xia22**, Theorem 0.10] that

$$\lim_{i\in I}\nu(\varphi^i, E) = \nu\left(\inf_{i\in I}\varphi^i, E\right).$$

We conclude (2.1). □

LEMMA 2.9. *Suppose that $\varphi^i \in \mathrm{PSH}(X,\theta)$ ($i \in I$) is a non-empty family of $\mathcal{I}$-good potentials, uniformly bounded from above. Then $\sup^*_{i\in I}\varphi^i$ is also $\mathcal{I}$-good.*

In the proof below, $V_\theta := \sup\{\varphi \in \mathrm{PSH}(X,\theta), \varphi \leq 0\}$.

PROOF. We may assume that X is smooth. Assume that this result is known when I is finite. By Choquet's lemma, we therefore assume that $I = \mathbb{Z}_{>0}$ and φ^i is an increasing sequence. This is proved in [**Xia24**, Chapter 6].

It remains to handle the case where I has two elements. Let us take $\varphi, \psi \in \mathrm{PSH}(X,\theta)$ be two $\mathcal{I}$-good potentials, we need to show that $\varphi \vee \phi$ is $\mathcal{I}$-good. By [**DX21**, Thorem 3.8], we may take sequences $\varphi_j, \psi_j \in \mathrm{PSH}(X,\theta)$ converging to φ, ψ respectively with respect to d_S. It remains to argue that $\varphi_j \vee \psi_j \xrightarrow{d_S} \varphi \vee \psi$. This clearly follows from the following assertion: if $\varphi, \psi, \eta \in \mathrm{PSH}(X,\theta)$, then

$$d_S(\varphi \vee \eta, \psi \vee \eta) \leq C_n d_S(\varphi, \psi)$$

for some constant $C_n > 0$ depending only on n. By [**DDNL21**, Proposition 3.5], we may assume that $\varphi \leq \psi$. Using the definition of d_S in [**DDNL21**], this would follow if we can show

$$d_1(\varphi \vee \eta \vee (V_\theta - C), \psi \vee \eta \vee (V_\theta - C)) \leq d_1(\varphi \vee (V_\theta - C), \psi \vee (V_\theta - C)),$$

which is just [**Xia23a**, Proposition 4.12] (Proposition 6.8 in the arXiv version). □

LEMMA 2.10. *Let ω be a semipositive closed real $(1,1)$-form on X. Let $\varphi^j \in \mathrm{PSH}(X, \theta + \epsilon_i\omega)$ be a decreasing sequence of model potentials, where ϵ_j is a decreasing sequence converging to 0. Define $\varphi := \inf_j \varphi^j$.*

(1) $\int_X \theta_\varphi^n = \lim_{j\to\infty}\int_X (\theta + \epsilon_j\omega + \mathrm{dd^c}\varphi^j)^n$.

(2) *Assume that φ has positive mass. Then for any prime divisor E over X,*

$$\lim_{j\to\infty}\nu(\varphi^j, E) = \nu(\varphi, E).$$

PROOF. We may assume that X is smooth.

In this case, the result is proved in [**DX24**, Proposition 2.1]. □

2.3. The trace operator. The results in this section are simple extensions of the results established in [**DX24**]. The proofs are almost identical, so we omit the details.

Assume that X is smooth. Let E be an irreducible reduced closed subspace of X of dimension m. Let $\tilde{E} \to E$ denote the normalization of E. It is well-known that $\tilde{E}$ is a normal Kähler space, see [**GK20**, Proposition 3.5].

Consider $\varphi \in \mathrm{PSH}(X,\theta)$ with generic Lelong number $\nu(\varphi, E) = 0$. The trace operator gives a canonical and non-trivial way to restrict φ to E, different from the naive restriction $\varphi|_E$.

The problem with the naive restriction is that even under the assumption $\nu(\varphi, E) = 0$ it can happen that $\varphi|_E \equiv -\infty$. Moreover, the naive restriction is usually poorly behaved with respect to sequences of quasi-psh functions. The trace operator is a novel way of solving these issues.

Let $\varphi \in \mathrm{PSH}(X,\theta)$ such that $\nu(\varphi, E) = 0$. We will define $\mathrm{Tr}_E(\varphi)$ as a qpsh function on the normalization $\tilde{E}$ of E, well-defined up to $\mathcal{I}$-equivalence. The definition is given in [**DX24**], which we recall now.

DEFINITION 2.11. Suppose that φ has analytic singularities, then we let $\mathrm{Tr}_E(\varphi) = \varphi|_{\tilde{E}}$. In general, fix a Kähler form ω on X and take a quasi-equisingular approximation $(\varphi_j)_j$ of φ in $\mathrm{PSH}(X, \theta + \omega)$. We define $\mathrm{Tr}_E(\varphi)$ as a d_S-limit of $\varphi_j|_{\tilde{E}}$.

Note that $\mathrm{Tr}_E(\varphi)$ is not identical to $\mathrm{Tr}^\theta_E(\varphi)$ in [**DX24**], but they are $\mathcal{I}$-equivalent whenever the latter has positive volume. Hence we can freely apply the results in [**DX24**].

The following result is proved in [**DX24**, Section 3].

PROPOSITION 2.12. *Given $\varphi \in \mathrm{PSH}(X,\theta)$ with $\nu(\varphi, E) = 0$, its trace operator $\mathrm{Tr}_E(\varphi)$ as a quasi-psh function on $\tilde{E}$ is well-defined up to $\mathcal{I}$-equivalence. Moreover, $\mathrm{Tr}_E(\varphi)$ depends only on the $\mathcal{I}$-equivalence class of φ.*

Moreover, consider a decreasing sequence of $\mathcal{I}$-good potentials $\varphi_j \in \mathrm{PSH}(X,\theta)$ with pointwise limit φ satisfying $\nu(\varphi, E) = 0$. Suppose that $\varphi_j \xrightarrow{d_S} \varphi$, then we have

$$\mathrm{Tr}_E(\varphi_j) \xrightarrow{d_S} \mathrm{Tr}_E(\varphi).$$

Here $\varphi_j \xrightarrow{d_S} \varphi$ means the same holds after pulling-back to an arbitrary projective resolution of singularities.

In general it is desirable to have a definition of $\mathrm{Tr}_E(\varphi)$ when X is just unibranch. Unfortunately, due to the lack of Demailly regularization on X, the author cannot define the trace operator when E is contained in the singular locus of X. Subsequently, in Section 4.7, we have to assume that our Kähler space is smooth.

3. The theories of non-Archimedean metrics

In this section, we briefly recall two different theories of non-Archimedean metrics.

3.1. Boucksom–Jonsson theory. In this section, we recall the basic concepts in Boucksom–Jonsson's theory as in [**BJ22b**].

Let X be an irreducible reduced variety over $\mathbb{C}$ of dimension n. Let X^{an} denote the Berkovich analytification X^{an} of X with respect to the trivial valuation on $\mathbb{C}$.

The set of real valuations on $\mathbb{C}(X)$ trivial on $\mathbb{C}$ is denoted by X^{val}. The center of a valuation v is the scheme-theoretic point $c = c(v)$ of X such that $v \geq 0$ on $\mathcal{O}_{X,c}$ and $v > 0$ on the maximal ideal $\mathfrak{m}_{X,c}$ of $\mathcal{O}_{X,c}$. The center is unique if exists. It exists if X is proper.

In the remaining of this section, we assume that X is projective.

As a set, X^{an} is the set of semi-valuations on X, in other words, real-valued valuations v on irreducible reduced subvarieties Y in X that is trivial on $\mathbb{C}$. We call Y the *support* of the semi-valuation v. In other words,

$$X^{\mathrm{an}} = \coprod_Y Y^{\mathrm{val}}.$$

We will write $v_{\mathrm{triv}} \in X^{\mathrm{an}}$ for the trivial valuation on X: $v_{\mathrm{triv}}(f) = 0$ for any $f \in \mathbb{C}(X)^{\times}$. See [**Ber93**] for more details.

We will be most interested in divisorial valuations. Recall that a divisorial valuation on X is a valuation of the form $c \operatorname{ord}_E$, where $c \in \mathbb{Q}_{>0}$ and E is a prime divisor over X. The set of divisorial valuations on X is denoted by X^{div}. This notation is compatible with that in Section 2.1.

Given any coherent ideal $\mathfrak{a}$ on X and any $v \in X^{\mathrm{an}}$, we define

$$v(\mathfrak{a}) := \min\{v(f) : f \in \mathfrak{a}_{c(v)}\} \in [0, \infty], \tag{3.1}$$

where $c(v)$ is the center of the valuation v on X.

Given any valuation v on X, the Gauss extension of v is a valuation $\sigma(v)$ on $X \times \mathbb{A}^1$:

$$\sigma(v)\left(\sum_i f_i t^i\right) := \min_i (v(f_i) + i).$$

Here t is the standard coordinate on $\mathbb{A}^1 = \operatorname{Spec} \mathbb{C}[t]$. The key property is that when v is a divisorial valuation, then so it $\sigma(v)$. See [**BHJ17**, Lemma 4.2].

Non-Archimedean plurisubharmonic functions. Let X be an irreducible complex projective variety of dimension n and L be a holomorphic pseudoeffective $\mathbb{Q}$-line bundle on X. Through the GAGA morphism $X^{\mathrm{an}} \to X$ of ringed spaces, L can be pulled-back to an analytic line bundle L^{an} on X.

Following [**BJ22b**, Definition 2.18], we define $\mathcal{H}^{\mathrm{gf}}_{\mathbb{Q}}(L)$, the set of *(rational) generically finite Fubini–Study functions* ϕ: $X^{\mathrm{an}} \to [-\infty, \infty)$, that are of the following form:

$$\phi = \frac{1}{m} \max_j \{\log |s_j| + \lambda_j\}. \tag{3.2}$$

Here $m \in \mathbb{Z}_{>0}$ is an integer such that $L^{\otimes m}$ is a line bundle, the s_j's are a finite collection of non-vanishing sections in $H^0(X, L^{\otimes m})$, and $\lambda_j \in \mathbb{Q}$. We followed the convention of Boucksom–Jonsson by writing $\log |s_j|(v) = -v(s_j)$.

Now we come to the main definition of this paragraph:

Definition 3.1 ([**BJ22b**, Definition 4.1]). A psh metric on L^{an} is a function $\phi : X^{\mathrm{an}} \to [-\infty, \infty)$ that is not identically $-\infty$, and is the pointwise limit of a decreasing net $(\phi_i)_{i\in I}$, where $\phi_i \in \mathcal{H}^{\mathrm{gf}}_{\mathbb{Q}}(L_i^{\mathrm{an}})$ for some $\mathbb{Q}$-line bundles L_i on X satisfying $c_1(L_i) \to c_1(L)$ in $\mathrm{NS}^1(X)_{\mathbb{R}}$.

3.2. The transcendental theory. Let X be a reduced irreducible unibranch compact Kähler space of dimension n and θ be a closed real $(1,1)$-form on X. Assume that $\mathrm{PSH}(X, \theta)$ is non-empty.

We briefly recall the transcendental theory of non-Archimedean metrics introduced in [**DXZ23**].

Definition 3.2. The space $\mathrm{PSH}^{\mathrm{NA}'}(X, \theta)$ is defined as the set consisting of maps Γ: $(-\infty, \Gamma_{\max}) \to \mathrm{PSH}(X, \theta)$ for some $\Gamma_{\max} \in \mathbb{R}$ satisfying the following conditions:

(1) for each $\tau \in (-\infty, \Gamma_{\max})$, Γ_τ is $\mathcal{I}$-model;
(2) the map $(-\infty, \Gamma_{\max}) \ni \tau \mapsto \Gamma_\tau$ is decreasing and concave.

We write

$$\Gamma_{-\infty} := \sup_{\tau \in (-\infty, \Gamma_{\max})}{}^{*} \Gamma_\tau.$$

We define $\mathrm{PSH}^{\mathrm{NA}}(X,\theta)_{>0}$ as the subset of $\mathrm{PSH}^{\mathrm{NA}'}(X,\theta)$ consisting of Γ satisfying furthermore

(3) $\int_X \theta^n_{\Gamma_{-\infty}} > 0$.

The curves $\Gamma \in \mathrm{PSH}^{\mathrm{NA}'}(X,\theta)$ are sometimes known as *test curves*. Here we choose the Greek letter Γ simply because it resembles the reflection of ℓ, the notation usually used to represent a geodesic ray, in view of their duality.

The object that we are eventually interested in is the space $\mathrm{PSH}^{\mathrm{NA}}(X,\theta)$ defined in Definition 3.5. The set $\mathrm{PSH}^{\mathrm{NA}'}(X,\theta)$ only plays an auxiliary role. In the notation $\mathrm{PSH}^{\mathrm{NA}}(X,\theta)_{>0}$, we did not put a prime in the index. Intuitively we would like to think of $\mathrm{PSH}^{\mathrm{NA}}(X,\theta)_{>0}$ as a subset of $\mathrm{PSH}^{\mathrm{NA}}(X,\theta)$ instead of a subset of $\mathrm{PSH}^{\mathrm{NA}'}(X,\theta)$. See Lemma 3.10.

REMARK 3.3. Sometimes it is convenient to extend the domain of definition of $\Gamma \in \mathrm{PSH}^{\mathrm{NA}'}(X,\theta)$ to the whole $\mathbb{R}$ as follows: for $\tau > \Gamma_{\max}$, we simply set $\Gamma_\tau = -\infty$ while for $\tau = \Gamma_{\max}$, we let

$$\Gamma_{\Gamma_{\max}} = \inf_{\tau < \Gamma_{\max}} \Gamma_\tau .$$

In particular, for each $x \in X$, the function $\mathbb{R} \ni \tau \mapsto \Gamma_\tau(x)$ is either constantly $-\infty$, or a proper usc concave function. In particular, it is a closed concave function and the Legendre–Fenchel duality applies. We will frequently make use of this convention without further explanation.

LEMMA 3.4. *Assume that* $\Gamma \in \mathrm{PSH}^{\mathrm{NA}}(X,\theta)_{>0}$, *then* $\int_X \theta^n_{\Gamma_\tau} > 0$ *for any* $\tau \in (-\infty, \Gamma_{\max})$.

PROOF. Fix $\tau \in (-\infty, \Gamma_{\max})$, we want to show that

$$\int_X \theta^n_{\Gamma_\tau} > 0. \tag{3.3}$$

We may assume that X is smooth.

By assumption, $\Gamma_{-\infty}$ has positive mass. By [**DDNL18**, Theorem 2.3], we have

$$\int_X \theta^n_{\Gamma_{-\infty}} = \lim_{\tau \to -\infty} \int_X \theta^n_{\Gamma_\tau}.$$

In particular, for a sufficiently small $\tau_0 < \tau$, we have

$$\int_X \theta^n_{\Gamma_{\tau_0}} > 0.$$

Now take $\tau' \in (\tau, \Gamma_{\max})$ and $t \in (0,1)$ so that

$$\tau = (1-t)\tau' + t\tau_0.$$

From the concavity of Γ, we find that

$$\Gamma_\tau \geq (1-t)\Gamma_{\tau'} + t\Gamma_{\tau_0}.$$

By the monotonicity theorem [**WN19**],

$$\int_X \theta^n_{\Gamma_\tau} \geq \int_X \theta^n_{(1-t)\Gamma_{\tau'} + t\Gamma_{\tau_0}} \geq t^n \int_X \theta^n_{\Gamma_{\tau_0}} > 0$$

and (3.3) follows. □

Definition 3.5. Given any Kähler form ω on X, we define transition maps

$$p_{\theta,\theta+\omega}\colon \mathrm{PSH}^{\mathrm{NA}'}(X,\theta) \to \mathrm{PSH}^{\mathrm{NA}'}(X,\theta+\omega),$$
$$p_{\theta,\theta+\omega}\colon \mathrm{PSH}^{\mathrm{NA}}(X,\theta)_{>0} \to \mathrm{PSH}^{\mathrm{NA}}(X,\theta+\omega)_{>0}$$

as follows: the image of $\Gamma\colon (-\infty,\Gamma_{\max}) \to \mathrm{PSH}(X,\theta)$ is given by

$$p_{\theta,\theta+\omega}(\Gamma)\colon (-\infty,\Gamma_{\max}) \to \mathrm{PSH}(X,\theta+\omega), \quad \tau \mapsto P_{\theta+\omega}[\Gamma_\tau]_{\mathcal{I}}.$$

Note that $\{\mathrm{PSH}^{\mathrm{NA}'}(X,\theta+\omega), p_{\theta,\theta+\omega}\}_\omega$ and $\{\mathrm{PSH}^{\mathrm{NA}}(X,\theta+\omega)_{>0}, p_{\theta,\theta+\omega}\}_\omega$ form projective systems, where the set of Kähler forms ω's is ordered by the reverse of the usual ordering. We define

$$\mathrm{PSH}^{\mathrm{NA}}(X,\theta) := \varprojlim_{\omega} \mathrm{PSH}^{\mathrm{NA}'}(X,\theta+\omega), \tag{3.4}$$

where the projective limit is taken in the category of sets.

We will denote the components of an element $\Gamma \in \mathrm{PSH}^{\mathrm{NA}}(X,\theta)$ by $\Gamma^{\theta+\omega} \in \mathrm{PSH}^{\mathrm{NA}'}(X,\theta+\omega)$. Observe that $\Gamma^{\theta+\omega}_{\max}$ is independent of the choice of ω, we will denote this common value by $\Gamma_{\max}$.

Remark 3.6. The extension in Remark 3.3 is compatible with transition maps in the following sense: Suppose that $\Gamma \in \mathrm{PSH}^{\mathrm{NA}}(X,\theta)$, then for any Kähler forms ω and ω' on X,

$$P_{\theta+\omega+\omega'}\left[\Gamma^{\theta+\omega}_{\Gamma_{\max}}\right]_{\mathcal{I}} = \Gamma^{\theta+\omega+\omega'}_{\Gamma_{\max}}.$$

Lemma 3.7. *The natural maps*

$$\varprojlim_{\omega} \mathrm{PSH}^{\mathrm{NA}}(X,\theta+\omega)_{>0} \to \mathrm{PSH}^{\mathrm{NA}}(X,\theta) \to \varprojlim_{\omega} \mathrm{PSH}^{\mathrm{NA}}(X,\theta+\omega)$$

are both bijective.

In the sequel, we will be constantly making use of these identifications without further mentioning.

Proof. It is clear that both maps are injective, so it suffices to verify that the map

$$\varprojlim_{\omega} \mathrm{PSH}^{\mathrm{NA}}(X,\theta+\omega)_{>0} \to \varprojlim_{\omega} \mathrm{PSH}^{\mathrm{NA}}(X,\theta+\omega)$$

is surjective. Consider an element $\Gamma \in \varprojlim_\omega \mathrm{PSH}^{\mathrm{NA}}(X,\theta+\omega)$ with components $\Gamma^{\theta+\omega} \in \mathrm{PSH}^{\mathrm{NA}}(X,\theta+\omega)$. For each ω, $\Gamma^{\theta+\omega}$ has components $\Gamma^{\theta+\omega,\theta+\omega+\omega'} \in \mathrm{PSH}^{\mathrm{NA}'}(X,\theta+\omega+\omega')$. It is easy to verify that Γ is the image of the following element in $\varprojlim_\omega \mathrm{PSH}^{\mathrm{NA}}(X,\theta+\omega)_{>0}$ with components $\Gamma^{\theta+\omega/2,\theta+\omega}$. □

Also note that the transition maps give natural injections

$$\mathrm{PSH}^{\mathrm{NA}}(X,\theta) \hookrightarrow \mathrm{PSH}^{\mathrm{NA}}(X,\theta+\omega). \tag{3.5}$$

In particular, we can define

$$\mathrm{QPSH}^{\mathrm{NA}}(X) := \varinjlim_{\omega} \mathrm{PSH}^{\mathrm{NA}}(X,\theta+\omega),$$

where ω runs over all Kähler forms on X. This definition is clearly independent of the choice of θ.

LEMMA 3.8. *The natural map*

$$\varprojlim_{\eta} \mathrm{PSH}^{\mathrm{NA}}(X,\theta+\eta)_{>0} \to \mathrm{PSH}^{\mathrm{NA}}(X,\theta) \tag{3.6}$$

is a bijection, where η runs over the set of semipositive closed real $(1,1)$-forms on X with positive total mass.

PROOF. The injectivity is trivial. In order to prove the surjectivity, let $\Gamma \in \mathrm{PSH}^{\mathrm{NA}}(X,\theta)$. We want to show that Γ can be extended to an element in the domain of (3.6). For this purpose, take a semipositive closed real $(1,1)$-form η on X with positive total mass. Fix a Kähler form ω on X. We define

$$\Gamma_\tau^{\theta+\eta} := \inf_{k\in\mathbb{Z}_{>0}} \Gamma_\tau^{\theta+\eta+k^{-1}\omega}$$

for any $\tau \in (-\infty, \Gamma_{\max})$. By Lemma 2.10,

$$\int_X (\theta+\eta+\mathrm{dd^c}\Gamma_\tau^{\theta+\eta})^n > 0$$

and

$$\nu\left(\Gamma_\tau^{\theta+\eta}, E\right) = \nu\left(\Gamma_\tau^{\theta+\eta+\omega}, E\right)$$

for any prime divisor E over X and any $\tau \in (-\infty, \Gamma_{\max})$. It is clear that $\Gamma^\eta \in \mathrm{PSH}^{\mathrm{NA}}(X,\theta+\eta)_{>0}$. The $(\Gamma^{\theta+\eta})_\eta$ defined in this way is clearly an element in the domain of (3.6) with image Γ. □

DEFINITION 3.9. The *non-pluripolar mass* of $\Gamma \in \mathrm{PSH}^{\mathrm{NA}}(X,\theta)$ is defined as the limit

$$\lim_{\omega} \int_X \left(\theta+\omega+\mathrm{dd^c}\Gamma_{-\infty}^{\theta+\omega}\right)^n,$$

where ω runs over the direct set of all Kähler forms on X. In other words, if we fix a Kähler form ω, the limit means

$$\lim_{k\to\infty} \int_X \left(\theta+k^{-1}\omega+\mathrm{dd^c}\Gamma_{-\infty}^{\theta+k^{-1}\omega}\right)^n,$$

Since the net is decreasing, the limit exists.

LEMMA 3.10. *The image of the canonical injection*

$$\mathrm{PSH}^{\mathrm{NA}}(X,\theta)_{>0} \hookrightarrow \mathrm{PSH}^{\mathrm{NA}}(X,\theta)$$

is given by the set of $\Gamma \in \mathrm{PSH}^{\mathrm{NA}}(X,\theta)$ with positive non-pluripolar mass.

PROOF. It is clear that the image of an element in $\mathrm{PSH}^{\mathrm{NA}}(X,\theta)_{>0}$ has positive non-pluripolar mass. Conversely, take $\Gamma \in \mathrm{PSH}^{\mathrm{NA}}(X,\theta)$ with positive non-pluripolar mass. We want to construct $\Gamma' \in \mathrm{PSH}^{\mathrm{NA}}(X,\theta)_{>0}$ representing Γ.

Fix a Kähler form ω on X. Using the same arguments as Lemma 3.4, we find that

$$\lim_{k\to\infty} \int_X \left(\theta+k^{-1}\omega+\mathrm{dd^c}\Gamma_\tau^{\theta+k^{-1}\omega}\right)^n > 0$$

for any $\tau < \Gamma_{\max}$. We define

$$\Gamma'_\tau := \lim_{k\to\infty} \Gamma_\tau^{\theta+k^{-1}\omega}$$

for any $\tau < \Gamma_{\max}$. It follows from Lemma 2.10 that Γ' represents Γ. □

3.3. Comparison between the two theories. Let X be a reduced irreducible unibranch projective variety over $\mathbb{C}$ and L be a pseudoeffective line bundle on X. Recall that in this case, the Zariski unibranchness is equivalent to the analytic unibranchness, as proved in [**Xia21a**], so there is no ambiguity in the adjective *unibranch.*

Choose a smooth Hermitian metric h_0 on L and let $\theta = c_1(L, h_0)$. In this case, both the Boucksom–Jonsson theory and the transcendental theory make sense, so we can consider the problem of comparison.

3.3.1. *The general case.*

DEFINITION 3.11. Given $\Gamma \in \mathrm{PSH}^{\mathrm{NA}}(X, \theta)$, we define its *associated non-Archimedean potential* $\Gamma^{\mathrm{an}} \colon X^{\mathrm{div}} \to \mathbb{R}$ as follows:

$$\Gamma^{\mathrm{an}}(c\,\mathrm{ord}_E) := \sup_{\tau < \Gamma_{\max}} \left(\Gamma_\tau^{\theta+\omega,\mathrm{an}}(c\,\mathrm{ord}_E) + \tau\right) \tag{3.7}$$

for any Kähler form ω on X. Clearly, this map is independent of the choice of ω.

The map $\Gamma \mapsto \Gamma^{\mathrm{an}}$ is compatible with the inclusions (3.5). In particular, given $\Gamma \in \mathrm{QPSH}^{\mathrm{NA}}(X)$, we can define Γ^{an} by choosing any representative of Γ.

It is clear that the map $\Gamma \mapsto \Gamma^{\mathrm{an}}$ is injective.

3.3.2. *On a smooth variety.* In the sequel we assume that X is smooth. In this case, we have the techniques of multiplier ideal sheaves.

A potential $\varphi \in \mathrm{PSH}(X, \theta)$ defines a potential $\varphi^{\mathrm{an}} \in \mathrm{PSH}(L^{\mathrm{an}})$ as follows:

$$\varphi^{\mathrm{an}}(v) := -\lim_{k\to\infty} \frac{1}{k} v\left(\mathcal{I}(k\varphi)\right)$$

for any $v \in X^{\mathrm{an}}$. As explained in [**BBJ21, DXZ23**], the limit exists and lies in $\mathrm{PSH}(L^{\mathrm{an}})$ and this function extends the non-Archimedean potential defined in Definition 2.2 on X^{div}.

We can then define a map

$$\mathrm{an} \colon \mathrm{PSH}^{\mathrm{NA}}(X, \theta) \to \mathrm{PSH}(L^{\mathrm{an}}) \tag{3.8}$$

as follows: take $\Gamma \in \mathrm{PSH}^{\mathrm{NA}}(X, \theta)$, its image is given by

$$\Gamma^{\mathrm{an}} \colon X^{\mathrm{an}} \to [-\infty, \infty), \quad v \mapsto \sup_{\tau < \Gamma_{\max}} \left(\Gamma_\tau^{\theta+\omega,\mathrm{an}}(v) + \tau\right)$$

for any Kähler form ω on X. Clearly, this map is independent of the choice of ω. The map Γ^{an} extends the corresponding definition in Definition 3.11.

THEOREM 3.12 ([**DXZ23**]). *Assume that X is smooth, then the map* (3.8) *is a bijection.*

Motivated by this bijection, we could translate the pluripotential-theoretic operations of $\mathrm{PSH}(L^{\mathrm{an}})$ to operations on the space $\mathrm{PSH}^{\mathrm{NA}}(X, \theta)$. Furthermore, in most cases, the resulting operations have natural extensions to the transcendental case as well. This is what we will carry out in the next section.

4. Pluripotential-theoretic operations

Let X be a reduced irreducible unibranch compact Kähler space of dimension n and θ be a closed real $(1,1)$-form on X. Assume that $\mathrm{PSH}(X, \theta) \neq \emptyset$. When necessary, we use θ' and θ'' to denote similar forms with the same properties as θ.

4.1. The partial order.

DEFINITION 4.1. We define a partial order $\leq$ on $\mathrm{PSH}^{\mathrm{NA}}(X,\theta)$ as follows: given $\Gamma,\Gamma' \in \mathrm{PSH}^{\mathrm{NA}}(X,\theta)$, we say $\Gamma \leq \Gamma'$ if for some (hence for every) Kähler form ω on X, we have $\Gamma^{\theta+\omega}_\tau \leq \Gamma'^{\theta+\omega}_\tau$ for all $\tau < \Gamma_{\max}$.

It is trivial to verify that this is indeed a partial order. Note that this partial order is compatible with the inclusions (3.5), so it defines a partial order on $\mathrm{QPSH}^{\mathrm{NA}}(X)$.

LEMMA 4.2. *Suppose that $\Gamma,\Gamma' \in \mathrm{PSH}^{\mathrm{NA}}(X,\theta)$, then the following are equivalent:*

(1) $\Gamma \leq \Gamma'$;
(2) $\Gamma^{\mathrm{an}}(c\,\mathrm{ord}_E) \leq \Gamma'^{\mathrm{an}}(c\,\mathrm{ord}_E)$ *for all* $c\,\mathrm{ord}_E \in X^{\mathrm{div}}$.

The proof is almost identical to [**DXZ23**, Theorem 3.12], we omit the details.

4.2. Addition by a constant.

DEFINITION 4.3. Given $\Gamma \in \mathrm{PSH}^{\mathrm{NA}}(X,\theta)$ and $C \in \mathbb{R}$, we define the addition $\Gamma + C \in \mathrm{PSH}^{\mathrm{NA}}(X,\theta)$ as the element with components

$$(\Gamma + C)^{\theta+\omega}\colon (-\infty, \Gamma_{\max} + C) \to \mathrm{PSH}(X, \theta+\omega)$$

given by

$$(\Gamma + C)^{\theta+\omega}_\tau := \Gamma^{\theta+\omega}_{\tau - C}.$$

The associated non-Archimedean potential behaves as expected:

PROPOSITION 4.4. *Given $\Gamma \in \mathrm{PSH}^{\mathrm{NA}}(X,\theta)$ and $C \in \mathbb{R}$, we have*

$$(\Gamma + C)^{\mathrm{an}}(c\,\mathrm{ord}_E) = \Gamma^{\mathrm{an}}(c\,\mathrm{ord}_E) + C \tag{4.1}$$

for all $c\,\mathrm{ord}_E \in X^{\mathrm{div}}$ *and*

$$(\Gamma + C)_{\max} = \Gamma_{\max} + C.$$

This follows trivially from (3.7). We will omit similar proofs in the sequel.

Of course, this operation has the obvious compatibility in itself: given another constant $C' \in \mathbb{R}$, we have

$$(\Gamma + C) + C' = \Gamma + (C + C').$$

It is compatible with the partial order in the following sense: given another $\Gamma' \in \mathrm{PSH}^{\mathrm{NA}}(X,\theta)$, the following are equivalent:

(1) $\Gamma \leq \Gamma'$;
(2) $\Gamma + C \leq \Gamma' + C$.

4.3. Maximum.

DEFINITION 4.5. Suppose that $\Gamma,\Gamma' \in \mathrm{PSH}^{\mathrm{NA}}(X,\theta)_{>0}$, we define their *maximum* $\Gamma \vee \Gamma' \in \mathrm{PSH}^{\mathrm{NA}}(X,\theta)_{>0}$ as the map $(-\infty, \Gamma_{\max} \vee \Gamma'_{\max}) \to \mathrm{PSH}(X,\theta)$:

$$\tau \mapsto P_\theta[\mathrm{EC}(\tau' \mapsto \Gamma_{\tau'} \vee \Gamma'_{\tau'})_\tau]_{\mathcal{I}},$$

where EC denotes the concave hull.

We need to verify that $\Gamma \vee \Gamma' \in \mathrm{PSH}^{\mathrm{NA}}(X,\theta)_{>0}$. The only non-trivial part is to show that the concave envelope $\mathrm{EC}(\tau' \mapsto \Gamma_{\tau'} \vee \Gamma'_{\tau'})$ is a curve of θ-psh functions. For this purpose, we may assume that X is smooth and connected. According to the Ross–Witt Nyström correspondence ([**RWN14**, Section 5, Section 6] and [**DXZ23**, Theorem 2.6]), Γ and Γ' correspond to geodesic rays ℓ and ℓ' in $\mathrm{PSH}(X,\theta)$: for any $t \geq 0$, we have

$$\ell_t = \sup_{\tau \in \mathbb{R}} (\Gamma_\tau + t\tau), \quad \ell'_t = \sup_{\tau \in \mathbb{R}} (\Gamma'_\tau + t\tau).$$

It follows that

$$\ell_t \vee \ell'_t = \sup_{\tau \in \mathbb{R}} (\Gamma_\tau \vee \Gamma'_\tau + t\tau).$$

Hence,

$$\mathrm{EC}(\tau \mapsto \Gamma_\tau \vee \Gamma'_\tau)_{\tau'} = \inf_{t \geq 0} (\ell_t \vee \ell'_t - t\tau')$$

and $\mathrm{EC}(\tau \mapsto \Gamma_\tau \vee \Gamma'_\tau)_{\tau'} \in \mathrm{PSH}(X,\theta) \cup \{-\infty\}$ by Kiselman's principle.

By a similar argument, we have

COROLLARY 4.6. *Suppose that $\Gamma, \Gamma' \in \mathrm{PSH}^{\mathrm{NA}}(X,\theta)_{>0}$. For any $c\,\mathrm{ord}_E \in X^{\mathrm{div}}$, we have*

$$-(\Gamma \vee \Gamma')^{\mathrm{an}}_\tau(c\,\mathrm{ord}_E) = \mathrm{EC}\left(\tau' \mapsto (-\Gamma^{\mathrm{an}}_{\tau'}(c\,\mathrm{ord}_E)) \vee (-\Gamma'^{\mathrm{an}}_{\tau'}(c\,\mathrm{ord}_E))\right)_\tau.$$

In particular,

$$(\Gamma \vee \Gamma')^{\mathrm{an}}(c\,\mathrm{ord}_E) = \Gamma^{\mathrm{an}}(c\,\mathrm{ord}_E) \vee \Gamma'^{\mathrm{an}}(c\,\mathrm{ord}_E). \tag{4.2}$$

It follows from Corollary 4.6 and Lemma 4.2 that the maximum operation is compatible with the projective system (3.4). Hence, we can make the following definition in general:

DEFINITION 4.7. Suppose that $\Gamma, \Gamma' \in \mathrm{PSH}^{\mathrm{NA}}(X,\theta)$, we define their *maximum* $\Gamma \vee \Gamma' \in \mathrm{PSH}^{\mathrm{NA}}(X,\theta)$ as the element with components

$$(\Gamma \vee \Gamma')^{\theta+\omega} = \Gamma^{\theta+\omega} \vee \Gamma'^{\theta+\omega}.$$

The associated non-Archimedean potential behaves as expected:

COROLLARY 4.8. *Suppose that $\Gamma, \Gamma' \in \mathrm{PSH}^{\mathrm{NA}}(\theta)$. For any $c\,\mathrm{ord}_E \in X^{\mathrm{div}}$, we have*

$$(\Gamma \vee \Gamma')^{\mathrm{an}}(c\,\mathrm{ord}_E) = \Gamma^{\mathrm{an}}(c\,\mathrm{ord}_E) \vee \Gamma'^{\mathrm{an}}(c\,\mathrm{ord}_E) \tag{4.3}$$

and

$$(\Gamma \vee \Gamma')_{\max} = \Gamma_{\max} \vee \Gamma'_{\max}.$$

The maximum operation is commutative and associative. In particular, finite (non-empty) maximum makes sense in the obvious way.

The maximum operation is compatible with the partial order in the following sense: given $\Gamma, \Gamma' \in \mathrm{PSH}^{\mathrm{NA}}(\theta)$, the following are equivalent

(1) $\Gamma \vee \Gamma' = \Gamma'$;
(2) $\Gamma \leq \Gamma'$.

The maximum operation is compatible with the addition by a constant: if $C \in \mathbb{R}$, then

$$(\Gamma \vee \Gamma') + C = (\Gamma + C) \vee (\Gamma' + C).$$

4.4. Decreasing limit along a net.

DEFINITION 4.9. Let $(\Gamma_i)_{i\in I}$ be a decreasing net in $\mathrm{PSH}^{\mathrm{NA}}(X,\theta)$. Assume that $\inf_{i\in I}\Gamma_{i,\max} > -\infty$. We define $\inf_{i\in I}\Gamma_i \in \mathrm{PSH}^{\mathrm{NA}}(X,\theta)$ as the element with components

$$\left(\inf_{i\in I}\Gamma_i\right)^{\theta+\omega}_{\tau} = \inf_{i\in I}\Gamma^{\theta+\omega}_{i,\tau}.$$

This gives an element in $\mathrm{PSH}^{\mathrm{NA}}(X,\theta)$ by Proposition 2.8 together with [**DX22**, Lemma 2.21(i)][2].

When $\inf_{i\in I}\Gamma_{i,\max} = -\infty$, we could still formally define $\inf_{i\in I}\Gamma_i$ as a symbol $-\infty$. As one could easily verify, all of our pluripotential-theoretic operations admit natural extensions to $-\infty$.

Using [**DXZ23**, Lemma 3.13], we have

LEMMA 4.10. *Let $(\Gamma_i)_{i\in I}$ be a decreasing net in $\mathrm{PSH}^{\mathrm{NA}}(X,\theta)$ with $\inf_{i\in I}\Gamma_i \in \mathrm{PSH}^{\mathrm{NA}}(X,\theta)$. Then for any $c\,\mathrm{ord}_E \in X^{\mathrm{div}}$,*

$$\left(\inf_{i\in I}\Gamma_i\right)^{\mathrm{an}}(c\,\mathrm{ord}_E) = \inf_{i\in I}\Gamma_i^{\mathrm{an}}(c\,\mathrm{ord}_E). \tag{4.4}$$

Moreover,

$$\left(\inf_{i\in I}\Gamma_i\right)_{\max} = \inf_{i\in I}\Gamma_{i,\max}.$$

This operation is compatible with itself: if I and J are two non-empty directed set and Γ_{ij} is a decreasing net in $\mathrm{PSH}^{\mathrm{NA}}(X,\theta)$ indexed by $I\times J$ with $\inf_{i,j}\Gamma_{ij} \in \mathrm{PSH}^{\mathrm{NA}}(X,\theta)$, then

$$\inf_{i,j}\Gamma_{ij} = \inf_{i\in I}\inf_{j\in J}\Gamma_{ij} = \inf_{j\in J}\inf_{i\in I}\Gamma_{ij}.$$

This operation is compatible with the partial order in the following senses: if $\Gamma_i \in \mathrm{PSH}^{\mathrm{NA}}(X,\theta)$ $(i\in I)$ is a decreasing net with $\inf_{i\in I}\Gamma_i \in \mathrm{PSH}^{\mathrm{NA}}(X,\theta)$. Then $\inf_{j\in I}\Gamma_j \leq \Gamma_i$ for all $i\in I$ and it is the biggest element in $\mathrm{PSH}^{\mathrm{NA}}(X,\theta)$ with this property. On the other hand, if $\Gamma'_i \in \mathrm{PSH}^{\mathrm{NA}}(X,\theta)$ $(i\in I)$ is another decreasing net with $\inf_{i\in I}\Gamma'_i \in \mathrm{PSH}^{\mathrm{NA}}(X,\theta)$ such that $\Gamma_i \leq \Gamma'_i$ for all $i\in I$, then

$$\inf_{i\in I}\Gamma_i \leq \inf_{i\in I}\Gamma'_i.$$

This operation is compatible with the addition by a constant: if $C\in\mathbb{R}$, then

$$\inf_{i\in I}(\Gamma_i + C) = \left(\inf_{i\in I}\Gamma_i\right) + C.$$

4.5. Regularized supremum.

DEFINITION 4.11. Let $(\Gamma^i)_{i\in I}$ be an increasing net in $\mathrm{PSH}^{\mathrm{NA}}(X,\theta)_{>0}$. We assume that $\sup_{i\in I}\Gamma^i_{\max} < \infty$. We define a map $\sup_{i\in I}\Gamma^i_{\max}\colon (-\infty, \sup_i \Gamma^i_{\max}) \to \mathrm{PSH}(X,\theta)$ as follows:

$$\left(\sup_{i\in I}{}^*\,\Gamma^i\right)_{\tau} = P_\theta\left[\sup_{i\in I}{}^*\,\Gamma^i_\tau\right].$$

By Lemma 2.9, it is easy to see that $\sup^*{}_i\,\Gamma^i \in \mathrm{PSH}^{\mathrm{NA}}(X,\theta)_{>0}$.

[2] The result is stated for a sequence, but the proof extends to a net.

More generally, suppose that $\Gamma^i \in \mathrm{PSH}^{\mathrm{NA}}(X,\theta)$ $(i \in I)$ is an increasing net with $\sup_i \Gamma^i_{\max} < \infty$. We define $\sup^*_i \Gamma^i \in \mathrm{PSH}^{\mathrm{NA}}(X,\theta)$ as the element with components

$$\left(\sup_{i\in I}{}^* \Gamma^i\right)^{\theta+\omega} = \sup_{i\in I}{}^* \Gamma^{i,\theta+\omega} \tag{4.5}$$

for any Kähler form ω on X. The compatibility follows from [**DX21**, Lemma 3.14].

DEFINITION 4.12. Let $(\Gamma^i)_{i\in I}$ be a (non-empty) family in $\mathrm{PSH}^{\mathrm{NA}}(X,\theta)$ such that $\sup_{i\in I} \Gamma^i_{\max} < \infty$. Let $\mathcal{J}$ denote the directed set of finite non-empty subsets of I ordered by inclusion. Then we define

$$\sup_{i\in I}{}^* \Gamma^i := \sup_{J\in\mathcal{J}}{}^* \left(\max_{j\in J} \Gamma^j\right).$$

Recall that the maximum is defined right after Corollary 4.8.

THEOREM 4.13. *Let $(\Gamma^i)_{i\in I}$ be a (non-empty) family in $\mathrm{PSH}^{\mathrm{NA}}(X,\theta)$ with $\sup_{i\in I} \Gamma^i_{\max} < \infty$. Then for any $c\,\mathrm{ord}_E \in X^{\mathrm{div}}$, we have*

$$\left(\sup_{i\in I}{}^* \Gamma^i\right)^{\mathrm{an}} (c\,\mathrm{ord}_E) = \sup_{i\in I} \Gamma^{i,\mathrm{an}}(c\,\mathrm{ord}_E).$$

Moreover,

$$\left(\sup_{i\in I}{}^* \Gamma^i\right)_{\max} = \sup_{i\in I} \Gamma^i_{\max}.$$

This result is the transcendental analogue of Boucksom–Jonsson's envelope conjecture. As explained in [**DXZ23**, Theorem 3.4], when X is smooth and projective, this result implies Boucksom–Jonsson's envelope conjecture.

PROOF. By Corollary 4.8, we may assume that $(\Gamma^i)_{i\in I}$ is an increasing net. Furthermore, it is easy to reduce to the case where $\Gamma^i \in \mathrm{PSH}^{\mathrm{NA}}(X,\theta)_{>0}$. In fact, by definition,

$$\left(\sup_{i\in I}{}^* \Gamma^i\right)^{\mathrm{an}} = \left(\sup_{i\in I}{}^* \Gamma^{i,\theta+\omega}\right)^{\mathrm{an}}$$

for any Kähler form ω on X. A similar equation holds for $(\bullet)_{\max}$ as well. So we could replace Γ^i by $\Gamma^{i,\theta+\omega}$ and assume that $\Gamma^i \in \mathrm{PSH}^{\mathrm{NA}}(X,\theta)_{>0}$.

In this case, the result follows from [**DXZ23**, Lemma 3.13]. □

In particular, this operation extends the finite maximum.

This operation is compatible with itself: if $(\Gamma^{i,j})_{i\in I, j\in J}$ is a (non-empty) family in $\mathrm{PSH}^{\mathrm{NA}}(X,\theta)$ with $\sup_{i\in I, j\in J} \Gamma^{i,j}_{\max} < \infty$. Then

$$\sup_{i\in I}{}^* \sup_{j\in J}{}^* \Gamma^{i,j} = \sup_{j\in J}{}^* \sup_{i\in I}{}^* \Gamma^{i,j}.$$

It is compatible with the partial order: Suppose that $(\Gamma^i)_{i\in I}, (\Gamma'^i)_{i\in I}$ are two families in $\mathrm{PSH}^{\mathrm{NA}}(X,\theta)$ with $\sup^*_{i\in I} \Gamma^i_{\max} < \infty$ and $\sup^*_{i\in I} \Gamma'^i_{\max} < \infty$ and $\Gamma^i \leq \Gamma'^i$ for each $i \in I$, then

$$\sup_{i\in I}{}^* \Gamma^i \leq \sup_{i\in I}{}^* \Gamma'^i.$$

It is compatible with addition by a constant: if $C \in \mathbb{R}$, then

$$\sup_{i\in I}{}^* (\Gamma^i + C) = \left(\sup_{i\in I}{}^* \Gamma^i\right) + C.$$

There is no non-trivial compatibility with the decreasing limits, as expected.

4.6. Rescaling and addition.

DEFINITION 4.14. Let $\Gamma \in \mathrm{PSH}^{\mathrm{NA}}(X,\theta)$ and $\lambda \in \mathbb{R}_{>0}$, we define the rescaling $\lambda\Gamma \in \mathrm{PSH}^{\mathrm{NA}}(X,\lambda\theta)$ as the element with components $(\lambda\Gamma)^{\lambda\theta+\omega}\colon (-\infty, \lambda\Gamma_{\max}) \to \mathrm{PSH}(X,\lambda\theta+\omega)$:

$$(\lambda\Gamma)^{\lambda\theta+\omega}_{\tau} := \lambda\Gamma^{\theta+\lambda^{-1}\omega}_{\lambda^{-1}\tau}$$

for any Kähler form ω on X.

DEFINITION 4.15. Let $\Gamma \in \mathrm{PSH}^{\mathrm{NA}}(X,\theta)_{>0}$ and $\Gamma' \in \mathrm{PSH}^{\mathrm{NA}}(X,\theta')_{>0}$. We define $\Gamma+\Gamma'\colon (-\infty, \Gamma_{\max}+\Gamma'_{\max}) \to \mathrm{PSH}(X,\theta+\theta')$ as

$$(\Gamma+\Gamma')_{\tau} := P_{\theta+\theta'}\left[\sup_{t\in\mathbb{R}}\left(\Gamma_t+\Gamma'_{\tau-t}\right)\right].$$

Observe that $\tau \mapsto \sup_{t\in\mathbb{R}}(\Gamma_t+\Gamma'_{\tau-t})$ is concave. In fact, this curve is just the concave involution of Γ and Γ'. From the general properties of infimal involution, we know that this map is concave and usc. For each fixed $x \in X$, $\tau \mapsto \sup_{t\in\mathbb{R}}(\Gamma_t(x)+\Gamma'_{\tau-t}(x))$ is either constantly $-\infty$ or proper. In particular, by [**Roc70**, Theorem 16.4] and Legendre–Fenchel duality, $\tau \mapsto \sup_{t\in\mathbb{R}}(\Gamma_t+\Gamma'_{\tau-t})$ is the same as the inverse Legendre transform of $\ell+\ell'$, where ℓ and ℓ' are the geodesic rays defined by Γ and Γ'. In particular, $\sup_{t\in\mathbb{R}}(\Gamma_t+\Gamma'_{\tau-t})$ is either in $\mathrm{PSH}(X,\theta)$ or constantly $-\infty$ by Kiselman's principle. By Lemma 2.9 and [**Xia22**, Proposition 2.8], we find that $\Gamma+\Gamma' \in \mathrm{PSH}^{\mathrm{NA}}(X,\theta)_{>0}$.

This argument also shows that for any $c\,\mathrm{ord}_E \in X^{\mathrm{div}}$, we have

$$(\Gamma+\Gamma')^{\mathrm{an}}(c\,\mathrm{ord}_E) = \Gamma^{\mathrm{an}}(c\,\mathrm{ord}_E)+\Gamma'^{\mathrm{an}}(c\,\mathrm{ord}_E). \tag{4.6}$$

DEFINITION 4.16. Let $\Gamma \in \mathrm{PSH}^{\mathrm{NA}}(X,\theta)$ and $\Gamma' \in \mathrm{PSH}^{\mathrm{NA}}(X,\theta')$. We define $\Gamma+\Gamma' \in \mathrm{PSH}^{\mathrm{NA}}(X,\theta+\theta')$ as the element with components

$$(\Gamma+\Gamma')^{\theta+\omega} = \Gamma^{\theta+\omega}+\Gamma'^{\theta+\omega}.$$

Note that $\Gamma+\Gamma' \in \mathrm{PSH}^{\mathrm{NA}}(X,\theta+\theta')$ by (4.6).

LEMMA 4.17. *Suppose that $\Gamma \in \mathrm{PSH}^{\mathrm{NA}}(X,\theta)$, $\Gamma' \in \mathrm{PSH}^{\mathrm{NA}}(X,\theta')$, $\lambda \in \mathbb{R}_{>0}$. Then for any $c\,\mathrm{ord}_E \in X^{\mathrm{div}}$, we have*

$$(\lambda\Gamma)^{\mathrm{an}}(c\,\mathrm{ord}_E) = \lambda\Gamma^{\mathrm{an}}(c\,\mathrm{ord}_E), \quad (\Gamma+\Gamma')^{\mathrm{an}}(c\,\mathrm{ord}_E) = \Gamma^{\mathrm{an}}(c\,\mathrm{ord}_E)+\Gamma'^{\mathrm{an}}(c\,\mathrm{ord}_E).$$

Moreover,

$$(\lambda\Gamma)_{\max} = \lambda\Gamma_{\max}, \quad (\Gamma+\Gamma')_{\max} = \Gamma_{\max}+\Gamma'_{\max}.$$

Both operations are compatible with themselves: if $\lambda,\lambda' \in \mathbb{R}_{>0}$ and $\Gamma \in \mathrm{PSH}^{\mathrm{NA}}(X,\theta)$, $\Gamma' \in \mathrm{PSH}^{\mathrm{NA}}(X,\theta')$ and $\Gamma'' \in \mathrm{PSH}^{\mathrm{NA}}(X,\theta'')$, then

$$\lambda(\lambda'\Gamma) = (\lambda\lambda')\Gamma$$

and

$$(\Gamma+\Gamma')+\Gamma'' = \Gamma+(\Gamma'+\Gamma'').$$

The addition is clear commutative. Moreover, these two operations are compatible with each other:

$$\lambda(\Gamma+\Gamma') = \lambda\Gamma+\lambda\Gamma', \quad (\lambda+\lambda')\Gamma = \lambda\Gamma+\lambda'\Gamma.$$

Both operations are compatible with the partial order: if $\Gamma,\Gamma' \in \mathrm{PSH}^{\mathrm{NA}}(X,\theta)$, $\lambda \in \mathbb{R}_{>0}$, then $\Gamma \leq \Gamma'$ if and only if $\lambda\Gamma \leq \lambda\Gamma'$. Suppose that $\Gamma^1,\Gamma^2 \in \mathrm{PSH}^{\mathrm{NA}}(X,\theta)$, $\Gamma'^1,\Gamma'^2 \in \mathrm{PSH}^{\mathrm{NA}}(X,\theta')$ and $\Gamma^1 \leq \Gamma^2$, $\Gamma'^1 \leq \Gamma'^2$, then $\Gamma^1+\Gamma'^1 \leq \Gamma^2+\Gamma'^2$.

Both operations are compatible with addition by a constant: if $C \in \mathbb{R}$, $\lambda \in \mathbb{R}_{>0}$, $\Gamma \in \mathrm{PSH}^{\mathrm{NA}}(X,\theta)$ and $\Gamma' \in \mathrm{PSH}^{\mathrm{NA}}(X,\theta')$, then

$$\lambda(\Gamma + C) = \lambda\Gamma + \lambda C$$

and

$$(\Gamma + \Gamma') + C = \Gamma + (\Gamma' + C).$$

Both operations satisfy obvious inequalities with respect to the limit operations. We leave the details to the readers.

4.7. The trace operator. We assume in addition that X is smooth. Recall that in this case, we have a natural bijection (3.8). Fix an irreducible reduced closed analytic subspace E of X.

DEFINITION 4.18. Assume that X is smooth. We say $\Gamma \in \mathrm{PSH}^{\mathrm{NA}}(X,\theta)$ has *non-trivial restriction* to E if there is a small enough $C \in \mathbb{R}$ so that $\nu(\Gamma_C^{\theta+\omega}, E) = 0$ for any Kähler form ω on X.

This definition is motivated by the following observation.

LEMMA 4.19. *Assume that X is smooth and projective. Let $\Gamma \in \mathrm{PSH}^{\mathrm{NA}}(X,\theta)$. Then the following are equivalent:*

(1) $\Gamma^{\mathrm{an}}(v_{E,\mathrm{triv}}) \neq -\infty$*;*
(2) $\Gamma^{\mathrm{an}}|_{E^{\mathrm{an}}} \not\equiv -\infty$*;*
(3) Γ *has non-trivial restriction to E.*

Here $v_{E,\mathrm{triv}}$ denotes the trivial valuation of $\mathbb{C}(E)$.

PROOF. The equivalence between (1) and (2) is a simple consequence of the maximum principle [**BJ22b**, Lemma 1.4(i)].

To see the equivalence between (1) and (3), it suffices to observe that for any $\varphi \in \mathrm{PSH}(X,\theta)$,

$$\varphi^{\mathrm{an}}(v_{E,\mathrm{triv}}) = \begin{cases} -\infty, & \text{if } \nu(\varphi, E) > 0; \\ 0, & \text{if } \nu(\varphi, E) = 0. \end{cases}$$

□

DEFINITION 4.20. Assume that X is smooth. Assume that $\Gamma \in \mathrm{PSH}^{\mathrm{NA}}(X,\theta)$ has non-trivial restriction to E, we define the *trace operator* of Γ along E as the element $\mathrm{Tr}_E(\Gamma) \in \mathrm{PSH}^{\mathrm{NA}}(\tilde{E}, \theta|_{\tilde{E}})$ defined as follows: for any Kähler form ω' on Y and any Kähler form ω on X satisfying $\omega' \geq \omega|_{\tilde{Y}}$, the component $\mathrm{Tr}_E(\Gamma)^{\theta|_{\tilde{E}}+\omega'} \in \mathrm{PSH}^{\mathrm{NA}'}(\tilde{E}, \theta|_{\tilde{E}} + \omega')$ is given by

$$\tau \mapsto P_{\theta|_{\tilde{E}}+\omega'}[\mathrm{Tr}_E(\Gamma_\tau^{\theta+\omega})],$$

where $\tau \in (-\infty, (\mathrm{Tr}_E(\Gamma))_{\max})$ and $(\mathrm{Tr}_E(\Gamma))_{\max} = \sup\{\tau \in \mathbb{R} : \nu(\Gamma_\tau^{\theta+\omega''}, E) = 0\}$ for any Kähler form ω'' on X.

We remind the readers that $\tilde{E}$ is the normalization of E. It is a normal Kähler space by [**GK20**, Proposition 3.5].

The non-Archimedean potential associated with $\mathrm{Tr}_E(\Gamma)$ is not very easy to describe. We will only do this in the following special but important case.

THEOREM 4.21. *Assume that X is smooth projective and $\{\theta\}$ is the first Chern class of a pseudoeffective line bundle L. Consider an element $\Gamma \in \mathrm{PSH}^{\mathrm{NA}}(X,\theta)$ with non-trivial restriction to E. Then*

$$\mathrm{Tr}_E(\Gamma)^{\mathrm{an}}|_{E^{\mathrm{div}}} = \Gamma^{\mathrm{an}}|_{E^{\mathrm{div}}}. \tag{4.7}$$

Observe that there is a canonical identification $E^{\mathrm{div}} = \tilde{E}^{\mathrm{div}}$.

In order to prove the theorem, we need to recall the following notion introduced in [**DXZ23**].

DEFINITION 4.22. A *piecewise linear curve* $\psi_\bullet$ in $\mathrm{PSH}(X,\theta)$ associated with $\psi_{\tau_j} \in \mathrm{PSH}(X,\theta)$, for a finite number of parameters $\tau_0 > \tau_1 > \cdots > \tau_N$ is the affine interpolation of these data:

(1) $\psi_\tau = \psi_{\tau_N}$ for $\tau \leq \tau_N$;
(2) For $t \in (0,1)$ and $i = 0, \ldots, N-1$, we have $\psi_{(1-t)\tau_i + t\tau_{i+1}} = (1-t)\psi_{\tau_i} + t\psi_{\tau_{i+1}}$;
(3) $\psi_\tau = -\infty$ for $\tau > \tau_0$.

The analytification ψ^{an} of a piecewise linear curve $\psi_\bullet$ as above is defined as

$$\psi^{\mathrm{an}}(v) := \sup_{\tau \leq \tau_0} (\psi_\tau^{\mathrm{an}}(v) + \tau) = \max_{\tau_i} (\psi_{\tau_i}^{\mathrm{an}}(v) + \tau_i) \quad \text{for } v \in X^{\mathrm{an}}. \tag{4.8}$$

PROOF OF THEOREM 4.21. We may assume that $\Gamma \in \mathrm{PSH}^{\mathrm{NA}}(X,\theta)_{>0}$. Let $\phi = \Gamma^{\mathrm{an}} \in \mathrm{PSH}(L^{\mathrm{an}})$. By Lemma 4.19, $\phi(v_{E,\mathrm{triv}}) \neq -\infty$.

Take an ample line bundle S on X, a Kähler form ω in $c_1(S)$. Write ϕ as the decreasing limit of a sequence ϕ^i of elements in $\mathcal{H}^{\mathrm{gf}}_{\mathbb{Q}}(L + i^{-1}S)$. This is possible by [**BJ22b**, Corollary 12.18].

Take $\Gamma^i \in \mathrm{PSH}^{\mathrm{NA}}(X, \theta + i^{-1}\omega)$ such that $\Gamma^{i,\mathrm{an}} = \phi^i$. Note that by Lemma 3.10, $\Gamma^i \in \mathrm{PSH}^{\mathrm{NA}}(X, \theta + i^{-1}\omega)_{>0}$.

It follows from Lemma 4.10 (applied to the images of Γ^i in $\mathrm{PSH}^{\mathrm{NA}}(X, \theta + \omega)$) that for any $\tau < \Gamma_{\max}$, we have

$$\lim_{i \to \infty} \Gamma^i_\tau = \Gamma_\tau.$$

In particular, by Lemma 2.10, the Lelong numbers of Γ^i_τ converge to those of Γ_τ. It follows that $\Gamma^i_\tau \xrightarrow{d_{S,\theta+\omega}} \Gamma_\tau$ for all $\tau < \Gamma_{\max}$.

By Lemma 4.19 again, each Γ^i has non-trivial restriction to E. By [**DX24**, Lemma 3.12], for any Kähler form ω' on $\tilde{E}$ satisfying $\omega' \geq \omega|_{\tilde{E}}$ we have

$$\mathrm{Tr}_E(\Gamma^i)_\tau^{\theta|_{\tilde{E}} + \omega'} \xrightarrow{d_S} \mathrm{Tr}_E(\Gamma)_\tau^{\theta|_{\tilde{E}} + \omega'}$$

for any $\tau < (\mathrm{Tr}_E(\Gamma))_{\max}$. Thanks to [**Xia22**, Theorem 0.10], the Lelong numbers of $\mathrm{Tr}_E(\Gamma^i)_\tau^{\theta|_{\tilde{E}} + \omega'}$ converge to those of $\mathrm{Tr}_E(\Gamma)_\tau^{\theta|_{\tilde{E}} + \omega'}$. Therefore, by [**DXZ23**, Lemma 3.13],

$$\mathrm{Tr}_E(\Gamma)^{\mathrm{an}}(c\,\mathrm{ord}_F) = \inf_i \mathrm{Tr}_E(\Gamma)^{i,\mathrm{an}}(c\,\mathrm{ord}_F)$$

for any $c\,\mathrm{ord}_F \in E^{\mathrm{div}}$. In particular, it suffices to prove (4.7) with Γ^i in place of Γ.

In other words, we have reduced to the case where $\phi \in \mathcal{H}^{\mathrm{gf}}_{\mathbb{Q}}(L)$ and L is big.

Under this extra assumption, as proved in the proof of [**DXZ23**, Theorem 3.12], there is a piecewise linear curve ψ connecting potentials lying in the space of potentials generated by functions of the form

$$m^{-1} \log \max_i (|s_i|^2_{h_0^m} + c) \tag{4.9}$$

using convex combinations and finite maxima, where $m \in \mathbb{Z}_{>0}$, the s_i's are finitely many holomorphic sections of $L^{\otimes m}$ and $c \in \mathbb{Q}$ such that Γ_τ is the $\mathcal{I}$-envelope of η_τ for any $\tau < \Gamma_{\max}$, where η is the concave envelope of ψ. Moreover, after adjusting ψ (c.f. [**Roc70**, Theorem 5.6]), we may guarantee that ψ itself is concave. In particular, $\Gamma_\tau \sim_{\mathcal{I}} \psi_\tau$ for all $\tau < \Gamma_{\max}$.

It follows that for any $c\operatorname{ord}_F \in E^{\mathrm{div}}$,

$$\begin{aligned}\phi|_{E^{\mathrm{an}}}(c\operatorname{ord}_F) &= \sup_{\tau<\Gamma_{\max}} \left(\psi_\tau^{\mathrm{an}}(c\operatorname{ord}_F) + \tau\right)\\ &= \sup_{\tau<\Gamma_{\max}} \left(\left(\psi_\tau|_{\tilde{E}}\right)^{\mathrm{an}}(c\operatorname{ord}_F) + \tau\right)\\ &= \sup_{\tau<\Gamma_{\max}} \left(\operatorname{Tr}_E(\psi_\tau)^{\mathrm{an}}(c\operatorname{ord}_F) + \tau\right)\\ &= \sup_{\tau<\Gamma_{\max}} \left(\operatorname{Tr}_E(\Gamma_\tau)^{\mathrm{an}}(c\operatorname{ord}_F) + \tau\right)\\ &= \operatorname{Tr}_E(\Gamma)^{\mathrm{an}}(c\operatorname{ord}_F).\end{aligned}$$

Here we adopt the convention that $(-\infty)^{\mathrm{an}}(c\operatorname{ord}_F) = -\infty$ to simplify the notations. The third equality follows from the linearity of the trace operator [**DX24**, Lemma 3.10] and the obvious fact that the trace operator preserves finite maxima. The fourth equality follows from [**DX24**, Lemma 3.9]. In order to justify the second equality, it suffices to show that for $\tau < (\operatorname{Tr}_E(\Gamma))_{\max}$, if ψ_τ has the form (4.9), then

$$\left(\psi_\tau|_{\tilde{E}}\right)^{\mathrm{an}}(\operatorname{ord}_F) = \psi_\tau^{\mathrm{an}}(\operatorname{ord}_F).$$

We could easily reduce to the case $m = 1$ and $\psi_\tau = \log|s|^2_{h_0}$ for a single holomorphic section s of L which does not vanish identically on E. In this case, the above equality is trivial. □

In particular, the trace operator is compatible with the partial order and the addition by a constant.

As an interesting corollary,

COROLLARY 4.23. *Assume the same assumptions as in Theorem 4.21. Suppose that $\phi \in \mathrm{PSH}(L^{\mathrm{an}})$ is a non-Archimedean metric represented by $\Gamma \in \mathrm{PSH}^{\mathrm{NA}}(X,\theta)$ and E is an irreducible reduced subvariety of X. Then*

$$\phi(v_{E,\mathrm{triv}}) = \sup\left\{\tau \in \mathbb{R} : \nu(\Gamma_\tau^{\theta+\omega}, E) = 0\right\}$$

for any Kähler form ω on X.

4.8. Functoriality. Let $\pi\colon Y \to X$ be a proper bimeromorphic morphism from a reduced irreducible unibranch compact Kähler space of dimension n. By Zariski's main theorem, the obvious pull-back map induces a bijection

$$\pi^*\colon \mathrm{PSH}(X,\theta) \xrightarrow{\sim} \mathrm{PSH}(Y,\pi^*\theta).$$

In particular, it induces a canonical bijection

$$\pi^*: \mathrm{PSH}^{\mathrm{NA}}(X,\theta)_{>0} \xrightarrow{\sim} \mathrm{PSH}^{\mathrm{NA}}(Y,\pi^*\theta)_{>0}$$

as follows: the image of $\Gamma \in \mathrm{PSH}^{\mathrm{NA}}(X,\theta)_{>0}$ is given by

$$\pi^*\Gamma\colon (-\infty,\Gamma_{\max}) \to \mathrm{PSH}(Y,\pi^*\theta), \quad (\pi^*\Gamma)_\tau := \pi^*\Gamma_\tau.$$

This induces a canonical bijection

$$\pi^*\colon \mathrm{PSH}^{\mathrm{NA}}(X,\theta) \xrightarrow{\sim} \mathrm{PSH}^{\mathrm{NA}}(Y,\pi^*\theta).$$

The injectivity is trivial and the surjectivity follows from Lemma 3.8.

Note that there is a canonical identification $Y^{\mathrm{div}} \xrightarrow{\sim} X^{\mathrm{div}}$, so that we can identify $(\pi^*\Gamma)^{\mathrm{an}}$ and Γ^{an}.

One trivially verifies that all the previous operations are functorial with respect to π^*.

5. Comparison with Boucksom–Jonsson's theory

Let X be a reduced irreducible unibranch projective variety and L be a pseudoeffective line bundle on X with $\xi = c_1(L)$. Fix a smooth Hermitian metric h_0 on L with $\theta = c_1(L, h_0)$.

Recall the following basic result:

THEOREM 5.1 ([**BJ22b**, Theorem 4.22]). *Let $\phi, \psi \in \mathrm{PSH}(L^{\mathrm{an}})$. Assume that $\phi \leq \psi$ on X^{div}, then the same holds on X^{an}.*

In particular, as we have computed the effect of our operations on X^{div}, their behaviours on X^{an} are in fact completely determined.

COROLLARY 5.2. *The order relation, addition by a constant, maximum, decreasing limit, rescaling and addition correspond to the corresponding operations in Boucksom–Jonsson's theory. Assume that the envelope conjecture holds for X, then the same holds for the supremum operation.*

Briefly speaking, we have established the correspondence between the non-Archimedean pluripotential theory and the convex geometry of complex potentials as in Table 1.

Finally, we have the following comparison result between our theory of non-Archimedean potentials and Boucksom–Jonsson's.

COROLLARY 5.3. *The following are equivalent:*

(1) *The envelope conjecture of Boucksom–Jonsson holds for (X, L);*
(2) *The comparison map $\Gamma \mapsto \Gamma^{\mathrm{an}}$ defined in Definition 3.11 admits an extension as a bijection $\mathrm{PSH}^{\mathrm{NA}}(X, \theta) \xrightarrow{\sim} \mathrm{PSH}(L^{\mathrm{an}})$.*

PROOF. By Theorem 5.1, if the map in (2) exists, it is necessarily unique.

Let $\pi \colon Y \to X$ be a projective resolution. Consider the following diagram:

$$\begin{array}{ccc} \mathrm{PSH}^{\mathrm{NA}}(X, \theta) & \xrightarrow{\text{an}} & \mathrm{PSH}(L^{\mathrm{an}}) \\ \downarrow{\scriptstyle \pi^*} & & \downarrow{\scriptstyle \pi^*} \\ \mathrm{PSH}^{\mathrm{NA}}(Y, \pi^*\theta) & \xrightarrow{\text{an}} & \mathrm{PSH}(\pi^* L^{\mathrm{an}}). \end{array}$$

Clearly, (2) means that we can complete the dotted map as a bijection so that the diagram is commutative. We already know that the left-vertical map is a bijection, see Section 4.8. The lower-horizontal map is also a bijection by Theorem 3.12.

Now assume (1), then the right-vertical map is also a bijection by [**BJ22b**, Lemma 5.13]. Therefore, (2) follows.

Conversely, assume that (2) holds. Take an increasing net $\phi_i \in \mathrm{PSH}^{\mathrm{NA}}(L^{\mathrm{an}})$ with $\phi_i \leq 0$. Recall that the envelope conjecture means that ϕ_i converges to some $\phi \in \mathrm{PSH}^{\mathrm{NA}}(L^{\mathrm{an}})$ with respect to the pointwise convergence topology on X^{div}. We write $\phi_i = \Gamma^{i,\mathrm{an}}$ for some $\Gamma^i \in \mathrm{PSH}^{\mathrm{NA}}(X, \theta)$. It suffices to take $\phi = (\sup^*{}_i \Gamma^i)^{\mathrm{an}}$ and apply Theorem 4.13. □

References

[BBJ21] Robert J. Berman, Sébastien Boucksom, and Mattias Jonsson, *A variational approach to the Yau-Tian-Donaldson conjecture*, J. Amer. Math. Soc. **34** (2021), no. 3, 605–652, DOI 10.1090/jams/964. MR4334189

[Ber93] Vladimir G. Berkovich, *Étale cohomology for non-Archimedean analytic spaces*, Inst. Hautes Études Sci. Publ. Math. **78** (1993), 5–161 (1994). MR1259429

[BFJ15] Sébastien Boucksom, Charles Favre, and Mattias Jonsson, *Solution to a non-Archimedean Monge-Ampère equation*, J. Amer. Math. Soc. **28** (2015), no. 3, 617–667, DOI 10.1090/S0894-0347-2014-00806-7. MR3327532

[BFJ16] Sébastien Boucksom, Charles Favre, and Mattias Jonsson, *Singular semipositive metrics in non-Archimedean geometry*, J. Algebraic Geom. **25** (2016), no. 1, 77–139, DOI 10.1090/jag/656. MR3419957

[BHJ17] Sébastien Boucksom, Tomoyuki Hisamoto, and Mattias Jonsson, *Uniform K-stability, Duistermaat-Heckman measures and singularities of pairs* (English, with English and French summaries), Ann. Inst. Fourier (Grenoble) **67** (2017), no. 2, 743–841. MR3669511

[BJ22a] Sébastien Boucksom and Mattias Jonsson, *Global pluripotential theory over a trivially valued field* (English, with English and French summaries), Ann. Fac. Sci. Toulouse Math. (6) **31** (2022), no. 3, 647–836, DOI 10.5802/afst.170. MR4452253

[BJ22b] Sébastien Boucksom and Mattias Jonsson, *Global pluripotential theory over a trivially valued field* (English, with English and French summaries), Ann. Fac. Sci. Toulouse Math. (6) **31** (2022), no. 3, 647–836, DOI 10.5802/afst.170. MR4452253

[CL06] Antoine Chambert-Loir, *Mesures et équidistribution sur les espaces de Berkovich* (French), J. Reine Angew. Math. **595** (2006), 215–235, DOI 10.1515/CRELLE.2006.049. MR2244803

[CLD12] Antoine Ducros, *Espaces de Berkovich, polytopes, squelettes et théorie des modèles* (French, with English summary), Confluentes Math. **4** (2012), no. 4, 1250007, 57, DOI 10.1142/S1793744212500077. MR3020334

[DDNL18] Tamás Darvas, Eleonora Di Nezza, and Chinh H. Lu, *Monotonicity of nonpluripolar products and complex Monge-Ampère equations with prescribed singularity*, Anal. PDE **11** (2018), no. 8, 2049–2087, DOI 10.2140/apde.2018.11.2049. MR3812864

[DDNL21] Tamás Darvas, Eleonora Di Nezza, and Hoang-Chinh Lu, *The metric geometry of singularity types*, J. Reine Angew. Math. **771** (2021), 137–170, DOI 10.1515/crelle-2020-0019. MR4234098

[Dem85] Jean-Pierre Demailly, *Mesures de Monge-Ampère et caractérisation géométrique des variétés algébriques affines* (French, with English summary), Mém. Soc. Math. France (N.S.) **19** (1985), 124. MR813252

[DX21] Tamás Darvas and Mingchen Xia, *The volume of pseudoeffective line bundles and partial equilibrium*, Geom. Topol. (2021).

[DX22] Tamás Darvas and Mingchen Xia, *The closures of test configurations and algebraic singularity types*, Adv. Math. **397** (2022), Paper No. 108198, 56, DOI 10.1016/j.aim.2022.108198. MR4366232

[DX24] Tamás Darvas and Mingchen Xia, *The trace operator of quasi-plurisubharmonic functions on compact Kähler manifolds*, 2024.

[DXZ23] Tamás Darvs, Mingchen Xia, and Kewei Zhang, *A transcendental approach to non-Archimedean metrics of pseudoeffective classes*, 2023.

[GK20] Patrick Graf and Tim Kirschner, *Finite quotients of three-dimensional complex tori* (English, with English and French summaries), Ann. Inst. Fourier (Grenoble) **70** (2020), no. 2, 881–914. MR4105953

[Gub16] Walter Gubler, *Forms and current on the analytification of an algebraic variety (after Chambert-Loir and Ducros)*, Nonarchimedean and tropical geometry, Simons Symp., Springer, [Cham], 2016, pp. 1–30. MR3700066

[Li20] Chi Li, *Geodesic rays and stability in the cscK problem* (English, with English and French summaries), Ann. Sci. Éc. Norm. Supér. (4) **55** (2022), no. 6, 1529–1574. MR4517682

[Roc70] R. Tyrrell Rockafellar, *Convex analysis*, Princeton Mathematical Series, No. 28, Princeton University Press, Princeton, NJ, 1970. MR274683

[RWN14] Julius Ross and David Witt Nyström, *Analytic test configurations and geodesic rays*, J. Symplectic Geom. **12** (2014), no. 1, 125–169, DOI 10.4310/JSG.2014.v12.n1.a5. MR3194078

[SD18] Zakarias Sjöström Dyrefelt, *K-semistability of cscK manifolds with transcendental cohomology class*, J. Geom. Anal. **28** (2018), no. 4, 2927–2960, DOI 10.1007/s12220-017-9942-9. MR3881961

[Thu05] Amaury Thuillier, *Théorie du potentiel sur les courbes en géométrie analytique non archimédienne. applications à la théorie d'arakelov*, Ph.D. thesis, Université Rennes 1, 2005.

[WN19] David Witt Nyström, *Monotonicity of non-pluripolar Monge-Ampère masses*, Indiana Univ. Math. J. **68** (2019), no. 2, 579–591, DOI 10.1512/iumj.2019.68.7630. MR3951074

[Xia21a] Mingchen Xia, *Mabuchi geometry of big cohomology classes*, J. Reine Angew. Math. **798** (2023), 261–292, DOI 10.1515/crelle-2023-0019. MR4579706

[Xia21b] Mingchen Xia, *Partial Okounkov bodies and Duistermaat–Heckman measures of non-Archimedean metrics*, 2021.

[Xia22] Ping Li, *Nonnegative Hermitian vector bundles and Chern numbers*, Math. Ann. **380** (2021), no. 1-2, 21–41, DOI 10.1007/s00208-020-01974-x. MR4263677

[Xia23a] Mingchen Xia, *Mabuchi geometry of big cohomology classes*, J. Reine Angew. Math. **798** (2023), 261–292, DOI 10.1515/crelle-2023-0019. MR4579706

[Xia23b] Mingchen Xia, *Pluripotential-theoretic stability thresholds*, Int. Math. Res. Not. IMRN (2023), no. 14, 12324–12382. MR4615232

[Xia24] Mingchen Xia, *Singularities in global pluripotential theory*, `https://mingchenxia.github.io/home/Lectures/SGPT.pdf`, 2024.

[Zha95] Shouwu Zhang, *Positive line bundles on arithmetic varieties*, J. Amer. Math. Soc. **8** (1995), no. 1, 187–221, DOI 10.2307/2152886. MR1254133

Institut de mathématiques de Jussieu—Paris Rive Gauche, 4 Place Jussieu, Paris
Email address: `mingchen@imj-prg.fr`